AF566614

# PARTICLES AND FIELDS

To learn more about AIP Conference Proceedings, including the Conference Proceedings Series, please visit the webpage
**http://proceedings.aip.org/proceedings**

# PARTICLES AND FIELDS

## Proceedings of the VI Latin American Symposium on High Energy Physics and the XII Mexican School of Particles and Fields

*Puerto Vallarta, México 1 – 8 November 2006*

*EDITORS*
Heriberto Castilla Valdez
Miguel A. Perez
*CINVESTAV*
*México City, México*

Juan Carlos D'Olivo
*ICN-UNAM*
*México City, México*

***SPONSORING ORGANIZATIONS***

Centro de Investigación y de Estudios Avanzados
Centro Latinoamericano de Física
Centro Latinoamericano de Física - México
Consejo Nacional de Ciencia y Tecnología
División de Partículas y Campos de la SMF
Department of Energy
International Center for Theoretical Physics
Instituto de Física - BUAP
Instituto de Física - UASLP
Instituto de Física y Matemáticas - UMSNH
Facultad de Ciencias Físico Matemáticas - BUAP
National Science Foundation
Instituto de Física de la Universidad de Guanajuato
Universidad Autónoma Metropolitana de Iztapalapa
Universidad de Guadalajara
Universidad Nacional Autónoma de México
Academia Mexicana de Ciencias
Sociedad Mexicana de Física

**Melville, New York, 2007**
**AIP CONFERENCE PROCEEDINGS ■ VOLUME 917**

**Editors:**

Heriberto Castilla Valdez
Miguel A. Perez

Physics Department
CINVESTAV
Av. IPN 2508
Col. San Pedro Zacatenco
México City CP 07360
México

E-mail: castilla@fnal.gov
mperez@fis.cinvestav.mx

Juan Carlos D'Olivo
Instituto de Ciencias Nucleares
Universidad Nacional Autónoma de México
Circuito Exterior C.U.
A. Postal 70-543
04510 México D.F.
México

E-mail: dolivo@nucleares.unam.mx



L.C. Catalog Card No. 2007928478
ISBN 978-0-7354-0427-4
ISSN 0094-243X
Printed in the United States of America

## CONTENTS



### PLENARY SESSION PRESENTATIONS

#### Section 1: Neutrino Physics

Section 2: Strings, Fields and Gravitation



Section 3: Astrophysics and Cosmology



Section 4: Heavy Ion Physics

Section 5: SM and Beyond



Section 6: Recent HEP Experimental Results



Section 7: Top Quark Physics

## POSTER SESSION

# PREFACE

The Division of Particles and Fields of the Mexican Physical Society (DPC-SMF) organized the VI Latin American Symposium on High Energy Physics (VI Simposio Latinoamericano de Fisica de Altas Energias, VI-Silafae), which was held in Puerto Vallarta, Mexico, on November 1-8, 2006. This meeting was celebrated jointly with the XII Mexican School of Particles and Fields (XII-MSPF) and the Panamerican Advanced Study Institute (PASI2006). While the scientific program of the VI-Silafae included 60 invited lectures, PASI2006 and XII-MSPF were organized in several courses for postdocs and advanced PhD students coming from USA and Latin America.

Silafae has become the main scientific meeting of the Latinamerica high energy physics community and in this occasion we had about 200 participants. Previous Silafaes were held in Merida (Mexico, 1996), Rio Piedras (Puerto Rico, 1998), Cartagena (Colombia, 2000), Aguas de Lindoaia (Brazil, 2002) and Lima (Peru, 2004). The following meetings have been arranged to take place in Bariloche (Argentina, 2008) and Santiago (Chile, 2010).

We had a special session (November 1-2) dedicated to celebrate the 50th anniversary of the discovery of the neutrino with the participation of Leon M. Lederman, James W. Cronin and Pier Oddone. On this subject, we had 22 review talks on the recent advances of neutrino physics, including experimental and phenomenological results. We have organized these proceedings according to the subject matter covered in the VI-Silafae sessions: Neutrino Physics, Experiments and Collider physics, Astrophysics and Cosmology, Standard Model and Beyond, Top Quark Physics, Strings and Field Theory.

**Mexico City,**
**May 2007**

**Heriberto Castilla Valdez**
**Juan Carlos D´Olivo**
**Miguel Angel Perez Angon**

## International Advisory Committee

José Bernabeu (UValencia)
Marcela Carena (Fermilab)
James W. Cronin (UChicago)
John Ellis (CERN)
Paolo Giubellino (INFN/Torino)
Benjamín Grinstein (UC/San Diego)
Jacobo Konigsberg (UFlorida)
León Lederman (IMSA/Fermilab)
Ernest Ma (UC/Riverside)
Luciano Maiani (URoma/La Sapienza)
Giora Mikenberg (Weizmann Inst.)
Juan Maldacena (UPrinceton)
Fernando Quevedo (UCambridge)
Oscar Saavedra (U.Torino)
Carlos Wagner (ANL)
Barton Zwiebach (MIT)

## Scientific Committee

Argentina: Tere Dova (UNLP)
Daniel de Florian (UBA)
Esteban Roulet (Bariloche)
Bolivia: Nicolás Martinic (ULP)
Brazil: Joao dos Anjos (CBPF)
Ma Beatriz Gay Ducati (UFRGS)
Vicente Pleitez (USP)
Chile: Jorge Alfaro (PUC)
Marco A. Díaz (PUC)
Ivan Schmidt (UTFSM)
Colombia: Marta Losada (UAN)
Roberto Martinez (UNal/Bogota)
Enrico Nardi (UAntioquia)
William Ponce (UAntioquia)
Cuba: Angelina Díaz (AENTA)
Guatemala: Edgar Cifuentes (USAC)
México: Juan Carlos D'Olivo (ICN-UNAM)
Miguel A. Pérez (Cinvestav), Chair
Luis M. Villaseñor (UMSNH)
Peru: Orlando Pereyra (UNI)
Puerto Rico: Angel López (UPR/Mayaguez)
Uruguay: Ramon Méndez Galain URU)
Venezuela: Anamaría Font (UCV)
Alejandra Melfo (ULA)
USA: Marleigh Sheaff (U. Wisconsin/Fermilab)
USA/Costa Rica: John Swain (Northeastern U)

## Local Organizing Committee

Alejandro Ayala (ICN-UNAM)
Adnan Bashir (IFM-UMSNH)
Guillermo Contreras (Cinvestav-UM)
Axel de la Macorra (IF-UNAM)
Jurgen Engelfried (IF-UASLP)
Julián Félix (IFUG)
José Luis García Luna (UdeG)
Rebeca Juárez (ESFM-IPN)
Gisela Mateos (CEIICH-UNAM)
Hilda Mercado Uribe (UdeG)
Omar Miranda (Cinvestav)
Hugo Morales (UAM-I)
Humberto Salazar (FCFM-BUAP)
Alberto Sánchez (Cinvestav)
Luis Urrutia (ICN-UNAM)
Juan Carlos D'Olivo (ICN-UNAM)
Heriberto Castilla (Cinvestav)
Luis Manuel Villaseñor (IFM-UMSNH)
Miguel A. Pérez (Cinvestav), Chair

## SPONSORS

The organizers of the VI Simposio Latinoamericano de Altas Energías and XII Escuela Mexicana de Particulas y Campos acknowledge financial support from the following instituions:

* Centro de Investigación y de Estudios Avanzados
* Centro Latinoamericano de Física
* Centro Latinoamericano de Física – México
* Consejo Nacional de Ciencia y Tecnología
* División de Partículas y Campos de la SMF
* Department of Energy
* Instituto de Ciencias Nucleares - UNAM
* International Center for Theoretical Physics
* Instituto de Física - BUAP
* Instituto de Física - UASLP
* Instituto de Física - UNAM
* Instituto de Física y Matemáticas - UMSNH
* Facultad de Ciencias Físico Matemáticas - BUAP
* National Science Foundation
* Instituto de Física de la Universidad de Guanajuato
* Universidad Autónoma Metropolitana de Iztapalapa
* Universidad de Guadalajara
* Academia Mexicana de Ciencias

## Foreword

Tuesday, November 7, 2006

SILAFAE The past few days I attended VI-SILAFAE, the Sixth Latin American Symposium on High Energy Physics, in Puerto Vallarta, Mexico. Held every two years in different Latin American countries, this symposium was organized by the Division of Particles and Fields of the Mexican Physical Society (DPF-MPS). More than 170 physicists mostly from Latin America attended the meeting. The meeting had more than 70 talks covering all aspects of HEP. I gave a talk on the program at Fermilab, present and future.

Pier Oddone

Leon Lederman, who is revered in the Latin American particle physics community for his years of support, opened the conference with a review of fifty years of neutrino physics. Later in the week, Jim Cronin presented results from the Pierre Auger observatory. Pierre Auger has become a major experiment for Latin American countries, and there are great expectations for the results coming in the next few years as the experiment accumulates statistics.

There is strong participation of Latin American physicists in the US program, especially at Fermilab, and in Europe, primarily at CERN. Leon Lederman 25 years ago was a prime mover in developing Latin American particle physics, making it possible to develop and train theorists and experimentalists at Fermilab. Since then the Latin American groups have grown and now contribute to many of the world's primary particle physics experiments. CERN starting with Carlo Rubbia also played a very active role in helping to develop Latin American participation in particle physics.

At this point there is not much involvement of Latin American physicists in the planning for ILC, although this machine is of great interest to them. In my discussions with Latin American colleagues it became clear that Fermilab can play an enabling role by providing the opportunity to participate in ILC planning and R&D while satisfying the needs of Latin American colleagues and their students for physics results by their simultaneous participation in the ongoing neutrino, particle astrophysics and energy frontier programs at Fermilab.

Pier Oddone
Fermilab Director

# PLENARY SESSION PRESENTATIONS

# Phenomenology with Massive Neutrinos

M. C. Gonzalez-Garcia

*Y.I.T.P., SUNY at Stony Brook, Stony Brook, NY 11794-3840, USA,*
*and*
*Institució Catalana de Recerca i Estudis Avançats (ICREA) & Departament d'Estructura i Constituents de la Matèria, Universitat de Barcelona, Diagonal 647, E-08028, Spain*

**Abstract.** In this talk I review the present status of neutrino masses and mixing and some of their implications for particle physics phenomenology.

**Keywords:** Neutrino, Oscillations, Phenomenology
**PACS:** 14.60Pq,14.60St

## INTRODUCTION: THE NEW MINIMAL STANDARD MODEL

The SM is a gauge theory based on the gauge symmetry $SU(3)_{\mathrm{C}} \times SU(2)_{\mathrm{L}} \times U(1)_{\mathrm{Y}}$ spontaneously broken to $SU(3)_{\mathrm{C}} \times U(1)_{\mathrm{EM}}$ by the the vacuum expectation value of a Higgs doublet field $\phi$. The SM contains three fermion generations which reside in chiral representations of the gauge group. Right-handed fields are included for charged fermions as they are needed to build the electromagnetic and strong currents. No right-handed neutrino is included in the model since neutrinos are neutral.

In the SM, fermion masses arise from the Yukawa interactions which couple the right-handed fermion singlets to the left-handed fermion doublets and the Higgs doublet. After spontaneous electroweak symmetry breaking these interactions lead to charged fermion masses but leave the neutrinos massless. No Yukawa interaction can be written that would give a tree level mass to the neutrino because no right-handed neutrino field exists in the model.

Furthermore, within the SM $G_{\mathrm{SM}}^{\mathrm{global}} = U(1)_B \times U(1)_e \times U(1)_\mu \times U(1)_\tau$ is an accidental global symmetry. Here $U(1)_B$ is the baryon number symmetry, and $U(1)_{e,\mu,\tau}$ are the three lepton flavor symmetries. Any neutrino mass term which could be built with the particle content of the SM would violate the $U(1)_L$ subgroup of $G_{\mathrm{SM}}^{\mathrm{global}}$ and therefore cannot be induced by loop corrections. Also, it cannot be induced by non-perturbative corrections because the $U(1)_{B-L}$ subgroup of $G_{\mathrm{SM}}^{\mathrm{global}}$ is non-anomalous.

It follows then that the SM predicts that neutrinos are *strictly* massless. Consequently, there is neither mixing nor CP violation in the leptonic sector.

We now know that this picture cannot be correct. Over several years we have accumulated important experimental evidence that neutrinos are massive particles and there is mixing in the leptonic sector. In particular we have learned that:

– Solar $\nu_e' s$ convert to $\nu_\mu$ or $\nu_\tau$ with confidence level (CL) of more than $7\sigma$ [1, 2].
– KamLAND find that reactor $\overline{\nu}_e$ disappear over distances of about 180 km and they observe a distortion of their energy spectrum. Altogether their evidence has more than $3\sigma$ CL [3].

CP917, *Particles and Fields,* edited by H. Castilla Valdez, J. C. D'Olivo, and M. A. Perez

- The evidence of atmospheric (ATM) $\nu_\mu$ disappearing is now at $> 15\sigma$, most likely converting to $\nu_\tau$ [1].
- K2K observe the disappearance of accelerator $\nu_\mu$'s at distance of 250 km and find a distortion of their energy spectrum with a CL of 2.5–4 $\sigma$ [1].
- MINOS observes the disappearance of accelerator $\nu_\mu$'s at distance of 735 km and find a distortion of their energy spectrum with a CL of $\sim 5\ \sigma$ [1].
- LSND found evidence for $\overline{\nu_\mu} \to \overline{\nu_e}$. This evidence has not been confirm by any other experiment so far and it is being tested by MiniBooNE.

These results imply that neutrinos are massive and the Standard Model has to be extended at least to include neutrino masses. This minimal extension is what I call *The New Minimal Standard Model.*

In the New Minimal Standard Model flavour is mixed in the CC interactions of the leptons, and a leptonic mixing matrix appears analogous to the CKM matrix for the quarks. However the discussion of leptonic mixing is complicated by two factors. First the number massive neutrinos ($n$) is unknown, since there are no constraints on the number of right-handed, SM-singlet, neutrinos. Second, since neutrinos carry neither color nor electromagnetic charge, they could be Majorana fermions. As a consequence the number of new parameters in the model depends on the number of massive neutrino states and on whether they are Dirac or Majorana particles.

In general, if we denote the neutrino mass eigenstates by $\nu_i$, $i = 1,2,\ldots,n$, and the charged lepton mass eigenstates by $l_i = (e,\mu,\tau)$, in the mass basis, leptonic CC interactions are given by

$$-\mathcal{L}_{\rm CC} = \frac{g}{\sqrt{2}} \overline{l_{iL}} \gamma^\mu U_{ij} \nu_j W_\mu^+ + \text{h.c.}. \tag{1}$$

Here $U$ is a $3 \times n$ matrix $U_{ij} = P_{\ell,ii} V_{ik}^{\ell\dagger} V_{kj}^{\nu} (P_{\nu,jj})$ where $V^\ell$ $(3\times 3)$ and $V^\nu$ $(n \times n)$ are the diagonalizing matrix of the charged leptons and neutrino mass matrix respectively $V^{\ell\dagger} M_\ell M_\ell^\dagger V^\ell = \text{diag}(m_e^2, m_\mu^2, m_\tau^2)$ and $V^{\nu\dagger} M_\nu^\dagger M_\nu V^\nu = \text{diag}(m_1^2, m_2^2, m_3^2, \ldots, m_n^2)$.

$P_\ell$ is a diagonal $3 \times 3$ phase matrix, that is conventionally used to reduce by three the number of phases in $U$. $P_\nu$ is a diagonal matrix with additional arbitrary phases (chosen to reduce the number of phases in $U$) only for Dirac states. For Majorana neutrinos, this matrix is simply a unit matrix, the reason being that if one rotates a Majorana neutrino by a phase, this phase will appear in its mass term which will no longer be real. Thus, the number of phases that can be absorbed by redefining the mass eigenstates depends on whether the neutrinos are Dirac or Majorana particles. In particular, if there are only three Majorana (Dirac) neutrinos, $U$ is a $3 \times 3$ matrix analogous to the CKM matrix for the quarks but due to the Majorana (Dirac) nature of the neutrinos it depends on six (four) independent parameters: three mixing angles and three (one) phases.

A consequence of the presence of the leptonic mixing is the possibility of flavour oscillations of the neutrinos. Neutrino oscillations appear because of the misalignment between the interaction neutrino eigenstates and the propagation eigenstates ( which for propagation in vacuum are the mass eigenstates). Thus a neutrino of energy $E$ produced in a CC interaction with a charged lepton $l_\alpha$ can be detected via a CC interaction with a charged lepton $l_\beta$ with a probability which presents an oscillatory behaviour, with oscillation lengths given by the phase difference between the different propagation

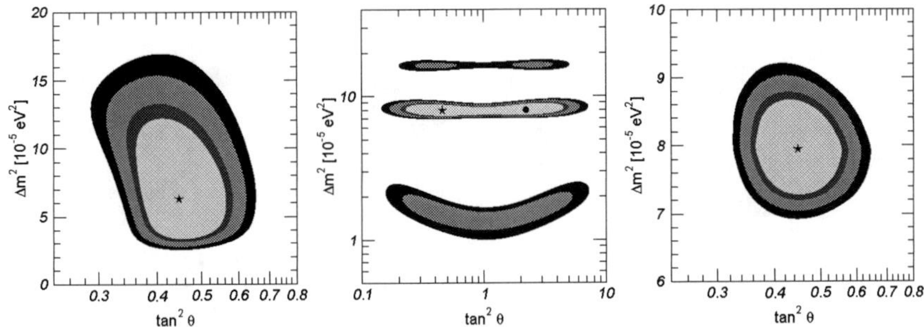


**FIGURE 1.** Allowed regions for 2-$\nu$ oscillations of solar $\nu_e$ and KamLAND $\overline{\nu}_e$ and for the combination of KamLAND and solar data under the hypothesis of CPT conservation. The different contours correspond to the allowed regions at 90%, 95%, 99% and $3\sigma$ CL.

eigenstates – which in the ultrarelativistic limit is $L^{\rm osc}_{0,ij} = \frac{4\pi E}{\Delta m^2_{ij}}$ – and amplitude that is proportional to elements in the mixing matrix.

It follows that neutrino oscillations are only sensitive to mass squared differences and do not give us information on the absolute value of thc masses. Also the Majorana phases do not affect oscillations because total lepton number is conserved in the process. Experimental information on absolute neutrino masses can be obtained from Tritium $\beta$ decay experiments and from its effect on the cosmic microwave background radiation and large structure formation data [4]. If neutrinos are Majorana particles their mass and also additional phases can be determined in $\nu$-less $\beta\beta$ decay experiments.

When neutrinos travel through regions of dense matter, they can undergo forward scattering with the particles in the medium. These interactions are, in general, flavour dependent and as a consequence the oscillation pattern is modified but it still depends only on the mass squared differences and it is independent of the Majorana phases.

The neutrino experiments described above have measured some non-vanishing $P_{\alpha\beta}$ and from these measurements we have inferred all the positive evidence that we have on the non-vanishing values of neutrino masses and mixing. In the following I will derive the allowed ranges for the mass and mixing parameters when the bulk of data is consistently combined. In Fig. 2 I show the results of our latest analysis of the ATM neutrino data which includes the full data set of Super-Kamiokande phases I+II.

## THE PARAMETERS OF THE NMSM: 3$\nu$ ANALYSIS

I describe here our present determination of the parameters of the model from the analysis which try to explain the evidences from solar, KamLAND, ATM and K2K experiments and assume that the LSND evidence will not be confirmed by MiniBoone.

In Fig. 1 I show the results from our latest analysis [5] of KamLAND $\overline{\nu}_e$ disappearance data, solar $\nu_e$ data and their combination under the hypothesis of CPT symmetry. The main features of these results are:

– In the analysis of solar data, only LMA is allowed at more than $3\sigma$ and maximal

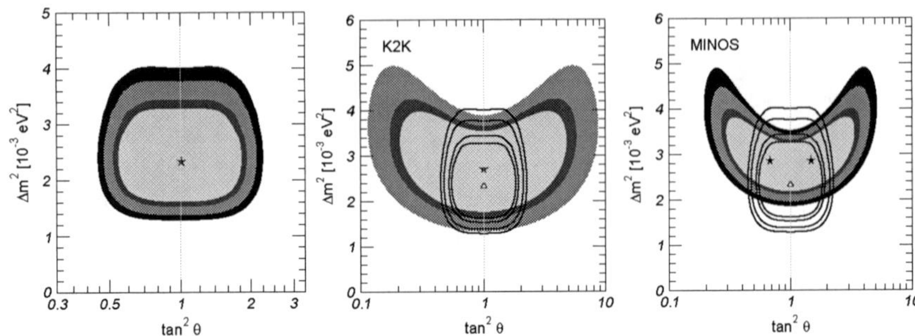


**FIGURE 2.** Allowed regions from the analysis of ATM data (left), K2K (central) and MINOS (right). The different contours correspond to at 90%, 95%, 99% and $3\sigma$ CL.

mixing is rejected by the solar analysis at more than $5\sigma$. This is so since the release of the SNO salt-data (SNOII) in Sep 2003.

- In the analysis of the KamLAND data the $3\sigma$ region does not extend to mass values larger than $\Delta m^2_{21} = 2\times10^{-4}$ eV$^2$ because for larger $\Delta m^2_{21}$ values, the predicted spectral distortions are too small to fit the spectral KamLAND data.
- the combined analysis allows only the LMA-I region at $3\sigma$.

The evidence of oscillation of ATM $\nu_\mu$ has been now confirmed by two long-baseline (LBL) experiments: K2K which first observed not a deficit of $\nu_\mu$'s at a distance of 250 km and in his final results also measured the distortion of their energy spectrum, and MINOS which has reported his first data in 2006 and which also observes an energy dependent deficit with a confidence level of about $\sim 5$ sigma. I show the results of our analysis of the K2K and MINOS data which graphically illustrate this agreement.

The minimum joint description of ATM, LBL, solar and reactor data requires that all the three known neutrinos take part in the oscillations. The mixing parameters are encoded in the $3\times3$ lepton mixing matrix which can be conveniently parametrized in the standard form $U = \begin{pmatrix} 1 & 0 & 0 \\ 0 & c_{23} & s_{23} \\ 0 & -s_{23} & c_{23} \end{pmatrix} \begin{pmatrix} c_{13} & 0 & s_{13}e^{i\delta} \\ 0 & 1 & 0 \\ -s_{13}e^{-i\delta} & 0 & c_{13} \end{pmatrix} \begin{pmatrix} c_{21} & s_{12} & 0 \\ -s_{12} & c_{12} & 0 \\ 0 & 0 & 1 \end{pmatrix}$ where $c_{ij} \equiv \cos\theta_{ij}$ and $s_{ij} \equiv \sin\theta_{ij}$. The angles $\theta_{ij}$ can be taken without loss of generality to lie in the first quadrant, $\theta_{ij} \in [0, \pi/2]$.

There are two possible mass orderings, which we denote as *Normal* and *Inverted*. In the normal scheme $m_1 < m_2 < m_3$ while in the inverted one $m_3 < m_1 < m_2$.

In total the 3-$\nu$ oscillation analysis involves six parameters: 2 mass differences (one of which can be positive or negative), 3 mixing angles, and the CP phase. Generic 3-$\nu$ oscillation effects include: (i) coupled oscillations with two different wavelengths; (ii) CP violating effects; (iii) difference between Normal and Inverted schemes.The strength of these effects is controlled by the values of the ratio of mass differences $\Delta m^2_{21}/|\Delta m^2_{31}|$, by the mixing angle $\theta_{13}$ and by the CP phase $\delta$.

From the previous $2\nu$ analysis we see that $\Delta m^2_{\odot} = \Delta m^2_{21} \ll |\Delta m^2_{31}| \simeq |\Delta m^2_{32}| = \Delta m^2_{\rm atm}$. As a consequence the joint 3-$\nu$ analysis simplifies as follows:

- for solar and KamLAND neutrinos, the oscillations with the $\Delta m^2_{31}$-driven oscillation length are completely averaged and the survival probability takes the form:

$$P^{3\nu}_{ee} = \sin^4\theta_{13} + \cos^4\theta_{13} P^{2\nu}_{ee} \tag{2}$$

  where in the Sun $P^{2\nu}_{ee}$ is obtained with the modified sun density $N_e \to \cos^2\theta_{13} N_e$. So the analyses of solar data constrain three of the six parameters: $\Delta m^2_{21}, \theta_{12}$ and $\theta_{13}$.
- for ATM and LBL neutrinos, the $\Delta m^2_{21}$-driven wavelength is too long and the corresponding oscillating phase is almost negligible. As a consequence, the ATM and LBL data analysis mostly restricts $\Delta m^2_{31} \simeq \Delta m^2_{32}$, $\theta_{23}$ and $\theta_{13}$, the latter being the only relevant parameter common to both solar+Kamland and ATM+LBL neutrino oscillations and which may potentially allow for some mutual influence. The effect of $\theta_{13}$ is to add a $\nu_\mu \to \nu_e$ contribution to the ATM and LBL oscillations;
- at CHOOZ the $\Delta m^2_{21}$-driven wavelength is unobservable and the relevant oscillation wavelength is determined by $\Delta m^2_{31}$ and its amplitude by $\theta_{13}$.

The CP phase is basically unobservable although there is somc marginal sensitivity in the present ATM neutrino analysis [5]. Normal versus Inverted orderings could be discriminated due to matter effects in the Earth for ATM neutrinos. However, this effect is controlled by the mixing angle $\theta_{13}$. Presently all data favour small $\theta_{13}$ with best fit point very near $\theta_{13} = 0$. The dominant constraint arises from the combined analysis of CHOOZ reactor and ATM data and it is further limited by the solar and KamLAND results. Consequently, the difference between Normal and Inverted orderings is too small to be statistically meaningful in the present analysis.

Altogether the derived ranges for the six parameters at $1\sigma$ ($3\sigma$) are:

$$\begin{aligned}
&\Delta m^2_{21} = 7.9^{+0.27}_{-0.28}\left(^{+1.1}_{-0.89}\right)\times 10^{-5}\,\mathrm{eV}^2, \quad |\Delta m^2_{31}| = 2.6 \pm 0.2\,(0.6)\times 10^{-3}\,\mathrm{eV}^2,\\
&\theta_{12} = 33.7 \pm 1.3\left(^{+4.3}_{-3.5}\right), \qquad \theta_{23} = 43.3^{+4.3}_{-3.8}\left(^{+9.8}_{-8.8}\right) \qquad \theta_{13} = 0^{+5.2}_{-0.0}\left(^{+11.5}_{-0.0}\right),\\
&\delta_{\rm CP} \in [0, 360].
\end{aligned} \tag{3}$$

These results can be translated into our present knowledge of the moduli of the mixing matrix $U$:

$$U_{3\sigma} = \begin{pmatrix} 0.79 \to 0.86 & 0.50 \to 0.61 & 0.00 \to 0.20 \\ 0.25 \to 0.53 & 0.47 \to 0.73 & 0.56 \to 0.79 \\ 0.21 \to 0.51 & 0.42 \to 0.69 & 0.61 \to 0.83 \end{pmatrix}. \tag{4}$$

## LEARNING ABOUT NATURAL NEUTRINO FLUXES

The expected number of solar and atmospheric neutrino events in an experiment depends on a variety of components: the neutrino fluxes, the neutrino oscillation parameters and the neutrino interaction cross section in the detector. Although the original goal of the experiments was the study of the solar and atmospheric neutrino fluxes, once it was found that the observed anomalies seemed to indicate that neutrinos oscillated, the main focus of the experiments changed to the determination of the neutrino masses and mixing. Consequently, in the standard analysis, such as the ones previously described, the remaining components of the event rate computation are inputs taken from other sources. In particular, the fluxes of solar neutrinos are taken from the results of Solar Model simulations [6] and the fluxes of atmospheric neutrinos are taken from the results of numerical calculations [7] based on the convolution of the primary cosmic ray spectrum with the expected yield of neutrinos per incident cosmic ray.

The oscillation of $\nu_e$ and $\nu_\mu$'s is now also tested at terrestrial facilities like KamLAND, K2K and MINOS and the relevant parameters will be further measured in upcoming experiments. The attainable accuracy in the independent determination of the relevant neutrino oscillation parameters from non-solar and non-atmospheric neutrino experiments makes it possible to attempt an inversion of the strategy: to use the oscillation parameters (independently determined in reactor and LBL neutrino experiments) as inputs in the solar and atmospheric neutrino analysis in order to extract the solar and atmospheric neutrino fluxes directly from the data. Alternatively one can perform global fits to natural and terrestrial neutrino data in which both the oscillation parameters and the natural fluxes are extracted simultaneously.

There are several motivations for such direct determination of the solar and atmospheric neutrino fluxes. First of all it would provide a cross-check of the standard flux calculations as well as of the size of the associated uncertainties (which, being mostly theoretical, are difficult to quantify). Also, such program quantitatively expands the physics potential of future solar and atmospheric neutrino experiments.

The application to this strategy to the solar and KamLAND neutrino data yields the following value for the ratio of the solar fluxes to the predictions from the BS05 [6] solar model at $1\sigma$:

$$\begin{array}{lll} f_{pp} = f_{pep} = 1.0 \pm 0.02\,, & f_{^8\mathrm{B}} = 0.88 \pm 0.04\,, & f_{^7\mathrm{Be}} = 1.03^{+0.24}_{-1.03}\,, \\ f_{^{13}\mathrm{N}} = 0.0^{+7.6}_{-0.0}\,, & f_{^{15}\mathrm{O}} = 0.0^{+5.0}_{-0.0}\,, & f_{^{17}\mathrm{F}} = 0.0^{+2.1}_{-0.0}\,. \end{array} \tag{5}$$

From these results it is possible to extract, for example, the allowed range of the fraction of the sun's luminosity that arises from CNO reactions:

$$\frac{L_{\mathrm{CNO}}}{L_\odot} = \sum_{i=\mathrm{N,O,F}} \left(\frac{\alpha_i}{10\ \mathrm{MeV}}\right) a_i f_i = 0.0^{+2.7}_{-0.0}\,(^{+7.3}_{-0.0})\,\% \tag{6}$$

at 1 (3)$\sigma$. $a_i$ is the ratio of the neutrino flux $i$ predicted by the standard solar model to the characteristic solar photon flux defined by $L_\odot/[4\pi(\mathrm{A.U.})^2(10\ \mathrm{MeV})]$. This constraint is consistent with the SSM prediction at $1\sigma$ which is an important empirical confirmation of the SSM and, in general, of our understanding of the Sun.

The determination of atmospheric neutrino fluxes directly from the atmospheric neutrino data is technically more involved than for solar neutrinos because there are four different fluxes to be determined: $\nu_e$, $\nu_\mu$, $\bar{\nu}_e$, and $\bar{\nu}_\mu$ which, after integration over the azimuthal angle, are a function of two variables: the zenith angle and the energy of the neutrino. Unlike for solar neutrinos, there is no simple physics which can determine the angular and energy dependence of the fluxes thus not only the normalization of the fluxes but also their functional dependence has to be extracted from the data. Consequently the fully empirically determination of the atmospheric fluxes from atmospheric

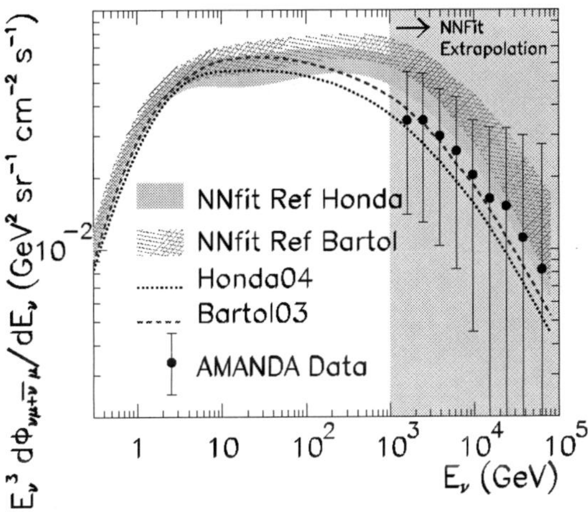


**FIGURE 3.** Results for the neural network fit for the angular averaged muon neutrino plus antineutrino flux and comparison with numerical computations. The fluxes are also shown extrapolated extrapolated to the high energy region and compared to the corresponding data from AMANDA [9]. The neural network fit was performed assuming oscillations with $\Delta m^2_{\rm atm} = 2.2 \times 10^{-3}\,{\rm eV}^2$ and $\tan^2\theta_{\rm atm} = 1$

neutrino data requires a generic parametrization of the energy and angular functional dependence of the fluxes which is valid in all the range of energies where there is available data. Such parametrization does not exist.

The problem of the unknown functional form for the neutrino flux can be bypassed by the use of neural networks as interpolants. However the precision of the available experimental data is not enough to allow for a separate determination of the energy, zenith angle and type dependence of the atmospheric flux. Consequently in Ref. [8] it was assumed that the zenith and type dependence of the flux is known with some precision and only its energy dependence was extracted from the data. The results from the fit are shown in Fig. 3. In the figure we show the neural network angular averaged muon neutrino and antineutrino fluxes obtained using as reference the fluxes of Honda and Bartol groups respectively. The results indicate that until about $E_\nu \sim 1$ TeV we have a good understanding of the normalization of the fluxes and the present accuracy from Super-Kamiokande neutrino data is comparable with the theoretical uncertainties from the numerical calculations.

## NEUTRINOS AS TESTS OF OTHER FORMS OF NEW PHYSICS

Using the good description of neutrino data in terms of neutrino oscillations, it is also possible to constraint other exotic forms of new physics. In my talk I discussed also some of the results in constraining the possibility of mass varying neutrinos [10,

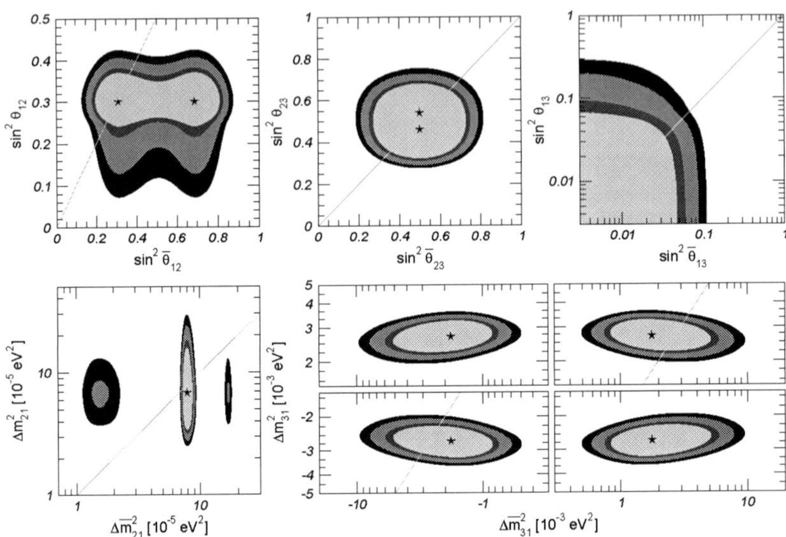


**FIGURE 4.** Allowed regions for neutrino and anti-neutrino mass splittings and mixing angles in the CPT violating scenario. Different contours correspond to the two-dimensional allowed regions at 90%, 95%, 99% and $3\sigma$ CL. The best fi t point is marked with a star.

11, 12] and the possibility of long-range leptonic forces [13]. In these proceedings I will summarize only the constraints which can be imposed in the violation of some fundamental symmetries. Examples of those are the violation of Lorentz Invariance (VLI) [14] induced by different asymptotic values of the velocity of the neutrinos, $c_1 \neq c_2$, or the violation of the equivalence principle (VEP) [15] due to non universal coupling of the neutrinos, $\gamma_1 \neq \gamma_2$ to the local gravitational potential. These forms of new physics, if non-universal, can also induce neutrino flavour oscillations whose main differentiating characteristic is a different energy dependence of the oscillation wavelength. For example for both VLI and VEP the oscillation wavelength decreases with energy unlike for mass oscillations. ATM neutrino events extend over several decades in energy. As a consequence they can test the presence of this effect even at the subdominant level. In Ref. [16] we performed an analysis of ATM and LBL neutrino data in terms of neutrino mass oscillations plus these new physics effects and we have concluded that the determination of mass and mixing parameters is robust under the presence of these unknown forms of new physics. Conversely, the analysis permits to impose strong constraints on the violations of these symmetries. For instance we find that at 90% CL the possible VLI a and VEP are limited to

$$\frac{|\Delta c|}{c} \leq 1.2 \times 10^{-24}, \qquad |\phi\,\Delta\gamma| \leq 5.9 \times 10^{-25}. \tag{7}$$

which constitute the strongest constraints on the violation of these symmetries.

As another example of the reach of the present experimental data in constraining exotic forms of new physics I comment here on alternative explanations to the LSND result which include the possibility of CPT [17] violation and imply that the masses and mixing angles of neutrinos may be different from those of antineutrinos. To test this possibility, in Ref. [18] we performed an analysis of the existing data from solar, ATM, LBL, reactor and SBL experiments in the framework of CPT violating oscillations. The outcome of the analysis is that, presently, the hypothesis of CPT violation is not supported by the data. This arises from two main facts: (i) KamLand finds that reactor $\overline{\nu}_e$ oscillate with wavelength and amplitude in good agreement with the expectations from the LMA solution of the solar $\nu_e$; (ii) both ATM neutrinos and antineutrinos have to oscillate with similar wavelengths and amplitudes to explain the ATM data. In general, as a result of these effects, the best fit to the data is very near CPT conservation as illustrated in Fig. 4.

Concerning LSND, the results show that values of $\Delta\overline{m}^2_{31} = \Delta\overline{m}^2_{\rm LSND}$ large enough to fit the LSND result do not appear as part of the $3\sigma$ CL allowed region of this all-but-LSND analysis. It is bounded to $\Delta\overline{m}^2_{31} < 0.01$ eV$^2$ with this upper bound being determined by atmospheric neutrino data. It is clear from these results that the CPT violation scenario cannot give a good description of the LSND data and simultaneously fit all-but-LSND data.

## ACKNOWLEDGMENTS

This work is supported by NSF grant PHY-0354776 and by Spanish MEC FPA-2004-00996.

## REFERENCES

1. See talk by Y. Suzuki in these proceedings
2. See talk by A. McDonald in these proceedings.
3. See talk by E. Berger in these proceedings.
4. See talk by S. Pastor in these proceedings.
5. M.C. Gonzalez-Garcia and M.Maltoni Physics Reports in preparation.
6. J. N. Bahcall, A. M. Serenelli and S. Basu, Astrophys. J. **621**, L85 (2005).
7. M. Honda, T. Kajita, K. Kasahara, and S. Midorikawa, Phys. Rev. **D70** (2004) 043008; G. D. Barr, T. K. Gaisser, P. Lipari, S. Robbins, and T. Stanev, Phys. Rev. **D70** (2004) 023006.
8. M. C. Gonzalez-Garcia, M. Maltoni and J. Rojo, JHEP **0610** (2006) 075
9. J. Ahrens *et al.* [IceCube Collaboration], Astropart. Phys. **20**, 507 (2004)
10. R. Fardon, A. E. Nelson and N. Weiner, JCAP **0410**, 005 (2004); D. B. Kaplan, A. E. Nelson and N. Weiner, Phys. Rev. Lett. **93**, 091801 (2004).
11. M. Cirelli, M. C. Gonzalez-Garcia and C. Pena-Garay, Nucl. Phys. B **719**, 219 (2005).
12. M.C. Gonzalez-Garcia M C, P.C. de Holanda and R. Zukanovich Funchal Phys. Rev. D **73** 033008 (2006).
13. M. C. Gonzalez-Garcia, P. C. de Holanda, E. Masso and R. Zukanovich Funchal, JCAP **0701** (2007) 005.
14. S. Coleman and S.L. Glashow, Phys. Lett. B **405**, 249 (1997); D. Colladay and V.A. Kostelecky, *Phys. Rev.* **D55**, 6760 (1997).
15. M. Gasperini, Phys. Rev. D **38** (1988) 2635; Phys. Rev. D **39**, 3606 (1989)
16. M. C. Gonzalez-Garcia and M. Maltoni, Phys. Rev. D **70**, 033010 (2004)
17. H. Murayama, and T. Yanagida, Phys. Lett. B **520**, 263 (2001); G. Barenboim *et al.*, JHEP **0210** (2002) 001.
18. M. C. Gonzalez-Garcia *et al.* Phys. Rev. D **68**, 053007 (2003).

# The Discovery and Nondiscovery of Neutrinos: The Reines Cowan Experiment and the 17Kev Neutrino

Allan Franklin

*Department of Physics*
*University of Colorado*
*Boulder, CO 80309-0390*

**Abstract.** In this paper I discuss the first observation of the neutrino by Frederick Reines, Clyde Cowan, and their collaborators. I also discuss how the physics community decided to reject the claim that a new, heavy, 17-keV neutrino had been observed. Both are examples of good science.

**Keywords:** Neutrino, 17-keV neutrino, Reines, Cowan, Simpson, liquid scintillator, magnetic spectrometer, solid-state detector.
**PACS:** 14.60.

The year 2006 marks the fiftieth anniversary of the first reported observation of the neutrino by Frederick Reines, Clyde Cowan, and their collaborators. It is altogether appropriate that we celebrate this significant contribution to physics. At the same time it is worth remembering that not all good science is correct. In this essay I will discuss not only the observation of the neutrino, but also the claim that there existed a heavy, 17-keV neutrino. It is now agreed that such a particle does not exist, but it is worth examining how the physics community reached that conclusion. Both of theses episodes are examples of good physics. Let us begin with the observation of the neutrino.

## THE OBSERVATION OF THE NEUTRINO

In 1927 Ellis and Wooster[1] demonstrated that the energy spectrum of electrons emitted in β decay was continuous. This posed a problem for the physics community because if β decay were a two-body process, as most physicists then believed, then the conservation of energy and of momentum required that the electron be monoenergetic. The conservation laws were under attack. In 1930, Wolfgang Pauli proposed what he called his "desperate way" out of the problem. This was a particle with very small or zero mass, zero electric charge, and spin ½ that was also emitted during β decay. This solved the problem of the conservation laws because for a three-body decay the energy spectrum of the electron is continuous.

CP917, *Particles and Fields,* edited by H. Castilla Valdez, J. C. D'Olivo, and M. A. Perez

In 1934 Enrico Fermi incorporated this new particle, which he named the neutrino, or little neutral one, into the first successful theory of β decay.[2] The success of Fermi's theory through the 1950s and the role of the neutrino in saving the conservation laws, persuaded most physicists of the neutrino's existence. Others, however, asked for more direct evidence of this elusive particle's existence. The problem was that the neutrino was expected to interact only very weakly with matter.

One possibility of directly observing the neutrino was to observe inverse β decay in which an antineutrino is absorbed by a proton giving rise to a neutron and a positron. ($\nu + p \ 6 \ n + e^+$). As Hans Bethe and Robert Bacher[3] pointed out, soon after Fermi's formulation of his theory, this reaction would be very difficult to observe. In order to detect such a rare interaction in a reasonable time one would need both a large amount of matter as a target and a very large flux of neutrinos. Developments during World War II made such an experiment feasible. The first experimental suggestion of what became known as Project Poltergeist, so named because of the properties of the elusive neutrino, was to use an atomic bomb as the source of antineutrinos. The explosion of an atomic bomb, a nuclear fission weapon, produces many short-lived fission fragments, each of which is radioactive. These fission fragments emit an electron and an antineutrino. Frederick Reines and Clyde Cowan, two scientists at the Los Alamos weapons laboratory, where the atomic bomb had first been developed and built, hypothesized that they could place a neutrino detector close enough to an atomic bomb explosion so that they would have a good chance of detecting that process. How would one detect such an interaction? It had been found that certain organic liquids would emit light, or scintillate, when an electron passed through them. The positron emitted by inverse β decay will quickly annihilate with an electron in the liquid producing two γ rays. These γ rays each have equal energy and are emitted in opposite directions, an important point for later experiments, and can be detected by the scintillator. Reines's and Cowan's first plan involved detecting only the γ rays resulting from the annihilation of the positron produced with an electron. The positron would be evidence that inverse β decay had occurred and for the presence of the neutrino. No other known reactions were expected to produce such positrons in any significant numbers. Not too surprisingly, the experiment was never performed. In the fall of 1952, following a suggestion by J.M.B. Kellogg, the director of the Physics Division at Los Alamos, that Reines and Cowan review the experimental plan to see if the flux of antineutrinos from a nuclear fission reactor could be used in place of that from an atomic bomb. Although the flux from a reactor was thousands of times lower than that expected from a 50 kiloton atomic bomb, such an experiment could run for months or a year in contrast to the one or two seconds expected for a bomb experiment. In addition, Reines and Cowan had devised a new method for detecting the antineutrino.

The new method for detecting the neutrino made use of both the positron and the neutron produced in inverse β decay ($\nu + p \rightarrow n + e^+$). The detector consisted of a large (10.7 cubic feet) tank of organic liquid scintillator, in which a cadmium compound had been dissolved. The cadmium was present because it had a high probability for absorbing neutrons. The tank was surrounded by ninety photomultiplier tubes, which would detect the γ rays produced. The positron from the reaction will almost immediately annihilate with an electron and produce two γ rays that are emitted back to back, each with an energy of 0.51 MeV. The neutron will wander through the scintillator for a time, approximately 5 microseconds on average, before it is absorbed by a cadmium nucleus. (The reaction is shown schematically in Figure 1. The apparatus sketched is from the second version of the experiment, discussed in detail below.) The excited nucleus produced will emit several $\gamma$ rays, which will be detected by the liquid scintillator. Thus, a signal for inverse $\beta$ decay will be an immediate $\gamma$ ray signal from the annihilation of the positron in delayed coincidence with the $\gamma$ rays from the neutron capture by the cadmium nucleus.

In principle this is a straightforward measurement. The apparatus is turned on and the number of delayed coincidences is counted. In practice things are very different. In almost every experiment there will be backgrounds, events or effects which mimic or mask the desired effect. This is a particularly serious problem when one is looking for such rare events, such as inverse $\beta$ decay. When the apparatus was turned on, the experimenters found that there was a large delayed-coincidence background of approximately 5 counts/minute. This was not only far larger than the expected signal for inverse $\beta$ decay, but was also independent of whether the nuclear reactor was on or off. It was not due to antineutrinos, or any other particles, coming from the reactor. The experimenters found that it was due to cosmic rays, particles passing through the apparatus from the atmosphere, that produced both neutrons and $\gamma$ rays in the shielding surrounding the experimental apparatus. Steps were taken to reduce the background. In particular, a bank of Geiger counters was placed around the apparatus, which served as a veto counter and reduced the background, although there was still a significant amount present. The results of this first experiment are shown in Table 1. A difference in delayed-coincidence rate of 0.41 ± 0.20 counts/minute was found between the reactor-on and reactor-off counting rates, over a background of approximately 2 counts/minute. This difference was attributed to inverse $\beta$ decay. It was in rough agreement with the approximately 0.2 counts per minute expected theoretically. The experimenters concluded cautiously, an experiment has been performed to detect the free neutrino. It appears probable that this aim has been accomplished although further confirmatory work is in progress.[4]

**Table 1. Data from the Hanford Experiment of Reines and Cowan (1953)**

| Run | Pile Status | Length of Run (Seconds) | Net Delayed Pair Rate (Counts/min) | Accidental Background Rate (Counts/min) |
|---|---|---|---|---|
| 1 | up | 4000 | 2.56 | 0.84 |
| 2 | up | 2000 | 2.46 | 3.54 |
| 3 | up | 4000 | 2.58 | 3.11 |
| 4 | down | 3000 | 2.20 | 0.45 |
| 5 | down | 2000 | 2.02 | 0.15 |
| 6 | Down | 1000 | 2.19 | 0.13 |

| | |
|---|---|
| Average: Pile up | 2.55 ± 0.15 |
| Pile down | 2.14 ± 0.13 |
| Difference due to pile | 0.41 ± 0.20 |

The first Reines-Cowan result was suggestive, but certainly not conclusive. The background effects were five times larger than the observed signal, and the signal itself was hardly statistically overwhelming. The final result, $0.41 \pm 0.20$ counts/minute, was only two standard deviations from zero, or no effect. There was a five percent chance that the result was, in fact, zero.

Reines, Cowan, and their collaborators improved their apparatus and performed a second search for the free antineutrino in an experiment at the Savannah River nuclear reactor. The new experiment had a much larger signal to background ratio and very careful arguments were presented to show that the signal was due to inverse β decay caused by antineutrinos from the reactor. Their results were first presented in Reference 6,[5] with a more detailed account in Reference 7.

The detection scheme is shown schematically in Figure [1]. An antineutrino (v) from the fission products in a powerful production reactor is incident on a water target in which $CdCl_2$ has been dissolved. By reaction (1) [$v + p \rightarrow n + e^+$], the incident v produces a positron ( $e^+$) and a neutron (n). The positron slows down and annihilates with an electron in a time short compared with the 0.2-μsec resolving time characteristic of our system, and the resulting two 0.5 Mev annihilation gamma rays penetrate the target and are detected in prompt coincidence by the two large scintillation detectors placed on opposite sides of the target. The neutron is moderated by the water and then captured by the cadmium in a time dependent on the cadmium concentration (in our experiment practically all the neutrons are captured within 10 μsec of their production). The multiple cadmium-capture gamma rays are detected in prompt coincidence by the two scintillation detectors, yielding a characteristic delayed-coincidence count with the preceding $e^+$ gammas (Ref. 4, p. 159).

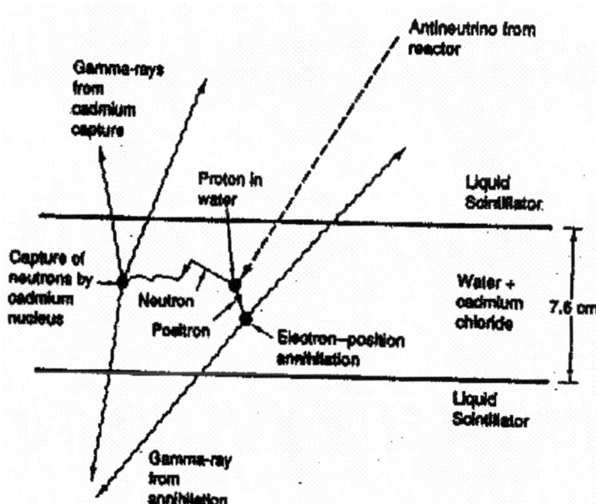


**Figure 1.** Schematic diagram of the antineutrino experiment of Reines and Cowan (Ref. 7).

The experimenters found a clear signal that depended on the reactor power. The signal was far larger with the reactor on than it was with the reactor off. These results are shown in Table 2. For runs (c) and (d) small changes were made in the operating conditions to reduce background. For the first part of the series [(a) and (b)] the total net rate (reactor on minus reactor off) was 1.23 ± 0.24 $hr^{-1}$. For the second part [(c) and (d)] (during which small changes were made) the net rate was 0.93 ± 0.22. From these data we conclude that there was a reactor-associated signal (Ref. 4, p. 163).

**Table 2. Results of the Savannah River Experiment (Ref. 7)**

| **Counter Triad** | **υ Flux Factor** | **Run length (Hours)** | **Total Counts** | **Calculated Accidentals** | **Net Rate ($hr^{-1}$)** |
|---|---|---|---|---|---|
| (a) Top | 1.03 | 192.7 | 283 | 144 | 0.88 ± 0.10 |
| Bottom | 1.04 | 171.8 | 224 | 95 | 0.75 ± 0.10 |
| (b) Top | 0 | 67.3 | 55 | 31.8 | 0.34 ± 0.14 |
| Bottom | 0 | 69.7 | 44 | 39.7 | 0.06 ± 0.13 |
| (c)Top | 0 | 63.2 | 48 | 27.6 | 0.32 ± 0.14 |
| Bottom | 0 | 63.2 | 38 | 35.3 | 0.04 ± 0.14 |
| (d) Top | 1.07 | 264.5 | 320 | 124.9 | 0.74 ± 0.08 |
| Bottom | 1.07 | 264.5 | 302 | 157.2 | 0.55 ± 0.08 |

The poltergeist had been observed.

## THE APPEARANCE AND DISAPPEARANCE OF THE 17-KEV NEUTRINO

The Reines-Cowan experiment was a great success. Let us now consider an episode in which a hypothesis turned out to be incorrect. This was the suggestion of the existence of a heavy, 17-keV neutrino. Although the consensus is that such a particle does not exist, it is also an example of good science because the method by which that decision was reached was appropriate. In 1985 John Simpson discovered a heavy, 17-keV neutrino. He had searched for such a neutrino by looking for a kink in the β-decay energy spectrum, or in the Kurie plot, at an energy equal to the maximum allowed decay energy minus the mass of the neutrino in energy units. If such a heavy neutrino existed then there would be a kink in the energy spectrum caused by the superposition of two energy spectra; one in which the ordinary low-mass neutrino was emitted and one in which the heavy neutrino was emitted. Simpson's original experimental result for the fractional change in the Kurie plot of tritium, a more sensitive measurement than the energy spectrum, is shown in Figure 2. A marked change in slope, a kink, is clearly seen at $T_\beta$, the electron energy, equal to 1.5 keV, corresponding to a neutrino mass of 17 keV (the maximum decay energy is 18.6 keV). Simpson concluded, in summary, the β spectrum of tritium recorded in the present experiment is consistent with the emission of a heavy neutrino of mass about 17.1 keV and a mixing probability [the fraction of heavy neutrinos] of about 3%.[6]

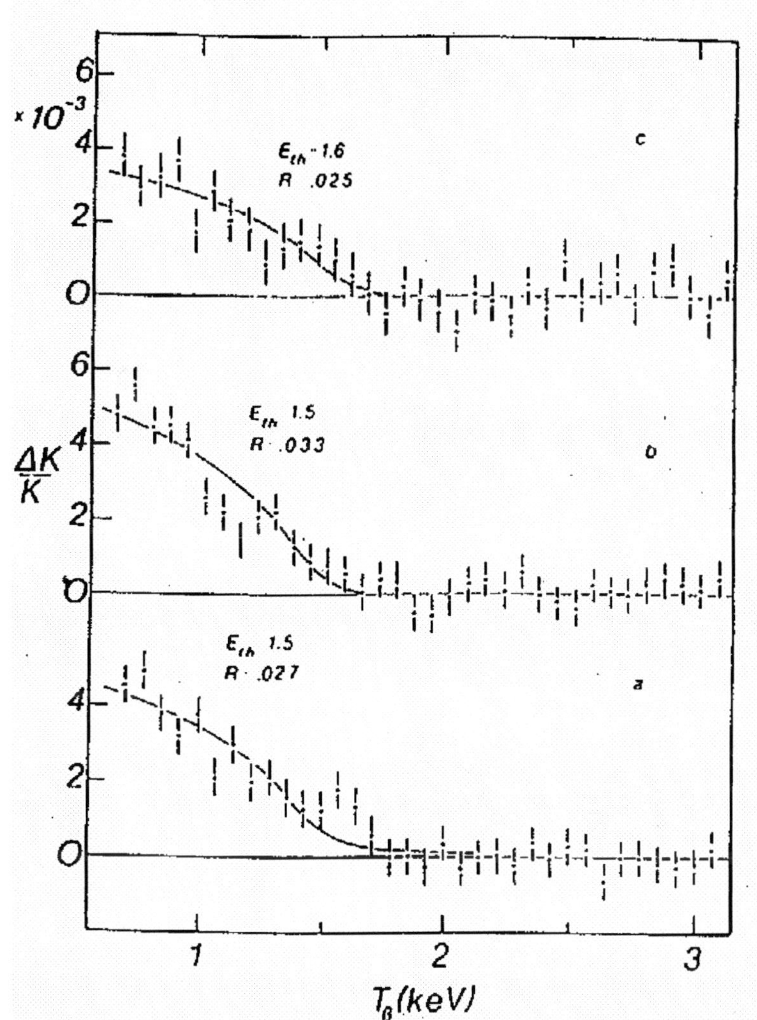


**Figure 2.** . The data of three runs presented as ΔK/K (the fractional change in the Kurie plot) as a function of the kinetic energy of the β particles. $E_{th}$ is the threshold energy, the difference between the endpoint energy and the mass of the heavy neutrino. A kink is clearly seen at $E_{th}$ = 1.5 keV, or at a mass of 17.1 keV. Run a included active pileup rejection, whereas runs *b* and *c* did not. *c* was the same as *b* except that the detector was housed in a soundproof box. No difference is apparent. From (Ref. 8)

Within a year of Simpson's first report there were five attempted replications of his experiment. These initial replications all gave negative results. Initial theoretical suggestions were made that attempted to explain Simpson's result using accepted physics, without the need for a heavy neutrino. Theoretical physicists took two broad approaches. One group rejected the heavy neutrino altogether. Among them, some attributed the positive observations to either systematic instrumental effects or to faulty analysis. Others thought that the observations required a deeper explanation, and they provided one, but without introducing a new particle. A second group accepted the positive observation. They also divided into two. Some in this group presumed the existence of the particle and sought to trace out its implications. Others proposed an altogether new theory, one that incorporated the particle at a fundamental level.

During the next few years the chilling effect of the negative results combined with possible theoretical explanations of the result which did not involve a heavy neutrino resulted in very little experimental work being done on the 17-keV neutrino. That which was done also gave negative results. The most stringent limits on the presence of the 17-keV neutrino were set by Hetherington et al.,[7] who set an upper limit of 0.3% for the mixing probability. This experiment was generally regarded as the most complete and convincing magnetic spectrometer experiment done to that point.[8]

In 1989 the tide started to turn in favor of the existence of the 17-keV neutrino. In April of that year Simpson and Hime reported positive results from two new experiments, one using a different tritium-implanted detector and the second using a $^{35}S$ source, the same element used in the original negative experiments.[9] It was deemed important to check the earlier result by measuring the $^{3}H$ $\beta$ spectrum in a different detector (Hime and Simpson, Ref. 11, p. 1837). These new results confirmed the existence of the 17-keV neutrino, but reported reduced mixing probabilities, (0.6 - 1.6)% and (0.73 ± 0.09)%, for $^{3}H$ and $^{35}S$, respectively. These were approximately a factor of three lower than Simpson's original result.

Hime and Jelley[10] reported a positive, high-statistics result using a $^{35}S$ source, the same as that used in the original negative experiments. They found a mass for the heavy neutrino of (17.0 ± 0.4) keV and a mixing probability of (0.84 ± 0.06 ± 0.05) %. This high-statistics extension of the Simpson technique with an improved instrumental geometry is, by consensus, the most compelling of the experimental results that claim to see the 17-keV neutrino.[11]

A second persuasive positive result came from a group at Berkeley.[12] This experiment was similar to that originally used by Simpson. It used radioactive $^{14}C$ embedded in a solid-state detector, and included an anticoincidence ring in the detector to veto events in which the full energy of the electron was not deposited in the detector, and thus distort the spectrum. They found (17 ± 2) keV and (1.4 ± 0.45 ± 0.14)% for the mass of the heavy neutrino and its mixing probability, respectively , Awhich supports the claim by Simpson that there is a 17-keV neutrino emitted with ~1% probability in $\beta$ decay (Ref. 14, p. 2447).

This new evidence persuaded many physicists that there was a 17-keV neutrino. Bert Schwarzschild wrote an article in the May 1991 issue of *Physics Today*, the semi-popular magazine of the American Physical Society entitled Four of Five New Experiments Claim Evidence for 17-keV Neutrinos (Ref. 13). He also injected a note of caution. On one thing everyone seems to agree. After six years, the experimenters must begin to resolve the stubborn discrepancy between the two different styles of beta-decay experiment [solid-state detectors and magnetic spectrometers] (Ref. 13, p. 19).

The year 1991 was the high point in the life of the 17-keV neutrino. Although the evidence for its existence was far from conclusive, its existence had been buttressed by the recent results of Simpson and Hime, of Hime and Jelley, and of the Berkeley group. From this point on, however, the evidence would be almost exclusively against it. Not only would there be high-statistics, extremely persuasive negative results, but serious questions would also be raised about its strongest support.

The uncertainty of the evidential situation with respect to the existence of the 17-keV neutrino was emphasized in the work of Bonvicini (Ref. 10). He supported Simpson's view that the early negative magnetic spectrometer results were not as decisive as claimed. In this work Bonvicini discussed the question of whether or not a kink in the energy spectrum due to an admixture of a 17-keV neutrino could be masked by the presence of unknown distortions, such as the shape correction factors used in magnetic spectrometer experiments or even mimicked by a combination of such distortions and statistical effects. He concluded that it could.

Support for the existence of the 17-keV neutrino began to erode in 1992. In early August, the Berkeley group presented a conference report which included a statistically improved result from $^{14}C$ with a neutrino mass of $17 \pm 1$ keV and a mixing probability of $(1.26 \pm 0.25)\%$.[13] This result was in agreement with their earlier result that had supported the existence of the 17-keV neutrino. They also reported, however, a high statistics measurement of the inner bremsstrahlung spectrum of $^{55}Fe$ and find no indication of the emission of a 17-keV neutrino (p. Ref. 15, p. 1123). They also cited three other recent negative results; two with magnetic spectrometers on $^{35}S$ and $^{63}Ni$, and one on $^{35}S$, that used a solid-state silicon detector. (Two of these experiments, those of Ohshima et al. and of Mortara et al., will be discussed below.) These negative results, along with their own result on $^{55}Fe$, led them to question their $^{14}C$ result and to state, We thus conclude that, whatever causes the kink in our $^{14}C$ spectrum, it is not a neutrino (Ref. 15, p. 1126). They also began a careful search for problems in their experiment.

The magnetic spectrometer experiment on $^{63}Ni$ by the Tokyo group was also presented at that conference.[14] The experimenters noted some of the problems of experiments that used wide energy regions and commented that, We have concentrated on performing a measurement of high statistical accuracy, in a narrow energy region, using very fine energy steps. Such a restricted energy scan ...also reduced the degree of energy-dependent corrections and other related systematic uncertainties (Kawakami et al., Ref. 16, p. 45). Their best value for the mixing probability of a 17-keV neutrino was ($-0.011 \pm 0.033$ (statistical) $\pm 0.030$ (systematic)) %, with an upper limit for the mixing probability of 0.073% at the 95% confidence level. This was the most stringent limit yet.

The 17-keV neutrino received another severe blow when Hime, following a suggestion of Piilonen and Abashian, extended his calculation of the electron response function of his detector to include electron scattering effects that had not been previously included. He found that he could fit the positive results of Hime and Jelley without the need for a 17-keV neutrino.[15] This seemed to remove one of the most persuasive pieces of evidence for the heavy neutrino. It will be shown that scattering effects are sufficient to describe the Oxford β-decay measurements .(Ref. 17, p. 166).

Further evidence against the 17-keV neutrino was provided by the Argonne group headed by Stuart Freedman.[16] The experiment used a solid-state, Si(Li), detector, the same type used by Simpson, an external $^{35}S$ source, and a solenoidal magnetic field to focus the decay electrons. The present experiment requires that we know the electron response function between 120 and 167 keV. Measurements of the conversion lines of $^{139}Cs$ at 127, 160, and 167 keV are the principal constraint on the model of the electron response function (Ref. 18, p. 395). Previous $^{35}S$ experiments had used an electron response function extrapolated from the lower energy $^{57}Co$ lines. Finally, and perhaps most importantly, this experiment required no arbitrary shape correction factor, which Bonvicini had shown could create problems.

The experimenters also demonstrated the sensitivity of their apparatus to a possible 17-keV neutrino. If there were a 17-keV neutrino then their apparatus would have detected it. The experimenters used a $^{35}S$ source with a one percent admixture of $^{14}C$. The two spectra have different end-point energies and their combination would produce a kink in the spectrum similar to that expected for a 17-keV neutrino.

To assess the reliability of our procedure, we introduced a known distortion into the $^{35}S$ beta spectrum and attempted to detect it. A drop of $^{14}C$-doped valine ($E_o$ - $m_e$ – 156 keV) was deposited on a carbon foil and a much stronger $^{35}S$ source was deposited over it. The data from the composite source were fitted using the $^{35}S$ theory, ignoring the $^{14}C$ contaminant. The residuals are shown in Figure [3]. The distribution is not flat; the solid curve shows the expected deviations from the single component spectrum with the measured amount of $^{14}C$. The fraction of decays from $^{14}C$ determined from the fit to the beta spectrum is (1.4 ± 0.1)%. This agrees with the value of 1.34% inferred from measuring the total decay rate of the $^{14}C$ alone while the source was being prepared. This exercise demonstrates that our method is sensitive to a distortion at the level of the positive experiments. Indeed, the smoother distortion with the composite source is more difficult to detect than the discontinuity expected from the massive neutrino.(Ref.18,p.396).

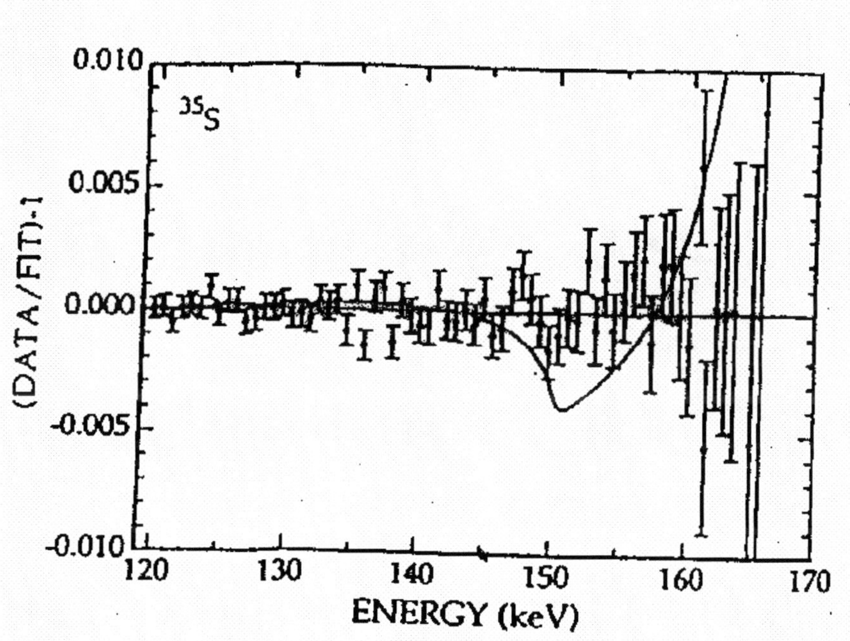


**Figure 3.** Residuals from fitting the beta spectrum of a mixed source of $^{14}$C and $^{35}$S with a pure $^{35}$S shape; the reduced $\chi^2$ of the data is 3.59. The solid curve indicates residuals expected from the known $^{14}$C contamination. The best fit yieldsa mixing of (1.4 ∀ 0.1)% and reduced $\chi^2$ of 1.06. From (Ref. 18).

Their final result, shown in Figure [4], was $\sin^2\theta$ = -0.0004 ± 0.0008 (statistical) ± 0.0008 (systematic), for the mixing probability of the 17-keV neutrino. In conclusion, we have performed a solid-state counter search for a 17 keV neutrino with an apparatus with demonstrated sensitivity. We find no evidence for a heavy neutrino, in serious conflict with some previous reports (Ref. 18, p. 396).

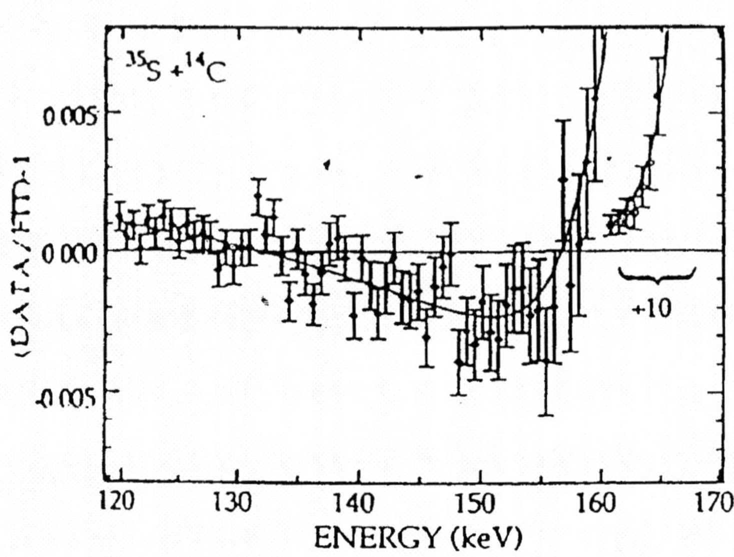


**Figure 4.** Residuals from a fit to the pile-up corrected $^{35}S$ data assuming no massive neutrino; the reduced $\chi^2$ for the fit is 0.88. The solid curve represents the residuals expected for decay with a 17-keV neutrino and $\sin^2\theta = 0.85\%$; the reduced $\chi^2$ of the data is 2.82. From (Ref. 18).

The Berkeley group had also found an error in their experiment. They had originally used an anticoincidence ring to veto events in which the full electron energy was not deposited in the detector. They had found that there was, in fact, energy sharing between the ring and the main detector, which caused a distortion in the spectrum that mimicked the presence of a 17-keV neutrino. There was virtually no evidence left that supported the existence of the 17-keV neutrino. The positive results of both Hime and Jelley and of the Berkeley group had been shown to be incorrect. There was, in addition, very persuasive evidence from both the Tokyo and Argonne groups that the 17-keV neutrino did not exist.

The consensus of the physics community was that the 17-keV neutrino does not exist. The kink was dead.

Thus, in both the observation of the neutrino and the failure to observe proposed a 17-keV neutrino we have seen examples of good science.

# REFERENCES


[1] C.D. Ellis and W. A. Wooster, *Proc. Roy. Soc. (London)* **A117,** 109-123 (1927).
[2] E. Fermi, *Nuovo Cimento* **11**, 1-21 (1934) and *Zeit. Phys* **88,** 161-177 (1934)
[3] H.A. Bethe and R. F. Bacher, *Rev. Mod. Phys.* **8,** 82-229 (1936).
[4] F. Reines and C. L. Cowan, *Phys. Rev.* **92,** 830-831 (1953).
[5] C.L. Cowan, F. Reines, F. B. Harrison, et al., *Science* **124**, 103-104 (1956).
[6] J.J. Simpson, *Phys. Rev. Letters* **54,** 1891-1893 (1985).
[7] D.W. Hetherington, R. L. Graham, M. A. Lone, et al., *Phys. Rev. C* **36,** 1504-1513 (1987).
[8] G. Bonvicini, *Zeit. Phys. A* **345,** 97-117 (1993).
[9] A. Hime and J. J. Simpson, *Phys. Rev. D* **39,** 1837-1850 (1989) and J.J. Simpson. and A. Hime, *Phys. Rev. D* **39,** 1825-1836 (1989).
[10] A. Hime and N. A. Jelley, *Phys. Letters* **257B,** 441-449 (1991).
[11] B. Schwarzschild, *Physics Today* **44**(5), 17-19 (1991).
[12] B. Sur, E. B. Norman, K. T. Lesko, et al., *Phys. Rev. Letters* **66,** 2444-2447 (1991).
[13] E.B. Norman, Y. Chan, M. T. F. Da Cruz, et al., XXVI International Conference on High Energy Physics, Dallas, American Institute of Physics (1992).
[14] H. Kawakami, H., S. Kato, T. Ohshima, et al., *Phys. Letters* **287B,** 45-50 (1992); T. Ohshima, *XXVI International Conference on High Energy Physics*, edited by J. R. Sanford. Dallas: American Institute of Physics. **1:** 1128-1135 (1993); and T. Ohshima, T., H. Sakamoto, T. Sato, et al., *Phys. Rev. D* **47,** 4840-4856 (1993).
[15] A. Hime, *Phys. Letters* **299B,** 165-173 (1993).
[16] J.L. Mortara, I. Ahmad, K. P. Coulter, et al., *Phys. Rev. Letters* **70,** 394-397 (1993).

# Physics Potential of Beta/EC Beams in regard to CP Violation in Neutrino Oscillations

Jose Bernabeu and Catalina Espinoza

*Universitat de València and IFIC, E-46100 Burjassot, València, Spain*

**Abstract.** The goal for future neutrino facilities is the determination of the $[U_{e3}]$ mixing and CP violation in neutrino oscillations. This will require precision experiments with a very intense neutrino source. With this objective the creation of neutrino beams from the radioactive decay of boosted ions by the SPS of CERN from either beta or electron capture transitions has been propossed. We discuss the capabilities of such facilities as a function of the energy of the boost and the baseline for the detector. We conclude that the SPS upgrade to 1000 GeV is crucial to have a better sensitivity to CP violation if it is accompanied by a longer baseline. We compare the physics potential for two different configurations. In the case of beta beams, with the same boost for both $\beta^+$ (neutrinos) and $\beta^-$ (antineutrinos), the two setups are: I) $\gamma = 120$, $L = 130$ Km (Frejus); II) $\gamma = 330$, $L = 650$ Km (Canfranc). In the case of monochromatic EC beams we exploit the energy dependence of neutrino oscillations to separate out the two parameters $U(e3)$ and the CP phase $\delta$. Setup I runs at $\gamma = 90$ and $\gamma = 195$ (maximum achievable at present SPS) to Frejus, whereas Setup II runs at $\gamma = 195$ and $\gamma = 440$ (maximum achievable at upgraded SPS) to Canfranc. The main conclusion is that, whereas the gain in the determination of $U(e3)$ is rather modest, setup II provides much better sensitivity to CP violation.

**Keywords:** Neutrino Oscillation, CP Violation, Beta/EC Beams
**PACS:** 14.60.Pq, 11.30.Er, 29.25.Rm

## 1. INTRODUCTION

Neutrinos are very elusive particles that are difficult to detect. Even so, physicists have over the last decades successfully studied neutrinos from a wide variety of sources, either natural, such as the sun and cosmic objects, or manmade, such as nuclear power plants or accelerated beams. Spectacular results have been obtained in the last few years for the flavour mixing of neutrinos obtained from atmospheric, solar, reactor and accelerator sources and interpreted in terms of the survival probabilities for the beautiful quantum phenomenon of neutrino oscillations [1, 2]. The weak interaction eigenstates $\nu_\alpha$ $(\alpha = e, \mu, \tau)$ are written in terms of mass eigenstates $\nu_k$ $(k = 1,2,3)$ as $\nu_\alpha = \sum_k U_{\alpha k}(\theta_{12}, \theta_{23}, \theta_{13}; \delta)\nu_k$, where $\theta_{ij}$ are the mixing angles among the three neutrino families and $\delta$ is the *CP* violating phase. Neutrino mass differences and the mixings for the atmospheric $\theta_{23}$ and solar $\theta_{12}$ sectors have thus been determined. The third connecting mixing $|U_{e3}|$ is bounded as $\theta_{13} \leq 10°$ from the CHOOZ reactor experiment [3]. The third angle, $\theta_{13}$, as well as the CP-violating phases $\delta$, remain thus undetermined. Besides the approved experiments Double CHOOZ [4], T2K [5] and NOVA [6], a number of experimental facilities to significantly improve on present sensitivity have been discussed in the literature: neutrino factories (neutrino beams from boosted-muon decays) [7, 8, 9], superbeams (very intense conventional neutrino beams) [10, 11, 12, 13] improved reactor experiments [14] and more recently $\beta$-beams [15]. The original

CP917, *Particles and Fields,* edited by H. Castilla Valdez, J. C. D'Olivo, and M. A. Perez

standard scenario for beta beams with lower $\gamma = 60/100$ and short baseline $L = 130$ Km from CERN to Frejus with $^{6}He$ and $^{18}Ne$ ions could be improved using an electron capture facility for monochromatic neutrino beams [16]. New proposals also include the high $Q$ value $^{8}Li$ and $^{8}Be$ isotopes in a $\gamma = 100$ facility [17]. In this paper we discuss the physics reach that a high energy facility for both beta beams [18] and EC beams may provide with the expected SPS upgrade at CERN. In Section 2 we discuss the virtues of the suppressed oscillation channel ($\nu_e \to \nu_\mu$) in order to have access to the parameters $\theta_{13}$ and $\delta$. The interest of energy dependence, as obtainable in the EC facility, is emphasized. In Section 3 we compare the beta beam capabilities at different energies and baselines using two ions, one for neutrinos, the other for antineutrinos. In Section 4 we present new results on the comparison between (low energies, short baseline) and (high energies, long baseline) configurations for an EC facility with a single ion. Section 5 gives our conclusions and outlook.

## 2. SUPPRESSED NEUTRINO OSCILLATION

The observation of *CP* violation needs an experiment in which the emergence of another neutrino flavour is detected rather than the deficiency of the original flavour of the neutrinos. At the same time, the interference needed to generate CP-violating observables can be enhanced if both the atmospheric and solar components have a similar magnitude. This happens in the suppressed $\nu_e \to \nu_\mu$ transition. The appearance probability $P(\nu_e \to \nu_\mu)$ as a function of the distance between source and detector ($L$) is given by [19]

$$P(\nu_e \to \nu_\mu) \simeq s_{23}^2 \sin^2 2\theta_{13} \sin^2\left(\frac{\Delta m_{13}^2 L}{4E}\right) + c_{23}^2 \sin^2 2\theta_{12} \sin^2\left(\frac{\Delta m_{12}^2 L}{4E}\right) + \tilde{J} \cos\left(\delta - \frac{\Delta m_{13}^2 L}{4E}\right) \frac{\Delta m_{12}^2 L}{4E} \sin\left(\frac{\Delta m_{13}^2 L}{4E}\right), \quad (1)$$

where $\tilde{J} \equiv c_{13} \sin 2\theta_{12} \sin 2\theta_{23} \sin 2\theta_{13}$. The three terms of Eq. (1) correspond, respectively, to contributions from the atmospheric and solar sectors and their interference. As seen, the *CP* violating contribution has to include all mixings and neutrino mass differences to become observable. The four measured parameters $(\Delta m_{12}^2, \theta_{12})$ and $(\Delta m_{23}^2, \theta_{23})$ have been fixed throughout this paper to their mean values [20].

Neutrino oscillation phenomena are energy dependent (see Fig.1) for a fixed distance between source and detector, and the observation of this energy dependence would disentangle the two important parameters: whereas $|U_{e3}|$ gives the strength of the appearance probability, the *CP* phase acts as a phase-shift in the interference pattern. These properties suggest the consideration of a facility able to study the detailed energy dependence by means of fine tuning of a boosted monochromatic neutrino beam. As shown below, in an electron capture facility the neutrino energy is dictated by the chosen boost of the ion source and the neutrino beam luminosity is concentrated at a single known energy which may be chosen at will for the values in which the sensitivity for the $(\theta_{13}, \delta)$

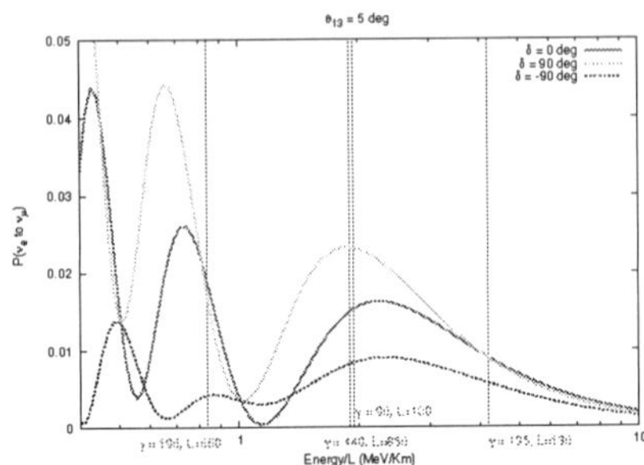


**FIGURE 1.** The appearance probability $P(\nu_e \to \nu_\mu)$ for neutrino oscillations as a function of the LAB energy E/L, with fixed connecting mixing. The three curves refer to different values of the CP violating phase $\delta$. The vertical lines are the energies of our simulation study in the EC facility.

parameters is higher. This is in contrast to beams with a continuous spectrum, where the intensity is shared between sensitive and non sensitive regions. Furthermore, the definite energy would help in the control of both the systematics and the detector background. In the beams with a continuous spectrum, the neutrino energy has to be reconstructed in the detector. In water-Cerenkov detectors, this reconstruction is made from supposed quasielastic events by measuring both the energy and direction of the charged lepton. This procedure suffers from non-quasielastic background, from kinematic deviations due to the nuclear Fermi momentum and from dynamical suppression due to exclusion effects [21].

From general arguments of CPT invariance and absence of absorptive parts the CP-odd probability is odd in time and then odd in the baseline L (formally). Vacuum oscillations are only a function of $E/L$ so that, at fixed $L$, the CP-odd probability is odd in the energy (formally). This proves that the study of neutrino oscillations in terms of neutrino energy will be able to separate out the CP phase $\delta$ from the mixing parameters. A control of this energy may be obtained from the choice of the boost in the EC facility with a single ion. In order for this concept to become operational it is necessary to combine it with the recent discovery of nuclei far from the stability line, having super allowed spin-isospin transitions to a giant Gamow-Teller resonance kinematically accessible [22]. Thus the rare-earth nuclei above $^{146}Gd$ have a small enough half-life for electron capture processes. This is in contrast with the proposal of EC beams with fully stripped long-lived ions [23]. We discuss the option of short-lived ions [16].

## 3. BETA-BEAM CAPABILITIES AT DIFFERENT ENERGIES AND BASELINES

A first question to be answered is: Is the sensitivity to CP violation and $\theta_{13}$ changing with energy at fixed baseline? Fixing the baseline to CERN-Frejus ($L = 130$ Km) [18], one notices that, for $\gamma > 80$, the sensitivity to both $\theta_{13}$ and $\delta$ changes rather slowly because the flux at low energies in the continuous spectrum does not reduce significantly. Then it is not advantageous to increase the neutrino energy unless the baseline is

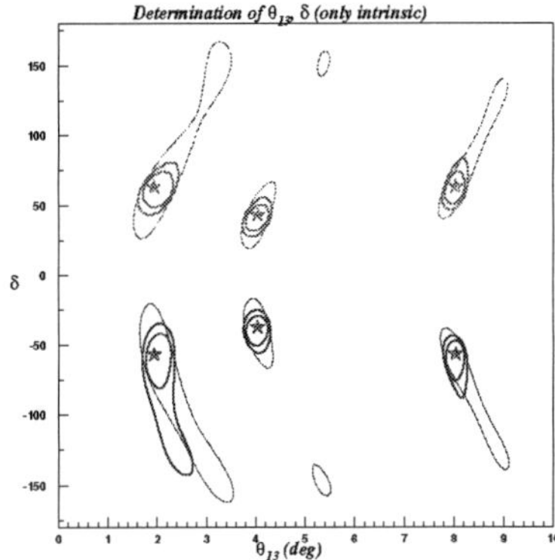


**FIGURE 2.** Determination of $(\theta_{13}, \delta)$ for the Setups I) and II), as explained in the text.

correspondingly scaled to remain close to the atmospheric oscillation maximum as suggested by the $E/L$ dependence.

With the present SPS of CERN the maximum energy reachable for the $^6He$ ion corresponds to $\gamma = 150$. Fixing this value of $\gamma$ for both $^6He$ and $^{18}Ne$ we may ask: Is the sensitivity to $\theta_{13}$ and $\delta$ changing with the baseline? Particularly for the CP phase $\delta$, $L = 300$ Km is clearly favoured [18]. There is neither an existing nor an envisaged laboratory at this particular distance from CERN.

Equipped with these previous results, it is of interest to make a comparison between the physics reach for two different Beta Beam Setups. With the same $\gamma$ for both neutrino and antineutrino sources, Setup I corresponds to $\gamma = 120$, $L = 130$ Km (Frejus), whereas Setup II is for $\gamma = 330$, $L = 650$ Km (Canfranc). Setup II needs the upgrade of the SPS until proton energies of 1000 GeV. The associated determinations of $\theta_{13}$ and $\delta$ are presented in Fig. 2. The main conclusion is that Setup II is clearly better for the CP violating phase. Not only the high energy Setup provides a better precision, but it is able to resolve the degeneracies. In [18] one may find the associated sensitivities of these Setups for each parameter $\theta_{13}$ and $\delta$. As a consequence, a *R&D* effort to design Beta Beams for the upgraded CERN SPS ($E_p = 1000$ GeV) appears justified.

## 4. EC-BEAM CAPABILITIES AT DIFFERENT ENERGIES AND BASELINES

Electron Capture is the process in which an atomic electron is captured by a proton of the nucleus leading to a nuclear state of the same mass number $A$, replacing the proton by a neutron, and a neutrino. Its probability amplitude is proportional to the atomic wavefunction at the origin, so that it becomes competitive with the nuclear $\beta^+$ decay at high $Z$. Kinematically, it is a two body decay of the atomic ion into a nucleus and the neutrino, so that the neutrino energy is well defined and given by the difference

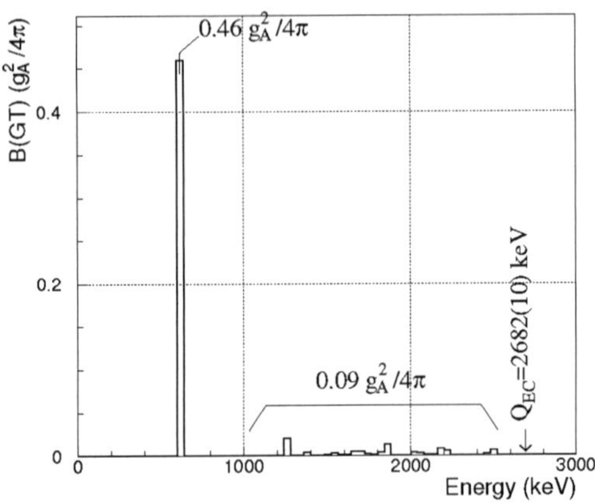


**FIGURE 3.** Gamow-Teller strength distribution in the $EC/\beta^+$ decay of $^{148}Dy$.

between the initial and final atomic masses $(Q_{EC})$ minus the excitation energy of the final nuclear state. In general, the high proton number $Z$ nuclear beta-plus decay $(\beta^+)$ and electron-capture $(EC)$ transitions are very "forbidden", i.e., disfavoured, because the energetic window open $Q_\beta/Q_{EC}$ does not contain the important Gamow-Teller strength excitation seen in (p,n) reactions. There are a few cases, however, where the Gamow-Teller resonance can be populated (see Fig.3) having the occasion of a direct study of the "missing" strength. For the rare-earth nuclei above $^{146}Gd$, the filling of the intruder level $h_{11/2}$ for protons opens the possibility of a spin-isospin transition to the allowed level $h_{9/2}$ for neutrons, leading to a fast decay. Our studies for neutrino beam preparation have used the $^{150}Dy$ ion with half live of 7.2 min, a Branching Ratio to neutrino channels of 64% (fully by EC) and neutrino energy of 1.4 MeV in the C.M. frame as obtained from its decay to the single giant Gamow-Teller resonance in the daugther $^{150}Tb^*$. A neutrino (of energy $E_0$) that emerges from radioactive decay in an accelerator will be boosted in energy. At the experiment, the measured energy distribution as a function of angle $(\theta)$ and Lorentz gamma $(\gamma)$ of the ion at the moment of decay can be expressed as $E = E_0/[\gamma(1-\beta\cos\theta)]$. The angle $\theta$ in the formula expresses the deviation between the actual neutrino detection and the ideal detector position in the prolongation of one of the long straight sections of the Decay Ring. The neutrinos are concentrated inside a narrow cone around the forward direction. If the ions are kept in the decay ring longer than the half-life, the energy distribution of the Neutrino Flux arriving to the detector in absence of neutrino oscillations is given by the Master Formula

$$\begin{aligned} \frac{d^2N_\nu}{dSdE} &= \frac{1}{\Gamma}\frac{d^2\Gamma_\nu}{dSdE}N_{ions} \\ &\simeq \frac{\Gamma_\nu}{\Gamma}\frac{N_{ions}}{\pi L^2}\gamma^2\delta(E-2\gamma E_0), \end{aligned} \tag{2}$$

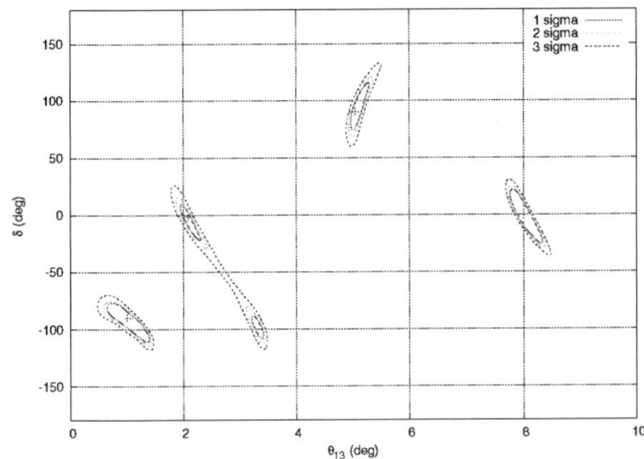


**FIGURE 4.** Setup I. Fit for $(\theta_{13}, \delta)$ from statistical distribution.

with a dilation factor $\gamma >> 1$. It is remarkable that the result is given only in terms of the branching ratio and the neutrino energy and independent of nuclear models. In Eq. (2), $N_{ions}$ is the total number of ions decaying to neutrinos. For an optimum choice with $E \sim L$ around the first oscillation maximum, Eq. (2) says that lower neutrino energies $E_0$ in the proper frame give higher neutrino fluxes. The number of events will increase with higher neutrino energies as the cross section increases with energy. To conclude, in the forward direction the neutrino energy is fixed by the boost $E = 2\gamma E_0$, with the entire neutrino flux concentrated at this energy. As a result, such a facility will measure the neutrino oscillation parameters by changing the $\gamma$'s of the decay ring (energy dependent measurement) and there is no need of energy reconstruction in the detector.

For the study of the physics reach associated with such a facility, we combine two different energies for the same $^{150}Dy$ ion using two Setups. In all cases we consider $10^{18}$ decaying ions/year, a water Cerenkov Detector with fiducial mass of 440 Kton and both appearance ($\nu_\mu$) and disappearance ($\nu_e$) events. Setup I is associated with a five year run at $\gamma = 90$ (close to the minimum energy to avoid atmospheric neutrino background) plus a five year run at $\gamma = 195$ (the maximum energy achievable at present SPS), with a baseline $L = 130$ Km from CERN to Frejus. The results for Setup I are going to be compared with those for Setup II, associated with a five year run at $\gamma = 195$ plus a five year run at $\gamma = 440$ (the maximum achievable at the upgraded SPS with Proton energy of 1000 GeV), with a baseline $L = 650$ Km from CERN to Canfranc.

For the Setup I we generate the statistical distribution of events from assumed values of $\theta_{13}$ and $\delta$. The corresponding fit is shown in Fig. 4 for selected values of $\theta_{13}$ from $6^o$ to $1^o$ and covering a few values of the CP phase $\delta$. As observed, the principle of an energy dependent measurement (illustrated here with two energies) is working to separate out the two parameters. With this configuration the precision obtainable for the mixing (even at 1 degree) is much better than that for the CP phase.

The corresponding exclusion plots which define the sensitivity to discover a non-vanishing mixing $\theta_{13} \neq 0$ and CP violation $\delta \neq 0$ are presented in Fig. 5 and Fig. 6

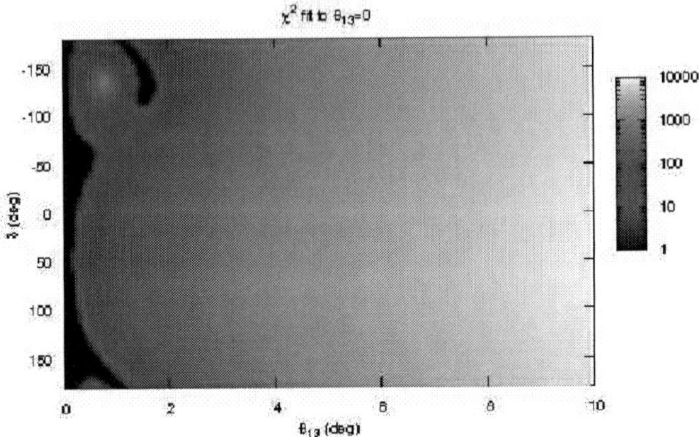


**FIGURE 5.** Setup I. $\theta_{13}$ sensitivity.

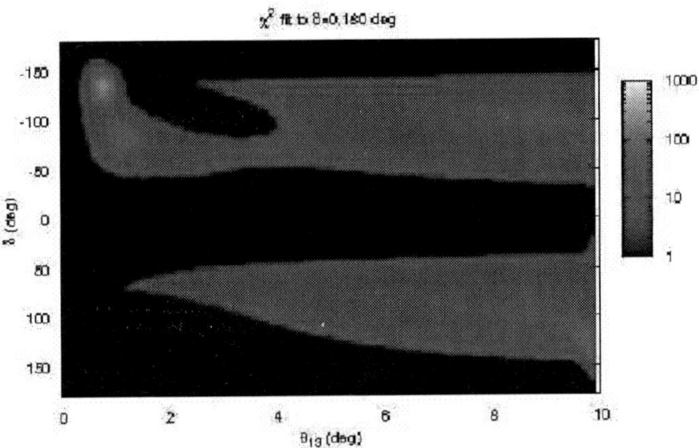


**FIGURE 6.** Setup I. CPV sensitivity.

for varying confidence levels. For 99% CL the sensitivity to a non-vanishing mixing is tipically around 1 degree. The corresponding sensitivity (99% CL) to see CP violation becomes significant for $\theta_{13} > 4^o$ with values of the phase $\delta$ around $30^o$ to be distinguished from zero.

In the case of Setup II the longer baseline for $\gamma = 195$ leads to a value of $E/L$ well inside the second oscillation. In that case the associated strip in the ($\theta_{13}$, $\delta$) plane has a more pronounced curvature, so that the two parameters can be better disantangled. The statistical distribution generated for some assumed values of ($\theta_{13}$, $\delta$) has been fitted and the $\chi^2$ values obtained. The results are given in Fig. 7. Qualitatively, one notices that the precision in the mixing is somewhat (but no much) better than that in Setup I. On the contrary, the precision reachable for the CP phase is much better than that for Setup

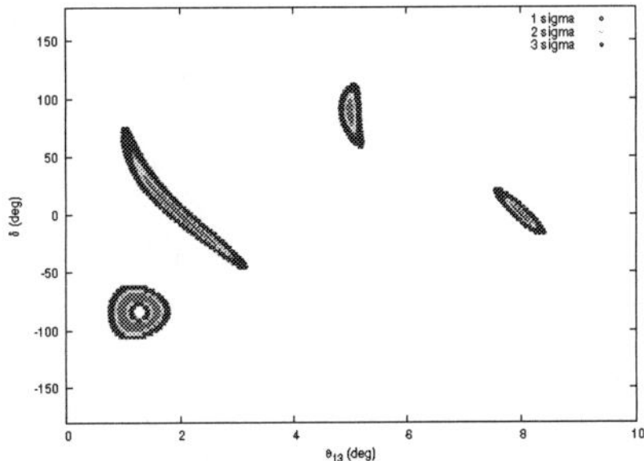


**FIGURE 7.** Setup II. Fit for $(\theta_{13}, \delta)$ from statistical distribution.

I. One should emphasize that this improvement in the CP phase has been obtained with the neutrino channel only, using two appropriate different energies.

The corresponding exclusion plots which define the sensitivities to discover a non-vanishing mixing $\theta_{13} \neq 0$ and CP violation $\delta \neq 0$ are presented in Fig. 8 and Fig. 9 for varying confidence levels. For 99% CL the sensitivity to a non-vanishing mixing is, as before, significant up to around 1 degree. The corresponding sensitivity (99% CL) to see CP violation for $\theta_{13} > 4^o$ becomes now significant with values of the phase $\delta$ around $20^o$ to be distinguished from zero.

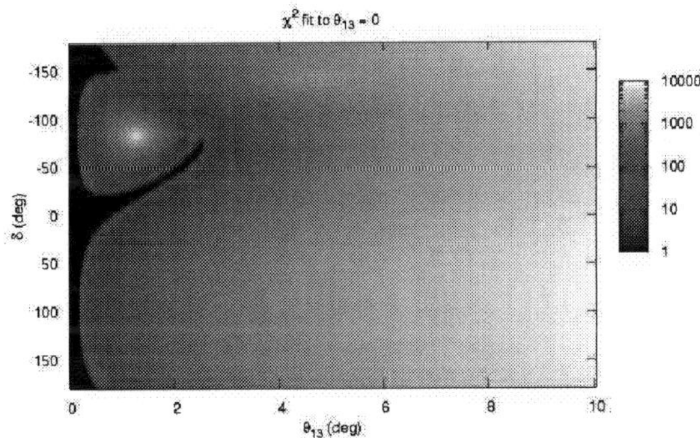


**FIGURE 8.** Setup II. $\theta_{13}$ sensitivity.

At the time of the operation of this proposed Facility in Setup II it could happen that the connecting mixing $\theta_{13}$ is already known from the approved experiments for second generation neutrino oscillations, like Double CHOOZ, T2K and NOVA. To illustrate the gain obtainable in the sensitivity to discover CP violation from the previous knowledge of $\theta_{13}$ we present in Fig. 10 the expected sensitivity with the distribution of events depending on a single parameter $\delta$ for a fixed known value of $\theta_{13}$. The result

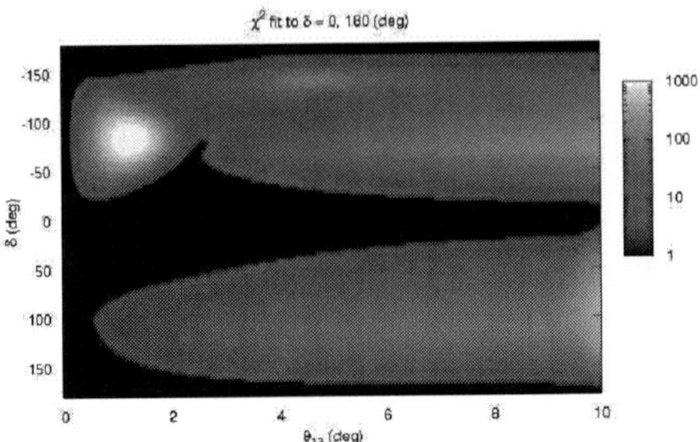


**FIGURE 9.** Setup II. CPV sensitivity for the statistical distribution depending on two parameters ($\theta_{13}$ and $\delta$).

is impressive: even for a mixing angle of one degree, the CP violation sensitivity at 99.7% CL reaches values around $10^o$.

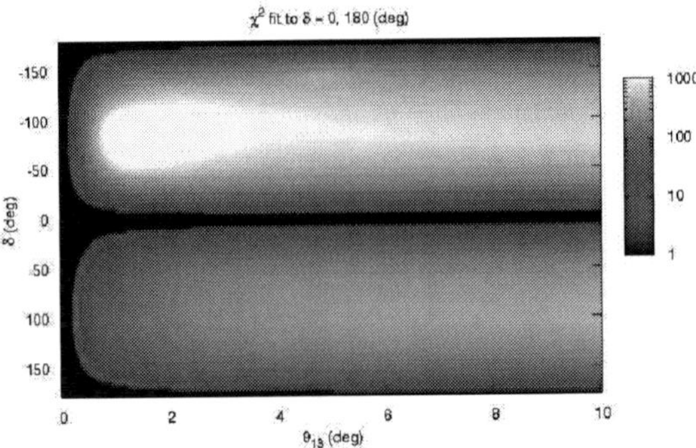


**FIGURE 10.** Setup II. CPV sensitivity for the statistical distribution depending on a single parameter $\delta$, assuming previous information on $\theta_{13}$.

## 5. CONCLUSIONS AND OUTLOOK

The simulations of the physics output for both Beta and EC beams indicate:

1) The upgrade to higher energy ($E_p = 1000$ GeV) is crucial to have a better sensitivity to CP violation, which is the main objective of the next generation neutrino oscillation experiments, iff accompanied by a longer baseline.

2) The best $E/L$ in order to have a higher sensitivity to the mixing $U(e3)$ is not the same than that for the CP phase. Like the phase-shifts, the presence of $\delta$ is easier to observe when the energy of the neutrino beams enters into the region of the second oscillation. The mixing is better seen around the first oscillation maximum, instead.

In particular, Setup II in EC beams, i.e., with $\gamma's$ between 195 and 440 and a baseline $L = 650$ Km (Canfranc), has an impressive sensitivity to CP violation, reaching precisions around $20^o$, for 99% CL, or better (if some knowledge on the value of $\theta_{13}$ is already established).

Besides the feasibility studies for the machine, most important for physics is the study of the optimal configuration by combining low energy with high energy neutrino beams, short baseline with long baseline and/or EC monochromatic neutrinos with $^6He$ $\beta^-$ antineutrinos.

Among the possible systematics associated with the proposed experiments one should define a program to determine independently the relevant cross sections of electron and muon neutrinos and antineutrinos with water in the relevant energy region from several hundreds of MeV's to 1 GeV or so.

The result of the synergy of Neutrino Physics with Nuclear Physics (EURISOL) and LHC Physics (SPS upgrade) for the Facility at CERN could be completed with the synergy with Astroparticle Physics for the Detector, which could be common to neutrino oscillation studies with terrestrial beams, atmospheric neutrinos (sensitive to the neutrino mass hierarchy through matter effects [24]), Supernova neutrinos and Proton decay.

The analysis shown in this paper shows that the proposals discussed here merit *R&D* studies in the immediate future for all their ingredients: Facility, Detector and Physics.

## REFERENCES


1. Y. Fukuda *et al.*, Phys. Rev. Lett. **81** (1998) 1562.
2. Q. R. Ahmad *et al.*, Phys. Rev. Lett. **89** (2002) 011301 and 011302.
3. M. Apollonio *et al.*, Phys. Lett. B **466** (1999) 415.
4. Double-Chooz, Th. Lasserre, Nucl. Phys. Proc. Suppl. **149** (2005) 163, [hep-ex/0409060].
5. T2K Collaboration, K. Kaneyuki, Nucl. Phys. Proc. Suppl. **145** (2005) 178-181.
6. D. S. Ayres *et al.* [NOvA Collaboration], [hep-ex/0503053].
7. S. Geer, Phys. Rev. **D57** (1998) 6989 [hep-ex/0210192]
8. A. De Rújula, M.B. Gavela and P. Hernández, Nucl. Phys. **B547** (1999) 21 [hep-ex/0210192].
9. For the extensive literature see for example the reviews M. Apollonio, *et al*, [hep-ex/0210192]; J. J. Gómez-Cadenas and D. A. Harris, Ann. Rev. Nucl. Part. Sci. **52** (2002) 253 and the annual proceedings of the International Nufact Workshop.
10. Y. Itow *et al*, Nucl. Phys. Proc. Suppl. **111** (2002) 146, [hep-ex/0106019].
11. A. Para and M. Szleper, [hep-ex/0110032]; D. Ayres *et al*, [hep-ex/0210005]; I. Ambats *et al.* [NOvA Collaboration], FERMILAB-PROPOSAL-0929
12. B. Autin *et al*, CERN/PS 2002-012.

13. H. Minakata and H. Nunokawa, Phys. Lett. B **495** (2000) 369 [arXiv:hep-ph/0004114]. J. J. Gómez-Cadenas *et al.* [CERN working group on Super Beams Collaboration], [hep-ph/0105297]. H. Minakata and H. Nunokawa, JHEP **0110** (2001) 001, [hep-ph/0108085]. P. Huber, M. Lindner and W. Winter, Nucl. Phys. B **645** (2002) 3 [hep-ph/0204352]. G. Barenboim *et al.*, [hep-ex/0206025].
14. V. Martemianov *et al*, Phys. Atom. Nucl. 66 (2003) 1934-1939; Yad.Fiz. 66 (2003) 1982-1987 [hep-ex/0211070]; H. Minakata *et al*, Phys. Rev. D **68** (2003) 033017 [Erratum-ibid. D **70** (2004) 059901] [hep-ph/0211111]. Phys. Lett. B **580** (2004) 216 [hep-ph/0309323]. P. Huber, M. Lindner, T. Schwetz and W. Winter, Nucl. Phys. B **665** (2003) 487 [hep-ph/0303232]. F. Ardellier *et al.*, [hep-ex/0405032].
15. P. Zucchelli, Phys. Lett. **B532** (2002) 166.
16. J. Bernabeu *et al.*, JHEP **0512** (2005) 014, [hep-ph/0505054].
17. C. Rubbia, A. Ferrari, Y. Kadi and V. Vlachoudis, Nucl. Instrum. Meth. A **568**, 475 (2006), [hep-ph/0602032].
18. J. Burguet-Castell, D. Casper, E. Couce, J. J. Gomez-Cadenas and P. Hernandez, Nucl. Phys. B **725**, 306 (2005), [hep-ph/0503021].
19. A. Cervera *et al.*, Nucl. Phys. B **579** (2000) 17; Erratum-ibid B **593** (2001) 731.
20. M. C. Gonzalez-Garcia, Global Analysis of Neutrino, in Nobel Symposium 2004: Neutrino Physics, Haga Slott, Enkoping, Sweden, 19-24 Aug 2004, [hep-ph/0410030].
21. J. Bernabeu, Nucl. Phys. B **49** (1972) 186.
22. A. Algora *et al.*, Phys. Rev. C **70** (2004) 064301.
23. J. Sato, Phys. Rev. Lett. **95** 131804 (2005), [hep-ph/0503144].
24. M. C. Banuls, G. Barenboim and J. Bernabeu, Phys. Lett. B **513**, 391 (2001) [arXiv:hep-ph/0102184]; J. Bernabeu, S. Palomares-Ruiz, A. Perez and S. T. Petcov, Phys. Lett. B **531**, 90 (2002); J. Bernabeu, S. Palomares Ruiz and S. T. Petcov, Nucl. Phys. B **669**, 255 (2003).

# Solar Neutrinos, SNO and SNOLAB

*A. B. McDonald*

*Queen's University, Kingston, Ontario, Canada K7L 3N6*

Abstract. The Sudbury Neutrino Observatory has completed operation in its third phase with an array of neutron detectors in 1000 tonnes of heavy water and Cherenkov light detection 2 km underground in INCO's Creighton mine near Sudbury, Ontario, Canada. Data from the third phase is now being analyzed. In the first two phases of the project reported previously, the neutral current reaction on deuterium was used to determine the total flux of active neutrinos and the charged current reaction on deuterium provided a measure of the flux and energy spectrum of solar electron neutrinos. The flux of electron neutrinos was found to be only about one third of the total flux, providing clear evidence of neutrino flavour change. The total flux of active neutrinos was found to be in agreement with solar model calculations. The underground laboratory is being expanded to create an international facility known as SNOLAB that will be completed at the end of 2007. Proposed future experiments for the detection of lower energy solar neutrinos, geo-neutrinos, dark matter and double beta decay are described.

Keywords: Neutrino physics, solar neutrinos, dark matter, double beta decay.
**PACS:** 14.60.Pq, 95.35.+d, 96.60.-j, 23.40.-s

## THE SNO EXPERIMENT

This presentation was made several weeks prior to the completion on Nov. 28, 2006 of operation of the SNO experiment [1] with 1000 tonnes of pure heavy water. The third and final phase of the experiment involved an array of $^3$He-filled, ultra-low radioactivity proportional counters placed in the heavy water for neutron detection. This measurement will have a very different set of systematic uncertainties from the previous measurement and is expected to provide improved overall accuracy for the measurement of the neutrino oscillation parameters

The detection reactions used in SNO for the measurement of solar neutrinos are as follows:

$$\nu_e + d \rightarrow p + p + e^- (CC)$$

$$\nu_x + d \rightarrow p + n + \nu_x (NC)$$

$$\nu_x + e^- \rightarrow \nu_x + e^- (ES)$$

The charged current (CC) reaction is sensitive only to electron neutrinos, while the neutral current (NC) reaction is sensitive to all active neutrino flavors (x = e, $\mu$,$\tau$) above the energy threshold of 2.2 MeV. The elastic scattering (ES) reaction on electrons is sensitive to all flavors as well, but with reduced sensitivity to $\nu_\mu$ and $\nu_\tau$. The Cherenkov light produced by the electrons in the final state is used to observe the CC and ES reactions. The NC reaction is observed through the detection of the neutron in the final state of the reaction using three different techniques. for the

CP917, *Particles and Fields,* edited by H. Castilla Valdez, J. C. D'Olivo, and M. A. Perez

separate phases of the experiment. For the final phase, the neutron signals for the NC reaction from the $^{3}$He-filled detectors are totally independent of the light generated by events from the CC reaction, thereby breaking any covariance between the two types of events.

During the first phase of SNO operation, neutrons from the NC reaction were detected from the Cerenkov light generated by a 6.25 MeV gamma ray emitted when the neutrons are captured on deuterium. During the second phase, with about 2 tonnes of NaCl added to the heavy water, the neutrons from the NC reaction captured predominantly on Cl with the emission of multiple gammas, creating events that were considerably more isotropic than the single electron Cerenkov events from the CC reaction.

A measurement of neutrino flavor change can be made by comparing the flux observed with the CC reaction (electron neutrinos) with the flux observed with the NC reaction (all active neutrino types). A comparison of the CC reaction with the ES reaction also provides a less sensitive measurement because the ES reaction is dominated by the contribution from electron neutrinos and the neutral current sensitivity is small. Because all reactions are sensitive to the flux of neutrinos from $^{8}$B decay in the Sun, the results are not sensitive to details of solar models. In fact, the flux of all neutrino types tests solar model calculations for $^{8}$B neutrino flux, independent of neutrino flavor change to active neutrinos. In the first phase, the SNO experiment provided direct evidence of neutrino flavor change via these flux comparisons, and through the NC reaction, provided the first "appearance" observation of oscillated neutrino flavors.

With salt in the heavy water, the strong correlation between the CC and NC event types can be broken because the pattern of light on the PMT's is different for those two reactions. An event angular distribution parameter $\beta_{14}$ (defined in reference [2]) was used to separate NC and CC events on a statistical basis. The fluxes determined from the salt phase data [2] (in units of $10^{6}$ cm$^{-2}$ sec$^{-1}$) are:

$$\phi_{CC} = 1.68 \ ^{+0.06}_{-0.06}(\text{stat.})^{+0.08}_{-0.09}(\text{syst.})$$
$$\phi_{NC} = 4.94 \ ^{+0.21}_{-0.21}(\text{stat.})^{+0.38}_{-0.34}(\text{syst.})$$
$$\phi_{ES} = 2.35 \ ^{+0.22}_{-0.22}(\text{stat.})^{+0.15}_{-0.15}(\text{syst.})$$

A comparison of the CC and NC fluxes shows that about two-thirds of the electron neutrinos have changed their flavor to other active neutrino types, violating a hypothesis test for no flavor change at greater than 7 $\sigma$. The observed total flux of active neutrinos (NC) is in excellent agreement with the flux of $^{8}$B neutrinos obtained from solar models:

$$\Phi_{e} = 5.82 \pm 1.3[3]; 5.31 \pm 0.6[4]$$

Oscillation purely to sterile neutrinos is strongly disfavored. By comparison of the active total flux of $^{8}$B solar neutrinos observed by the NC reaction with the $^{8}$B flux calculated with solar models, restrictions are placed on sub-dominant oscillations to sterile neutrinos.

The data for solar neutrinos can be analyzed in terms of neutrino oscillations and the MNSP [5] mixing matrix. If the neutrinos travel in a region of high electron density, such as in regions within the Sun or Earth, the interaction of neutrinos with electrons can produce matter enhancement of the oscillation process, via the MSW effect [6]. The MSW process produces adjustments to the effective values for the parameters $\theta_{12}$ and $\Delta m_{12}^2$, depending on the electron density. The sign of $\Delta m_{12}$ can be determined if matter interactions are substantial and distortions of the energy spectrum can occur. Interactions in the Earth can produce differences in the measured fluxes at detectors for day and night time periods.

Following the SNO evidence for neutrino flavor change, the KAMLAND collaboration reported their results [7] showing the oscillation of reactor neutrinos with parameters that overlap with the solar neutrino results for part of their allowed region. The agreement between the solar neutrino oscillation parameters and the reactor anti-neutrino oscillation parameters confirms MNSP mass oscillations as the dominant process for flavor change. The agreement provides a confirmation of CPT for neutrinos. If full CPT invariance is assumed, the agreement confirms matter enhancement for the solar neutrinos and restricts the possibility of flavor change arising from other sub-dominant processes.

The best parameter values for an analysis for $\nu_1\nu_2$ mixing [2], following the SNO salt data and including other solar neutrino measurements and the Kamland data is:

$\Delta m^2 = 8.0^{+0.6}_{-0.4} \times 10^{-5} eV^2$ and $\theta_{12} = 33.9^{+2.4}_{-2.2}$ degrees.

The mixing is found to be non-maximal (mixing angle less than $\pi/4$) with a confidence level corresponding to 5.5 standard deviations. The LMA region, involving matter enhancement of the oscillation via the MSW effect, is the preferred solution and that provides a determination of the sign of $\Delta m_{12}$. It is found that $m_2$ is greater than $m_1$.

Analysis is in progress for the third phase of the SNO experiment in which the array of $^3$He-filled, ultra-low radioactivity proportional counters replaced the salt in the pure heavy water for neutron detection. In this measurement the covariance between the CC and NC events is broken and there is a very different set of systematic uncertainties from the previous measurement. Another analysis is also being carried out on a combination of the first two phases of the experiment with a lower energy threshold. This will also break the correlation and improve the accuracy. These analyses are expected to provide improved overall accuracy for the measurement of the neutrino oscillation parameters, particularly $\sin^2 \theta_{12}$. For the LMA solution of the MSW effect, solar electron neutrinos are very nearly in a pure $\nu_2$ eigenstate when emerging from the Sun, and the survival probability is almost exactly proportional to $\sin^2 \theta_{12}$. Other analyses in progress will provide improved measurements of atmospheric neutrinos (including above horizon events accessible because SNO is deep beneath a horizontal overburden).

## THE NEW SNOLAB

A new international laboratory for underground science, SNOLAB, is being constructed on the same underground level as the SNO site, with space for 5 or more next-generation experiments. Over 20 letters of interest have been received from international collaborations. An Experiment Advisory Committee has designated a number of projects in double beta decay, dark matter and solar neutrinos for development space and/or priority for large scale deployment when the laboratory is completed at the end of 2007. The entire SNOLAB laboratory is being constructed with clean room conditions and the depth of 2 km will provide substantially lower fluxes of penetrating muons than other existing underground laboratories. Figure 1 shows the proposed SNOLAB layout. The cylindrical chamber in the upper left is the subject of an imminent funding decision. The remainder of the new areas will be completed at the end of 2007.

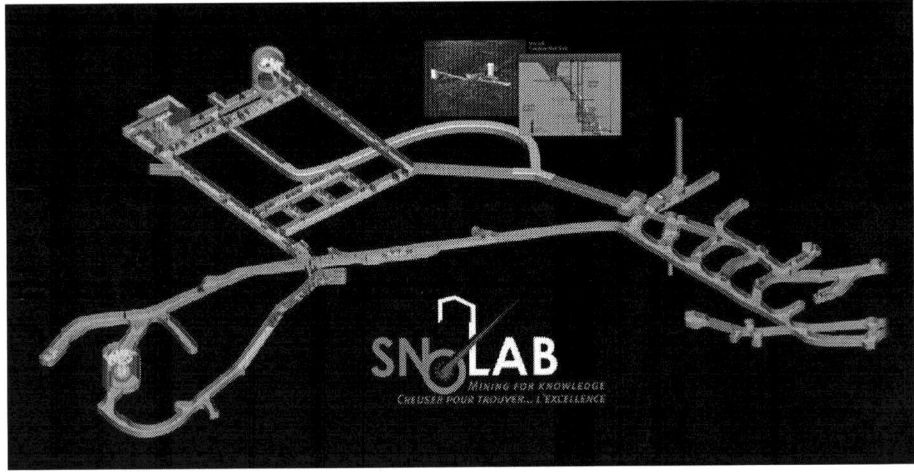


**FIGURE 1.** SNOLAB layout. The area to the lower left is the existing SNO detector. The area to the upper left contains the new experimental locations.

## FUTURE EXPERIMENTS AT SNOLAB

The Experiment Advisory Committee for SNOLAB has identified a number of experiments for future development space at SNOLAB, including Super-CDMS, EXO, MAJORANA, SNO+, DEAP, CLEAN, PICASSO and ZEPLIN. Several other proposals were encouraged to submit more detailed technical proposals in future. I will emphasize three of these projects in this presentation: SNO+, DEAP/CLEAN and PICASSO.

## The SNO+ Experiment

Upon completion of the SNO project and removal of the heavy water in 2007, a new international collaboration (SNO+) is proposing [8] to replace the heavy water with liquid scintillator and proceed with measurements of neutrino-less double beta decay, lower energy solar neutrinos, supernova neutrinos, geo-neutrinos and reactor neutrinos. The addition of the scintillator Linear Alkyl Benzene (LAB) could provide up a factor of 100 greater light output than Cerenkov light from heavy water, enabling measurements of double beta decay, and low energy neutrinos from the $^{7}$Be, pep and CNO reactions in the Sun. The latter two reactions could be accessible because the background from $^{11}$C produced by penetrating muons is so much lower at SNO than it is at other sites with less effective depth (SNOLAB has more than 40 times lower muon flux than the Gran Sasso location). Of course, extreme care must be taken to reduce normal radioactive backgrounds, such as those from the U and Th chain and potassium. The LAB is not expected to cause a corrosion problem for the acrylic containment sphere but some engineering adjustments must be made to hold down the sphere to deal with the lower density of the LAB relative to the surrounding ultra-pure light water.

Development work to date is very promising for the addition of the neutrino-less double beta decay candidate $^{150}$Nd as a metallo-organic compound [9]. The neutrino-less double beta transition for $^{150}$Nd is very favorable as it has a transition energy of 3.3 MeV, resulting in a high decay rate for a given effective neutrino mass and placing the summed two-electron peak above background transitions from the Uranium decay chain. Measurements of light attenuation for Nd [9] indicate that up to 500 kg of isotope could be added while maintaining light sensitivity similar to or better than existing large scale liquid scintillation detectors. Preliminary simulations indicate that effective neutrino mass sensitivities better than 0.1 eV could be obtained in SNO+ with a few hundred kg of $^{150}$Nd. ( See figure 2). Discussions with the Super-NEMO collaboration have led to the development of a design study to convert an existing laser isotope enrichment facility in France for enrichment of hundreds of kilograms of $^{150}$Nd. The tuning parameters are already known from previous small scale studies, so the participants are optimistic about the availability of large amounts of enriched isotope in the near future.

The pep reaction provides a very interesting test of the MSW energy dependence in the intermediate region between pp neutrinos exhibiting vacuum oscillations and the $^{8}$B neutrinos that exhibit strong matter effects. Several authors [10] have indicated the sensitivity in this region to other effects such as non-standard interactions, sterile neutrinos and even mass-varying neutrinos. The pep flux is predicted with an accuracy of about 1.5% by solar models [3,4]. A measurement of the CNO flux is also possible and would provide the first explicit measurement of this important solar reaction.

SNO+ would also allow sensitive measurements of geo-neutrinos with fluxes predicted to be higher than at Kamland and reactor backgrounds as much as 4 times lower [8]. Measurements of reactor neutrinos would provide an interesting addition to the Kamland oscillation measurements with a baseline about 50% longer. Supernova detection capabilities and continued participation in the SNEWS supernova early warning collaboration will also be maintained with the SNO+ detector.

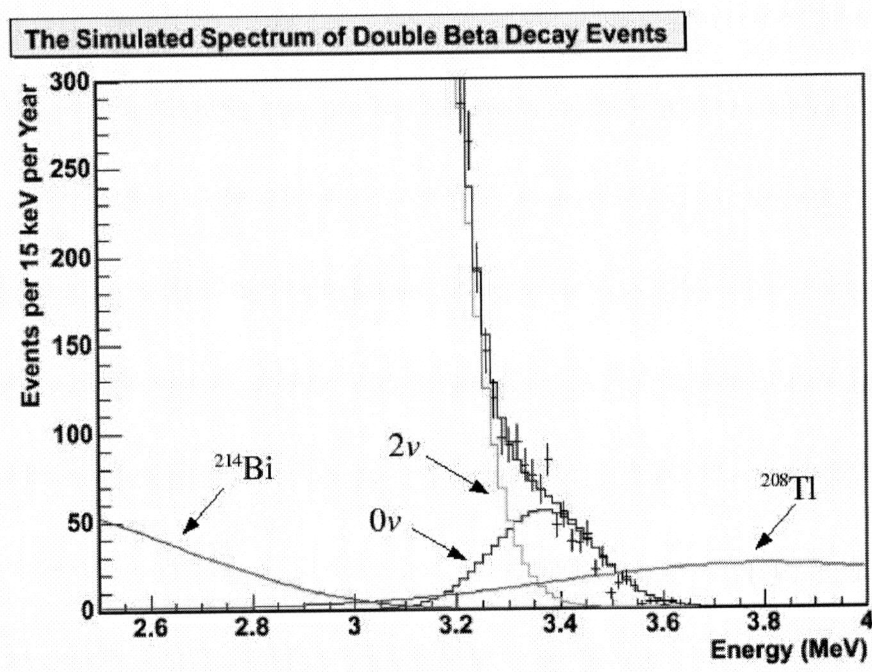


**FIGURE 2.** Simulation of the SNO+ detector response for 500 kg of $^{150}Nd$, assuming an effective neutrino mass of 0.15 eV, matrix elements from ref [11] and backgrounds as achieved at the Kamland experiment.

## The DEAP/CLEAN Experiment

As pointed out in reference [12] the scintillation decay times of nuclear recoils and ionizing particles in liquid argon are very different and can be used for a sensitive discrimination process using scintillation light alone. This discrimination is adequate to eliminate the background from $^{39}Ar$ and $^{42}Ar$ and enables a sensitive measurement of dark matter through nuclear recoils caused by collisions with particles forming the dark matter halo in our galaxy. The fraction of the prompt light within the first 100 nsec to the total light in a 15 microsecond period (the prompt fraction $F_{prompt}$) is used to discriminate nuclear recoils from ionizing background. A discrimination ratio of $10^5$ has been demonstrated in a laboratory at ground level, limited by coincidences from room background. Simulations of the technique [12] indicate that a discrimination factor of up to $10^8$ could be obtained in an underground environment (See Figure 3). A 7 kg detector (DEAP-1) has been run for several months in a surface laboratory at Queen's University, Canada and is being prepared for deployment in SNOLAB in the spring of 2007. This detector should be capable of determining discrimination factors of $10^9$ without interference from background events. Members of the DEAP and CLEAN collaborations from Canada and the US have joined to together to form the DEAP/CLEAN collaboration with the principal objective of developing a 3.5 tonne liquid argon detector with 1 tonne central fiducial volume.. Preliminary simulations indicate that backgrounds can be controlled to provide [12] a cross section sensitivity of about $10^{-46}$ $cm^2$ for a broad WIMP mass range near 100 Gev.

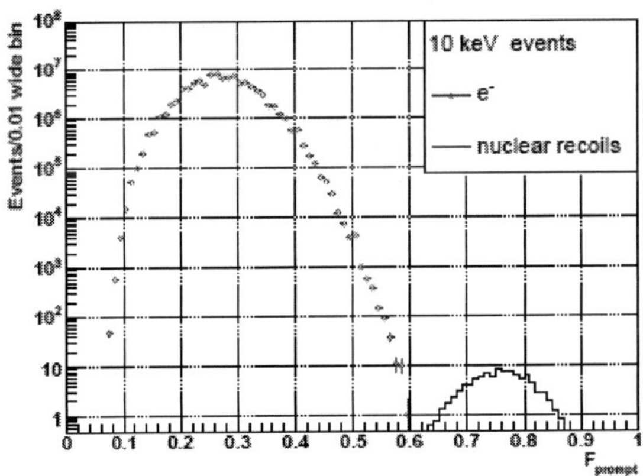


**FIGURE 3.** Discrimination of nuclear recoils from electrons in liquid argon. Shown are the Fprompt distributions for $10^8$ simulated electron events and for 100 simulated nuclear recoil events with effective energies of 10 keV.

## The PICASSO Experiment

The PICASSO collaboration has developed dark matter detectors [13] using supersaturated gels containing Freon to discriminate nuclear recoils from ionizing radiation. By observing interaction rates as a function of temperature, the ionizing backgrounds can be discriminated against by a large factor. By using acoustic detection of bubble formation, the PICASSO collaboration have made some of the most sensitive measurements to date [14] of dark matter using the spin dependent interaction, for which fluorine is calculated to have much higher sensitivity than other nuclei. An improved version of the experiment, scaled up by a factor of more than fifty and employing reduced background techniques is being deployed in SNOLAB for initial operation during 2006. This technique could also be scaled up to much larger mass at reasonable cost and future plans call for larger detectors as detector background reduction techniques progress.

## REFERENCES


1. SNO Collaboration, J. Boger et al., Nucl. Instrum. Meth. **A449**, 172 (2000)
2. SNO Collaboration, B. Aharmim et al, Phys. Rev. **C72**, 055502 (2005)
3. J. N. Bahcall and M. H. Pinsonneault, *Phys. Rev. Lett.* **93** 121301 (2004); J. N. Bahcall and C. Pena-Garay hep-ph/0305159 (2003)

4. S. Turck-Chieze et al, Phys. Rev. Lett. **93**, 211102 (2004)
5. Z. Maki, N. Nakagawa and S. Sakata, Prog. Theor. Phys **28,** 870 (1962); B. Pontecorvo, J. Expt.Theoret. Phys **33**, 549 (1957); J. Expt. Theoret. Phys **34,** 247 (1958); V. Gribov and B. Pontecorvo, Phys. Lett.**B28,** 493 (1969).
6. S. P. Mikheyev and A. Y. Smirnov in Massive Neutrinos in Astrophysics and in Particle Physics, Proceedings of the Moriond Workshop, edited by O. Fackler and J. Tran Thanh Van, Editions Frontieres, Gif-sur-Yvette, 335 (1986); S. P. Mikheyev and A. Y. Smirnov, Sov. J. Nucl. Phys. **42,** 1441 (1985); L. Wolfenstein, Phys. Rev. **D 17,** 2369 (1978)
7. T. Araki et al, hep-ex/0406035 (2004).
8. M. C. Chen, *Nuclear Physics B (Proc. Suppl.)* **145** 65–68 (2005).
9. R. Hahn, M. Yeh, Brookhaven National Laboratory, 2006, private communication.
10. A. Friedland A et al, hep-ph/0402266 (2004); O. Miranda et al, hep-ph/0406280 (2004); V. Barger et al, hep-ph/0502196 (2005); P. De Holanda and A. Smirnov, hep-ph/0307266 (2003).
11. S. R. Eliott and P. Vogel, Ann.Rev.Nucl.Part.Sci. **52** 115-151 (2002)
12. M. Boulay and A. Hime, astro-ph/0411358 (2004)
13. M. Barnabé Heider et al, hep-ex/0508098 (2005)
14. M. Barnabé Heider et al, hep-ex/0502028 (2005)

# The Fermilab Main Injector Neutrino Program

Jorge G. Morfín

*Fermi National Accelerator Laboratory,P.O. Box 500, Batavia, IL,60510*

**Abstract.** The NuMI Facility at Fermilab provides an extremely intense beam of neutrinos making it an ideal place for the study of neutrino oscillations as well as high statistics (anti)neutrino-nucleon/nucleus scattering experiments. The MINOS neutrino oscillation $\nu_\mu$ disappearance experiment is currently taking data and has published first results. The NO$\nu$A $\nu_e$ appearance experiment is planning to begin taking data at the start of the next decade.

For the study of neutrino scattering, the MINER$\nu$A experiment at Fermilab is a collaboration of elementary-particle and nuclear physicists planning to use a fully active fine-grained solid scintillator detector. The overall goals of the experiment are to measure absolute exclusive cross-sections, nuclear effects in $\nu$ - A interactions, a systematic study of the resonance-DIS transition region and the high-$x_{Bj}$ - low $Q^2$ DIS region.

**Keywords:** neutrino, scattering, oscillation
**PACS:** 13.15, 14.6, 25.3, 95.55

## 1. THE FERMILAB NUMI FACILITY

The Fermilab NuMI (Neutrinos at the Main Injector) facility consists of the technical beamline components including target, two magnetic focusing horns, evacuated decay pipe, monitoring devices, shielding and the underground facilities to contain the beamline components. A large, on-site experimental detector hall $\sim$ 100 meters underground currently contains the MINOS near detector. It will also house the MINER$\nu$A detector, just upstream of the MINOS near detector, and the NO$\nu$A near detector at an upstream off-axis location.

Two parabolic magnetic horns are pulsed with 200 kA of current to focus charged hadrons ($\pi^+$ and $K^+$) into a 670 m long decay pipe that ends with an aluminum and steel absorber. Just downstream of the absorber, 240 m of Dolomite is used to range out muons before the $\nu$ beam enters the Near Detector Hall.

The neutrino energy distribution of the NuMI beam can be chosen by changing the distance of the target and second horn with respect to the first horn, as in a zoom lens, or, with reduced intensity but quicker tuning time, by simply varying the distance of target from the first horn and leaving the second horn in a fixed position. Depending on the chosen configuration, event rates in the near hall detectors will vary from 60K in the low energy (LE) configuration to 520K in the high energy (HE) configuration per ton of detector and $10^{20}$ protons on target (POT). At the far detector site in Soudan, Minnesota, the expected event rate in the 5.4 kT MINOS far detector is (assuming no oscillations) $\sim$ 275 per $10^{20}$ protons on target (POT) in the LE configuration. For the MINOS experiment and the start of the MINER$\nu$A experiment, the beamline will be operating mainly in its lowest possible neutrino energy configuration to be able to reach desired low values of $\delta m^2$ for MINOS. For the proposed NO$\nu$A experiment, the beam

CP917, *Particles and Fields,* edited by H. Castilla Valdez, J. C. D'Olivo, and M. A. Perez

will be operating in the full ME configuration.

The Main Injector is now delivering protons to MINOS at a rate equivalent to around 2.0-2.5 x$10^{20}$ POT/year or $\sim$ 200kW. Upgrades to the Main Injector and other components of the Fermilab accelerator complex will increase the NuMI beam power before and during the planned operation of MINER$\nu$A and NO$\nu$A. The current phase, Proton Plan I, includes upgrades to the MI RF system and other components leading to a maximum power of 430 kW. Further upgrades proposed for after the end of TeVatron collider operations use the existing anti-proton Recycler and Accumulator rings as proton accumulators in the MI injection phase, to achieve up to 1 MW of beam power.

## 2. THE MINOS EXPERIMENT [2]

The MINOS experiment [1] tests the oscillation hypothesis by making two measurements of a beam of $\nu_\mu$ produced in the NuMI beam at Fermilab. The first measurement occurs at the Near Detector (ND) located onsite at Fermilab and the second measurement is made at the Far Detector (FD) located 735 km away in the Soudan Underground Mine in Soudan, Minnesota, USA. MINOS extracts the oscillation parameters by comparing the reconstructed energy spectra of the $\nu_\mu$ at the ND and FD.

### 2.1. The MINOS Detectors

The MINOS detectors are steel-scintillator tracking calorimeters with toroidal magnetic fields averaging 1.3 T. The steel planes are 2.54 cm thick and the scintillator is mounted to the steel. The scintillator planes are made of 4.1 cm wide and 1 cm thick strips. The strips in each plane are rotated 45° from the vertical and the strips in successive planes are rotated 90° from each other. The light produced in the scintillator is collected in wavelength shifting fibers embedded in the scintillator. The fibers transport the light to multi-anode photomultiplier tubes (PMTs).

The detectors were made as similar as possible in order to cancel the majority of the uncertainties in the neutrino interaction modeling and detector response. The main design differences between the two detectors are due to the much higher rate ($\sim$ 100 times higher) in the ND than in the FD. The FD is 705 m below the surface, has a mass of 5.40 kton. The scintillator is read out at both ends of the strips and the front end readout electronics are designed to provide high precision timing information. The ND is 103 m below the surface and has a mass of 0.98 kton. It uses special electronics to handle the increased rate compared to the FD. The geometry of the planes optimizes containment of hadronic showers and allows for the magnetic field to be similar to that in the FD.

---

[1] My thanks to Brian Rebel, Fermilab, for providing much of the information in this section

## 2.2. Event Definition and Reconstruction

The FD data were blinded until the procedures for event selection and energy spectrum prediction were defined and understood. The blinding procedure hid a substantial and unknown fraction of the events in the FD. The data in the ND were not blinded.

The energy of each neutrino interaction is found in the same way for both detectors. Muon tracks are found and their curvature in the magnetic field is fit to determine their energy. The hadronic showers are also found and their energy is determined. The events selected in both detectors were required to have visible energy, Evis, less than 30 GeV and the events had to have a negatively charged track, a requirement chosen to select only $\nu_\mu$ interactions. A fiducial volume was defined to contain the hadronic energy of the event and reject background cosmic ray muons. The events were also required to occur within a 50 $\mu$ s window surrounding the spill time.

## 2.3. The Expected Far Detector Neutrino Energy Distribution

The measured energy spectrum in the ND is used to predict the unoscillated spectrum in the FD. The method used by MINOS to predict the FD spectrum uses the ND data to measure effects such as beam modeling, neutrino interactions and detector response that are common to both detectors. The beam simulation is used to derive a transfer matrix that relates $\nu_\mu$ in the two detectors via their parent hadrons. The matrix element Mij gives the relative probability that the distribution of secondary hadrons producing the observed $\nu_\mu$ of energy $E_i$ in the ND will produce the observed $\nu_\mu$ of energy $E_j$ in the FD. .

## 2.4. Results

The FD data set contains a total of 215 events with Evis < 30 GeV compared to the unoscillated expectation of 336.0 $\pm$ 14.4. The uncertainty is due to systematic uncertainties associated with (a) the fiducial mass calculation and POT counting accuracy (4 %), (b) the hadronic energy scale (11 %) and (c) the NC component (50 %).

The data were fit to the hypothesis of $\nu_\mu$ to $\nu_\tau$ oscillations, and the fit has $\sin^2(2\theta_{23}) \geq 0.87$ at the 68 % C.L. The best fit for the mass squared difference is $\Delta m^2_{23} = 2.74^{+0.44}_{-0.26}x10^{-3}eV^2/c^4$ with the fit probability of 8.9 %.

# 3. THE NO$\nu$A EXPERIMENT [1]

The proposed NO$\nu$A (NuMI Off-axis $\nu_e$ Appearance) experiment [2] will search primarily for the oscillation of muon neutrinos into electron neutrinos and the corresponding mixing angle $\theta_{13}$. NO$\nu$A will use the Fermilab NuMI beam in the ME configuration

[2] My thanks to Peter Shanahan, Fermilab, for providing much of the information in this section

over a baseline of 810 km, and detectors at Fermilab and the Ash River site in northern Minnesota. The detector design is optimized for the identification of electrons in the final state of $\nu_e$ charged current (CC) interactions. The Far detector will be located 12 km from the central axis of the beam to suppress backgrounds from intrinsic beam $\nu_e$ and high-energy neutral current interactions

## 3.1. The NO$\nu$A Detectors

The design of the NO$\nu$A detector enhances identification of $\nu_e$ CC events by the separation of electromagnetic and hadronic showers. This is achieved with nearly totally active detector of relatively low Z/A ratio, allowing a high number of samples per radiation length that is not cost-prohibitive for a detector of large mass. The detectors will be composed of liquid scintillator contained in planes of an extruded PVC cell structure, read out on one end of each cell via a wavelength-shifting fiber. The basic active cell unit will be approximately 6 cm deep along the beam direction, and 3.8 cm wide along the measurement coordinate. The wavelength shifting fibers will be read out by 32-pixel Avalanche Photo-Diodes (APDs).

The Far Detector will be situated below grade, with an overburden of between 10 and 20 radiation lengths to reduce backgrounds due to cosmic rays. A Far Detector of $\sim$ 20 kT total mass will comprise roughly 1300 planes approximately 15.7 m on a side. The Near Detector will have the same structure, although with smaller and fewer planes, and a muon ranger to compensate.

## 3.2. The NO$\nu$A Physics Goals

NO$\nu$A will have a greatly improved sensitivity to $\nu_\mu$ to $\nu_e$ oscillations over current experiments, in part due to its low Z/A ratio and off-axis location. NO$\nu$A will have a unique level of sensitivity to matter effects, and therefore the neutrino mass hierarchy, among present and approved experiments due to its uniquely long baseline. Depending on the size of the remaining unmeasured mixing angle, $\theta_{13}$, the detection of CP violation in the lepton sector could be within the reach of NO$\nu$A

# 4. THE MINER$\nu$A EXPERIMENT [3]

The MINER$\nu$A (Main Injector ExpeRiment: $\nu$ A) experiment, a collaboration of elementary-particle and nuclear physicists [4], will install a fully active fine-grained solid scintillator detector in the NuMI beam. The overall goals of the experiment are to measure absolute exclusive cross-sections, study nuclear effects in $\nu$ - A interactions (with A varying from He to Pb), perform a systematic study of the resonance-DIS transition region and the lower $Q^2$ DIS region including the extraction of high-$x_{Bj}$ parton distribution functions.

## 4.1. The MINERνA Detector

The MINERνA detector is a hybrid of a fully-active fine-grained detector and a traditional calorimeter and is made up of a number of sub-detectors with distinct functions in reconstructing neutrino interactions. The fiducial volume for most analyses is the inner "Active Target" where all the material of the detector is the scintillator strips themselves. The scintillator detector does not fully contain events due to its low density and low Z, and therefore, the MINERνA design surrounds the scintillator fiducial volume with sampling detectors. To construct these sampling detectors, the scintillator strips are intermixed with absorbers. For example, the side and downstream (DS) electromagnetic calorimeters (ECALs) have lead foil absorbers. Surrounding the ECALs are the hadronic calorimeter (HCAL) where the absorbers are steel plates. On the sides of the detector the outer detector (OD) plays the role of the HCAL. In the upstream end of the detector are the nuclear targets of pure C, Fe and Pb as well as a LHe target. The He target vessel is directly upstream of the main MINERνA detector. Upstream of the detector and LHe target vessel is a veto of steel and scintillator strips to shield MINERνA from incoming soft particles produced upstream in the hall. A complete description of MINERνA is found in the proposal [3] and TDR [5].

The core active element will be extruded triangular-shaped scintillator strips read out *via* wavelength-shifting fibers. Readout of the fibers will be done with multi-anode photomultiplier tubes (MAPMTs), connected to the wavelength shifting fibers *via* an optical cable system.

There are three distinct orientations of strips in the inner detector and veto, separated by 60°, and labeled X, U, V. A single module of MINERνA has two X layers to seed two-dimensional track reconstruction, and one each of the U and V layers to reconstruct three-dimensional tracks.

Monte Carlo studies of this detector and subsequent prototype studies have confirmed that light-sharing with the triangular-shaped scintillator extrusions (3.1 cm base and 1.7 cm height) yield reconstructed point resolution of just under 3 mm. The electromagnetic ($\pi^0$) energy resolution is 6%/sqrt($E_{em}$) while the hadron energy resolution is 4% + 18%/sqrt($E_{had}$).

## 4.2. Overview of the MINERνA Physics Program

For a four-yeart run with 1-year of LE running parasitically with MINOS and 3-years of ME running parasitically with NOνA we expect a total of **9.0M** in the 3-ton fiducial volume of the active scintillator target and a total of another **5.5M** total events on the four nuclear targets of the MINERνA experiment.

The high-statistics studies listed below are important for both the particle and nuclear physics communities, providing information complementary to JeffersonLab charged lepton studies in the same kinematic range

- Precision measurement of the quasi-elastic neutrino–nucleus cross-section, including its $E_\nu$ and $q^2$ dependence, and study of the nucleon axial form factors. Over **800 K** events are expected in the fiducial volume during the four-year MINERνA run.

- Determination of cross-sections in the resonance-dominated region for both neutral-current (NC) and charged-current (CC) interactions, including study of isospin amplitudes, measurement of pion angular distributions, isolation of dominant form factors, and measurement of the effective axial mass. A total of **1.7M** one-pion events make up the low-W resonance sample.
- Clarification of the W ($\equiv$ mass of the hadronic system) transition region where resonance production merges with neutrino deep-inelastic scattering, including tests of phenomenological characterizations of this transition such as quark/hadron duality. A sample of **2.1 M** multi-pion events is expected with W $\leq$ 2.0 GeV.
- Precision measurement of coherent single-pion production cross-sections, with particular attention to target A dependence. Coherent $\pi^0$ production, via the neutral current, is a significant background for next-generation neutrino oscillation experiments seeking to observe $\nu_\mu \rightarrow \nu_e$ oscillation. A sample of **89 K** CC events is expected off carbon. The expected NC sample is roughly half the CC sample.
- Examination of nuclear effects in neutrino interactions, including final-state modifications in heavy nuclei, by employing helium, carbon, iron and lead targets. These effects play a significant role in neutrino oscillation experiments measuring $\nu_\mu$ disappearance as a function of $E_\nu$. It has recently been suggested [**?** ] that, for a given $Q^2$, shadowing can occur at much lower energy transfer ($\nu$) for neutrinos than for charged leptons. This effect is unaccounted for in neutrino event generators. With sufficient $\overline{\nu}$ running, a study of flavor-dependent nuclear effects can also be performed. Due to the different mix of quark flavors, this is another way in which neutrino and charged-lepton nuclear effects differ. MINER$\nu$A will collect over **0.6, 0.4, 2.0 and 2.5 M** CC events off He, C, Fe and Pb targets respectively in addition to the carbon of the scintillator.
- Study of nuclear effects on $\sin^2\theta_W$ measurements, and the NC/CC ratio for different nuclear targets.
- With a sample of over **4.3 M** CC DIS events, a much-improved measurement of the parton distribution functions, particularly at large $x_{Bj}$, will be possible using a measurement of all three $\nu$ structure functions. Although we expect over **150 K** CC $\overline{\nu}$ events in the four year MINER$\nu$A $\nu$ run, an additional dedicated $\overline{\nu}$ run would be required to measure the three $\overline{\nu}$ structure functions with similar precision.
- Examination of the leading exponential contributions of perturbative QCD.
- With nearly **240 K** fully reconstructed exclusive events, precision measurement of exclusive strange-production channels near threshold. This will significantly improve our knowledge of backgrounds in nucleon-decay searches. Also, determination of $V_{us}$, and searches for strangeness-changing neutral-currents and candidate pentaquark resonances will be undertaken Measurement of hyperon-production cross-sections, including hyperon polarization, is feasible with exposure of MINER$\nu$A to $\overline{\nu}$ beams

In addition to these being extremely interesting and challenging research topics, improved knowledge in most is essential to minimizing systematic uncertainties in neutrino-oscillation experiments.

# 5. CONCLUSIONS

The Fermilab Main Injector Neutrino Program covers the entire contemporary study of neutrino physics. There is the disappearance oscillation experiment MINOS that is currently taking data and will make the most accurate measurement of $\Delta m^2_{23}$. The NO$\nu$A appearance oscillation experiment will atempt to measure $\sin^2(2\theta_{13})$ and, if possible, the sign of the mass hierarchy. The MINER$\nu$A experiment is a high-statistics study of the neutrino interactions that both MINOS and NO$\nu$A use to study oscillations and will help the MI oscillation experiments minimize their systematic errors.

# REFERENCES


2. E. Ables *et al.* [MINOS Collaboration], FERMILAB-PROPOSAL-0875 and A. Marchionni [MINOS Collaboration], FERMILAB-CONF-05-429-AD-E
1. D. S. Ayres *et al.* [NOvA Collaboration], arXiv:hep-ex/0503053.
3. D. Drakoulakos *et al.* [Minerva Collaboration], arXiv:hep-ex/0405002. http://minerva.fnal.gov/
4. The MINER$\nu$A Collaboration consists of groups from the following institutions: U Athens, U California/Irvine, CBPF/Rio de Janeiro, U Dortmund, Fermilab, Hampton U, IL Inst. Tech., Inst. for Nuc. Research - Moscow, James Madison U, U Minnesota-Duluth, Jefferson Lab, U Nacional Ingeneria de Lima, N. Illinois U, Northwestern U, Pontifica U Catolica de Lima, U Pittsburgh, U Rochester, Rutgers U, U Texas-Austin, Tufts U, William and Mary U.
5. The MINER$\nu$A Technical Design Report, 1 December 2006, http://minerva-docdb.fnal.gov/cgi-bin/ShowDocument?docid=700

# Application of Finite Groups to Neutrino Mass Matrices

Ernest Ma

*Physics and Astronomy Department, University of California, Riverside, California 92521, USA*

**Abstract.** Recent progress in the application of finite groups to neutrino mass matrices is reviewed, with special emphasis on the tetrahedral symmetry $A_4$.

**Keywords:** Lepton Family Symmetry, Neutrino Masses and Mixing
**PACS:** 11.30.Hv, 14.60.Pq

## INTRODUCTION

Using present data from neutrino oscillations, the $3 \times 3$ neutrino mixing matrix is largely determined, together with the two mass-squared differences [1]. In the Standard Model of particle interactions, there are 3 lepton families. The charged-lepton mass matrix linking left-handed $(e, \mu, \tau)$ to their right-handed counterparts is in general arbitrary, but may always be diagonalized by 2 unitary transformations:

$$M_l = U_L^l \begin{pmatrix} m_e & 0 & 0 \\ 0 & m_\mu & 0 \\ 0 & 0 & m_\tau \end{pmatrix} (U_R^l)^\dagger. \tag{1}$$

Similarly, the neutrino mass matrix may also be diagonalized by 2 unitary transformations if it is Dirac:

$$M_\nu^D = U_L^\nu \begin{pmatrix} m_1 & 0 & 0 \\ 0 & m_2 & 0 \\ 0 & 0 & m_3 \end{pmatrix} (U_R^\nu)^\dagger, \tag{2}$$

or by just 1 unitary transformation if it is Majorana:

$$M_\nu^M = U_L^\nu \begin{pmatrix} m_1 & 0 & 0 \\ 0 & m_2 & 0 \\ 0 & 0 & m_3 \end{pmatrix} (U_L^\nu)^T. \tag{3}$$

Notice that whereas the charged leptons have individual names, the neutrinos are only labeled as $1,2,3$, waiting to be named. The observed neutrino mixing matrix is the mismatch between $U_L^l$ and $U_L^\nu$, i.e.

$$U_{l\nu} = (U_L^l)^\dagger U_L^\nu \simeq \begin{pmatrix} 0.83 & 0.56 & < 0.2 \\ -0.39 & 0.59 & -0.71 \\ -0.39 & 0.59 & 0.71 \end{pmatrix} \simeq \begin{pmatrix} \sqrt{2/3} & 1/\sqrt{3} & 0 \\ -1/\sqrt{6} & 1/\sqrt{3} & -1/\sqrt{2} \\ -1/\sqrt{6} & 1/\sqrt{3} & 1/\sqrt{2} \end{pmatrix}. \tag{4}$$

CP917, *Particles and Fields,* edited by H. Castilla Valdez, J. C. D'Olivo, and M. A. Perez

This approximate pattern has been dubbed tribimaximal by Harrison, Perkins, and Scott [2]. Notice that the 3 vertical columns are evocative of the mesons $(\eta_8, \eta_1, \pi^0)$ in their $SU(3)$ decompositions.

Historically, once the third lepton $\tau$ was established, it was speculated by Cabibbo [3] and Wolfenstein [4] that

$$U_{l\nu}^{CW} = \frac{1}{\sqrt{3}} \begin{pmatrix} 1 & 1 & 1 \\ 1 & \omega & \omega^2 \\ 1 & \omega^2 & \omega \end{pmatrix}, \tag{5}$$

where $\omega = \exp(2\pi i/3) = -1/2 + i\sqrt{3}/2$. Note now

$$U_{l\nu}^{HPS} = (U_{l\nu}^{CW})^\dagger \begin{pmatrix} 1 & 0 & 0 \\ 1 & 1/\sqrt{2} & -1/\sqrt{2} \\ 0 & 1/\sqrt{2} & 1/\sqrt{2} \end{pmatrix} \begin{pmatrix} 0 & 1 & 0 \\ 1 & 0 & 0 \\ 0 & 0 & i \end{pmatrix}. \tag{6}$$

Comparing this to Eq. (4), it tells us that if $U_L^l$ is in fact $U_{l\nu}^{CW}$, then $U_{l\nu}^{HPS}$ can be obtained if maximal mixing occurs in the $2-3$ submatrix of $M_\nu$.

How can $U_{l\nu}^{HPS}$ be derived from a symmetry? The difficulty comes from the fact that any symmetry defined in the basis $(\nu_e, \nu_\mu, \nu_\tau)$ is automatically applicable to $(e, \mu, \tau)$ in the complete Lagrangian. To do so, usually one assumes the canonical seesaw mechanism and studies the Majorana neutrino mass matrix

$$M_\nu = -M_\nu^D M_N^{-1} (M_\nu^D)^T \tag{7}$$

in the basis where $M_l$ is diagonal; but the symmetry apparent in $M_\nu$ (such as $\nu_\mu - \nu_\tau$ interchange) is often incompatible with a diagonal $M_l$ with 3 very different eigenvalues. Obviously a more sophisticated approach is needed. To obtain $U_{l\nu}^{HPS}$, the non-Abelian discrete symmetry $A_4$ turns out to be very useful. In this talk, I will focus mainly on this approach, but first I will discuss $S_3$ which is the smallest non-Abelian finite group. I will also mention $S_4$ and $\Delta(27)$ at the end.

## PERMUTATION SYMMETRY $S_3$

$S_3$ is the permutation group of 3 objects, which is also the symmetry group of the equilateral triangle. It has 6 elements divided into 3 equivalence classes, with the irreducible representations $\underline{1}$, $\underline{1}'$, and $\underline{2}$, and the multiplication rule $\underline{2} \times \underline{2} = \underline{1} + \underline{1}' + \underline{2}$. Its character table is given below.

**TABLE 1.** Character table of $S_3$.

| class | $n$ | $h$ | $\chi_1$ | $\chi_{1'}$ | $\chi_2$ |
|---|---|---|---|---|---|
| $C_1$ | 1 | 1 | 1 | 1 | 2 |
| $C_2$ | 2 | 3 | 1 | 1 | $-1$ |
| $C_3$ | 3 | 2 | 1 | $-1$ | 0 |

Let me discuss briefly 4 recent $S_3$ models.

- Kubo, Mondragon, Mondragon, and Rodriguez-Jauregui, [5] (recently updated by Felix, Mondragon, Mondragon, and Peinado [6]): The symmetry used is actually

$S_3 \times Z_2$, with the assignments

$$(\nu,l),\ l^c,\ N,\ (\phi^+,\phi^0) \sim \underline{1}+\underline{2}, \tag{8}$$

and $v_1 = v_2$. The $Z_2$ symmetry serves to eliminate 4 Yukawa couplings otherwise allowed by $S_3$, resulting in an inverted ordering of neutrino masses with

$$\theta_{23} \simeq \pi/4, \quad \theta_{13} \simeq 0.0034, \quad m_{ee} \simeq 0.05 \text{ eV}, \tag{9}$$

where $m_{ee}$ is the effective Majorana neutrino mass measured in neutrinoless double beta decay. This model relates $\theta_{13}$ to the ratio $m_e/m_\mu$.

- Chen, Frigerio, and Ma [7]: The symmetry here is $S_3$ only, with the assignments

$$(\nu,l) \sim \underline{1}+\underline{2}, \quad l^c \sim \underline{1}+\underline{1}+\underline{1}', \quad (\phi^+,\phi^0) \sim \underline{1}+\underline{2}, \quad (\xi^{++},\xi^+,\xi^0) \sim \underline{2}, \tag{10}$$

and $v_1 = v_2$ but $u_1 \neq u_2$. This results in a normal ordering of neutrino masses with

$$\theta_{23} \simeq \pi/4, \quad 0.008 < \theta_{13} < 0.032, \quad m_{ee} < 0.01 \text{ eV}. \tag{11}$$

This model relates $\theta_{13}$ to $\theta_{12}$ and the ratio $\Delta m^2_{sol}/\Delta m^2_{atm}$.

- Grimus and Lavoura [8]: The symmetry is $S_3 \times Z_2$, with the assignments

$$\begin{aligned}&(\nu,l) \sim (\underline{1},+)+(\underline{2},+), \quad l^c \sim (\underline{1},-)+(\underline{2},+), \quad N \sim (\underline{1},-)+(\underline{2},-),\\ &(\phi^+,\phi^0) \sim (\underline{1},-)+(\underline{1},+)+(\underline{1}',+), \quad (\chi,\chi^*) \sim (\underline{2},+),\end{aligned} \tag{12}$$

and $\langle\chi\rangle^3 = $ real, resulting in a diagonal $M_l$ and a $\mu-\tau$ symmetric $M_\nu$, i.e. $\theta_{23} = \pi/4$ and $\theta_{13} = 0$, whereas $m_{ee}$ is not predicted.

- Mohapatra, Nasri, and Yu [9]: The symmetry $S_3$ is extended to include 3 $Z_3$ transformations which do not commute with $S_3$, so it is not really $S_3$. For $M_\nu$, the assignments are

$$(\nu,l) \sim \underline{1}+\underline{2}, \quad N \sim \underline{1}'+\underline{2}, \quad (\phi^+,\phi^0) \sim \underline{1}, \quad (\xi^{++},\xi^+,\xi^)) \sim \underline{1}, \tag{13}$$

but the extended $S_3$ is grossly broken by $M_N$ in a very special way, resulting then in the tribimaximal form of $M_\nu$. This is not what I would consider a *bona fide* derivation of $U^{HPS}_{l\nu}$.

## TETRAHEDRAL SYMMETRY $A_4$

For 3 families, we should look for a group with a $\underline{3}$ representation, the simplest of which is $A_4$, the group of the even permutation of 4 objects, which is also the symmetry group of the tetrahedron. The tetrahedron is one of five perfect geometric solids known to the ancient Greeks. In order to match them to the 4 elements (fire, air, earth, and water) already known, Plato invented a fifth (quintessence) as that which pervades the cosmos and presumably holds it together. In terms of symmetry, since a cube (hexahedron) may be embedded inside an octahedron and vice versa, the two must have the same

**TABLE 2.** Perfect geometric solids in 3 dimensions.

| solid | faces | vertices | Plato | group |
|---|---|---|---|---|
| tetrahedron | 4 | 4 | fire | $A_4$ |
| octahedron | 8 | 6 | air | $S_4$ |
| cube | 6 | 8 | earth | $S_4$ |
| icosahedron | 20 | 12 | water | $A_5$ |
| dodecahedron | 12 | 20 | quintessence | $A_5$ |

group structure and are thus dual to each other. The same holds for the icosahedron and dodecahedron. The tetrahedron is self-dual. For amusement, compare this first theory of everything to today's contender, i.e. string theory. (A) There are 5 consistent string theories in 10 dimensions. (B) Type I is dual to Heterotic $SO(32)$, Type IIA is dual to Heterotic $E_8 \times E_8$, and Type IIB is self-dual.

$A_4$ has 12 elements divided into 4 equivalence classes, with the irreducible representations $\underline{1}$, $\underline{1}'$, $\underline{1}''$, and $\underline{3}$, and the fundamental multiplication rule

$$\begin{aligned} \underline{3} \times \underline{3} &= \underline{1}(11+22+33) + \underline{1}'(11+\omega^2 22+\omega 33) + \underline{1}''(11+\omega 22+\omega^2 33) \\ &+ \underline{3}(23,31,12) + \underline{3}(32,13,21). \end{aligned} \tag{14}$$

Its character table is given below, where $\omega = \exp(2\pi i/3) = -1/2 + i\sqrt{3}/2$ is exactly

**TABLE 3.** Character table of $A_4$.

| class | $n$ | $h$ | $\chi_1$ | $\chi_{1'}$ | $\chi_{1''}$ | $\chi_3$ |
|---|---|---|---|---|---|---|
| $C_1$ | 1 | 1 | 1 | 1 | 1 | 3 |
| $C_2$ | 4 | 3 | 1 | $\omega$ | $\omega^2$ | 0 |
| $C_3$ | 4 | 3 | 1 | $\omega^2$ | $\omega$ | 0 |
| $C_4$ | 3 | 2 | 1 | 1 | 1 | -1 |

what we saw before in Eq. (5). Note that $\underline{3} \times \underline{3} \times \underline{3} = \underline{1}$ is possible in $A_4$, i.e. $a_1 b_2 c_3 +$ permutations, and $\underline{2} \times \underline{2} \times \underline{2} = \underline{1}$ is possible in $S_3$, i.e. $a_1 b_1 c_1 + a_2 b_2 c_2$.

Other useful sets of finite groups are subgroups of $SU(3)$. The series $\Delta(3n^2)$ has $\Delta(3) \equiv Z_3$, $\Delta(12) \equiv A_4$, $\Delta(27)$, etc. The series $\Delta(3n^2 - 3)$ has $\Delta(9) \equiv Z_3 \times Z_3$, $\Delta(24) \equiv S_4$, etc.

**TABLE 4.** Representations of $SU(3)$ and its subgroups.

| $SU(3)$ | $A_4$ | $S_4$ | $\Delta(27)$ |
|---|---|---|---|
| 1 | 1 | 1 | $1_1$ |
| 3 | 3 | $3'$ | 3 |
| $\bar{3}$ | 3 | $3'$ | $\bar{3}$ |
| 6 | $1+1'+1''+3$ | 1+2+3 | $\bar{3}+\bar{3}$ |
| 8 | $1'+1''+3+3$ | $2+3+3'$ | $\sum_{i=2,9} 1_i$ |
| 10 | $1+3+3+3$ | $1'+3'+3'+3'$ | $1_1 + \sum_{i=1,9} 1_i$ |

Using $A_4$, there are two ways to obtain $U_{lv}^{CW}$ as the unitary matrix which diagonalizes $M_l$: (I) the original proposal of Ma and Rajasekaran [10] and (II) the recent one by Ma [11].

**(I)** Let $(\nu_i, l_i) \sim \underline{3}$, $l_i^c \sim \underline{1}, \underline{1}', \underline{1}''$, then with $(\phi_i^0, \phi_i^-) \sim \underline{3}$,

$$
\begin{aligned}
M_l &= \begin{pmatrix} h_1 v_1 & h_2 v_1 & h_3 v_1 \\ h_1 v_2 & h_2 v_2 \omega & h_3 v_2 \omega^2 \\ h_1 v_3 & h_2 v_3 \omega^2 & h_3 v_3 \omega \end{pmatrix} \\
&= \frac{1}{\sqrt{3}} \begin{pmatrix} 1 & 1 & 1 \\ 1 & \omega & \omega^2 \\ 1 & \omega^2 & \omega \end{pmatrix} \begin{pmatrix} h_1 & 0 & 0 \\ 0 & h_2 & 0 \\ 0 & 0 & h_3 \end{pmatrix} \sqrt{3} v, \qquad (15)
\end{aligned}
$$

for $v_1 = v_2 = v_3 = v$.

**(II)** Let $(\nu_o, l_i) \sim \underline{3}$, $l_i^c \sim \underline{3}$, then with $(\phi_i^0, \phi_i^-) \sim \underline{1}, \underline{3}$,

$$
\begin{aligned}
M_l &= \begin{pmatrix} h_0 v_0 & h_1 v_3 & h_2 v_2 \\ h_2 v_3 & h_0 v_0 & h_1 v_1 \\ h_1 v_2 & h_2 v_1 & h_0 v_0 \end{pmatrix} \\
&= \frac{1}{\sqrt{3}} \begin{pmatrix} 1 & 1 & 1 \\ 1 & \omega & \omega^2 \\ 1 & \omega^2 & \omega \end{pmatrix} \begin{pmatrix} m_e & 0 & 0 \\ 0 & m_\mu & 0 \\ 0 & 0 & m_\tau \end{pmatrix} \frac{1}{\sqrt{3}} \begin{pmatrix} 1 & 1 & 1 \\ 1 & \omega^2 & \omega \\ 1 & \omega & \omega^2 \end{pmatrix}, \qquad (16)
\end{aligned}
$$

for $v_1 = v_2 = v_3$ with $m_e = h_0 v_0 + (h_1 + h_2) v$, $m_\mu = h_0 v_0 + (h_1 \omega + h_2 \omega^2) v$, and $m_\tau = h_0 v_0 + (h_1 \omega^2 + h_2 \omega) v$.

In either case, $U_{l\nu}^{CW}$ has been derived. Each allows arbitrary values of the charged-lepton masses, and yet retains a symmetry for us to consider $M_\nu$. Let

$$
M_\nu = \begin{pmatrix} a+b+c & f & e \\ f & a+b\omega+c\omega^2 & d \\ e & d & a+b\omega^2+c\omega \end{pmatrix} \qquad (17)
$$

be the Majorana neutrino mass matrix in question. Under $A_4$, $a$ comes from $\underline{1}$, $b$ from $\underline{1}'$, $c$ from $\underline{1}''$, and $(d, e, f)$ from $\underline{3}$. Since there are 6 free parameters, this is the most general symmetric mass matrix. To proceed further, these 6 parameters must be restricted.

## SELECTED $A_4$ MODELS

Using **(I)**, the first two proposed $A_4$ models start with only $a \neq 0$, yielding thus 3 degenerate neutrino masses. In Ma and Rajasekaran [10], the degeneracy is broken sofly by $N_i N_j$ terms, allowing $b, c, d, e, f$ to be nonzero. In Babu, Ma, and Valle [12], the degeneracy is broken radiatively through flavor-changing supersymmetric scalar lepton mass terms. In both cases, $\theta_{23} \simeq \pi/4$ is predicted. In the latter, maximal CP violation in $U_{l\nu}$ is also predicted. Consider the case $b = c$ and $e = f = 0$ [13], then

$$
M_\nu = \begin{pmatrix} a+2b & 0 & 0 \\ 0 & a-b & d \\ 0 & d & a-b \end{pmatrix}, \qquad (18)
$$

which is diagonalized by

$$\begin{pmatrix} 1 & 0 & 0 \\ 1 & 1/\sqrt{2} & -1/\sqrt{2} \\ 0 & 1/\sqrt{2} & 1/\sqrt{2} \end{pmatrix} \begin{pmatrix} 0 & 1 & 0 \\ 1 & 0 & 0 \\ 0 & 0 & i \end{pmatrix}, \tag{19}$$

with eigenvalues $a-b+d$, $a+2b$, and $-a+b+d$. Comparing this with Eq. (6), we see that tribimaximal mixing has been achieved. However, since $\underline{1}'$ and $\underline{1}''$ are unrelated in $A_4$, $b=c$ is rather *ad hoc*. A very clever solution by Altarelli and Feruglio [14, 15] is to eliminate both, then $b=c=0$ naturally. This results in a normal ordering of neutrino masses with the prediction [16]

$$|m_{\nu_e}|^2 \simeq |m_{ee}|^2 + \Delta m^2_{atm}/9. \tag{20}$$

A closely related model by Babu and He [17] has $e=f=0$, $b=c$, and $d^2=3b(b-a)$. Here both normal and inverted ordering of neutrino masses are allowed. The technical challenge in this common approach is to break $A_4$ spontaneously along two incompatible directions: (1,1,1) and (1,0,0). One recent proposal [18] is to add $Z_3$ in a supersymmetric model, with singlets carrying the $A_4$ symmetry at a high scale, and require the breaking of $A_4$ without breaking the supersymmetry.

As for possible deviations from tribimaximal mixing, although $b \neq c$ would allow $U_{e3}$ to be different from zero, the assumption $e=f=0$ means that $\nu_2 = (\nu_e+\nu_\mu+\nu_\tau)/\sqrt{3}$ remains an eigenstate. The experimental bound $|U_{e3}| < 0.16$ then implies [13] $0.5 < \tan^2\theta_{12} < 0.52$, whereas experimentally, $\tan^2\theta_{12} = 0.45 \pm 0.05$.

**(III)** A third $A_4$ scenario [19] is to have $(\nu_i, l_i) \sim \underline{3}$, $l_i^c \sim \underline{3}$, but with $(\phi_i^0, \phi_i^-) \sim \underline{1}, \underline{1}', \underline{1}''$. The charged-lepton mass matrix is now diagonal and $M_\nu^{(e,\mu,\tau)} = M_\nu$ already. Using again Eq. (17) but with $d=e=f$,

$$M_\nu = \begin{pmatrix} a+b+c & d & d \\ d & a+b\omega+c\omega^2 & d \\ d & d & a+b\omega^2+c\omega \end{pmatrix}. \tag{21}$$

Assume $b=c$ and rotate to the basis $[\nu_e, (\nu_\mu+\nu_\tau)/\sqrt{2}, (-\nu_\mu+\nu_\tau)/\sqrt{2}]$, then

$$M_\nu = \begin{pmatrix} a+2b & \sqrt{2}d & 0 \\ \sqrt{2}d & a-b+d & 0 \\ 0 & 0 & a-b-d \end{pmatrix}, \tag{22}$$

i.e. maximal $\nu_\mu - \nu_\tau$ mixing and $U_{e3}=0$. The solar mixing angle is now given by $\tan 2\theta_{12} = 2\sqrt{2}d/(d-3b)$. For $b << d$, $\tan 2\theta_{12} \to 2\sqrt{2}$, i.e. $\tan^2\theta_{12} \to 1/2$, but $\Delta m^2_{sol} << \Delta m^2_{atm}$ implies $2a+b+d \to 0$, so that $\Delta m^2_{atm} \to 6bd \to 0$ as well. Therefore, $b \neq 0$ is required, and $\tan^2\theta_{12} \neq 1/2$, but should be close to it, because $b=0$ enhances the symmetry of $M_\nu$ from $Z_2$ to $S_3$. Here $\tan^2\theta_{12} < 1/2$ implies inverted ordering and $\tan^2\theta_{12} > 1/2$ implies normal ordering.

## $S_4$ AND $\Delta(27)$

In the above (III) application of $A_4$, approximate tribimaximal mixing involves the *ad hoc* assumption $b=c$. This problem is overcome by using $S_4$ in a supersymmetric seesaw model [20], yielding the result

$$M_\nu(S_4) = \begin{pmatrix} a+2b & e & e \\ e & a-b & d \\ e & d & a-b \end{pmatrix}. \tag{23}$$

Here $b=0$ and $d=e$ are related limits. A more recent proposal [21] uses $\Delta(27)$, resulting in

$$M_\nu(\Delta(27)) = \begin{pmatrix} fa & c & b \\ c & fb & a \\ b & a & fc \end{pmatrix}. \tag{24}$$

The permutation group of 4 objects is $S_4$. It contains both $S_3$ and $A_4$. It is also the symmetry group of the hexahedron (cube) and the octahedron. It has 24 elements divided into 5 equivalence classes, with 5 irreducible representations $\underline{1},\underline{1}',\underline{2},\underline{3},\underline{3}'$. The fundamental multiplication rules are

$$\begin{aligned} \underline{3}\times\underline{3} &= \underline{1}(11+22+33)+\underline{2}(11+\omega^2 22+\omega 33, 11+\omega 22+\omega^2 33) \\ &+ \underline{3}(23+32,31+13,12+21)+\underline{3}'(23-32,31-13,12-21), \end{aligned} \tag{25}$$

$$\underline{3}'\times\underline{3}' = \underline{1}+\underline{2}+\underline{3}_S+\underline{3}'_A, \quad \underline{3}\times\underline{3}' = \underline{1}'+\underline{2}+\underline{3}'_S+\underline{3}_A. \tag{26}$$

Note that both $\underline{3}\times\underline{3}\times\underline{3}=\underline{1}$ and $\underline{2}\times\underline{2}\times\underline{2}=\underline{1}$ are possible in $S_4$. Let $(\nu_i,l_i),l_i^c,N_i\sim\underline{3}$ under $S_4$. Assume singlet superfields $\sigma_{1,2,3}\sim\underline{3}$ and $\zeta_{1,2}\sim\underline{2}$, then

$$M_N = \begin{pmatrix} M_1 & h\langle\sigma_3\rangle & h\langle\sigma_2\rangle \\ h\langle\sigma_3\rangle & M_2 & h\langle\sigma_1\rangle \\ h\langle\sigma_2\rangle & h\langle\sigma_1\rangle & M_3 \end{pmatrix}, \tag{27}$$

where $M_1=A+f(\langle\zeta_2\rangle+\langle\zeta_1\rangle)$, $M_2=A+f(\langle\zeta_2\rangle\omega+\langle\zeta_1\rangle\omega^2)$, and $M_3=A+f(\langle\zeta_2\rangle\omega^2+\langle\zeta_1\rangle\omega)$. The most general $S_4$-invariant superpotential of $\sigma$ and $\zeta$ is given by

$$\begin{aligned} W &= M(\sigma_1\sigma_1+\sigma_2\sigma_2+\sigma_3\sigma_3)+\lambda\sigma_1\sigma_2\sigma_3+m\zeta_1\zeta_2+\rho(\zeta_1\zeta_1\zeta_1+\zeta_2\zeta_2\zeta_2) \\ &+ \kappa[(\sigma_1\sigma_1+\sigma_2\sigma_2\omega+\sigma_3\sigma_3\omega^2)\zeta_2+(\sigma_1\sigma_1+\sigma_2\sigma_2\omega^2+\sigma_3\sigma_3\omega)\zeta_1]. \end{aligned} \tag{28}$$

The resulting scalar potential has a minimum at $V=0$ (thus preserving supersymmetry) only if $\langle\zeta_1\rangle=\langle\zeta_2\rangle$ and $\langle\sigma_2\rangle=\langle\sigma_3\rangle$, so that $M_N$ is of the form given by Eq. (23). To obtain $M_\nu$ of the same form, $M_l$ should be diagonal and $M_{\nu N}$ proportional to the identity. These are both possible with $\phi^l_{1,2,3}\sim\underline{1}+\underline{2}$, $\phi^N_{1,2,3}\sim\underline{1}+\underline{2}$, but with zero vacuum expectation value for $\phi^N_{2,3}$.

$\Delta(27)$ has 27 elements divided into 11 equivalence classes. There are 9 one-dimensional irreducible representations $\underline{1}_i$ and 2 three-dimensional ones $\underline{3},\underline{\bar{3}}$, with the multiplication rules

$$\underline{3}\times\underline{3}=\underline{\bar{3}}+\underline{\bar{3}}+\underline{\bar{3}}, \quad \underline{3}\times\underline{\bar{3}}=\sum_{i=1,9}\underline{1}_i. \tag{29}$$

For the product $\underline{3} \times \underline{3} \times \underline{3}$, there are 3 invariants: $123+231+312-213-321-132$ which is invariant under $SU(3)$, $123+231+312+213+321+132$ which is also invariant under $A_4$, and $111+222+333$. Let $(\nu_i, l_i) \sim \underline{3}$, $l_i^c \sim \underline{\bar{3}}$, $(\phi_i^0, \phi_i^-) \sim \underline{1}_{1,2,3}$, $(\xi_i^{++}, \xi_i^{+}, \xi_i^{0}) \sim \underline{3}$, then Eq. (24) is obtained. Again let $b=c$, then two solutions for example are $f=1.1046$ and $f=-0.5248$, for both of which $\tan^2\theta_{12}=0.45$ and $m_{ee}=0.05$ eV.

## CONCLUSION

With the application of the non-Abelian discrete symmetry $A_4$, a plausible theoretical understanding of the tribimaximal form of the neutrino mixing matrix has been achieved. Other symmetries such as $S_4$ and $\Delta(27)$ are beginning to be studied. They share some of the properties of $A_4$ and may help to extend our understanding of possible discrete family symmetries, with eventual links to grand unification [22].

## ACKNOWLEDGEMENTS

I thank Miguel Perez and the other organizers of VI-Silafae for their great hospitality and a stimulating symposium in Puerto Vallarta. This work was supported in part by the U. S. Department of Energy under Grant No. DE-FG03-94ER40837.

## REFERENCES


1. See for example M. C. Gonzalez-Garcia, these proceedings.
2. P. F. Harrison, D. H. Perkins, and W. G. Scott, Phys. Lett. **B530**, 167 (2002).
3. N. Cabibbo, Phys. Lett. **B72**, 333 (1978).
4. L. Wolfenstein, Phys. Rev. **D18**, 958 (1978).
5. J. Kubo, A. Mondragon, M. Mondragon, and E. Rodriguez-Jauregui, Prog. Theor. Phys. **109**, 795 (2003).
6. O. Felix, A. Mondragon, M. Mondragon, and E. Peinado, hep-ph/0610061.
7. S.-L. Chen, M. Frigerio, and E. Ma, Phys. Rev. **D70**, 073008 (2004).
8. W. Grimus and L. Lavoura, JHEP **0508**, 013 (2005).
9. R. N. Mohapatra, S. Nasri, and H.-B. Yu, Phys. Lett. **B639**, 318 (2006).
10. E. Ma and G. Rajasekaran, Phys. Rev. **D64**, 113012 (2001).
11. E. Ma, hep-ph/0607190.
12. K. S. Babu, E. Ma, and J. W. F. Valle, Phys. Lett. **B552**, 207 (2003).
13. E. Ma, Phys. Rev. **D70**, 031901(R) (2004).
14. G. Altarelli and F. Feruglio, Nucl. Phys. **B720**, 64 (2005).
15. G. Altarelli and F. Feruglio, Nucl. Phys. **B741**, 215 (2006).
16. E. Ma, Phys. Rev. **D72**, 037301 (2005).
17. K. S. Babu and X.-G. He, hep-ph/0507217.
18. E. Ma, hep-ph/0610342.
19. M. Hirsch, E. Ma, A. Villanova del Moral, and J. W. F. Valle, Phys. Rev. **D72**, 091301(R) (2005); Erratum-ibid. **D72**, 119904 (2005).
20. E. Ma, Phys. Lett. **B632**, 352 (2006).
21. E. Ma, Mod. Phys. Lett. **A21**, 1917 (2006).
22. C. Hagedorn, M. Lindner, and R. N. Mohapatra, JHEP **0606**, 042 (2006); Y. Cai and H.-B. Yu, hep-ph/0608022; I. de Medeiros Varzielas, S. F. King, and G. G. Ross, hep-ph/0607054.

# Sterile neutrinos

Alexander Kusenko

*Department of Physics and Astronomy, University of California, Los Angeles, CA 90095-1547*

**Abstract.** Neutrino masses are usually described by adding to the Standard Model some SU(2)-singlet fermions that have Yukawa couplings, as well as some Majorana mass terms. The number of such fields and the scales of their Majorana masses are not known. Several independent observations point to the possibility that some of these singlets may have masses well below the electroweak scale. A sterile neutrino with mass of a few keV can account for cosmological dark matter. The same particle would be emitted anisotropically from a cooling neutron star in the event of a supernova. This anisotropy can be large enough to explain the observed velocities of pulsars. A lighter sterile neutrino, with mass of the order of eV, is implied by the LSND results; it can have profound implications for cosmology. We review the physics of sterile neutrinos and the roles they may play in astrophysics and cosmology.

**Keywords:** sterile neutrinos, dark matter
**PACS:** 14.60.St,13.15.+g,14.60.Pq,95.35.+d

## STERILE NEUTRINOS IN PARTICLE PHYSICS

The name *sterile neutrino* was coined by Bruno Pontecorvo, who hypothesized the existence of the right-handed neutrinos in a seminal paper [1], in which he also considered vacuum neutrino oscillations in the laboratory and in astrophysics, the lepton number violation, the neutrinoless double beta decay, some rare processes, such as $\mu \to e\gamma$, and several other questions that have dominated the neutrino physics for the next four decades. Most models of the neutrino masses introduce sterile (or right-handed) neutrinos to generate the masses of the ordinary neutrinos via the seesaw mechanism [2]. The seesaw lagrangian

$$\mathscr{L} = \mathscr{L}_{\rm SM} + \bar{N}_a \left(i\gamma^\mu \partial_\mu\right) N_a - y_{\alpha a} H \bar{L}_\alpha N_a - \frac{M_a}{2} \bar{N}_a^c N_a + h.c., \tag{1}$$

where $\mathscr{L}_{\rm SM}$ is the lagrangian of the Standard Model, includes some number $n$ of singlet neutrinos $N_a$ ($a = 1,...,n$) with Yukawa couplings $y_{\alpha a}$. Here $H$ is the Higgs doublet and $L_\alpha$ ($\alpha = e, \mu, \tau$) are the lepton doublets. Theoretical considerations do not constrain the number $n$ of sterile neutrinos. In particular, there is no constraint based on the anomaly cancellation because the sterile fermions do not couple to the gauge fields. The experimental limits exist only for the larger mixing angles [3]. To explain the neutrino masses inferred from the atmospheric and solar neutrino experiments, $n = 2$ singlets are sufficient [4], but a greater number is required if the lagrangian (1) is to explain the LSND [5], the r-process nucleosynthesis [6], the pulsar kicks [7, 8, 9], dark matter [10, 11, 12, 13], and the formation of supermassive black holes [14].

The scale of the right-handed Majorana masses $M_a$ is unknown; it can be much greater than the electroweak scale [2], or it may be as low as a few eV [5, 13, 15]. Even if some of

CP917, *Particles and Fields,* edited by H. Castilla Valdez, J. C. D'Olivo, and M. A. Perez

the right-handed Majorana masses are much larger than others, for example, if some of the $M_a$ ($a = 1, ..., n_l$) are smaller than 100 GeV, while some others ($a = n_l, ..., n$) are much greater than 100 GeV, both classes can have a non-negligible contribution to the active neutrino masses. Obviously, this does not contradict the usual decoupling theorems, because the heavy states decouple from all the physical processes at low energies, but they can still contribute to the values of the active neutrino masses if the corresponding Yukawa couplings are large enough.

## Are they natural?

The seesaw mechanism [2] can explain the smallness of the neutrino masses in the presence of the Yukawa couplings of order one if the Majorana masses $M_a$ are much larger than the electroweak scale. Indeed, in this case the masses of the lightest neutrinos are suppressed by the ratios $\langle H \rangle / M_a$.

However, the origin of the Yukawa couplings remains unknown, and there is no experimental evidence to suggest that these couplings must be of order 1. In fact, the Yukawa couplings of the charged leptons are much smaller than 1. For example, the Yukawa coupling of the electron is as small as $10^{-6}$. One can ask whether some theoretical models are more likely to produce the numbers of order one or much smaller than one. The two possibilities are, in fact, realized in two types of theoretical models. If the Yukawa couplings arise as some topological intersection numbers in string theory, they are generally expected to be of order one [16], although very small couplings are also possible [17]. If the Yukawa couplings arise from the overlap of the wavefunctions of fermions located on different branes in extra dimensions, they can be exponentially suppressed and are expected to be very small [18].

In the absence of the fundamental theory, one may hope to gain some insight about the size of the Yukawa couplings using 't Hooft's naturalness criterion [19], which states essentially that a number can be naturally small if setting it to zero increases the symmetry of the lagrangian. A small breaking of the symmetry is then associated with the small non-zero value of the parameter. This naturalness criterion has been applied to a variety of theories; it is, for example, one of the main arguments in favor of supersymmetry. (Setting the Higgs mass to zero does not increase the symmetry of the Standard Model. Supersymmetry relates the Higgs mass to the Higgsino mass, which is protected by the chiral symmetry. Therefore, the light Higgs boson, which is not natural in the Standard Model, becomes natural in theories with softly broken supersymmetry.) In view of 't Hooft's criterion, the *small* Majorana mass is natural because setting $M_a$ to zero increases the symmetry of the lagrangian (1) [20, 5].

One can ask whether cosmology can provide any clues as to whether the mass scale of sterile neutrinos should be above or below the electroweak scale. It is desirable to have a theory that could generate the matter–antimatter asymmetry of the universe. In both limits of large and small $M_a$ one can have a successful leptogenesis: in the case of the high-scale seesaw, the baryon asymmetry can be generated from the out-of-equilibrium decays of heavy neutrinos [21], while in the case of the low-energy seesaw, the matter-antimatter asymmetry can be produced by the neutrino oscillations [22]. The Big-Bang nucleosynthesis (BBN) can provide a constraint on the number of light relativistic species in equilibrium [23, 24], but the sterile neutrinos with the small mixing angles may never be in equilibrium in the early universe, even at the highest temperatures [10]. Indeed, the effective mixing angle of neutrinos at high temperature is suppressed due to the interactions with plasma [26], and, therefore, the sterile neutrinos may never thermalize. High-precision measurements of the primordial abundances may probe the existence of sterile neutrinos and the lepton asymmetry of the universe in the future [50].

While many seesaw models assume that the sterile neutrinos have very large masses, which makes them unobservable, it is worthwhile to consider light sterile neutrinos in view of the above arguments, and also because they can explain several experimental results. In particular, sterile neutrinos can account for cosmological dark matter [10], they can explain the observed velocities of pulsars [7, 8, 9], the x-ray photons from their decays can affect the star formation [28]. Finally, sterile neutrinos can explain the LSND result [5, 30, 31], which is currently being tested by the MiniBooNE experiment.

## EXPERIMENTAL STATUS

Laboratory experiments are able to set limits or discover sterile neutrinos with a large enough mixing angle. Depending on the mass, they can be searched in different experiments.

The light sterile neutrinos, with masses below $10^2$ eV, can be discovered in one of the neutrino oscillations experiments [32]. In fact, LSND has reported a result [33], which, in combination with the other experiments, implies the existence of at least one sterile neutrino, more likely, two sterile neutrinos [5, 30]. It is also possible that sterile neutrino decays, rather than oscillations, are the explanation of the LSND result [31].

In the eV to MeV mass range, the "kinks" in the spectra of beta-decay electrons can be used to set limits on sterile neutrinos mixed with the electron neutrinos [34]. Neutrinoless double beta decays can probe the Majorana neutrino masses [35]. An interesting proposal is to search for sterile neutrinos in beta decays using a complete kinematic reconstruction of the final state [36].

For masses in the MeV–GeV range, peak searches in production of neutrinos provide the limits. The massive neutrinos $\nu_i$, if they exist, can be produced in meson decays, e.g. $\pi^{\pm} \to \mu^{\pm}\nu_i$, with probabilities that depend on the mixing in the charged current. The energy spectrum of muons in such decays should contain monochromatic lines [34] at $T_i = (m_\pi^2 + m_\mu^2 - 2m_\pi m_\mu - m_{\nu_i}^2)/2m_\pi$. Also, for the MeV–GeV masses one can set a number of constraints based on the decays of the heavy neutrinos into the "visible" particles, which would be observable by various detectors. These limits are discussed in Ref. [3].

## STERILE NEUTRINOS IN ASTROPHYSICS AND COSMOLOGY

Sterile neutrinos can be produced in the early universe, as well as in supernova explosions. The light sterile neutrino, consistent with the LSND result, is consistent with the existing bounds on the big-bang nucleosynthesis [23, 24] and large-scale structure, es-

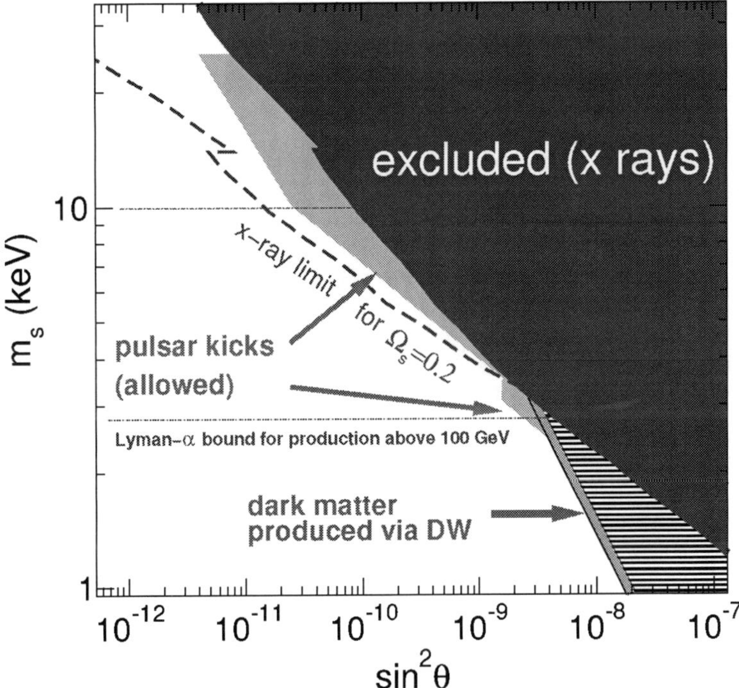


**FIGURE 1.** Sterile neutrinos with masses 2–25 keV can explain the pulsar kicks if the mixing angles are large enough (as shown). In the region marked *excluded (x-rays)*, the relic sterile neutrinos produced in neutrino oscillations via the Dodelson–Widrow (DW) mechanism would have a density inconsistent with the existing x-ray bounds. If the sterile neutrinos constitute all the dark matter, their masses and mixings should fall below the dashed line. Note that DW mechanism is not sufficient to produce enough dark matter for points on the dashed line: a large lepton asymmetry [12] or a new production mechanism [44, 38] is required. The Lyman-$\alpha$ bound for dark-matter sterile neutrinos produced at temperatures $T > 100$ GeV is $m_s > 2.7$ keV (see text or Ref. [38] for discussion). The cosmological and the x-ray bounds do not apply if the universe was never reheated above $T \sim$ MeV [37].

pecially if the mixing lepton asymmetry of the universe is larger than the baryon asymmetry [25]. A heavier sterile neutrino, with mass in the keV range is an appealing dark-matter candidate, as discussed below.

## Dark matter in the form of sterile neutrinos

The sterile neutrinos can be the cosmological dark matter [10, 11, 12, 13]. The interactions already present in the lagrangian (1) allow for the production of relic sterile neutrinos via the Dodelson-Widrow (DW) mechanism [10] in the right amount to account for all dark matter, i.e. $\Omega_s \approx 0.2$, if one of the Majorana masses is of the order of a keV. We will denote the dark-matter sterile neutrino $\nu_s$ and its mass $m_s$. By assumption, the mixing angles are small, and

$$m_s \approx M_1 \tag{2}$$

$$\sin\theta \approx \frac{y\langle H\rangle}{M_1}. \tag{3}$$

The mass and mixing angle are subject to the x-ray limits on the photons from the decays of the relic sterile neutrinos [39], as well as the Lyman-$\alpha$ bound [40] discussed below (see Fig. 1).

As was mentioned above, the relic sterile neutrinos can decay into the lighter neutrinos and an the x-ray photons [41], which can be detected by the x-ray telescopes [39]. The flux of x-rays depends on the sterile neutrino abundance. If all the dark matter is made up of sterile neutrinos, $\Omega_s \approx 0.2$, then the limit on the mass and the mixing angle is given by the dashed line in Fig. 1. However, the interactions in the lagrangian (1) cannot produce such an $\Omega_s = 0.2$ population of the sterile neutrinos for the masses and mixing angles along this dashed line, unless the universe has a relatively large lepton asymmetry [12]. If the lepton asymmetry is small, the interactions in eq. (1) can produce the relic sterile neutrinos via the neutrino oscillations off-resonance at some sub-GeV temperature [10]. This mechanism provides the lowest possible abundance (except for the low-temperature cosmologies, in which the universe is never reheated above a few MeV after inflation [37]). The model-independent bound [38] based on this scenario is shown as a solid (purple) region in Fig. 1. It is based on the flux limit from Ref. [39] and the analytical fit to the numerical calculation of the sterile neutrino production by Abazajian [42]. This calculation may have some hadronic uncertainties [43], but they appear to be under control for the mass and mixing angle in the range of interest [27].

If the lepton asymmetry of the universe is relatively large, the resonant oscillations can produce the requisite amount of dark matter even for smaller mixing angles [12], for which the x-ray limits are weak. (The x-ray flux is proportional to the square of the mixing angle.) It is also possible that some additional interactions, not present in eq. (1) can be responsible for the production of dark-matter sterile neutrinos [44, 38]. We will discuss this possibility in more detail below.

The x-ray photons from sterile neutrino decays in the early universe could have affected the star formation. Although these x-rays alone are not sufficient to reionize the universe, they can catalyze the production of molecular hydrogen and speed up the star formation [28], which, in turn, could cause the reionization. Molecular hydrogen is a very important cooling agent necessary for the collapse of primordial gas clouds that gave birth to the first stars. The fraction of molecular hydrogen must exceed a certain minimal value for the star formation to begin [46]. The reaction $\mathrm{H+H \to H_2} + \gamma$ is very slow in comparison with the combination of reactions

$$\mathrm{H^+ + H} \rightarrow \mathrm{H_2^+} + \gamma, \tag{4}$$

$$\mathrm{H_2^+ + H} \rightarrow \mathrm{H_2 + H^+}, \tag{5}$$

which are possible if the hydrogen is ionized. Therefore, the ionization fraction determines the rate of molecular hydrogen production. If dark matter is made up of sterile neutrinos, their decays produce a sufficient flux of photons to increase the ionization fraction by as much as two orders of magnitude [28]. This has a dramatic effect on the production of molecular hydrogen and the subsequent star formation.

Decays of the relic sterile neutrinos during the dark ages could produce an observable signature in the 21-cm background [29]. It can be detected and studied by such instruments as the Low Frequency Array (LOFAR), the 21 Centimeter Array (21CMA), the Mileura Wide-field Array (MWA) and the Square Kilometer Array (SKA).

### *New physics at the electroweak scale, and the Lyman-α bounds*

One can ask whether the mass $M \sim$ keV in equation (1) is a fundamental constant of nature, or whether it could arise from some symmetry breaking via the Higgs mechanism. For example, let us consider the following modification of the lagrangian (1) following Ref. [38]:

$$\mathscr{L} = \mathscr{L}_0 + \bar{N}_a \left( i\gamma^\mu \partial_\mu \right) N_a - y_{\alpha a} H \bar{L}_\alpha N_a - \frac{h_a}{2} S \bar{N}_a^c N_a + V(H,S) + h.c., \tag{6}$$

where $\mathscr{L}_0$ includes the gauge and kinetic terms of the Standard Model, $H$ is the Higgs doublet, $S$ is the real boson, which is SU(2)-singlet, $L_\alpha$ $(\alpha = e, \mu, \tau)$ are the lepton doublets, and $N_a$ $(a = 1, ..., n)$ are the additional singlet neutrinos. Let us consider the following scalar potential:

$$V(H,S) = m_1^2 |H|^2 + m_2^2 S^2 + \lambda_3 S^3 + \lambda_{HS} |H|^2 S^2 + \lambda_S S^4 + \lambda_H |H|^4. \tag{7}$$

After the symmetry breaking, the Higgs doublet and singlet fields each develop a VEV, $\langle H \rangle = v_0 = 247$ GeV, $\langle S \rangle = v_1$, and the singlet neutrinos acquire the Majorana masses $M_a = h_a v_1$. The mass of the $S$ boson in after the symmetry breaking is $\tilde{m}_S \sim v_1$.

This modification makes no difference in the low-energy theory, for example, in its application to the masses of active neutrinos. However, the new coupling opens a new channel for production of sterile neutrinos in the early universe. Indeed, if the couplings of $S$ to $H$ are large enough, while $h < 10^{-6}$, the $S$ boson can be in equilibrium at temperatures above its mass, while the sterile neutrino with a small mixing angle can be out of equilibrium at all times. This is the case, as long as the annihilations $NN \to NN$, $NN \to$ scalars, etc. are not fast enough to keep the sterile neutrinos in equilibrium. Now, since $S$ is in thermal equilibrium at high temperatures, some amount of sterile neutrinos can be produced through decays $S \to NN$. The amount of sterile neutrinos produced this way is determined by the $h$ coupling (and is independent of the active-sterile neutrino mixing angle):

$$\Omega_s = 0.2 \left( \frac{33}{\xi} \right) \left( \frac{h}{1.4 \times 10^{-8}} \right)^3 \left( \frac{\langle S \rangle}{\tilde{m}_S} \right), \tag{8}$$

where $\xi$ is the change in the number density of sterile neutrinos relative to $T^3$ due to the dilution taking place as the universe cools. For example, in the Standard Model, the reduction in the number of effective degrees of freedom that occurs during the cooling from the temperature $T \sim 100$ GeV to a temperature below 1 MeV causes the entropy increase and the dilution of any species out of equilibrium by factor $\xi \approx 33$.

At the same time, the sterile neutrino mass is determined by the VEV of $S$:

$$h\langle S\rangle \sim \text{keV} \quad \Longrightarrow \quad \langle S\rangle \sim \frac{\text{keV}}{h} \sim 10^2\text{GeV} \tag{9}$$

Based on the required values of $\Omega_s$ and the mass, we conclude that the Higgs singlet should have a VEV at the electroweak scale. In this case most of the dark matter is produced at temperature above 100 GeV.

At a lower temperature, some sterile neutrinos are also produced via the Dodelson–Widrow mechanism. This mechanism cannot be turned off. However, for small mixing angles, left of the yellow line in Fig. 1, the dominant contribution comes from the $S$-boson decays.

This has dramatic implications for the Lyman-$\alpha$ bounds because the relation between the mass and the average momentum is very different from the Dodelson-Widrow case. The Lyman-$\alpha$ forest [40] constrains the free-streaming length of the dark-matter particles, but the relation between this length and the particle mass depends on the production mechanism. One can approximately relate the free-streaming length to the mass $m_s$ and the average momentum of the sterile neutrino:

$$\lambda_{FS} \approx 1\text{Mpc}\left(\frac{\text{keV}}{m_s}\right)\left(\frac{\langle p_s\rangle}{3.15T}\right)_{T\approx 1\text{keV}} \tag{10}$$

The sterile neutrinos produced in the $S$-boson decays have an almost thermal spectrum at the time of production. However, as the universe cools down, the number of effective degrees of freedom decreases from $g_*(T_S) = 110.5$ to $g_*(0.1\,\text{MeV}) = 3.36$. Then $\xi = g_*(T_S)/g_*(0.1\,\text{MeV}) \approx 33$. This causes the redshifting of $\langle p_s\rangle$ by the factor $\xi^{1/3}$:

$$\langle p_s\rangle_{(T\ll 1\text{MeV})} = 0.76\,T\left[\frac{110.5}{g_*(\tilde{m}_S)}\right]^{1/3} \tag{11}$$

Comparing eq. (11) with the DW case, one concludes that, as long as the population of sterile neutrinos is dominated by those produced at a high temperature (large enough $h$, small $\theta$), the Lyman-$\alpha$ limit changes from 10 keV to

$$m_s > 2.7\,\text{keV} \tag{12}$$

This lower bound is shown in Fig. 1.

## Sterile neutrinos and the supernova

Sterile neutrinos with masses below several MeV can be produced in the supernova explosion; they can play an important role in the nucleosynthesis, as well as in generating the supernova asymmetries and the pulsar kicks.

The keV masses of sterile neutrinos coinside with the Mikheev-Smirnov-Wolfenstein [47] (MSW) resonance in nuclear matter for typical momenta of the supernova neutrinos [7]. The position of this resonance is affected by the magnetic field due to the D'Olivo-Nieves-Pal-Semikoz effect [48], which plays an important role in generating the pulsar velocities, as discussed below. Since the sterile neutrinos interact with nuclear matter very weakly, they can be very efficient at transporting the heat in the cooling proto-neutron star, altering the dynamics of the supernova [49]. This could lead to an enhancement of the supernova explosion. An additional enhancement can come from the increase in convection in front of the neutron star propelled by the asymmetric emission of sterile neutrinos [51].

In addition to playing an important role in the primordial nucleosynthesis [24, 50], sterile neutrinos can affect the $r$-process and the synthesis of heavy elements in the supernova [6].

## *The pulsar kicks*

The observations of neutrinos from SN1987A constrain the amount of energy that the sterile neutrinos can take out of the supernova, but they are still consistent with the sterile neutrinos that carry away as much as a half of the total energy of the supernova. A more detailed analysis shows that the emission of sterile neutrinos from a cooling newly born neutron star is anisotropic due to the star's magnetic field [7, 8]. The anisotropy of this emission can result in a recoil velocity of the neutron star as high as $\sim 10^3$km/s. While both the active and the sterile neutrinos are produced with some anisotropy, the asymmetry in the amplitudes of active neutrinos is quickly washed out in multiple scatterings as these neutrinos diffuse out of the star in the approximate thermal equilibrium [52]. In contrast, the sterile neutrinos are emitted from the supernova with the asymmetry equal to their production asymmetry. Hence, they give the recoiling neutron star a momentum, large enough to explain the pulsar kicks for the neutrino emission anisotropy as small as a few per cent [7, 8]. This mechanism can be the explanation of the observed pulsar velocities [9]. The range of masses and mixing angles required to explain the pulsar kicks is shown in Fig. 1.

The pulsar kick mechanism based on the sterile neutrino emission has several additional predictions [9]:

- the kick velocities are expected to correlate with the axis of rotation
- the kick should last 10 to 15 seconds, while the protoneutron star is cooling by the emission of neutrinos, but the onset of the kick can be delayed by a few seconds, depending on the mass and mixing angles [8].
- neutrino-driven kicks can deposit additional energy behind the supernova shock [51, 49], and they are expected to produce asymmetric jets with the stronger jet pointing *in the same direction* as the neutron star velocity [51].

Some of these predictions can already be tested statistically using the pulsar data [53].

## CONCLUSIONS

The underlying physics responsible for the neutrino masses is likely to involve the additional SU(2)-singlet fermions, or sterile neutrinos. The Majorana masses of these states can range from a few eV to values well above the electroweak scale. Theoretical arguments have been made in favor of both the high-scale and the low-scale seesaw mechanisms: the high-scale seesaw may be favored by the connection with the Grand Unified Theories, while the low-scale seesaw is favored by 't Hooft's naturalness criterion. Cosmological considerations are consistent with a vast range of mass scales. The laboratory bounds do not provide significant constraints on the sterile neutrinos, unless they have a large mixing with the active neutrinos. The atmospheric and solar neutrino oscillation results cannot be reconciled with the LSND result, unless sterile neutrinos (or other new physics) exist.

There are several indirect astrophysical hints in favor of sterile neutrinos at the keV scale. Such neutrinos can explain the observed velocities of pulsars, they can be the dark matter, and they can play a role in star formation and reionization of the universe.

The preponderance of indirect astrophysical hints may be a precursor of a major discovery, although it may also be a coincidence. One can hope to discover the sterile neutrinos in the X-ray observations. The mass around 3 keV and the mixing angle $\sin^2\theta \sim 3\times10^{-9}$ appear to be particularly interesting because the sterile neutrino with such parameters could simultaneously explain the pulsar kicks and dark matter (assuming these sterile neutrinos are produced at the electroweak scale). However, it is worthwhile to search for the signal from sterile dark matter in other parts of the allowed parameter space shown in Fig. 1. The existence of a much lighter sterile neutrino, with a much greater mixing angle can be established experimentally if MiniBooNE confirms the LSND result.

## ACKNOWLEDGMENTS

This work was supported in part by the DOE grant DE-FG03-91ER40662 and by the NASA ATP grants NAG 5-10842 and NAG 5-13399.

## REFERENCES

1. B. Pontecorvo, JETP, 53, 1717 (1967).
2. P. Minkowski, Phys. lett. **B67** , 421 (1977); M. Gell-Mann, P. Ramond, and R. Slansky, *Supergravity* (P. van Nieuwenhuizen et al. eds.), North Holland, Amsterdam, 1980, p. 315; T. Yanagida, in *Proceedings of the Workshop on the Unified Theory and the Baryon Number in the Universe* (O. Sawada and A. Sugamoto, eds.), KEK, Tsukuba, Japan, 1979, p. 95; S. L. Glashow, *The future of elementary particle physics*, in *Proceedings of the 1979 Cargèse Summer Institute on Quarks and Leptons* (M. Lévy et al. eds.), Plenum Press, New York, 1980, pp. 687; R. N. Mohapatra and G. Senjanović, Phys. Rev. Lett. **44**, 912 (1980).
3. A. Kusenko, S. Pascoli and D. Semikoz, JHEP **0511**, 028 (2005).
4. P. H. Frampton, S. L. Glashow and T. Yanagida, Phys. Lett. B **548**, 119 (2002).
5. A. de Gouvea, Phys. Rev. D **72**, 033005 (2005).
6. G. C. McLaughlin, J. M. Fetter, A. B. Balantekin and G. M. Fuller, Phys. Rev. C **59**, 2873 (1999); D. O. Caldwell, G. M. Fuller and Y. Z. Qian, Phys. Rev. D **61**, 123005 (2000); J. Fetter,

G. C. McLaughlin, A. B. Balantekin and G. M. Fuller, Astropart. Phys. **18**, 433 (2003).
7. A. Kusenko and G. Segrè, Phys. Lett. B **396**, 197 (1997); A. Kusenko and G. Segre, Phys. Rev. D **59**, 061302 (1999).
8. G. M. Fuller, A. Kusenko, I. Mocioiu, and S. Pascoli, Phys. Rev. D **68**, 103002 (2003); M. Barkovich, J. C. D'Olivo and R. Montemayor, Phys. Rev. D **70**, 043005 (2004);
9. For review, see, *e.g.*, A. Kusenko, Int. J. Mod. Phys. D **13**, 2065 (2004).
10. S. Dodelson and L. M. Widrow, Phys. Rev. Lett. **72**, 17 (1994).
11. K. Abazajian, G. M. Fuller and M. Patel, Phys. Rev. D **64**, 023501 (2001); A. D. Dolgov and S. H. Hansen, Astropart. Phys. **16**, 339 (2002).
12. X. d. Shi and G. M. Fuller, Phys. Rev. Lett. **82**, 2832 (1999). C. T. Kishimoto, G. M. Fuller and C. J. Smith, arXiv:astro-ph/0607403.
13. T. Asaka, S. Blanchet and M. Shaposhnikov, Phys. Lett. B **631**, 151 (2005)
14. F. Munyaneza, P.L. Biermann, P. L., Astron and Astrophys., 436, 805 (2005)
15. A. de Gouvea, J. Jenkins and N. Vasudevan, arXiv:hep-ph/0608147.
16. P. Candelas and S. Kalara, Nucl. Phys. B **298**, 357 (1988). D. Gepner, Nucl. Phys. B **311**, 191 (1988); W. Buchmuller, K. Hamaguchi, O. Lebedev, S. Ramos-Sanchez and M. Ratz, arXiv:hep-ph/0703078.
17. O. J. Eyton-Williams and S. F. King, JHEP **0506**, 040 (2005).
18. E. A. Mirabelli and M. Schmaltz, Phys. Rev. D **61**, 113011 (2000).
19. G. 't Hooft, *Naturalness, chiral symmetry, and spontaneous chiral symmetry breaking*, in Recent Developments in Gauge Theories, Cargese 1979, eds. G. 't Hooft et al., Plenum Press, New York, 1980.
20. K. Fujikawa, Prog. Theor. Phys. **113**, 1065 (2005).
21. M. Fukugita and T. Yanagida, Phys. Lett. B **174** (1986) 45.
22. E. K. Akhmedov, V. A. Rubakov and A. Y. Smirnov, Phys. Rev. Lett. **81**, 1359 (1998); T. Asaka and M. Shaposhnikov, Phys. Lett. B **620**, 17 (2005).
23. R. Barbieri and A. Dolgov, Phys. Lett. B **237**, 440 (1990); K. Kainulainen, Phys. Lett. B **244**, 191 (1990); K. Enqvist, K. Kainulainen and M. J. Thomson, Nucl. Phys. B **373**, 498 (1992); D. P. Kirilova and M. V. Chizhov, Phys. Rev. D **58**, 073004 (1998); A. D. Dolgov, Phys. Lett. B **506**, 7 (2001); M. Cirelli, G. Marandella, A. Strumia and F. Vissani, Nucl. Phys. B **708**, 215 (2005).
24. C. Y. Cardall and G. M. Fuller, Phys. Rev. D **54**, 1260 (1996); X. D. Shi and G. M. Fuller, Phys. Rev. D **59**, 063006 (1999);
25. Y. Z. Chu and M. Cirelli, Phys. Rev. D **74**, 085015 (2006).
26. L. Stodolsky, Phys. Rev. D **36**, 2273 (1987); R. Barbieri and A. Dolgov, Nucl. Phys. B **349** (1991) 743.
27. T. Asaka, M. Laine and M. Shaposhnikov, JHEP **0701**, 091 (2007) [arXiv:hep-ph/0612182].
28. P. L. Biermann and A. Kusenko, Phys. Rev. Lett. **96**, 091301 (2006); M. Mapelli, A. Ferrara and E. Pierpaoli, Mon. Not. Roy. Astron. Soc. **369**, 1719 (2006); J. Stasielak, P. L. Biermann and A. Kusenko, Astrophys. J. **654**, 290 (2007); E. Ripamonti, M. Mapelli and A. Ferrara, Mon. Not. Roy. Astron. Soc. **375**, 1399 (2007)
29. M. Valdes, A. Ferrara, M. Mapelli and E. Ripamonti, arXiv:astro-ph/0701301.
30. M. Sorel, J. M. Conrad and M. Shaevitz, Phys. Rev. D **70**, 073004 (2004).
31. S. Palomares-Ruiz, S. Pascoli and T. Schwetz, JHEP **0509**, 048 (2005) [arXiv:hep-ph/0505216].
32. A. Y. Smirnov and R. Zukanovich Funchal, Phys. Rev. D **74**, 013001 (2006).
33. C. Athanassopoulos *et al.* [LSND Collaboration], Phys. Rev. Lett. **77**, 3082 (1996); Phys. Rev. C **58**, 2489 (1998); Phys. Rev. Lett. **81**, 1774 (1998).
34. R. E. Shrock, Phys. Rev. D **24**, 1232 (1981).
35. S. R. Elliott and P. Vogel, Ann. Rev. Nucl. Part. Sci. **52**, 115 (2002).
36. G. Finocchiaro and R. E. Shrock, Phys. Rev. D **46**, 888 (1992); F. Bezrukov and M. Shaposhnikov, Phys. Rev. D **74**, 053005 (2007).
37. G. Gelmini, S. Palomares-Ruiz and S. Pascoli, Phys. Rev. Lett. **93**, 081302 (2004).
38. A. Kusenko, Phys. Rev. Lett. **97**, 241301 (2006).
39. K. Abazajian, G. M. Fuller and W. H. Tucker, Astrophys. J. **562**, 593 (2001); A. Boyarsky, A. Neronov, O. Ruchayskiy and M. Shaposhnikov, arXiv:astro-ph/0512509; A. Boyarsky, A. Neronov, O. Ruchayskiy and M. Shaposhnikov, JETP Lett. **83**, 133 (2006); A. Boyarsky, A. Neronov, A. Neronov, O. Ruchayskiy and M. Shaposhnikov, arXiv:astro-ph/0603368; A. Boyarsky, A. Neronov, O. Ruchayskiy, M. Shaposhnikov and I. Tkachev, arXiv:astro-ph/0603660;

S. Riemer-Sorensen, S. H. Hansen and K. Pedersen, arXiv:astro-ph/0603661; K. Abazajian and S. M. Koushiappas, arXiv:astro-ph/0605271; C. R. Watson, J. F. Beacom, H. Yuksel and T. P. Walker, arXiv:astro-ph/0605424; K. N. Abazajian, M. Markevitch, S. M. Koushiappas and R. C. Hickox, arXiv:astro-ph/0611144. S. Riemer-Sorensen, K. Pedersen, S. H. Hansen and H. Dahle, arXiv:astro-ph/0610034. A. Boyarsky, J. W. den Herder, A. Neronov and O. Ruchayskiy, arXiv:astro-ph/0612219.
40. M. Viel, et al., Phys. Rev. D **71**, 063534 (2005); U. Seljak, A. Makarov, P. McDonald and H. Trac, arXiv:astro-ph/0602430; M. Viel, J. Lesgourgues, M. G. Haehnelt, S. Matarrese and A. Riotto, Phys. Rev. Lett. **97**, 071301 (2006).
41. P. B. Pal and L. Wolfenstein, Phys. Rev. D **25**, 766 (1982).
42. K. Abazajian, Phys. Rev. D **73**, 063506 (2006).
43. T. Asaka, M. Laine and M. Shaposhnikov, JHEP **0606**, 053 (2006).
44. M. Shaposhnikov and I. Tkachev, Phys. Lett. B **639**, 414 (2006).
45. Y. Chikashige, R. N. Mohapatra and R. D. Peccei, Phys. Lett. B **98**, 265 (1981).
46. M. Tegmark, J. Silk, M. J. Rees, A. Blanchard, T. Abel and F. Palla, Astrophys. J. **474**, 1 (1997).
47. S. P. Mikheev and A. Yu. Smirnov, Yad. Fiz. **42**, 1441 (1985) [Sov. J. Nucl. Phys. **42**, 913 (1985)]; L. Wolfenstein, Phys. Rev. **D 17**, 2369 (1978).
48. V. B. Semikoz, Yad. Fiz. **46**, 1592 (1987) [Sov. J. Nucl. Phys. 46, 946 (1987)]; J. C. D'Olivo, J. F. Nieves and P. B. Pal, Phys. Rev. D **40**, 3679 (1989).
49. J. Hidaka and G. M. Fuller, Phys. Rev. D **74**, 125015 (2006).
50. C. J. Smith, G. M. Fuller, C. T. Kishimoto and K. N. Abazajian, arXiv:astro-ph/0608377.
51. C. L. Fryer and A. Kusenko, Astrophys. J. Suppl. **163**, 335 (2006).
52. A. Kusenko, G. Segre and A. Vilenkin, Phys. Lett. B **437**, 359 (1998).
53. C. Y. Ng and R. W. Romani, arXiv:astro-ph/0702180.

# GeV neutrinos due to $n\bar{n}$ oscillation in Gamma-Ray Burst Fireballs

Sarira Sahu

*Instituto de Ciencias Nucleares, Universidad Nacional Autónoma de México, Circuito Exterior, C. U., A. Postal 70-543, 04510 Mexico DF, Mexico*

**Abstract.** The long and short gamma-ray bursts are believed to be produced due to collapse of massive stars and merger of compact binaries respectively. All these objects are rich in neutron and the jet outflow from these objects must have a neutron component in it. By postulating the $n\bar{n}$ oscillation in the gamma-ray burst fireball, we show that, 19-38 GeV neutrinos and anti-neutrinos can be produced due to annihilation of anti-neutrons with the background neutrons. These neutrinos and anti-neutrinos will be produced before the 5-10 GeV neutrinos due to dynamical decoupling of neutrons from the rest of the fireball. Observation of these neutrinos will shed more light on the nature of the GRB progenitors and also be a unique signature of physics beyond the standard model.

**Keywords:** neutron, GRB, neutrino

Gamma Ray Bursts (GRBs) are flashes of non-thermal bursts of low energy ($\sim$ 100 KeV-1 MeV) photons and release about $10^{51}$-$10^{53}$ erg in a few seconds making them the most luminous object in the universe. They are known to occur at cosmological distance[1] and fall into two classes; short-hard bursts ($\leq 2\ s$) and long-soft bursts. It is now widely accepted that long duration bursts are produced due to the core collapse of massive stars the so called hypernovae. The origin of short-duration bursts are still a mystery, but recently there has been tremendous progress due to accurate localization of many short bursts by the Swift[2] and HETE-2[3] satellites which seem to support the coalescing of compact binaries as the progenitor for the short-hard bursts.

It is believed that, gamma-ray emission arrises from the collision of different internal shocks (shells) due to relativistic outflow from the source. A class of models call *fireball model* seems to explain the temporal structure of the bursts and the non-thermal nature of their spectra[1]. Sudden release of copious amount of $\gamma$ rays into a compact region with a size $c\delta t \sim$ 100-1000 km creates an opaque $\gamma - e^{\pm}$ fireball due to the process $\gamma + \gamma \rightarrow e^+ + e^-$. The average optical depth of this process is $\tau_{\gamma\gamma} \simeq 10^{13}$. This optical depth is very large and even if there are no pairs to begin with, they will form very rapidly and will Compton scatter to lower energy photons. Because of the huge optical depth, photons can not escape freely. In the fireball the $\gamma$ and $e^{\pm}$ pairs will thermalize with a temperature of about 3-10 MeV. The fireball expands relativistically with a Lorentz factor $\Gamma \sim 100 - 1000$ under its own pressure and cools adiabatically due to the expansion. The radiation emerges freely to the inter galactic medium (ISM), when the optical depth is $\tau_{\gamma\gamma} \simeq 1$.

In addition to $\gamma$, $e^{\pm}$ pairs, fireballs may also contain some baryons, both from the progenitor and the surrounding medium. These baryons can be either free or in the form of nuclei. If the fireball temperature is high enough (more than 0.7 MeV), then it will

CP917, *Particles and Fields,* edited by H. Castilla Valdez, J. C. D'Olivo, and M. A. Perez

be mostly in the form of neutrons and protons. The baryonic load has to be very small, otherwise, the expansion of the fireball will be Newtonian, which is inconsistent with the present observations. In both the long and the short GBRs, neutrons are the main ejected materials from the central engine. So it seems obvious that, the outflow from the central engine, which later on give rise to GRBs, has neutron as a principal component. The dynamical decoupling of neutron from the rest of the relativistic shell will give rise to inelastic n-p scattering and lead to emission of observable multi-GeV neutrinos[4, 5].

It is believed that, baryon number is not a good symmetry of nature, due to the fact that, our universe is matter dominant and this requires baryon number violating interactions. Thus the breaking of exact-baryon number conservation in unified theories of the fundamental gauge interactions[6] has lead to searches for proton decay and $n\bar{n}$ oscillation, but so far with no success. The $n\bar{n}$ oscillation occurs due to $\Delta B = 2$ transition. The present experimental limit for unbound neutrons is $\tau_{n\bar{n}} \sim 10^9\ s$[7].

Here we propose that, if $n\bar{n}$ oscillations take place within the neutron rich GRB fireball, $19-38\ GeV$ neutrinos will be produced prior to the production of 5-10 GeV neutrinos due to dynamical neutron decoupling from the rest of the fireball as predicted by Bahcall and Mészáros[4]. By using earth as the detector and assuming the magnetic field in the jet outflow from the central engine to be $\sim 10^{-6}\ G$, about 4 events can be recorded per year. The future international project Extreme Universe Space Observatory (EUSO) may be able to detect these neutrinos[8].

In the presence of a magnetic field, the neutron energy levels are split by an amount $\Delta E = g\mu B$ and this is responsible for the oscillation of neutron to anti-neutron, where $g = -1.91$ is the anomalous magnetic moment of neutron, $\mu = e\hbar/2m_p c = 3.152 \times 10^{-12}\ eV\ G^{-1}$ and $B$ is the magnetic field in Gauss. Due to $n\bar{n}$ oscillation, the number of anti-neutron is given by[9]

$$N_{\bar{n}} \simeq \frac{1}{2} N_n \left(\frac{\delta m}{\Delta E}\right)^2 = 0.6 \times 10^{-26} N_n \left(\frac{B}{G}\right)^{-2}, \tag{1}$$

where $\delta m = \tau_{n\bar{n}}{}^{-1}$ and we take $\tau_{n\bar{n}} \simeq 10^9 s$.

In the fireball, a substantial fraction of the baryon kinetic energy is transfered to a non thermal population of electrons through Fermi acceleration at the shock and these accelerated electrons cool via synchrotron emission and/or inverse Compton scattering to produce the observed prompt (due to internal shocks) and afterglow (due to external shocks) emission. For the synchrotron emission, strong magnetic field is needed to fit the observed data. But it is difficult to estimate the strength of the magnetic field from the first principle. One would expect large magnetic field if the progenitors are highly magnetized. Also a relatively small pre-existing magnetic field can be amplified due to turbulent dynamo mechanism, compression or shearing. Despite all these, there is no satisfactory explanation for the existence of strong magnetic field in the fireball. Also even if some magnetic flux is carried by the outflow, it will decrease due to the expansion of the fireball at a larger radius. Recently, it has been proposed that, the emission in the GRBs can also be explained through Compton-drag process and magnetic field is not necessary for the production of gamma rays.

If we include the neutron decay and the $n\bar{n}$ oscillation, the number of anti-neutrons at a distance $r$ from the source are given by

$$
\begin{aligned}
N_{\bar{n}} &\simeq \frac{1}{2}\left(\frac{\xi}{1+\xi}\right)\frac{E}{\eta m_p}e^{-r/r_\beta}\left(\frac{\delta m}{\Delta E}\right)^2 \\
&= 2\times 10^{27}\left(\frac{B}{G}\right)^{-2}\left(\frac{2\xi}{1+\xi}\right)\frac{E_{53}}{\eta_{100}}e^{-r/r_\beta}, \qquad (2)
\end{aligned}
$$

where $\xi$ is the neutron to proton ratio at the source and for simplicity we take $\xi \sim 1$, the dimensionless entropy $\eta = E/M \sim 10^2 - 10^3$ where $E$ is the initial radiation energy produced due to $e^{\pm}, \gamma$ and $M$ is the rest mass of the fireball due to baryon load. Also $r_\beta$ is the neutron decay radius.

The protons, neutrons and electrons are coupled to the thermal radiation in the expanding jet outflow from the central engine until the Compton scattering time scale $t'_{Th} \simeq (n'_p\sigma_{Th})^{-1}$ and the elastic n-p scattering time scale $t'_{np} \simeq (n'_p\sigma_{np})^{-1}$ are shorter than the comoving plasma expansion time scale $t'_{exp} \simeq r/\Gamma$, where $\sigma_{np} \sim 3\times 10^{-26}\ cm^2$ is the n-p elastic scattering cross section, $n'_p$ is the comoving number density of protons in the fireball. In fact when all the components in the jet are coupled, they have a common bulk Lorentz factor $\Gamma \sim 300$. The neutrons and protons are coupled until the opacity $\tau_{np} > 1$, which corresponds to the $r_{np}$ radius

$$
r_{np} < \frac{L\sigma_{np}}{(1+\xi)4\pi m_p\Gamma^2\eta} \sim 6\times 10^{10}\frac{L_{52}}{(1+\xi)\Gamma^2_{300}\eta_{100}}\ cm. \qquad (3)
$$

So the radius $r$ for the n-p to be coupled should be smaller than $r_{np}$, i.e. $r < r_{np}$. Also the mean beta decay radius is $r_\beta \sim 10^{16}\ cm >> r$. So $exp(-r/r_\beta) \sim 1$ and we can comfortably neglect the effect of beta decay in Eq.(2) when we are in the regime $r < r_{np} < r_\beta$.

In the jet frame protons and neutrons have the same Lorentz factor $\Gamma$ and the $n\bar{n}$ oscillation takes place with in the equilibrium plasma. The produced anti-neutrons can annihilate with the background neutrons to produce pions and also can have elastic scattering with the background protons. The following annihilation processes can take place within the comoving plasma,

$$
\begin{aligned}
\bar{n}+n &\rightarrow \pi^+ + \pi^- + \pi^0 \\
\bar{n}+n &\rightarrow \pi^+ + \pi^- \\
\bar{n}+n &\rightarrow \pi^0 + \pi^0. \qquad (4)
\end{aligned}
$$

The $\pi^0$ will decay to $2\gamma$ and the charge pions will decay as,

$$
\pi^{\pm} \rightarrow \mu^{\pm} + \nu_\mu(\bar{\nu}_\mu) \rightarrow e^{\pm} + \nu_e(\bar{\nu}_e) + \nu_\mu(\bar{\nu}_\mu) + \bar{\nu}_\mu(\nu_\mu). \qquad (5)
$$

In Eq.(4) the energy carried by each pion in the c.m. frame is 1.88 $GeV/k_\pi$ , where $k_\pi$ is the multiplicity of pion. The produced pions are relativistic and in the first step of Eq.(5), the charged pions will decay to $\mu^{\pm} + \nu_\mu(\bar{\nu}_\mu)$. In these decays, the average energy carried by muon is $\sim 80\%$ of the pion energy and rest is carried by $\nu_\mu(\bar{\nu}_\mu)$. In the second step of the decay muon will decay as shown in Eq.(5) and the average energy carried by each particle is about 1/3 of the muon energy. So on an average the energy carried by each neutrino or anti-neutrino due to muon decay is $\varepsilon'_{\nu,\bar{\nu}} \simeq 0.5GeV/k_\pi$ in the comoving frame. In the observer frame this will be translated to

$$\varepsilon_{\nu,\bar{\nu}} \simeq 0.5 \frac{GeV}{k_\pi} \Gamma \frac{1}{(1+z)}. \tag{6}$$

By normalizing the burst at redshift $z = 1$, we obtain

$$\varepsilon_{\nu,\bar{\nu}} \simeq 75 \frac{GeV}{k_\pi} \Gamma_{300} \frac{2}{1+z}. \tag{7}$$

For the pion multiplicity $k_\pi = 3$ and 2, this gives $\varepsilon_{\nu,\bar{\nu}} = 25\ GeV$ and $37.5\ GeV$ respectively. Similarly the energy of the muon neutrinos due to pion decay $\pi^\pm \to \mu^\pm + \nu_\mu(\bar{\nu}_\mu)$ is $\varepsilon_{\nu_\mu,\bar{\nu}_\mu} \simeq 56\ GeV/k_\pi\ \Gamma_{300}\ \frac{2}{1+z}$, which is 19 $GeV$ and 28 $GeV$ respectively for $k_\pi = 3$ and 2. So the neutrinos and anti-neutrinos produced due to $n\bar{n}$ annihilation are having energies in the range 19 $GeV$ to 38 $GeV$.

The neutral pions will decay to photons and have energies 0.3 $GeV$ and 0.47 $GeV$ respectively in the c.m. frame. In the observer frame, these will be boosted to 45 $GeV$ and 71 $GeV$ respectively. These photons are produced at a radius $r < r_{np}$ and $\tau_{\gamma\gamma} > 1$, they will degrade their energy by producing $e^\pm$ pairs and thermalize within the fireball medium. But the neutrinos and anti-neutrinos which are produced due to $n\bar{n}$ annihilation will stream away from the fireball before the decoupling of neutrons. So these neutrinos are different from the one that are produced due to neutron decoupling in the later stage of the fireball evolution.

By considering both the long and short GRBs rate within a Hubble radius of $R_b \sim 10^5 R_{b5}$[10], the number of events per year in a detector with $N_t$ target protons is $R_{\nu\bar{\nu}} \sim (N_t/4\pi D^2) R_b N_{\bar{n}} \bar{\sigma}_{\nu\bar{\nu}}$, where $D = 2.64 \times 10^{28} h_{70}^{-1}(1 - 1/\sqrt{1+z})\ cm$ is the proper distance out to red shift $z$ with $h_{70} = H_0/70\ km/s/Mpc$ and $\bar{\sigma}_{\nu\bar{\nu}} \sim 5 \times 10^{-40} \Gamma (1+z)^{-1}\ cm^2$ is the total detection cross section per neutron. If we consider the whole earth as the detector with $N_t \sim 10^{51}$ and taking a small magnetic field $B \sim 10^{-6}\ G$ in the jet outflow, the event rate is

$$\begin{aligned} R_{\nu\bar{\nu}} \sim\ & 4\, h_{70}^2 R_{b5} \frac{E_{53}}{\eta_{100}} N_{t51} \Gamma_{300} \left(\frac{B}{10^{-6} G}\right)^{-2} \\ & \times \left(\frac{2\xi}{1+\xi}\right) \left(\frac{3-2\sqrt{2}}{2+z-2\sqrt{1+z}}\right) year^{-1}. \end{aligned} \tag{8}$$

So the expected rate of multi-GeV neutrino event is about 4 per year and the increase in the initial neutron to proton ratio $\xi$ will slightly increase the event rate. May be in future, the EUSO will be able to detect these neutrinos by emission of fluorescence light of nitrogen due to the extensive air showers through the interaction of the multi-GeV neutrinos with the atmosphere[8]. The production of neutrinos and anti-neutrinos due to $n\bar{n}$ annihilation is inversely proportional to the square of the magnetic field. If the magnetic field in the fireball is very weak, the number of anti-neutrons will be more as can be seen from Eq.(1). The produced anti-neutrons will annhilate with the background neutrons through the processes in Eq.(4) to produce neutrinos, anti-neutrons and photons.

Apart from the $n\bar{n}$ annihilation, if beta decay of anti-neutrons take place within the region $r < r_{np}$ with a very small magnetic field, the excess amount of baryon, anti-baryon annihilation will result in decreasing the baryon content of the fireball further and making it cleaner, which can probably explain the almost baryon free environment of the fireball. On the other hand, strong magnetic field will suppress the $n\bar{n}$ oscillation and ultimately produce very less number of neutrinos and anti-neutrinos. So if the $n\bar{n}$ oscillation exist in GRB fireballs, magnetic field can not be arbitrarily weak or strong. The produced neutrinos and anti-neutrinos are unique signature of $n\bar{n}$ oscillation as well as the presence of magnetic field in the GRB fireballs. It is unique in the sense that, there is no other process which can produce these neutrinos in between the production point (the source at $r = r_0$ ) and $r = r_{np}$. Flux of these neutrinos depends on the original neutron contain as well as the magnetic field in the fireball. Due to very low flux, these neutrinos will be very difficult to detect. The future project EUSO may be able to detect these neutrinos. However, if at all detected, it will be before the 5-10 GeV neutrinos due to inelastic scattering of decoupling neutrons with the protons in the fireball. So the neutrinos within the energy range 19-38 GeV have different physics (production mechanism) as compared to 5-10 GeV neutrinos. Observation of these multi-GeV neutrinos will not only tell more about the nature of the GRB progenitors but also shed more light on the physics beyond the standard model.

## ACKNOWLEDGMENTS

This material is based upon work supported by DGAPA-UNAM under PAPIIT grant number IN119405.

## REFERENCES

1. T. Piran, *Phys. Rep.* **314**, 575 (1999).
2. N. Gehrels, et al., *Nature* **437**, 851 (2005).
3. J. S. Villasenor, et al., *Nature* **437**, 855 (2005).
4. J. N. Bahcall and P. Mészáros, *Phys. Rev. Lett.* **85**, 1362 (2000).
5. P. Mészáros and M. J. Rees, *Astro. Phys. J.* **541**, L5 (2000).
6. J. C. Pati and A. Salam, *Phys. Rev. Lett.* **31**, 661 (1973); H. Georgi and S. L. Glashow, *Phys. Rev. Lett.* **32**, 438 (1974).
7. Yu. Kamyshkov, *Nucl. Phys. B* (Proc. Suppl.) **52A**, 263 (1997).
8. See EUSO Web page, http://www.euso-mission.org/.
9. O. Sawada and M. Fukugita, *Astro. Phys. J.* **248**, 1162 (1981).
10. P. Mészáros and E. Waxman, *Phys. Rev. Lett.* **87**, 171101 (2001).

# Leptogenesis from a Flavour Perspective

Marta Losada

*Centro de Investigaciones, Universidad Antonio Nariño, Cll. 58A No. 37-94, Bogotá, Colombia*

**Abstract.** Leptogenesis is a successful mechanism interlacing particle physics and cosmology to describe the observed matter-antimatter asymmetry of the Universe. We briefly review the standard scenario and describe the novel features that appear when a more accurate analysis including flavour is performed.

**Keywords:** Leptogenesis, neutrino physics, Baryon Asymmetry of the Universe
**PACS:** 13.35Hb, 11.30Fs, 14.60Pq

## THE BASICS

Observational evidence confirms that we are surrounded by matter, in comparison the presence of antimatter is extremely suppressed. Although this may seem natural to us in fact known fundamental physical laws indicate that we should have the same proportion of both matter and antimatter. Thus a special mechanism, denoted baryogenesis, must have occured during the evolution of the early Universe to provide consistency with current experimental results. The necessary features that any given mechanism must include were identified by Sakharov [1]: baryon number violation, C and CP violation and a departure from thermal equilibrium. Throughout the years various models of baryogenesis have been proposed and analyzed. After many efforts and detailed analyses it has been shown that successful baryogenesis in solely the context of the SM is not possible for two main reasons:

- there is not enough CP violation in the quark sector to account for the necessary CP asymmetry
- one must invoke special scenarios [3] or extensions of the SM so that the produced Baryon Asymmetry of the Universe (BAU) is not erased after the electroweak phase transition given current experimental constraints on the scalar sector of the SM.

The above discussion implies exciting and challenging issues to both particle physics and cosmology. The first important conclusion is that the observed BAU is a *CP violating observable* and consequently is a strong indication of the existence of new sources of CP violation. A second important point is that new physics is absolutely called for to have successful modified scenarios or in the case of additional particles/interactions.

Amongst the various baryogenesis mechanisms the simple proposal of Fukugita and Yanagida [2] in which the out-of-equilibrium decay of a heavy Standard Model (SM) singlet (s)fermion in a lepton number and CP violating way has received much attention. The produced lepton asymmetry is then converted into a baryon asymmetry via the non-perturbative interactions which violate B+L present in the SM that are in equilibrium at high temperatures. This is in a nutshell the leptogenesis mechanism.

CP917, *Particles and Fields,* edited by H. Castilla Valdez, J. C. D'Olivo, and M. A. Perez

There is another strong motivation for the existence of new physics which arises from experiments in neutrino physics, currently indicating that neutrinos have masses. In the simplest leptogenesis model, adding 3 right-handed (RH) neutrinos to the particle content of the SM it is also possible via the see-saw mechanism to account simultaneously for solar and atmospheric neutrino oscillations. Thus in this model neutrinos acquire a mass through a different mechanism that the other fermions of the SM. In addition the Yukawa-type coupling of the left-handed doublets, right-handed singlets and Higgs fields contain additional CP violating phases. The existence of these phases in the lepton sector is crucial for successful leptogenesis.

## CP Violation in the Lepton Sector

Let us consider the number of parameters in the lepton sector in the case of the SM plus 3 RH neutrinos at the leptogenesis scale, which is on the order of $\sim M_{N_1}$, the mass of the lightest RH neutrino. There are 3 charged fermion masses, 3 RH neutrino masses and 15 parameters from the complex Yukawa matrix. Instead at low energies, we have 12 parameters: 6 masses, 3 mixing angles and 2 Majorana phases and 1 Dirac phase. Thus it is not simple to disentagle the contributions from the left and right hand sectors respectively. It is important to note that the Dirac phase is related to flavour à la CKM, and it is not related to *lepton number violation.* This phase is physically meaningful if the quantity

$$\Delta m^2_{\tau\mu}\Delta m^2_{\tau e}\Delta m^2_{\mu e}\Delta m^2_{32}\Delta m^2_{31}\Delta m^2_{21}J_{MNS} \neq 0, \tag{1}$$

where $\Delta m^2_{ij} = m^2_i - m^2_j$, and $m_{1,2,3}$ are the masses of the light neutrinos. $J_{MNS}$ is the Jarlskog invariant constructed from the mixing matrix of the lepton sector. In particular, the existence of this flavour-associated CP violated can be tested experimentally via differences in neutrino and antineutrino oscillation probabilities.

The relevance for leptogenesis can be summarized as follows:

- the necessary CP violation arises through phases in lepton sector.
- the CP asymmetry is produced through interference of tree and one-loop contributions to the decay rate of the lightest singlet.
- the CP asymmetry is determined by the particle physics model that produces couplings and masses for the RH neutrinos. Thus, the connection with low scale CP violating observables is model dependent.

## Out of equilibrium Decay

For the number densities of matter and antimatter to differ at some stage of the Universe's evolution there must be a deviation from thermal equilibrium. Leptogenesis makes use of the fact that the decay rate of the lightest right-handed neutrino will at some temperature become smaller that the Hubble expansion rate of the Universe,

$$\Gamma_D < H|_{T=M_{N_1}}. \qquad (2)$$

For temperatures below this scale the interactions which involve the RH neutrino $N_1$ are too slow to catch up with the expanding Universe, implying that the $N_1$s start to become overabundant. The standard numerical anaylses are usually performed using Boltzmann equations for the number density of the $N_1$

$$\frac{dY_{N_1}}{dt} + 3HY_{N_1} = \sum_{j,l,m} \int \Pi f_l f_m \cdots (1 \pm f_{N_1})(1 \pm f_j) \cdots W(l+m+\cdots \to N_1 + j \cdots)$$
$$- f_{N_1} f_j \cdots (1 \pm f_l)(1 \pm f_m) \cdots W(N_1 + j + \cdots \to l + m \cdots)$$

where the R.H.S. denotes the variation of the number density of the $N_1$, normalized to the entropy, due to *all* elementary processes involving $N_1$, inverse decays, scatterings, annihilations.

## SUMMARY OF UNFLAVOURED RESULTS

The Lagrangian of new physics necessary for leptogenesis in the minimal model is

$$L_{NP} = h_{ik} L_k \phi N_i + h.c. + M_\nu N_i N_i. \qquad (3)$$

These few new terms however imply that at high temperatures there are many relevant processes involving the RH neutrinos such as decays, inverse decays, 2-2 scatterings, $\Delta L = 1,2$ interactions, etc., all of which should be included in the BE. The BEs can be written as

$$\frac{dY_{N_1}}{dz} = -\frac{z}{sH(M_1)}(\gamma_D + \gamma_{\Delta L=1})(\frac{Y_{N_1}}{Y_{N_1}^{eq}} - 1) \qquad (4)$$

$$\frac{dY_L}{dz} = -\frac{z}{sH(M_1)}\left[\varepsilon\gamma_D(\frac{Y_{N_1}}{Y_{N_1}^{eq}} - 1) - (\gamma_D + \gamma_{\Delta L=1} + \gamma_{\Delta L=2})\frac{Y_L}{Y_L^{eq}}\right], \qquad (5)$$

where the quantities $\gamma_X$ correspond to the thermally averaged rates of the interactions named above. We next summarize the important quantities for leptogenesis:

- The zero temperature decay rate:

$$\Gamma = \frac{1}{8\pi}(hh^\dagger)_{11} M_{N_1} = \frac{\tilde{m}_1 M_{N_1}^2}{8\pi v^2} \qquad (6)$$

with $\tilde{m}_1 = \frac{(m_D^\dagger m_D)_{11}}{M_1}$ and $\bar{m}^2 = m_1^2 + m_2^2 + m_3^3$.

- The parameter which verifies the out of equilibrium condition:

$$K \equiv \frac{\Gamma}{H}|_{T=M_1} = \frac{\tilde{m}_1}{\tilde{m}_*} < 1. \qquad (7)$$

- The total CP asymmetry $\varepsilon = \frac{\Gamma(N_1->\phi+\bar{l})-\Gamma(N_1->l+\bar{\phi})}{\Gamma(N_1->\phi+\bar{l})+\Gamma(N_1->l+\bar{\phi})}$, which is bounded from above by

$$\varepsilon < \frac{3M_{N_1}\Delta m^2_{atm}}{8\pi v^2 m_3}. \tag{8}$$

- The final total lepton asymmetry $Y_L \sim \frac{1}{g_*}\varepsilon\delta$. $\delta$ incorporates washout effects by L-violating interactions after the RH neutrinos decay. For strong washout $\delta \propto \frac{H}{\Gamma}$.

One of the most important results that have been obtained in the literature [5] have been the connections with experimental results of low energy physics. In fact it can be shown that there is an upper bound on the CP asymmetry which scales as $1/m_3$ which in turn restricts the value of the light neutrino mass scale to be below $0.2-0.3$eV depending on whether an initial abundance of $N_1$s exists.

Precise studies have shown that leptogenesis is an interesting mechanism that can explain the BAU and simultaneously accomodate the neutrino oscillation scenario. In the SM + 3 RH $\nu$'s (or SUSY version) all requirements for successful leptogenesis can be/are natural. However, strong constraints on neutrino masses are implied. Similarly to the electroweak baryogenesis scenario for the production of the BAU it becomes a quantitative question [4].

## FLAVOUR MATTERS

Flavour physics refers to interactions that are not universal for the different generations. If one is in the interaction basis, we would discuss the physics of Yukawa coupling constants. Instead in the mass basis we refer to fermion masses and mixings. In the context of leptogenesis what should be relevant intuitively is the direction in flavour space that the lightest RH neutrino is decaying into. As the charged Yukawa coupling constants are small and the total lepton asymmetry is obtained via a trace over flavour, it is quite remarkable that flavour *matters* [6, 7, 8, 14, 16, 15, 17].

The main observation is quite simple. Flavour distinguishes mass eigenstates, so for temperatures in which the tau and muon Yukawa couplings are in equilibrium although they have no effect on the produced CP asymmetry, the lepton asymmetry produced by the decayof $N_1$ is distributed among *distinguishable* flavours. Each one of these flavours is then washed out in a different amount, thus the contribution to the total asymmetry is weighted by different factors depending on each flavour.

For simplicity we consider the system describing two flavours, say the electron and muon asymmetries,

$$\begin{aligned}\frac{d}{dz}\begin{bmatrix} Y_L^{ee} & Y_L^{e\mu} \\ Y_L^{\mu e} & Y_L^{\mu\mu}\end{bmatrix} &= \frac{z}{sH(M_1)}\left(\gamma_D\left(\frac{Y_{N_1}}{Y_{N_1}^{eq}}-1\right)\begin{bmatrix}\varepsilon^{ee} & \varepsilon^{e\mu} \\ \varepsilon^{\mu e} & \varepsilon^{\mu\mu}\end{bmatrix}\right. \\ &- \left.\frac{1}{2Y_L^{eq}}\left\{\begin{bmatrix}\gamma_D^{ee} & \gamma_D^{e\mu} \\ \gamma_D^{\mu e} & \gamma_D^{\mu\mu}\end{bmatrix},\begin{bmatrix} Y_L^{ee} & Y_L^{e\mu} \\ Y_L^{\mu e} & Y_L^{\mu\mu}\end{bmatrix}\right\}\right) - i\left[[\Lambda_\omega],\begin{bmatrix} Y_L^{ee} & Y_L^{e\mu} \\ Y_L^{\mu e} & Y_L^{\mu\mu}\end{bmatrix}\right]\end{aligned}$$

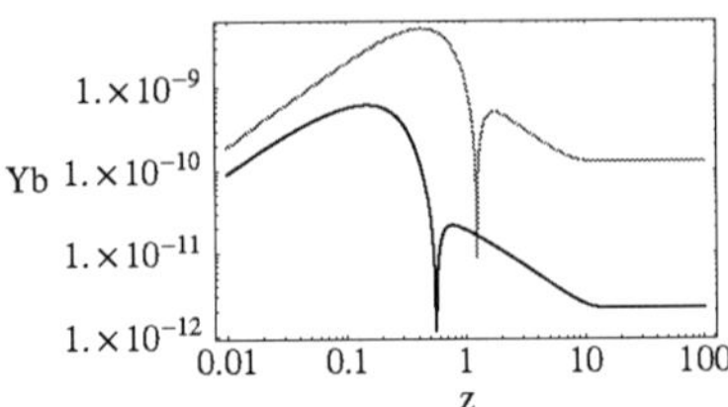


**FIGURE 1.** Total asymmetry in the strong washout case with and without flavour

$$-\frac{1}{2}\left\{[\Lambda_d],\left[\begin{array}{cc} Y_L^{ee} & Y_L^{e\mu} \\ Y_L^{\mu e} & Y_L^{\mu\mu} \end{array}\right]\right\}+[\Lambda_d]Y_L^{\mu\mu} \tag{9}$$

where $z = M_{N_1}/T$, and the parenthesis $[\cdot,\cdot]$ and $\{\cdot,\cdot\}$ stand for commutators and anti-commutators, respectively. The matrix $[\gamma_D]$ is defined as

$$\gamma_D^{ij} = \gamma_D \frac{\lambda_{1i}\lambda_{1j}^*}{\sum_k |\lambda_{1k}|^2}, \tag{10}$$

where $\gamma_D = \sum_i \gamma_D^{ii}$ is the total (thermally averaged) decay rate. For $i = j$, $\gamma_D^{ii}$ is the decay rate of $N_1$ to the $i$-th flavour. The CP asymmetry in the $i$-th flavour is $\varepsilon^{ii}$, $\varepsilon = \sum_i \varepsilon^{ii}$, and notice that $\varepsilon^{ij}$ is normalised by the total decay rate The matrix $[\varepsilon]$ is now defined as

$$\begin{aligned} \varepsilon_{ij} &= \frac{1}{(16\pi)}\frac{1}{[\lambda\lambda^\dagger]_{11}}\sum_J \Im\left\{(\lambda_{1i})[\lambda\lambda^\dagger]_{1J}\lambda_{Jj}^* - (\lambda_{1j}^*)[\lambda^*\lambda^T]_{1J}\lambda_{Ji}\right\}\left[f\left(\frac{M_J^2}{M_I^2}\right)\right] \\ &\simeq \frac{3}{(16\pi v^2)}\frac{M_1}{[\lambda\lambda^\dagger]_{11}}\mathrm{Im}\{\lambda_i\lambda_\sigma[m_\nu^*]_{\sigma j} - \lambda_\sigma^*[m_\nu]_{\sigma i}\lambda_j^*\} \qquad (\text{for } M_1 \ll M_2, M_3) \end{aligned} \tag{11}$$

where $f(x) = \sqrt{x}\left[1-(1+x)\ln\frac{1+x}{x}+\frac{1}{1-x}\right]$ is the loop function.

The matrix $\Lambda$ is given by

$$\Lambda = \left[\begin{array}{cc} \Lambda^{ee} & 0 \\ 0 & \Lambda^{\mu\mu} \end{array}\right], \tag{12}$$

where $\Lambda^{ee}$ and $\Lambda^{\mu\mu}$ indicate the *complex* thermal masses due to the electron and muon Yukawa couplings:

$$\Lambda^{ii} \equiv \Lambda_\omega^{ii} - i\,\Lambda_{\mathrm{d}}^{ii} = \left.\frac{\omega_{ii} - i\,\Gamma_{ii}}{H(M_1)}\right|_{T=M_1}, \quad i = e, \mu. \tag{13}$$

The terms in eq.(9) in which the $\Lambda$'s appear will cause the appearance of flavour oscillations of the asymmetries but additionally they guarantee that the oscillations are damped.

Alternatively [10], we write eq. (9) by expanding the various matrices on the basis provided by the $\sigma$-matrices, $\sigma^\mu = (I, \vec{\sigma})$. Using

$$A = A^\mu \sigma_\mu, \quad A^\mu = \frac{1}{2}\mathrm{Tr}\{A\sigma^\mu\}, \tag{14}$$

the Boltzmann equation (9) becomes a system of equations of the form

$$
\begin{aligned}
\frac{dY_L^0}{dz} &= \frac{z}{sH(M_1)}\left(\gamma_D\left(\frac{Y_{N_1}}{Y_{N_1}^{eq}}-1\right)\varepsilon^0-\frac{1}{Y_L^{eq}}\gamma_D^0 Y_L^0-\frac{1}{Y_L^{eq}}\vec{\gamma}_D\cdot\vec{Y}_L\right), \\
\frac{d\vec{Y}_L}{dz} &= \frac{z}{sH(M_1)}\left(\gamma_D\left(\frac{Y_{N_1}}{Y_{N_1}^{eq}}-1\right)\vec{\varepsilon}-\vec{\gamma}_D\frac{Y_L^0}{Y_L^{eq}}-\frac{1}{Y_L^{eq}}\gamma_D^0\vec{Y}_L\right) \\
& +\vec{\Lambda}_\omega\times\vec{Y}_L+\hat{\Lambda}_d\times\vec{\Lambda}_d\times\vec{Y}_L.
\end{aligned}
\tag{15}
$$

Here $Y_L^0=(1/2)\mathrm{Tr}Y_L$ is half of the total lepton asymmetry while $Y_L^z=(1/2)\left(Y_{ee}-Y_{\mu\mu}\right)$ is half of the difference of the asymmetries in the electron and muon flavour densities. The components $Y_L^x$ and $Y_L^y$ parametrize the off-diagonal entries encoding the quantum correlations between the lepton asymmetries.

The oscillations among the different flavour asymmetries are quickly damped. The timescale of the oscillations and the timescale of decoherence effect on the plasma are roughly on the same order. This implies that only if leptogenesis occurs and the muon Yukawa coupling enters equilibrium at roughly the same temperature there will be any net effect on the final lepton asymmetry, as the mass basis is changing.

However the point is that if leptogenesis occurs once the charged lepton Yukawa coupling is in equilibrium the flavour basis is fixed and it is not possible to rotate the basis arbitrarily. Specific consequences of flavour are reviewed next.

## Decay rate, CP asymmetry and BEs revisited

We can now give the expressions of the relevant quantities when the flavour of the decay products is considered. The decay rate is $\Gamma_{\alpha\alpha}=\frac{1}{8\pi}|h_{1\alpha}|^2M_1$ and the generalization of the quantity parametrizing out of equilibrium $K_{\alpha\alpha}=K\frac{\tilde{m}_{\alpha\alpha}}{\tilde{m}_*}<1$, where

$$
K=\sum_\alpha K_{\alpha\alpha}. \tag{16}
$$

From $N_1$ decays into flavour $\alpha$ an asymmetry will be produced $Y_{\alpha\alpha}$ which is preserved once inverse decays from flavour $\alpha$ are switched off. The corresponding flavour CP asymmetry is

$$
\varepsilon_{\alpha\alpha}=\frac{3}{8\pi[hh^\dagger]_{11}}Im\left(h_{1\beta}\frac{[m^*]_{\beta\alpha}}{v^2}h_{1\alpha}\right)\leq\frac{\sqrt{3}M_1\tilde{m}}{8\pi v^2} \tag{17}
$$

Note that the upper bound on the CP asymmetry is weaker than in unflavoured case in the case of degenerate light neutrinos. We can rewrite the CP asymmetry in each flavour as

$$
\varepsilon_{\alpha\alpha}\leq\frac{3M_1m_{max}}{8\pi v^2}\sqrt{\frac{K_\alpha}{K}} \tag{18}
$$

which grows as the scale of the heaviest of the light LH neutrinos increases. In terms of the Casas-Ibarra parametrization,

$$\varepsilon_{\alpha\alpha} = -\frac{3M_1}{16\pi v^2} \frac{Im(\sum_{\beta\rho} m_\beta^{1/2} m_\rho^{3/2} U^*_{\alpha\beta} U_{\alpha\rho} R_{\beta 1} R_{\rho 1})}{\sum_\beta m_\beta |R_{1\beta}|^2} \tag{19}$$

Where $U$ is the neutrino mixing matrix and $R$ is an orthogonal matrix expressed as $R = vM^{-1/2}\tilde{h}^\dagger_{ij} m^{-1/2}$. Then in the specific case in which $R$ is real, if there is CP violation in $U$ then

$$\varepsilon_{\alpha\alpha} \neq 0, \tag{20}$$

but total CP asymmetry which has been shown to depend in this parametrization only on $R$ and not on $U$ gives $\varepsilon = 0$. A real $R$ matrix implies that there is no CP violation from RH sector and also that possible CP violation at low energy could be correlated to the CP violation that produced the BAU. Something quite difficult to establish in the general cases.

The BE for the sum over flavour is

$$\sum_i \frac{dY_L^{ii}}{dz} = \frac{z}{sH}\left(\left(\frac{Y_{N_1}}{Y_{N_1}^{eq}} - 1\right)\varepsilon\gamma_D - \sum_i \gamma_D^{ii} \frac{Y_L^{ii}}{Y_L^{eq}}\right) \tag{21}$$

where $\sum_i \varepsilon^{ii} = \varepsilon$ is the total CP asymmetry. Like $\gamma_D$ it is the trace of a matrix in flavour space, see eq. (10). In the one-flavour approximation, the trace of the product of matrices in the last term on the RHS of eq. (21) is approximated as the product of the traces: this term is replaced by $\sum_j Y_L^{jj} \sum_i \gamma_D^{ii}$. This is correct in the absence of charged lepton Yukawa couplings; but incorrect when these interactions are in equilbrium as we are now not free to rotate in the direction of flavour space into which the RH neutrino is decaying. The flavour basis is fixed.

So if 1/3 of the asymmetry is in each flavour, and for each flavour, the inverse decay rate is 1/3 of the total. Summing over flavour, leptogenesis is described by the usual equation with the washout reduced by a factor of 1/3.

We present in Fig. (1) the results of the final baryon asymmetry with and without taking flavour into account. The results are for parameters values of: $\varepsilon_{ee} = 2.5\times 10^{-6}$, $\varepsilon_{\mu\mu} = -2\times 10^{-6}$, $\varepsilon_{\tau\tau} = 10^{-7}$ $M_1 = 10^{10}$ GeV $\frac{\Gamma_{ee}}{H} \simeq 10$, $\frac{\Gamma_{\mu\mu}}{H} \simeq 30$, $\frac{\Gamma_{\tau\tau}}{H} \simeq 30$

## CONCLUSIONS

We have discussed the existence of quantum oscillations of the flavour asymmetries which are relevant when the temperature of leptogenesis and the equilibrium of the charged Yukawa interaction occur simultaneously. Another consequence of flavour is that the lepton asymmetry in each flavour is modified and the washout term in the Boltzmann equation is reduced. The CP asymmetry is enhanced and implies modifications of the bounds on both the light left-handed neutrinos and the heavy right-handed neutrinos. Specifically the light neutrino mass scale can increase and there is a smaller lower bound on the reheat temperature. In addition, the careful analysis with flavour may provide more insight into the connection between the high scale and low scale parameters of CP violation and lepton flavour violation.

## ACKNOWLEDGMENTS

It is a pleasure to acknowledge my collaborators Asmaa Abada, Sacha Davidson, Alejandro Ibarra, Francois-Xavier Josse Michaux and Toni Riotto in the development of this work. Research supported in part by Colciencias.

## REFERENCES

1. A.D. Sakharov. *JETP Lett.*, 5:24, 1967.
2. M. Fukugita and T. Yanagida. *Phys. Lett.*, B174:45, 1986.
3. For example, C. Grojean, G. Servant and J. Wells *Phys.Rev.* D71 (2005) 036001
4. See, *e.g.* G. F. Giudice, A. Notari, M. Raidal, A. Riotto and A. Strumia, Nucl. Phys. B **685**, 89 (2004) [arXiv:hep-ph/0310123]. W. Buchmuller, P. Di Bari and M. Plumacher, Annals Phys. **315** (2005) 305 [arXiv:hep-ph/0401240].
5. A partial list: W. Buchmuller, P. Di Bari and M. Plumacher, Nucl. Phys. B **643** (2002) 367 [arXiv:hep-ph/0205349]; W. Buchmuller, P. Di Bari and M. Plumacher, Phys. Lett. B **547** (2002) 128 [arXiv:hep-ph/0209301]; G. C. Branco, R. Gonzalez Felipe, F. R. Joaquim, I. Masina, M. N. Rebelo and C. A. Savoy, Phys. Rev. D **67** (2003) 073025 [arXiv:hep-ph/0211001]; J. R. Ellis, M. Raidal and T. Yanagida, Phys. Lett. B **546** (2002) 228 [arXiv:hep-ph/0206300]; G. C. Branco, R. Gonzalez Felipe, F. R. Joaquim and M. N. Rebelo, Nucl. Phys. B **640** (2002) 202 [arXiv:hep-ph/0202030]; G. C. Branco, T. Morozumi, B. M. Nobre and M. N. Rebelo, Nucl. Phys. B **617** (2001) 475 [arXiv:hep-ph/0107164]; R. N. Mohapatra and S. Nasri, Phys. Rev. D **71** (2005) 033001 [arXiv:hep-ph/0410369]; R. N. Mohapatra, S. Nasri and H. B. Yu, Phys. Lett. B **615** (2005) 231 [arXiv:hep-ph/0502026]; A. Broncano, M. B. Gavela and E. Jenkins, Nucl. Phys. B **672** (2003) 163 [arXiv:hep-ph/0307058]; Phys. Lett. B **552** (2003) 177 [arXiv:hep-ph/0210271];A. Pilaftsis, Phys. Rev. D **56** (1997) 5431 [arXiv:hep-ph/9707235]; G. C. Branco, R. Gonzalez Felipe, F. R. Joaquim and B. M. Nobre, arXiv:hep-ph/0507092.
6. R. Barbieri, P. Creminelli, A. Strumia and N. Tetradis, Nucl. Phys. B **575** (2000) 61 [arXiv:hep-ph/9911315].
7. T. Endoh, T. Morozumi and Z. h. Xiong, Prog. Theor. Phys. **111**, 123 (2004) [arXiv:hep-ph/0308276].
8. T. Fujihara, S. Kaneko, S. Kang, D. Kimura, T. Morozumi and M. Tanimoto, Phys. Rev. D **72**, 016006 (2005) [arXiv:hep-ph/0505076].
9. G. Sigl and G. Raffelt, Nucl. Phys. B **406** (1993) 423.
10. L. Stodolsky, Phys. Rev. D **36** (1987) 2273.
11. G. Raffelt, G. Sigl and L. Stodolsky, Phys. Rev. Lett. **70** (1993) 2363 [arXiv:hep-ph/9209276].
12. L. Covi, E. Roulet and F. Vissani, Phys. Lett. B **384** (1996) 169 [arXiv:hep-ph/9605319].
13. W. Buchmuller, P. Di Bari and M. Plumacher, Nucl. Phys. B **665** (2003) 445 [arXiv:hep-ph/0302092].
14. E. Nardi, Y. Nir, J. Racker and E. Roulet, *JHEP* 0601 (2006) 068 hep-ph/0512052.
15. A. Abada, S. Davidson, F. X. Josse-Michaux, M. Losada and A. Riotto, JCAP **0604** (2006) 004 [arXiv:hep-ph/0601083].
16. E. Nardi, Y. Nir, J. Racker and E. Roulet, *JHEP* 0601 (2006) 164 hep-ph/0601084.
17. A. Abada, S. Davidson, A.Ibarra,F. X. Josse-Michaux, M. Losada and A. Riotto, JHEP 0609 (2006) 0104 [arXiv:hep-ph/0605281].

# Topics in Leptogenesis[1]

Enrico Nardi

*INFN, Laboratori Nazionali di Frascati, C.P. 13, 100044 Frascati, Italy*
*and*
*Instituto de Física, Universidad de Antioquia, A.A.1226, Medellín, Colombia*

**Abstract.** Baryogenesis via leptogenesis provides an appealing mechanism to explain the observed baryon asymmetry of the Universe. Recent refinements in the understanding of the dynamics of leptogenesis include detailed studies of the effects of lepton flavor and of the role possibly played by the lepton asymmetries generated in the decays of the heavier singlet neutrinos $N_{2,3}$. In this talk I present short a review of these recent developments in the theory of leptogenesis.

**Keywords:** Baryogenesis, Cosmology of Theories beyond the SM
**PACS:** 14.60.St 11.30.Fs 12.60.-i

## INTRODUCTION

The possibility that the Cosmic Baryon Asymmetry (BAU) could originate from a lepton number asymmetry generated in the *CP* violating decays of the heavy seesaw Majorana neutrinos was put forth twenty years ago by Fukugita and Yanagida [1]. Their proposal came shortly after Kuzmin, Rubakov and Shaposhnikov pointed out that above the electroweak phase transition $B+L$ is violated by fast electroweak anomalous interactions [2]. This implies that any lepton asymmetry would be in part converted into a baryon asymmetry. However, the discovery that at $T \gtrsim 100\,\mathrm{GeV}$ electroweak interactions do not conserve baryon number, also suggested the exciting possibility that baryogenesis could be a purely standard model phenomenon, and opened the way to electroweak baryogenesis [3]. Indeed, in the early 90's electroweak baryogenesis attracted more interest than leptogenesis. Still, a few remarkable papers appeared that put the first basis for *quantitative* studies of leptogenesis. Here I will just mention two important contributions: the 1992 Luty's paper [4] in which the rates for several processes relevant for the Boltzmann equations for leptogenesis were first presented, and the 1996 paper of Covi, Roulet and Vissani [5] that computed the correct expression for the *CP* asymmetry in the decays of the lightest Majorana neutrino.

Around year 2000 a flourishing of detailed studies of leptogenesis begins, with a corresponding burst in the number of papers dealing with this subject (see fig. 1). This raise of interest in leptogenesis can be traced back to two main reasons: firstly, by this time it became clear that within the standard model, electroweak baryogenesis fails to reproduce the correct BAU by many orders of magnitude, and that even in supersymmetric models this scenario quite likely is not viable. Secondly, the experimental confirmation (from oscillation experiments) that neutrinos have nonvanishing masses strengthened the

[1] Based on work done in collaboration with G. Engelhard, Y. Grossman, Y. Nir, J. Racker and E. Roulet.

CP917, *Particles and Fields,* edited by H. Castilla Valdez, J. C. D'Olivo, and M. A. Perez

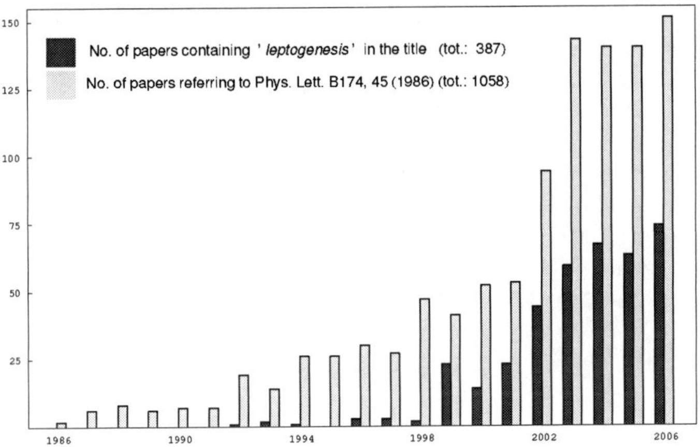


**FIGURE 1.** Barchart showing the number of publications in each year containing a reference to the Fukugita and Yanagida paper [1] (yellow) and containing the word 'leptogenesis' in the title (blue).

case for the seesaw mechanism, that in turn implies the existence, at some large energy scale, of lepton number violating ($\not{L}$) interactions.

The number of important papers and the list of people that contributed to the development of leptogenesis studies and to understand the various implications for the low energy neutrino parameters is too large to be recalled here. Let me just mention the remarkable paper of Giudice *et al.* [6] that appeared in 2003: in this paper a whole set of thermal corrections for the relevant leptogenesis processes were carefully computed, a couple of mistakes common to previous studies were pointed out and corrected, and a detailed numerical analysis was presented both for the SM and the MSSM cases. Eventually, it was claimed that the residual numerical uncertainties would probably not exceed the 10%-20% level [6]. A couple of years later, Nir, Roulet, Racker and myself [7] carried out a detailed study of additional effects that were not accounted for in the analysis of ref. [6]. This included electroweak and QCD sphaleron effects, the effects of the asymmetry in the Higgs number density, as well as the constraints on the particles asymmetry-densities implied by the spectator reactions that are in thermal equilibrium in the different temperature ranges relevant for leptogenesis [7]. Indeed, we found that the largest of theses new effects would barely reach the level of a few tens of percent.

However, two important ingredients that had been overlooked in practically all previous leptogenesis studies, had still to be accounted for. These were the role of the light lepton flavors, and the role of the heavier seesaw Majorana neutrinos. One remarkable exception was the 1999 paper of Barbieri *et al.* [8] that, besides addressing as the main topic the issue of flavor effects in leptogenesis, pointed out that the lepton number asymmetries generated in the decays of the heavier seesaw neutrinos should also be taken into account in computing the BAU.[2] However, these important results did not have much impact on subsequent analyses. The reason might be that these were thought to be just order one effects on the final value of the BAU, with no other major consequences for leptogenesis. As I will discuss in the following, the size of the effects could easily reach the level of one order of magnitude, and, most importantly, they can spoil the leptogenesis constraints on the neutrino low energy parameters, and in particular the limit on the absolute scale of neutrino masses [10]. This is important, since it was thought that this limit was a firm prediction of leptogenesis with hierarchical seesaw neutrinos, and that the discovery of a neutrino mass $m_\nu \gtrsim 0.2\,\mathrm{eV}$ would have strongly disfavored leptogenesis, or hinted to different scenarios (as e.g. resonant leptogenesis [11]).

## THE STANDARD SCENARIO

Let us start by writing the first few terms of the leptogenesis Lagrangian:

$$\mathscr{L} = \frac{1}{2}\left[\bar{N}_1(i\,\partial\!\!\!/)N_1 - M_1 N_1 N_1\right] - (\lambda_1 \bar{N}_1 \ell_1 H + \mathrm{h.c.}). \tag{1}$$

Here $N_1$ is the lightest right-handed Majorana neutrino with mass $M_1$, $H$ is the Higgs field, and $\ell_1$ is the lepton doublet to which $N_1$ couples, that when expressed on a complete orthogonal basis $\{\ell_i\}$ reads

$$|\ell_1\rangle = (\lambda\lambda^\dagger)_{11}^{-1/2} \sum_i \lambda_{1i} |\ell_i\rangle. \tag{2}$$

In practice it is always convenient to use the basis that diagonalizes the charged lepton Yukawa couplings (the flavor basis) that also has well defined $CP$ conjugation properties $CP(\{\ell_i\}) = \{\bar{\ell}_i\}$ with $i = e, \mu, \tau$. Note that in the first and third term in (1) a lepton number can be assigned to $N_1$, that is however violated by two units by the mass term. Then eq. (1) implies processes that violate $L$ like inverse-decays followed by $N_1$ decays $\ell_1 \leftrightarrow N_1 \leftrightarrow \bar{\ell}_1$ or of-shell $\ell_1 H \leftrightarrow \bar{\ell}_1 \bar{H}$ scatterings. At large temperatures $T \gg M_1$, $N_1$ decays and inverse decays are blocked because of thermal effects[3] and off-shell $\not{L}$ processes are suppressed with respect to $L$-conserving ones as $M_1^2/T^2$. At $T \ll M_1$ decays and inverse decays are Boltzmann suppressed, while $\ell_1 \leftrightarrow \bar{\ell}_1$ scatterings are suppressed as $T^2/M_1^2$. Therefore the temperature range in which $\not{L}$ processes can be important for leptogenesis is around $T \sim M_1$. The possibility of generating an asymmetry between the number of leptons $n_{\ell_1}$ and antileptons $n_{\bar{\ell}_1}$ is due to a non-vanishing $CP$ asymmetry in $N_1$ decays:

$$\varepsilon_1 \equiv \frac{\Gamma(N_1 \to \ell_1 H) - \Gamma(N_1 \to \bar{\ell}_1 \bar{H})}{\Gamma(N_1 \to \ell_1 H) + \Gamma(N_1 \to \bar{\ell}_1 \bar{H})} \neq 0. \tag{3}$$

---

[2] Lepton flavor effects were also considered by Endoh, Morozumi and Xiong in their 2003 paper [9], in the context of the minimal seesaw model with just two right handed neutrinos.

[3] $CP$ violating Higgs decays can kinematically occur when $T \gtrsim 7M_1$ [6] however, because of a large time dilation factor, only a small fraction of the Higgses can actually decay before thermal phase space closes.

In order that a macroscopic $L$ asymmetry can build up, the condition that $\not{L}$ reactions are (at least slightly) out of equilibrium at $T \sim M_1$ must also be satisfied. This condition can be expressed in terms of two dimensionful parameters, defined in terms of the Higgs vev $v \equiv \langle H \rangle$ and of the Plank mass $M_P$ as:

$$\tilde{m}_1 = \frac{(\lambda \lambda^\dagger)_{11} v^2}{M_1}, \qquad m_* \approx 10^3 \frac{v^2}{M_P} \approx 10^{-3}\,\text{eV}. \tag{4}$$

The first parameter ($\tilde{m}_1$) is related to the rates of $N_1$ processes (like decays and inverse decays) while the second one ($m_*$) is related to the expansion rate of the Universe at $T \sim M_1$. When $\tilde{m}_1 < m_*$ $\not{L}$ processes are slower than the expansion and leptogenesis can occur. As $\tilde{m}_1$ increases to values larger than $m_*$, $\not{L}$ reactions approach thermal equilibrium thus rendering leptogenesis inefficient because of the back-reactions that tend to erase any macroscopic asymmetry. However, even for $\tilde{m}_1$ as large as $\sim 100 m_*$ a lepton asymmetry sufficient to explain the BAU can be generated. It is customary to refer to the condition $\tilde{m}_1 > m_*$ as to the *strong washout regime* since washout reactions are rather fast. This regime is considered more likely than the *weak washout regime* $\tilde{m}_1 < m_*$ in view of the experimental values of the light neutrino mass-squared differences (that are both $> m_*^2$) together with the theoretical lower bound $\tilde{m}_1 \geq m_{\nu_1}$, where $m_{\nu_1}$ is the mass of the lightest neutrino. The strong washout regime is also theoretically more appealing since the final value of the lepton asymmetry is independent of the particular value of the $N_1$ initial abundance and also of a possible asymmetry $Y_{\ell_1} = (n_{\ell_1} - n_{\bar{\ell}_1})/s \neq 0$ ($s$ is the entropy density) preexisting the $N_1$ decay era. This last fact has been often used to argue that for $\tilde{m}_1 > m_*$ only the dynamics of the lightest Majorana neutrino $N_1$ is important, since asymmetries generated in the decays of the heavier $N_{2,3}$ would be efficiently erased by the strong $N_1$-related washouts. As we will see below, the effects of $N_1$ interactions on the $Y_{\ell_{2,3}}$ asymmetries are subtle, and the previous argument is incorrect. The result of numerical integration of the Boltzmann equations for $Y_{\ell_1}$ can be conveniently expressed in terms of an efficiency factor $\eta_1$, that ranges between 0 and 1:

$$Y_{\ell_1} = 3.9 \times 10^{-3}\, \eta_1 \varepsilon_1, \qquad \eta_1 \approx \frac{m_*}{\tilde{m}_1}. \tag{5}$$

The second relation gives a rough approximation for $\eta_1$ in the strong washout regime that will become useful in analyzing the impact of flavor effects. The possibility of deriving an upper limit for the the light neutrino masses [10] follows from the existence of a theoretical bound on the maximum value of the $CP$ asymmetry $\varepsilon_1$ (that holds when $N_{1,2,3}$ are sufficiently hierarchical) and relates $M_1$, $m_{\nu_3}$ and the washout parameter $\tilde{m}_1$:

$$|\varepsilon_1| \leq \left[\frac{3}{16\pi}\frac{M_1}{v^2}(m_{\nu_3} - m_{\nu_1})\right]\sqrt{1 - \frac{m_{\nu_1}^2}{\tilde{m}_1^2}}. \tag{6}$$

The term in square brackets is the so called Davidson-Ibarra limit [12] while the square root is a correction that was first given in [13]. When $m_{\nu_3} \gtrsim 0.1\,\text{eV}$, the light neutrinos are quasi-degenerate and $m_{\nu_3} - m_{\nu_1} \sim \Delta m^2_{atm}/2m_{\nu_3} \to 0$ so that, to keep $\varepsilon_1$ finite, $M_1$ is pushed to large values $\gtrsim 10^{13}\,\text{GeV}$. Since at the same time $\tilde{m}_1$ must remain larger than

$m_{\nu_{1,3}}$ the washout effects also increase, until the surviving lepton-asymmetry is too small to explain the BAU.[4] The interesting limit $m_{\nu_3} \lesssim 0.15\,\text{eV}$ results.

## LEPTON FLAVOR EFFECTS

In the Lagrangian (1) the terms involving the charged lepton Yukawa couplings have not been included. Since all these couplings are rather small, if leptogenesis occurs at temperatures $T \gtrsim 10^{12}\,\text{GeV}$, when the Universe is still very young, not many of the related (slow) processes could have occurred during its short lifetime, and leptogenesis has essentially no knowledge of lepton flavors. At $T \lesssim 10^{12}\,\text{GeV}$ the reactions mediated by the tau Yukawa coupling $h_\tau$ become important, and at $T \lesssim 10^9\,\text{GeV}$ also $h_\mu$-reactions have to be accounted for. Including the Yukawa terms for the leptons yields the Lagrangian:

$$\mathscr{L} = \frac{1}{2}[\bar{N}_1(i\,\partial\!\!\!/)N_1 - M_1 N_1 N_1] - (\lambda_{1i}\bar{N}_1\,\ell_i H + h_i \bar{e}_i \ell_i H^\dagger + \text{h.c.}), \tag{7}$$

where, since we are using the flavor basis, the matrix $h$ of the Yukawa couplings is diagonal. The flavor content of the (anti)lepton doublets $\ell_1$ ($\bar{\ell}'_1$) to which $N_1$ decays is now important, since these states do not remain coherent, but are effectively resolved into their flavor components by the fast Yukawa interactions $h_i$ [8, 14, 15]. Note that in general, due to $CP$ violating loop effects, $CP(\bar{\ell}'_1) \neq \ell_1$, that is the antileptons produced in $N_1$ decays are not the $CP$ conjugated of the leptons, implying that the flavor projections $K_i \equiv |\langle \ell_i | \ell_1 \rangle|^2$ and $\bar{K}_i \equiv |\langle \bar{\ell}_i | \bar{\ell}'_1 \rangle|^2$ differ: $\Delta K_i = K_i - \bar{K}_i \neq 0$. The fact that $\ell_1$ and $\bar{\ell}'_1$ can differ in their flavor content implies that even when $N_1$ decays with equal branchings into leptons and antileptons (yielding $\varepsilon_1 = 0$) the $CP$ asymmetries for the decays into single flavors can still be non-vanishing. The flavor $CP$ asymmetries are defined as [15]:

$$\varepsilon_1^i = \frac{\Gamma(N_1 \to \ell_i H) - \Gamma(N_1 \to \bar{\ell}_i \bar{H})}{\Gamma_{N_1}} = K_i \varepsilon_1 + \Delta K_i/2. \tag{8}$$

The factor $\Delta K_i$ in the second equality accounts for the flavor mismatch between leptons and antileptons, while the factor $K_i$ in front of $\varepsilon_1$ accounts for the reduction in the strength of the $N_1$-$\ell_i$ coupling with respect to $N_1$-$\ell_1$. The same factor also reduces the strength of the washouts for the $i$-flavor, yielding an efficiency factor $\eta_1^i = \min(\eta_1/K_i, 1)$. Assuming for illustration $\eta_1/K_i < 1$ the resulting asymmetry is

$$Y_L \approx \sum_i \varepsilon_1^i \eta_1^i \approx n_f Y_{\ell_1} + \sum_i \frac{\Delta K_i}{2K_i}\frac{m_*}{\tilde{m}_1}. \tag{9}$$

In the first term on the r.h.s. $n_f$ represents the number of flavors effectively resolved by the charged lepton Yukawa interactions ($n_f = 2$ or 3), while $Y_{\ell_1}$ is the asymmetry that would have been obtained by neglecting the decoherence of $\ell_1$. The second term, that

[4] $\Delta L = 2$ washout processes, that depend on a different parameter than $\tilde{m}_1$, and that can become important when $M_1$ is large, also play a role in establishing the limit.

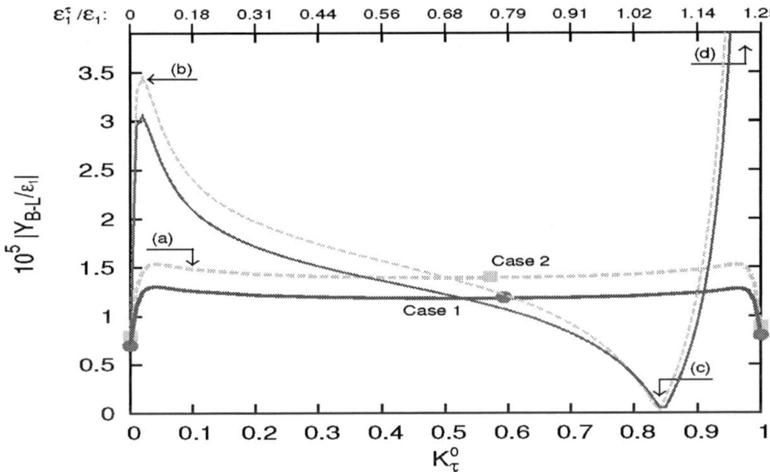


**FIGURE 2.** $|Y_{B-L}|$ (in units of $10^{-5}|\varepsilon_1|$) as a function of $K_\tau$ in two two-flavor regimes. The thick lines correspond to the special case when $K_\tau = \bar{K}_\tau$. The thin lines give an example of the results for $K_\tau \neq \bar{K}_\tau$. The filled circles and squares at $K_\tau = 0, 1$ correspond to the aligned cases without flavor effects.

is controlled by the 'flavor mismatch' factor $\Delta K_i$, can become particularly large in the cases when the flavor $i$ is almost decoupled from $N_1$ ($K_i \ll 1$). This situation is depicted in fig. 2 for the two-flavor case and for two different temperature regimes. The two flat curves give $|Y_{B-L}|$ as a function of the flavor projector $K_\tau$ assuming $\Delta K_\tau = 0$, and show rather clearly the enhancement of a factor $\approx 2$ with respect to the aligned cases (the points at $K_\tau = 0$, 1) for which flavor effects are irrelevant. The other two curves, that peak at values close to the boundaries, when $\ell_\tau$ or a combination orthogonal to $\ell_\tau$ are almost decoupled from $N_1$, show that $\ell_1$-$\ell'_1$ flavor mismatch can produce much larger enhancements. In conclusion, the relevance of flavor effects is at least twofold:

1. The BAU resulting form leptogenesis can be several times larger than what would be obtained neglecting flavor effects.
2. If leptogenesis occurs at temperatures when flavor effects are important, the limit on the light neutrino masses does not hold [14, 16]. This is because there is no analogous of the Davidson-Ibarra bound eq. (6) for the flavor asymmetries $\varepsilon_1^i$.

## THE EFFECTS OF THE HEAVIER MAJORANA NEUTRINOS

What about the possible effects of the heavier Majorana neutrinos $N_{2,3}$ that we have so far neglected ? Some recent studies analyzed the so called "$N_1$-decoupling" scenario, in which the Yukawa couplings of $N_1$ are simply too weak to washout an asymmetry generated in $N_2$ decays (and $\varepsilon_1$ is too small to explain the BAU) [17]. This is a consistent scenario in which $N_2$ leptogenesis could successfully explain the BAU. However, in the opposite situation when the Yukawa couplings of $N_1$ are very large, it was generally assumed that the asymmetries related to $N_{2,3}$ are irrelevant for the computation of the BAU, since they would be washed out during $N_1$ leptogenesis. In contrast to this, in ref. [8] (and more recently also in ref. [18]) it was stated that part of the asymmetry from $N_{2,3}$ decays does in general survive, and must be taken into account when computing the BAU. In ref. [19] Engelhard, Grossman, Nir and myself carried out a detailed study of the fate of a lepton asymmetry preexisting $N_1$ leptogenesis, and we reached conclusions that agree with these statements. I will briefly describe the reasons for this and the importance of the results. Including also $N_{2,3}$ the leptogenesis Lagrangian reads:

$$\mathscr{L} = \frac{1}{2}[\bar{N}_\alpha(i\,\partial\!\!\!/)N_\alpha - N_\alpha M_\alpha N_\alpha] - (\lambda_{\alpha i}\bar{N}_\alpha \ell_i H + h_i \bar{e}_i \ell_i H^\dagger + \text{h.c.}), \tag{10}$$

where $\alpha = 1,2,3$ is a heavy neutrinos index, and the heavy neutrinos are written in the mass basis. It is convenient to define the three (in general non-orthogonal) combinations of lepton doublets $\ell_\alpha$ to which the corresponding $N_\alpha$ decay:

$$|\ell_\alpha\rangle = (\lambda\lambda^\dagger)^{-1/2}_{\alpha\alpha}\sum_i \lambda_{\alpha i}|\ell_i\rangle. \tag{11}$$

Let us discuss for definiteness the case where a sizeable asymmetry originates from $N_2$ decays, while $N_1$ leptogenesis is instead rather inefficient. That is, let us assume that $N_2$-related washouts are not too strong, while $N_1$-related washouts are so strong that by itself $N_1$ leptogensis would not be successful:

$$\tilde{m}_2 \not\gg m_*, \qquad\qquad \tilde{m}_1 \gg m_*. \tag{12}$$

To simplify the arguments, let us also impose two additional conditions: thermal leptogenesis, that is a vanishing initial $N_1$ abundance $n_{N_1}(T \gg M_1) \approx 0$, and a strong hierarchy $M_2/M_1 \gg 1$. From this it follows that there are no $N_1$ related washout effects during $N_2$ leptogenesis and, because $n_{N_2}(T \approx M_1)$ is Boltzmann suppressed, there are no $N_2$ related washouts during $N_1$ leptogenesis. Thus $N_2$ and $N_1$ dynamics are decoupled. Now, the second condition in (12) implies that already at $T \gtrsim M_1$ the interactions mediated by the $N_1$ Yukawa couplings are sufficiently fast to quickly destroy the coherence of the state $\ell_2$ produced in $N_2$ decays. Then a statistical mixture of $\ell_1$ and of the states orthogonal to $\ell_1$ builds up, and it can be described by a suitable *diagonal* density matrix. On general grounds one expects that decoherence effects proceed faster than washout. In the relevant range, $T \gtrsim M_1$, this is also ensured by the fact that because of thermal effects the dominant $\mathscr{O}(\lambda^2)$ washout process (the inverse decay $\ell H \to N_1$) is blocked [6], and only scatterings with top-quarks and gauge bosons, that have additional suppression factors of $h_t^2$ and $g^2$, contribute to the washout.

Let us consider the case where both $N_2$ and $N_1$ decay at $T \gtrsim 10^{12}\,\text{GeV}$ and flavor effects are irrelevant. In this regime a convenient choice for the orthogonal lepton basis is $(\ell_1, \ell_0, \ell_0')$ where, without loss of generality, $\ell_0'$ can be chosen to satisfy $\langle \ell_0'|\ell_2\rangle = 0$. Then the asymmetry $Y_{\ell_2}$ produced in $N_2$ decays decomposes in the two components:

$$Y_{\ell_0} = c^2 Y_{\ell_2}, \qquad\qquad Y_{\ell_1} = s^2 Y_{\ell_2}, \tag{13}$$

where $c^2 \equiv |\langle \ell_0|\ell_2\rangle|^2$ and $s^2 = 1 - c^2$. The crucial point here is that in general we expect $c^2 \neq 0$, and since $\ell_0$ is orthogonal to $\ell_1$, $Y_{\ell_0}$ is protected against $N_1$ washouts. Then a finite part of the asymmetry $Y_{\ell_2}$ from $N_2$ decays survives through $N_1$ leptogenesis. A more detailed study [19] reveals also some unexpected features. For example, $Y_{\ell_1}$ is not driven to zero in spite of the strong $N_1$-related washouts, rather, only the sum of $Y_{\ell_1}$ and of the Higgs asymmetry $Y_H$ vanishes, but not the two separately. (This can be traced back to the presence of a conserved charge related to $Y_{\ell_0}$.) As a result we obtain $Y_{\ell_1} = -Y_{\ell_0}/4$ and $Y_H = Y_{\ell_0}/2$, while the total lepton asymmetry is $Y_L = (3/2)Y_{\ell_0} = (3/2)\,c^2 Y_{\ell_2}$.

For $10^9 \lesssim M_1 \lesssim 10^{12}\,$GeV the $\ell_{2,1}$ flavor structures are only partially resolved during $N_1$ leptogenesis, and a similar result is obtained. However, when $M_1 \lesssim 10^9\,$GeV and the full flavor basis $(\ell_e, \ell_\mu, \ell_\tau)$ is resolved, there are no directions in flavor space where an asymmetry can remain protected, and then the whole $Y_{\ell_2}$ can be erased. In conclusion, the common assumption that when $N_1$ leptogenesis occurs in the strong washout regime the final BAU is independent of initial conditions, does not hold in general, and is justified only in the following cases [19]: *i)* Vanishing decay asymmetries and/or efficiency factors for $N_{2,3}$ ($\varepsilon_2\eta_2 \approx 0$ and $\varepsilon_3\eta_3 \approx 0$); *ii)* $N_1$-related washouts are still significant at $T \lesssim 10^9\,$GeV; *iii)* Reheating occurs at a temperature in between $M_2$ and $M_1$. In all other cases the $N_{2,3}$-related parameters cannot be ignored when calculating the BAU, and any constraint inferred from analyses based only on $N_1$ leptogenesis are not reliable.

## ACKNOWLEDGMENTS

I thank Guy Engelhard, Yuval Grossman, Yosef Nir, Juan Racker and Esteban Roulet for the enjoyable collaboration, and Diego Aristizabal, Marta Losada, Luis Alfredo Muñoz, Jorge Noreña and Antonio Riotto for useful discussions. This work was supported in part by INFN in Italy and by Colciencias in Colombia.

## REFERENCES

1. M. Fukugita and T. Yanagida, *Phys. Lett. B***174**, 45 (1986).
2. V. A. Kuzmin, V. A. Rubakov and M. E. Shaposhnikov, *Phys. Lett. B***155**, 36 (1985).
3. G. R. Farrar and M. E. Shaposhnikov, *Phys. Rev. Lett.* **70**, 2833 (1993) [Erratum-ibid. **71**, 210 (1993)]
4. M. A. Luty, *Phys. Rev. D***45**, 455 (1992).
5. L. Covi, E. Roulet and F. Vissani, *Phys. Lett. B***384**, 169 (1996)
6. G. F. Giudice, A. Notari, M. Raidal, A. Riotto and A. Strumia, *Nucl. Phys. B***685**, 89 (2004)
7. E. Nardi, Y. Nir, J. Racker and E. Roulet, *JHEP***0601**, 068 (2006)
8. R. Barbieri, P. Creminelli, A. Strumia and N. Tetradis, *Nucl. Phys. B***575**, 61 (2000)
9. T. Endoh, T. Morozumi and Z. h. Xiong, *Prog. Theor. Phys.* **111**, 123 (2004)
10. See *e.g.*: W. Buchmuller, P. Di Bari and M. Plumacher, Phys. Lett. B **547**, 128 (2002); Nucl. Phys. B **665**, 445 (2003).
11. A. Pilaftsis and T. E. J. Underwood, Nucl. Phys. B **692**, 303 (2004); Phys. Rev. D **72**, 113001 (2005).
12. S. Davidson and A. Ibarra, Phys. Lett. B **535**, 25 (2002).
13. T. Hambye, Y. Lin, A. Notari, M. Papucci and A. Strumia, Nucl. Phys. B **695** (2004) 169.
14. A. Abada, S. Davidson, F. X. Josse-Michaux, M. Losada and A. Riotto, *JCAP***0604**, 004 (2006).
15. E. Nardi, Y. Nir, E. Roulet and J. Racker, *JHEP***0601**, 164 (2006).
16. A. De Simone and A. Riotto, arXiv:hep-ph/0611357.
17. O. Vives, Phys. Rev. D **73**, 073006 (2006); P. Di Bari, Nucl. Phys. B **727**, 318 (2005); S. Blanchet and P. Di Bari, arXiv:hep-ph/0603107.
18. A. Strumia, arXiv:hep-ph/0608347.
19. G. Engelhard, Y. Grossman, E. Nardi and Y. Nir, arXiv:hep-ph/0612187.

# CP violation in Semi-Leptonic $\tau$ decays

David Delepine

*Instituto de Fisica de la Universidad de Guanajuato, Loma del Bosque 103, 37150 Leon, Gto, Mexico*

**Abstract.** We study CP violation in semi-leptonic $\tau$ decays and we determine the conditions necessary to be able to define a observable CP asymmetry. We apply these conditions in both models, the standard model for the electroweak interactions and its supersymmetric extensions. In the first case, the leading order contribution to the direct CP asymmetry in $\tau^{\pm} \to K^{\pm}\pi^{0}\nu_{\tau}$ decay rates is evaluated. In the second case,we compute the SUSY effective hamiltonian that describes the $|\Delta S| = 1$ semileptonic decays of tau leptons. We show that SUSY contributions may enhance the CP asymmetry of $\tau \to K\pi\nu_{\tau}$ decays by several orders of magnitude compared to the standard model expectations.

**Keywords:** CP violation, $\tau$ decay, supersymmetry
**PACS:** 11.30.Er, 11.30.Pb, 12.60.Jv,13.35.Dx

## INTRODUCTION

Experimental searches for CP violating asymmetries in tau lepton semileptonic decays have been carried out in the $\tau \to \pi\pi\nu_{\tau}$ [1] and $\tau \to K_s\pi\nu_{\tau}$ [2] modes. Motivation for these searches in the context of beyond the Standard Model approaches were provided in refs. [3, 4]. In ref. [2], the missing evidence for a non-zero CP asymmetry was interpreted in terms of a (CP-violating) coupling $\Lambda$ due to a charged scalar exchange and the limit $-0.172 < Im(\Lambda) < 0.067$ (at 90% c.l.) has been derived. The CP-odd observable studied in [2] depends upon two variables of a particular kinematical distribution of semileptonic tau decays as long as this effect is assumed to have its origin in the interference of scalar and vector form factors In this talk, we shall first determine the general conditions to get observable CP asymmetry. Then we should apply it to Standard Model and its supersymmetric extensions

The general amplitude for $\tau^{-}(p) \to K^{-}(k)\pi^{0}(k')\nu_{\tau}(p')$is given by

$$\begin{aligned}\mathcal{M} &= \frac{G_F V_{us}}{\sqrt{2}}\Big\{\bar{u}(p')\gamma^{\mu}(1-\gamma_5)u(p)F_V(t)\left[(k-k')_{\mu} - \frac{\Delta^2}{t}q_{\mu}\right] \\ &\quad + \bar{u}(p')(1+\gamma_5)u(p)m_{\tau}\Lambda F_S(t)\frac{\Delta^2}{t} \\ &\quad + F_T\langle K\pi|\bar{s}\sigma_{\mu\upsilon}u|0\rangle\bar{u}(p')\sigma^{\mu\upsilon}(1+\gamma_5)u(p)\Big\},\end{aligned}$$

where $q = k + k'$ ($t = q^2$) is the momentum transfer to the hadronic system, $\Delta^2 \equiv m_K^2 - m_{\pi}^2$ and $F_{V,S,T}(t)$ are the *effective* form factors describing the hadronic matrix elements (In Standard Model, $F_T = 0, \Lambda = 1$)

CP917, *Particles and Fields,* edited by H. Castilla Valdez, J. C. D'Olivo, and M. A. Perez

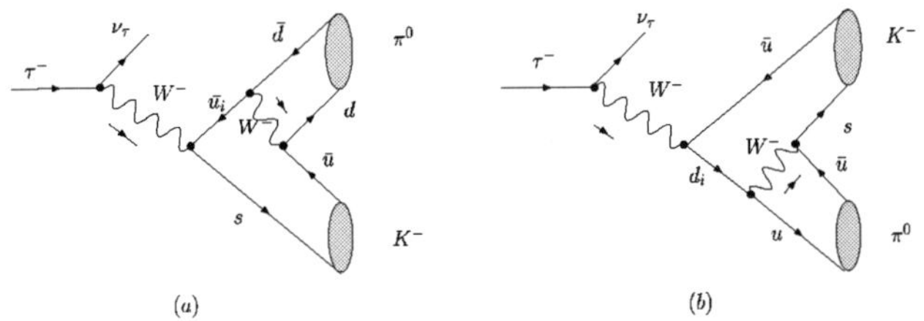


**FIGURE 1.** Higher order contributions to CP asymmetry in Standard Model

$$\sum_{pols} |\mathcal{M}|^2 \sim |F_V|^2(2p.Qp'.Q - p.p'Q^2) + |\Lambda|^2|F_S(t)|^2 M^2 p.p'$$
$$+2Re\Lambda \cdot Re(F_S F_V^*) M m_\tau p'.Q - 2Im\Lambda \cdot Im(F_S F_V^*) M m_\tau p'.Q$$

where $Q_\mu = (k-k')_\mu - \frac{\Delta^2}{t} q_\mu$and $F_T$ contributions have been neglected. The last term is odd under a CP transformation but the last two terms disappear once we integrate on the kinematical variable $u$.

It is not possible to generate a CP asymmetry in total decay rates corresponding to this process unless:

- $F_{V,S} = f_{V,S} + a_{V,S}$ which $a_{V,S}$ containing a weak CP violating phase and a different strong phase than $f_{V,S}$
- Interference between tensorial and vector contribution $\Rightarrow$ we need to go beyond SM.
- to look for the double differential distribution $(d^2\Gamma/dudt)$ or a variant of it as CLEO collaboration did [2].

## CP ASYMMETRY IN STANDARD MODEL

The form factor $f_{V,S}(t)$ are dominated at tree level by a single vector or scalar strange resonance:

$$f_i(t) = \frac{f_i(0) m_i^2}{m_i^2 - t - i m_i \Gamma_i}, \quad i = V,\ S, \tag{1}$$

where $(m_i,\ \Gamma_i)$ denote the mass and width of the resonance in the vector or scalar configuration (respectively the$K^*(892)$ or $K_0^*(1430)$).

$\Rightarrow$ the tree-level strong phase fixed by the decay width of these resonances, while the weak phase is absent at the tree-level. Higher order contributions can induce weak phase contributions $(a_{V,S})$ (see figure 1) [1].

[1] for details of the computation, see reference [5]

$$A_{CP} = \frac{\Gamma(\tau^+ \to K^+\pi^0\bar{\nu}_\tau) - \Gamma(\tau^- \to K^-\pi^0\nu_\tau)}{\Gamma(\tau^+ \to K^+\pi^0\bar{\nu}_\tau) + \Gamma(\tau^- \to K^-\pi^0\nu_\tau)}$$

$$\approx -\frac{\sqrt{2}G_F^3 m_\tau^5 \text{Im}(V_{us}V_{ud}^* V_{td} V_{ts}^*) f_K f_\pi}{768\pi^3 \Gamma(\tau^+ \to K^+\pi^0\bar{\nu}_\tau)} \times I_{CP} ,$$

where

$$I_{CP} = \frac{1}{m_\tau^6} \int_{(m_K+m_\pi)^2}^{m_\tau^2} \frac{dt}{t^3} (m_\tau^2 - t)^2 h(t) \left(1 + \frac{2t}{m_\tau^2}\right) \lambda^{3/2}(t, m_K^2, m_\pi^2) \quad (2)$$

where $\lambda(x,y,z) = x^2 + y^2 + z^2 - 2xy - 2xz - 2yz$

$$A_{CP} \approx -\frac{\sqrt{2}G_F \text{Im}(V_{us}V_{ud}^* V_{td} V_{ts}^*) f_K f_\pi}{20 B(\tau^+ \to K^+\pi^0\bar{\nu}_\tau)} \times I_{CP} \approx 2.3 \times 10^{-12},$$

with $B(\tau^+ \to K^+\pi^0\bar{\nu}_\tau) = (4.5 \pm 0.3) \times 10^{-3}$ and $m_c = 1.35$ GeV,

## CP ASYMMETRY IN SUPERSYMMETRIC MODELS

Strangeness-changing $|\Delta S| = 1$ decays of tau leptons are driven by the $\tau^- \to \bar{u}s\nu_\tau$ elementary process. The effective Hamiltonian $H_{eff}$ derived from SUSY can be expressed as

$$H_{eff} = \frac{G_F}{\sqrt{2}} V_{us} \sum_i C_i(\mu) Q_i(\mu), \quad (3)$$

where $C_i$ are the Wilson coefficients and $Q_i$ are the relevant local operators at low energy scale $\mu \simeq m_\tau$. The operators are given by

$$Q_1 = (\bar{\nu}\gamma^\mu L\tau)(\bar{s}\gamma_\mu Lu), \quad (4)$$
$$Q_2 = (\bar{\nu}\gamma_\mu L\tau)(\bar{s}\gamma_\mu Ru), \quad (5)$$
$$Q_3 = (\bar{\nu}R\tau)(\bar{s}Lu), \quad (6)$$
$$Q_4 = (\bar{\nu}R\tau)(\bar{s}Ru), \quad (7)$$
$$Q_5 = (\bar{\nu}\sigma_{\mu\nu}R\tau)(\bar{s}\sigma^{\mu\nu}Ru). \quad (8)$$

where $L, R$ are as defined in the previous section and $\sigma^{\mu\nu} = \frac{i}{2}[\gamma^\mu, \gamma^\nu]$. The Wilson coefficients $C_i$, at the electroweak scale, can be expressed as $C_i = C_i^{SM} + C_i^{SUSY}$ where $C_i^{SM}$ are given by

$$C_1^{SM} = 1,$$
$$C_{2,3,4,5}^{SM} = 0. \quad (9)$$

SUSY contributions to the Hamiltonian of $\tau^- \to \bar{u}s\nu_\tau$ transitions can be generated through two topological box diagrams as shown in Figs. (2,3). Other SUSY contributions (vertex corrections) are suppressed either due to small Yukawa couplings of light quarks or because they have the same structure as the SM in the hadronic vertex.

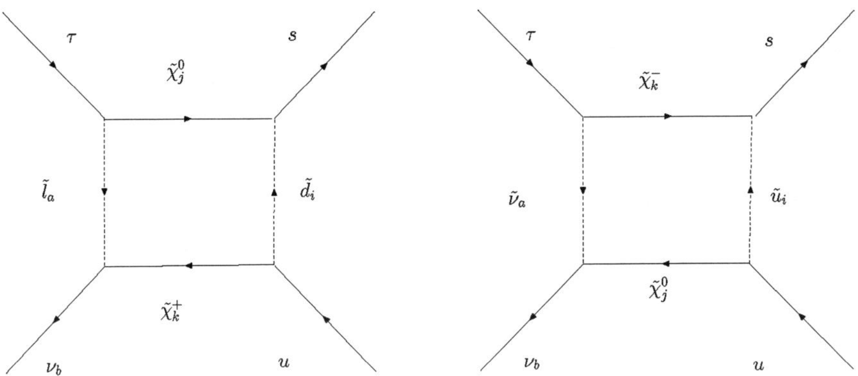


**FIGURE 2.** SUSY box contributions to $\tau^- \to \bar{u}s\nu_\tau$ transition.

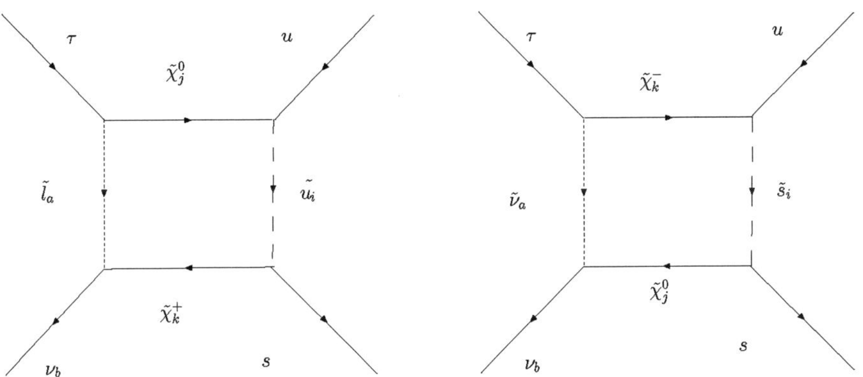


**FIGURE 3.** Crossed diagrams of Figure 2.

In our computations of Wilson coefficients, we will work in the mass insertion approximation (MIA), where gluino and neutralino are flavor diagonal. Denoting by $(\Delta^f_{AB})_{ab}$ the *off*-diagonal terms in the sfermion mass matrices where $A,B$ indicate chirality, $A,B=(L,R)$, the $A-B$ sfermion propagator can be expanded as

$$\langle \tilde{f}^a_A \tilde{f}^{b*}_B \rangle = i(k^2 I - \tilde{m}^2 I - \Delta^f_{AB}))^{-1}_{ab} \simeq \frac{i\delta_{ab}}{k^2-\tilde{m}^2} + \frac{i(\Delta^f_{AB})_{ab}}{(k^2-\tilde{m}^2)^2} + O(\Delta^2), \tag{10}$$

where $\tilde{f}$ denotes any scalar fermion, $a,b=(1,2,3)$ are flavor indices, $I$ is the unit matrix, and $\tilde{m}$ is the average sfermion mass. It is convenient to define a dimensionless quantity $(\delta^f_{AB})_{ab} \equiv (\Delta^f_{AB})_{ab}/\tilde{m}^2$. As long as $(\Delta^f_{AB})_{ab}$ is smaller than $\tilde{m}^2$ we can consider only the first order term in $(\delta^f_{AB})_{ab}$ of the sfermion propagator expansion. In our analysis we will keep only terms proportional to the third generation Yukawa couplings and terms of order $\lambda$ where $\lambda = V_{us}$.

The complete expressions for the Wilson coefficients $C_i$ at $m_W$ scale induced by SUSY computed from Figs. (2,3) can be found in reference [6]. As can be seen in ref.[6], the $C_i$ are given in terms of several mass insertions that represent the flavor transitions between different generations of quarks or leptons. In general, these mass insertions are complex and of order one. However, the experimental limits of several flavor changing neutral currents impose severe constraints on most of these mass insertions. In the following, we summarize all the important constraints on the relevant mass insertions for our process.

1. From the experimental measurements of $BR(\mu \to e\gamma) < 1.2 \times 10^{-11}$, the following bounds on $|(\delta^l_{12})_{AB}|$ and $|(\delta^\nu_{12})_{AB}|$ are obtained [7]: For $M_1 \sim M_2 = 100$ GeV and $\mu = \tilde{m}_l = 200$ GeV,

$$|(\delta^l_{12})_{LL}| \lesssim 10^{-3}, \qquad |(\delta^l_{12})_{LR}| \lesssim 10^{-6}, \tag{11}$$

$$|(\delta^\nu_{12})_{LL}| \lesssim 6 \times 10^{-4}, \quad |(\delta^\nu_{13})_{LL}| \lesssim 4 \times 10^{-4}, \quad |(\delta^\nu_{23})_{LL}| \lesssim 7 \times 10^{-4}. \tag{12}$$

2. From $BR(\tau \to \mu\gamma) < 1.1 \times 10^{-6}$, one gets the following constraint on $|(\delta^l_{23})_{LR})|$ [8]:

$$|(\delta^l_{23})_{LR}| \lesssim 2 \times 10^{-2}, \tag{13}$$

and from $BR(\tau \to e\gamma) < 2.7 \times 10^{-6}$, one finds [8]:

$$|(\delta^l_{13})_{LR}| \lesssim 1 \times 10^{-1}. \tag{14}$$

3. The mass insertions $(\delta^d_{12})_{AB}$ are constrained by the $\Delta M_K$ and $\varepsilon_K$ as follows [8]:

$$|(\delta^d_{12})_{LL}| \lesssim 4 \times 10^{-2}, \qquad |(\delta^d_{12})_{LR}| \lesssim 4 \times 10^{-3}, \tag{15}$$

$$\sqrt{|\mathrm{Im}\left[(\delta^{\mathrm{d}}_{12})_{\mathrm{LL}}\right]^2|} \lesssim 3 \times 10^{-3}, \quad \sqrt{|\mathrm{Im}\left[(\delta^{\mathrm{d}}_{12})_{\mathrm{LR}}\right]^2|} \lesssim 3 \times 10^{-4}. \tag{16}$$

4. The mass insertion $(\delta^u_{12})_{AB}$ are constrained by the $\Delta M_D$ as follows [9]:

$$|(\delta^u_{12})_{LL}| \lesssim 1.7 \times 10^{-2}, \qquad |(\delta^u_{12})_{LR}| \lesssim 2.4 \times 10^{-2}. \tag{17}$$

Here, three comments are in order. *i*) Due to the Hermiticity of the $LL$ sector in the sfermion mass matrix, $(\delta^f_{AB})_{LL} = (\delta^f_{AB})^\dagger_{LL} = (\delta^f_{BA})^*_{LL}$, where $A,B = 1,2,3$. *ii*) The above constraints imposed on the mass insertions $(\delta^{q,l}_{AB})_{LL,LR}$ are derived from supersymmetric contributions through exchange of gluino or neutralino which preserves chirality, therefore same constraints are also imposed on the mass insertions $(\delta^{q,l}_{AB})_{RR,RL}$. *iii*) The mass insertions $(\delta^f_{AB})_{LR(RL)}$ are not, in general, related to the mass insertions $(\delta^f_{BA})_{LR(RL)}$. Taking the above constraints into account, one finds that the dominant contribution to the $\tau^- \to u\bar{s}\nu_\tau$ is given in terms of $(\delta^\nu_{32})_{LR}$, $(\delta^\nu_{32})_{RL}$, $(\delta^d_{21})_{RL}$ and $(\delta^u_{21})_{LR}$. Notice that the effective Hamiltonian (eq. 3) derived in this section can induce supersymmetric effects in all the $|\Delta S| = 1$ exclusive $\tau$ lepton decay.

The total amplitude (SM and SUSY) of the $\tau(p) \to K(q)\pi(q')\nu_\tau(p')$decay as

$$\mathscr{A}_T(\tau \to K\pi\nu) = \frac{G_F V_{us}}{\sqrt{2}}\Big[(1+C_1)\langle K\pi|\bar{s}\gamma_\mu u|0\rangle \bar{\nu}(p')\gamma^\mu L\tau(p)$$
$$+ (C_3+C_4)\langle K\pi|\bar{s}u|0\rangle \bar{\nu}(p')R\tau(p) + C_5\langle K\pi|\bar{s}\sigma_{\mu\upsilon}u|0\rangle \bar{\nu}(p')\sigma^{\mu\upsilon}R\tau(p)\Big],$$

where $C_i$ stand for $C_i^{SUSY}$ We consider the following two interesting scenarios:

*i*) The case of $C_3$ or $C_4$ gives relevant contributions while $C_5$ is negligible. In this case, SUSY induces a relative weak phase between the vector and scalar form factors describing this process.

*ii*) The case of $C_5$ gives relevant contributions while $C_{3,4}$ are negligible. In this case, SUSY induces a relative weak phase between the vector and tensor form factors. A CP asymmetry in decay rate could be measured.

## First scenario

$$\langle K\pi|\bar{s}\gamma_\mu u|0\rangle = f_V(t)Q_\mu + f_S(t)(q+q')_\mu\ , \Rightarrow \langle K\pi|\bar{s}u|0\rangle = \frac{t}{m_s - m_u} f_S(t)\,, \tag{18}$$

$$\mathscr{A}_T(\tau \to K\pi\nu) = \frac{G_F V_{us}}{\sqrt{2}}(1+C_1)\times$$
$$\left\{ f_V Q^\mu \bar{u}(p')\gamma_\mu L u(p) + \left[m_\tau + \left(\frac{C_3+C_4}{1+C_1}\right)\frac{t}{m_s-m_u}\right] f_S \bar{u}(p')Ru(p)\right\}.$$

Using CLEO limit, we can translate this bound into:

$$-0.010 \le Im\left(\frac{C_3+C_4}{1+C_1}\right) \le 0.004\,, \tag{19}$$

where we have used $m_s - m_u = 100$ MeV, and the average value $\langle t\rangle \approx (1332.8\ \text{MeV})^2$. Using $M_1 = 100$ and $M_2 = 200$ GeV and $\mu = M_{\tilde{q}} = 400$ GeV and $\tan\beta = 20$, one gets

$$Im\left(\frac{C_3+C_4}{1+C_1}\right) \simeq 1.3\times 10^{-5} Im(\delta^d_{21})_{RL} \tag{20}$$

## Second scenario

$$\mathscr{A}_T(\tau \to K\pi\nu) = \frac{G_F V_{us}}{\sqrt{2}}(1+C_1)\left\{f_V(t)Q_\mu \bar{u}(p')\gamma^\mu L u(p)\right.$$

$$+ \quad \frac{C_5}{1+C_1}\langle K\pi|\bar{s}\sigma_{\mu\upsilon}u|0\rangle \bar{u}(p')\sigma^{\mu\upsilon}Ru(p)\Big\}$$

$$\langle K\pi|\bar{s}\sigma_{\mu\upsilon}u|0\rangle = \frac{ia}{m_K}[(p_\pi)^\mu(p_K)^\nu - (p_\pi)^\nu(p_K)^\mu] \quad (21)$$

where $a$ is a dimensionless quantity which fixes the scale of the hadronic matrix element.

$$a_{CP} = \frac{\Gamma(\tau^- \to K^-\pi^0\nu_\tau) - \Gamma(\tau^+ \to K^+\pi^0\nu_\tau)}{\Gamma(\tau^- \to K^-\pi^0\nu_\tau) + \Gamma(\tau^+ \to K^+\pi^0\nu_\tau)} \simeq \frac{a}{2}Im\, C_5$$
$$\simeq 1.4\times 10^{-7} a\, Im(\delta^u_{21})_{LR}$$

## CONCLUSION

The goal of this work is to illustrate the difficulties to define an observable CP asymmetry in semi-leptonic $\tau$ decays. We show that in standard model, a CP asymmetry in the total decay rate can be induced through higher order contributions but the absolute value of this CP asymmety is too small to be once accessible to experiments. In the supersymmetric case, we have computed the effective hamiltonian derived from SUSY for $|\Delta S| = 1$ tau lepton decays using the mass insertion approximation. Supersymmetric extensions of the SM could induce CP violating asymmetry in the double differential distribution as CLEO collaboration did and could also induce CP asymmetry in total decay rate due to interference between $O_5$ and $O_1$ operators. A direct consequence of our computation is that any CP asymmetry in the channel under consideration bigger than $10^{-6}$ will be a clear evidence of not only Physics beyond Standard Model but also an evidence of Physics beyond SUSY extensions of the SM.

## ACKNOWLEDGMENTS

D.D. wants to thank Gabriel Lopez Castro, Shaaban Khalil, Gaber Faisel and L.T. Lopez Lozano for their collaborations in the realization of this work. This work was financially supported by PROMEP PTC project, Conacyt Project numero 46195.

## SUSY CONTRIBUTIONS TO WILSON COEFFICIENTS

Here we provide the complete expressions for the supersymmeric contributions, at leading order in MIA, for the Wilson coefficients of $\tau^- \to s\bar{u}\nu_\tau$ transition, $C_i(M_W)$, $i = 1,..,5$. The dominant SUSY contributions are given by chargino-neutralino box diagram exchanges, as illustrated in Fig. 2.

The effective Hamiltonian $H_{eff}$ derived from SUSY can be expressed as

$$H_{eff} = \frac{G_F}{\sqrt{2}}V_{us}\sum_i C_i(\mu)Q_i(\mu), \quad (22)$$
$$= \sum_i \tilde{C}_i(\mu)Q_i(\mu), \quad (23)$$

where $C_i$ are the dimensionless Wilson coefficients and $Q_i$ are the relevant local operators at low energy scale $\mu \simeq m_\tau$. In terms of the vertex, one can write the complete vertex as a product of the vertex coming from leptonic sector and of the vertex coming from hadronic sector. In this respect we can also write the Wilson coefficients as

$$\begin{aligned}\tilde{C}_i &= C^l_{i(\tau-\chi^-)}\left(C^q_{i(\tau-\chi^--s)}+C^q_{i((\tau-\chi^--u)}\right)\\ &\quad +C^l_{i(\tau-\chi^0)}\left(C^q_{i(\tau-\chi^0-s)}+C^q_{i((\tau-\chi^0-u)}\right)\end{aligned}$$

where the $C^l_i$ is due to the leptonic vertex and $C^q_i$ is from the quark sectors. If we expand $C^{l,q}_i$ in terms of the mass insertions, one finds that the leading contributions are given by

$$\begin{aligned}\tilde{C}_3 &= C^{l(0)}_{3(\tau-\chi^-)}C^{q(1)}_{3(\tau-\chi^--s)}I_n(x_i,x_j)+C^{l(0)}_{3(\tau-\chi^0)}C^{q(1)}_{3(\tau-\chi^0-s)}I_n(x_i,x_j)\\ &\quad +O(\delta^2),\end{aligned} \tag{24}$$

where $I_n(x_i,x_j)$ is defined below and $x_i = m^2_{\chi^\pm_i}/\tilde{m}^2$ and $x_j = m^2_{\chi^0_j}/\tilde{m}^2$.

$$C^{l(0)}_{3(\tau-\chi^0)} = g(h_e)_{33}N_{i3}U^*_{j1}(U^*_{MNS})_{33} \tag{25}$$

$$-g\sqrt{2}\tan\theta_w N_{i1}U^*_{j2}(h_e)_{33}(U^*_{MNS})_{33}, \tag{26}$$

$$C^{l(0)}_{3(\tau-\chi^-)} = -(h_e)_{33}\frac{g}{\sqrt{2}}(N_{i2}-\tan\theta_w N_{i1})U_{j2}(U^*_{MNS})_{33}, \tag{27}$$

$$\begin{aligned}C^{q(1)}_{3(\tau-\chi^--s)} &= (-\frac{1}{8})\left(\frac{g^2}{\sqrt{2}}\frac{2}{3}\tan\theta_w N^*_{i1}U^*_{j1}(V^*_{CKM})_{1a}(\delta^d_{RL})_{2a}\right.\\ &\quad \left.-\frac{g}{\sqrt{2}}\frac{2}{3}\tan\theta_w N^*_{i1}U^*_{j2}(h_d)_{33}(V^*_{CKM})_{13}(\delta^d_{RR})_{23}\right),\end{aligned} \tag{28}$$

$$\begin{aligned}C^{q(1)}_{3(\tau-\chi^0-s)} &= (-\frac{1}{8})\left(\frac{2}{3}\frac{g^2}{\sqrt{2}}\tan\theta_w U^*_{j1}N^*_{i1}(V^*_{CKM})_{1a}(\delta^d_{RL})_{2a}\right.\\ &\quad \left.-\frac{2}{3}\frac{g}{\sqrt{2}}\tan\theta_w U^*_{j2}N^*_{i1}(V^*_{CKM})_{13}(\delta^d_{RR})_{23}(h_d)_{33}\right).\end{aligned} \tag{29}$$

$$\tilde{C}_4 = C^{l(0)}_{4(\tau-\chi^-)}C^{q(1)}_{4(\tau-\chi^--u)}\tilde{I}_n(x_i,x_j)+C^{l(0)}_{4(\tau-\chi^0)}C^{q(1)}_{4((\tau-\chi^0-u)}\tilde{I}_n(x_i,x_j)+O(\delta^2), \tag{30}$$

where $\tilde{I}_n(x_i,x_j)$ is given below.

$$\begin{aligned}C^{l(0)}_{4(\tau-\chi^-)} &= -(h_e)_{33}\frac{g}{\sqrt{2}}(N_{i2}-\tan\theta_w N_{i1})U_{j2}(U^*_{MNS})_{33}\\ &= C^{l(0)}_{3(\tau-\chi^-)},\end{aligned} \tag{31}$$

$$C^{l(0)}_{4(\tau-\chi^0)} = g(h_e)_{33}N_{i3}U^*_{j1}(U^*_{MNS})_{33}$$
$$-g\sqrt{2}\tan\theta_w N_{i1}U^*_{j2}(h_e)_{33}(U^*_{MNS})_{33}, \quad (32)$$

$$C^{q(1)}_{4(\tau-\chi^0-u)} = (-\frac{1}{8})\left(-\frac{4}{3}\frac{g^2}{\sqrt{2}}\tan\theta_w N_{i1}V_{j1}(V^*_{CKM})_{a2}(\delta^u_{LR})_{a1}\right.$$
$$\left.+\frac{4}{3}\frac{g}{\sqrt{2}}\tan\theta_w N_{i1}V_{j2}(h_u)_{33}(V^*_{CKM})_{32}(\delta^u_{RR})_{31}\right), \quad (33)$$

$$C^{q(1)}_{4(\tau-\chi^--u)} = (-\frac{1}{8})\left(-\frac{4}{3}\frac{g^2}{\sqrt{2}}\tan\theta_w V_{j1}N_{i1}(V^*_{CKM})_{a2}(\delta^u_{LR})_{a1}\right.$$
$$\left.+\frac{4}{3}\frac{g}{\sqrt{2}}\tan\theta_w V_{j2}N_{i1}(h_u)_{33}(V^*_{CKM})_{32}(\delta^u_{RR})_{31}\right). \quad (34)$$

$$\tilde{C}_{5(\tau-\chi^0-u)} = -\frac{1}{4}\tilde{C}_{4(\tau-\chi^0-u)}, \quad (35)$$

$$\tilde{C}_{5(\tau-\chi^--u)} = \frac{1}{4}\tilde{C}_{4(\tau-\chi^--u)}. \quad (36)$$

The loop integrals $I_n(x_i,x_j)$ and $\tilde{I}_n(x_i,x_j)$ are defined as follows:

$$I(x_i,x_j) = \frac{1}{16\pi^2\tilde{m}^2}\left(\frac{1}{x_i-x_j}\right)\left(\frac{x_i^2-x_i-x_i^2 logx_i}{(1-x_i)^2}-(x_i\leftrightarrow x_j)\right),$$
$$\tilde{I}(x_i,x_j) = \frac{\sqrt{x_ix_j}}{16\pi^2\tilde{m}^2}\left(\frac{1}{x_i-x_j}\right)\left(\frac{x_i^2-x_i-x_i logx_i}{(1-x_i)^2}-(x_i\leftrightarrow x_j)\right),$$
$$I_n(x_i,x_j) = \frac{1}{32\pi^2\tilde{m}^2}\left(\frac{1}{x_i-x_j}\right)\left(\frac{2x_i^2-2x_i-2x_i logx_i}{(x_i-1)^2}-\frac{x_i^3-4x_i^2+3x_i+2x_i logx_i}{(x_i-1)^3}\right.$$
$$\left.-(x_i\leftrightarrow x_j)\right),$$
$$\tilde{I}_n(x_i,x_j) = \frac{-\sqrt{x_ix_j}}{32\pi^2\tilde{m}^2}\left(\frac{1}{x_i-x_j}\right)\left(\frac{x_i^3-4x_i^2+3x_i+2x_i logx_i}{(x_i-1)^3}-(x_i\leftrightarrow x_j)\right). \quad (37)$$

The expression for the other Wilson coefficients can be found in ref.[6].

## REFERENCES


1. P. Avery et al (CLEO Collaboration), Phys. Rev. **D64**, 092005 (2001)
2. G. Bonvicini et al (CLEO Collaboration), Phys. Rev. Lett. **89**, 111803 (2002).
3. J. H. Kühn and E. Mirkes, Phys. Lett. **B398**, 407 (1997);

4. Y. S. Tsai, Nuc. Phys. (PS)**55C**, 293 (1997).
5. D. Delepine, G. Lopez Castro and L. T. Lopez Lozano, Phys. Rev. D **72**, 033009 (2005) [arXiv:hep-ph/0503090].
6. D. Delepine, G. Faisl, S. Khalil and G. L. Castro, Phys. Rev. D **74**, 056004 (2006) [arXiv:hep-ph/0608008].
7. G. C. Branco, D. Delepine and S. Khalil, Phys. Lett. B **567** (2003) 207 [arXiv:hep-ph/0304164].
8. F. Gabbiani, E. Gabrielli, A. Masiero and L. Silvestrini, Nucl. Phys. B **477** (1996) 321.
9. S. Khalil, arXiv:hep-ph/0604118

# Robustness of solar neutrino oscillations in the presence of non–standard physics[1]

O. G. Miranda*, M. A. Tórtola† and J. W. F. Valle**

*Departamento de Física, Centro de Investigación y de Estudios Avanzados del IPN
Apdo. Postal 14-740 07000 Mexico, DF, Mexico.
†Centro de Fíisica Teórica de Partículas. Departamento de Física. Instituto Superior Técnico.
Avenida Rovisco Pais, 1049-001 Lisboa, Portugal.
**AHEP Group, Instituto de Física Corpuscular – C.S.I.C./Universitat de València
Edificio Institutos de Paterna, Apt. 22085, E–46071 València, Spain

**Abstract.** We have reconsidered the status of the large mixing angle (LMA) oscillation interpretation of the solar neutrino data in a more general framework where non-standard neutrino interactions are present. Using the latest data from all solar neutrino experiments and KamLAND we have found the existence of three LMA solutions, instead of the unique solution which holds in the absence of non-standard interactions, LMA-I. In addition to LMA-I, there is another solution with smaller value of $\Delta m^2$ (LMA-0), and a new "dark-side" solution (LMA-D) with $\sin^2\theta = 0.70$. We comment on the complementary role of atmospheric, laboratory, reactor and future solar neutrino experiments in lifting the degeneracy between this three solutions. We also mention that establishing the issue of robustness of the oscillation picture in the most general case will require further experiments, such as those involving low energy solar neutrinos.

**Keywords:** Neutrino masses and mixings, Solar neutrinos, Neutrino interactions
**PACS:** 14.60.Pq, 13.15.+g

## INTRODUCTION

The first data of the KamLAND collaboration [2] have been enough to isolate neutrino oscillations as the correct mechanism explaining the solar neutrino problem, indicating also that large mixing angle (LMA) was the right solution. The 766.3 ton-yr KamLAND data sample strengthens the validity of the LMA oscillation interpretation of the data [3]. With neutrino experiments now entering the precision age [4], the determination of neutrino parameters and their theoretical impact have become one of the main goals in astroparticle and high energy physics [5]. Now the main efforts should be devoted to the precision determination of the oscillation parameters and to test for sub-leading non-oscillation effects such as spin-flavour conversions [6, 7] or non-standard neutrino interactions (NSI) [8].

Here we focus on the case of neutrinos endowed with non-standard interactions. These are a natural outcome of many neutrino mass models [9] and can be of two types: flavour-changing (FC) and non-universal (NU). Non–standard interactions may in principle affect neutrino propagation properties in matter as well as detection cross sections. Thus their existence can modify the solar neutrino signal observed at experiments. They may

[1] Based on the results of Ref. [1]

CP917, *Particles and Fields*, edited by H. Castilla Valdez, J. C. D'Olivo, and M. A. Perez

be parametrized with the effective low–energy four–fermion operator:

$$\mathscr{L}_{NSI} = -\varepsilon_{\alpha\beta}^{fP} 2\sqrt{2} G_F \left(\bar{\nu}_\alpha \gamma_\mu L \nu_\beta\right)\left(\bar{f}\gamma^\mu P f\right), \tag{1}$$

where P = L, R and $f$ is a first generation fermion: $e, u, d$. The coefficients $\varepsilon_{\alpha\beta}^{fP}$ denote the strength of the NSI between the neutrinos of flavours $\alpha$ and $\beta$ and the P–handed component of the fermion $f$. In the present work, for definiteness, we take for $f$ the down-type quark. However, one can also consider the presence of NSI with electrons and up and down quarks simultaneously. Current limits and perspectives in the case of NSI with electrons have been reported in the literature [10]. While strong constraints exist from $\nu_\mu$ interactions with a down-type quark ($\varepsilon_{e\mu}^{dP} \lesssim 10^{-3}$, $\varepsilon_{\mu\mu}^{dP} \lesssim 10^{-3} - 10^{-2}$) from CHARM and NuTeV [11], the constraints for all other NSI couplings, including those involved in solar neutrino physics, are rather loose [11, 12]. Therefore, in our analysis we consider $\varepsilon_{\alpha\mu}^{dP} = 0$ and we concentrate our efforts in the rest of NSI parameters.

For our solar neutrino analysis, we will consider the simplest approximate two–neutrino picture, which is justified in view of the stringent limits on $\theta_{13}$ [5] that follow mainly from reactor neutrino experiments [13]. In this approximation, the Hamiltonian describing solar neutrino evolution in the presence of NSI contains, in addition to the standard oscillations term

$$\begin{pmatrix} -\frac{\Delta m^2}{4E}\cos 2\theta + \sqrt{2}G_F N_e & \frac{\Delta m^2}{4E}\sin 2\theta \\ \frac{\Delta m^2}{4E}\sin 2\theta & \frac{\Delta m^2}{4E}\cos 2\theta \end{pmatrix} \tag{2}$$

a term $H_{\rm NSI}$, accounting for an effective potential induced by the NSI with matter, which may be written as:

$$H_{\rm NSI} = \sqrt{2}G_F N_d \begin{pmatrix} 0 & \varepsilon \\ \varepsilon & \varepsilon' \end{pmatrix}. \tag{3}$$

Here $\varepsilon$ and $\varepsilon'$ are two effective parameters that, according to the current bounds discussed above ($\varepsilon_{\alpha\mu}^{fP} \sim 0$), are related with the fundamental couplings of Eq. (1) by:

$$\varepsilon = -\sin\theta_{23}\,\varepsilon_{e\tau}^{dV} \qquad \varepsilon' = \sin^2\theta_{23}\,\varepsilon_{\tau\tau}^{dV} - \varepsilon_{ee}^{dV} \tag{4}$$

The quantity $N_d$ in Eq. (3) is the number density of the down-type quark along the neutrino path. It is important to note that the neutrino evolution inside the Sun and the Earth is sensitive only to the vector component of the NSI, $\varepsilon_{\alpha\beta}^{dV} = \varepsilon_{\alpha\beta}^{dL} + \varepsilon_{\alpha\beta}^{dR}$. The effect of NSI with down-type quarks on the neutrino detection has been discussed at Ref. [1].

## ANALYSIS OF SOLAR AND KAMLAND DATA

Here we reanalyse the robustness of the oscillation interpretation of the solar neutrino data in the presence of non–standard interactions. We perform a complete analysis of the most recent solar and KamLAND neutrino data using a numerical computation for the survival probabilities in the light as well as in the dark side of the mixing angle, for values of $\Delta m^2$ in the range of $10^{-6}$ to $10^{-3}$ eV$^2$ and running also the effective

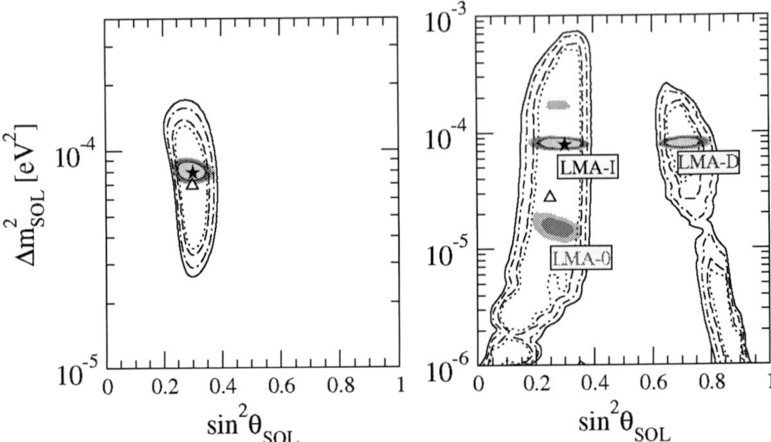


**FIGURE 1.** Left panel: 90%, 95%, 99% and 99.73% C.L. allowed regions of the neutrino oscillation parameters from the analysis of the latest solar data (hollow lines), and from the combined analysis of solar and KamLAND data (colored regions). Right panel: allowed regions for the generalized OSC + NSI case, corresponding to a solar only analysis (hollow lines) and to a combined solar+KamLAND analysis (colored regions).

**TABLE 1.** Best fit solar neutrino oscillation points with and without non-standard neutrino interactions.

| | $\sin^2\theta_{SOL}$ | $\Delta m^2$ [eV$^2$] | $\varepsilon$ | $\varepsilon'$ | $\chi^2$ |
|---|---|---|---|---|---|
| | | OSC analysis | | | |
| LMA-I | 0.29 | $8.1\times10^{-5}$ | – | – | 79.9 |
| | | OSC+NSI analysis | | | |
| LMA-I | 0.30 | $7.9\times10^{-5}$ | 0 | -0.05 | 79.7 |
| LMA-D | 0.70 | $7.9\times10^{-5}$ | -0.15 | 0.90 | 80.2 |
| LMA-0 | 0.25 | $1.6\times10^{-5}$ | 0.10 | 0.30 | 86.8 |

NSI couplings $\varepsilon$ and $\varepsilon'$ in the range $[-1,1]^2$. Our results for the pure oscillation case ($\varepsilon = \varepsilon' = 0$) are shown in the left panel of Fig. 1. The best fit point for this global analysis is given by $\sin^2\theta_{SOL} = 0.29$ and $\Delta m^2 = 8.1\times10^{-5}$ eV $^2$. This is in excellent agreement with the results obtained in [5] for the solar case. Concerning the generalized OSC+NSI picture, our results are shown in the right panel of Fig. 1 and Table 1. One sees that, in the light side, we obtain a region of allowed oscillation parameters larger than in the pure oscillation case, but more restricted than those obtained in previous

[2] More details about the statistical analysis performed can be found at Ref.[1]

OSC+NSI analysis of Refs. [14, 15] due to the effect of the recent KamLAND data, visible mainly in $\Delta m^2$. The table gives the parameter best fit values for the OSC and OSC+NSI fits. For the OSC+NSI analysis the best fit occurs for $\varepsilon = 0.0$ and $\varepsilon' = -0.05$. Clearly the quality of the fit obtained with and without NSI is comparable, as seen from the $\chi^2$ values given in the last column of the table. The most remarkable result is, however, the appearance of an additional solution in the dark side region, LMA-D. This solution has $\sin^2\theta_{\rm SOL} = 0.70$ and the same $\Delta m^2$ value as the LMA-I solution and is significantly better than the LMA-0 OSC+NSI solution of Refs. [14, 15], as shown in the table. On the other hand, it is nearly degenerate with the LMA-I solution, as seen by the $\chi^2$ value. This solution is characterized by $\varepsilon' = 0.90$, although lower values $\sim 0.75$ are allowed at $3\sigma$. Even if embarrassingly large, one sees that such large NSI strength values are perfectly compatible with all existing solar and reactor neutrino data, including the small values of the neutrino masses indicated by current oscillation data. This opens a potentially physics challenge for upcoming low energy solar neutrino experiments, such as Borexino. Note that large NSI values could affect also solar neutrino detection, as considered in [16]. In what follows we give a discussion of the role of other experiments in probing neutrino properties at the level implied by the LMA-D solution.

## CONSTRAINTS ON NSI: PRESENT AND FUTURE

As we just saw there are constraints on non-standard neutrino interaction strength parameters that follow from current solar and KamLAND data. The existence of NSI could also affect neutrino-nucleon scattering and there are laboratory data that potentially constrain their allowed strength. Moreover, one must check restrictions that follow from atmospheric data. Here we discuss their complementarity.

### Solar and KamLAND

We can derive limits on NSI parameters from solar and KamLAND data by displaying our $\chi^2$ as a function of the NSI parameters $\varepsilon$ or $\varepsilon'$ and marginalizing with respect to the remaining three parameters. Figure 2 gives the $\Delta\chi^2$ profiles with respect to $\varepsilon$ and $\varepsilon'$. From here one can determine the corresponding constraints on $\varepsilon$ and $\varepsilon'$. We can see that at 90% C.L. $-0.93 \leq \varepsilon \leq 0.30$ while for $\varepsilon'$ the only forbidden region is $[0.20, 0.78]$. The dashed lines in Fig. 2 denote the ultimate reach of this method of constraining NSI parameters (through their effect in solar neutrino propagation), namely they correspond to the case where solar neutrino oscillation parameters $\Delta m^2$ and $\theta_{\rm SOL}$ are determined with infinite precision. One sees that in this ideal case the allowed range narrows down mainly for negative NSI parameter values. We conclude that there is substantial room still left for sub–leading non-standard neutrinos conversions in matter and, moreover, that the determination of solar neutrino oscillation parameters, especially the solar mixing angle, is currently ambiguous. It is unlikely that more precise reactor measurements by KamLAND will resolve this mixing angle ambiguity, as they are expected to constrain mainly $\Delta m^2$.

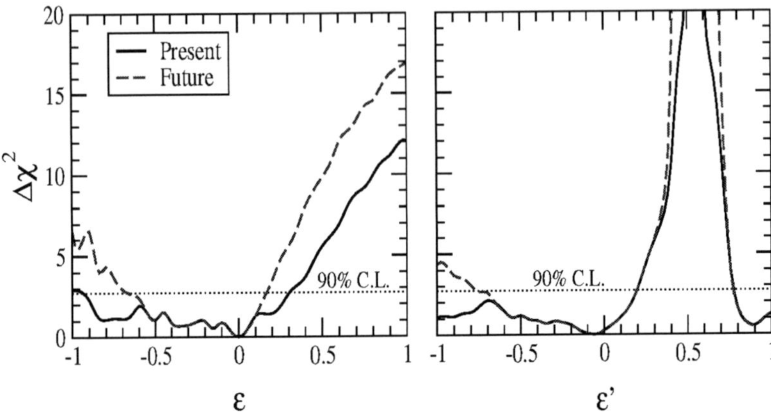


**FIGURE 2.** Constraining NSI parameters with solar and KamLAND neutrino data: dependence of $\Delta\chi^2$ with respect to $\varepsilon$ and $\varepsilon'$, illustrating the current limits.

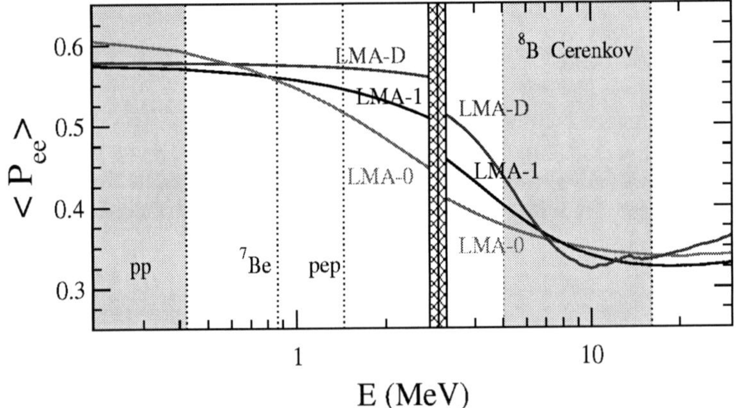


**FIGURE 3.** Predicted neutrino survival probability for low-energy neutrinos (left) and boron neutrinos (right) at the best fit points of the LMA-I, LMA-D and LMA-0 solutions.

In Fig. 3 we present the predicted neutrino survival probabilities versus energy, from the region of pp neutrinos up to the high energy solar neutrinos, for the three best–fit points of the allowed regions found above. One sees that the solutions predict different rates for the low energy neutrinos, so that future low energy solar neutrino experiments may have a hope of disentangling these solutions. Similarly, in the region of boron neutrinos our LMA-D solution also predicts a distortion in the spectrum that might be detectable at future water Cerenkov experiments such as UNO or Hyper-K [17], given the high statistics expected.

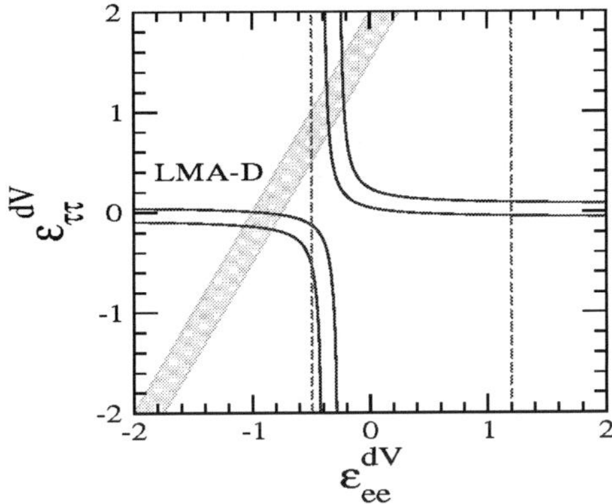


**FIGURE 4.** Consistency between the $\varepsilon'$ coupling required for our LMA-D solution (shaded band) and the regions allowed by atmospheric data in the analytic approximation of Ref. [19] for $\varepsilon^{dV}_{e\tau} = 0.21$ (solid lines). The laboratory constraints are also shown (dashed lines).

## Laboratory experiments

The laboratory bounds on the neutrino non-standard interactions with down-type quarks can be summarized as $|\varepsilon^{dP}_{\tau e}| < 0.5$, $|\varepsilon^{dR}_{\tau\tau}| < 6$, $|\varepsilon^{dL}_{\tau\tau}| < 1.1$, $-0.6 < \varepsilon^{dR}_{ee} < 0.5$, $-0.3 < \varepsilon^{dL}_{ee} < 0.3$, $0.6 < \varepsilon^{dL}_{ee} < 1.1$, see e. g. Ref [11]. Here we are interested in vector-like NSI couplings. For the case of $\varepsilon^{dV}_{ee}$, these bounds can be translated to $-0.5 < \varepsilon^{dV}_{ee} < 1.2$, while for $\varepsilon^{dV}_{\tau\tau}$ one finds a much wider range. However, we stress that these bounds have been obtained assuming that only one parameter is effective at a time. Relaxing this assumption opens more freedom. Assuming maximal mixing in the 2–3 sector in Eq. (4), one has

$$\varepsilon = -0.15 \quad \rightarrow \quad \varepsilon^{dV}_{e\tau} = 0.21 \tag{5}$$

$$\varepsilon' = \quad 0.90 \quad \rightarrow \quad \varepsilon^{dV}_{\tau\tau} = 2\,(\varepsilon^{dV}_{ee} + 0.90) \tag{6}$$

From this one can see explicitly that, even taking the above constraints at face value, they still leave room for our degenerate dark-side solution with $\varepsilon' = 0.90$.

## Atmospheric data

Concerning the atmospheric neutrino data, it is known that a large NSI strengths can originate a suppression of the neutrino oscillation amplitude. This has indeed been used in a two-neutrino analysis [18] in order to obtain relatively strong bounds on the NSI strength. However, in a 3–neutrino analysis of atmospheric data [19] it has been explicitly shown that large NSI strengths are not excluded. In particular, these authors have found two specific scenarios where somewhat large NSI strengths can fit well the experimental data, because their effect will be indistinguishable from the standard oscillation case, at least at high and low energies. Adapting their definitions to our notation, and using their analytical description, we obtain the two branches indicated in Fig. 4. One sees that the shaded band corresponding to our LMA-D solution at 90% C.L. (with $\varepsilon_{e\tau}^{dV} = 0.21$) intersects these branches in two disjoint regions, suggesting that, indeed, the NSI couplings required by the LMA-D solution are compatible with the atmospheric neutrino data. However, in a more complete numerical analysis of atmospheric neutrino data [20], it has been shown that values of $\varepsilon_{\tau\tau}^{dV}$ in the right region are not allowed by atmospheric data: only the left disjoint region is compatible with atmospheric neutrino data. As indicated by the dashed lines in Fig. 4 one can see that $\varepsilon_{ee}^{dV}$ values in this region lie outside the range allowed by current laboratory data. This leads us to conclude that the LMA-D solution induced by the simplest non-standard interactions of neutrinos with only down-type quarks is ruled out by its incompatibility with atmospheric and laboratory data. However, one can verify that for the general case where neutrinos have other NSI couplings one can reconcile the above laboratory bounds with the parameters required by the LMA-D solution.

## CONCLUSIONS

We have reanalysed the status of the LMA oscillation interpretation of the solar neutrino data in a more general framework where non-standard neutrino interactions are present. We have seen that combining the solar neutrino data, including the latest SNO fluxes of the salt phase with the full KamLAND data sample still leaves room for a degenerate determination of solar neutrino oscillation parameters. To this extent the solar neutrino oscillation parameters extracted from the experiments may be regarded as non-robust. In addition to the lower LMA-0 solution, we have found a LMA-D solution characterized by values of the solar mixing angle larger than $\pi/4$. This solution requires large non-universal neutrino interactions on down-type quarks. While the LMA-0 solution is already disfavored, and will soon be in conflict with further data, e.g. future KamLAND reactor data, the degeneracy implied by LMA-D solution will not be resolved by more precise KamLAND reactor measurements. This shows that the determination of solar neutrino parameters only from solar and KamLAND data is not fully robust. It is crucial to consider other data samples, such as atmospheric and laboratory data, since these bring complementary information. In the present case they allow one to rule out the LMA-D solution induced by the simplest NSI between neutrinos and down-type-quarks-only, given the large values of the non–universal NSI couplings required by that solution. It is therefore important to perform similar analyses for the more general case of non-standard interactions involving electrons and/or up-type quarks. Only in such scenario (NSI with u-type, d-type and electrons) we can confidently establish the robustness of the oscillation interpretation. Further experiments, like low-energy solar neutrino experiments are therefore required in order to clear up the situation.

## ACKNOWLEDGMENTS

Work supported by the European Commission under the RTN contract MRTN-CT-2004-503369.

## REFERENCES

1. O. G. Miranda, M. A. Tortola and J. W. F. Valle, JHEP **0610**, 008 (2006) [hep-ph/0406280].
2. K. Eguchi *et al.*, [KamLAND Collaboration], Phys. Rev. Lett. **90**, 021802 (2003) [hep-ex/0212021].
3. T. Araki *et al.*, [KamLAND Collaboration], Phys. Rev. Lett. **94**, 081801 (2005) [hep-ex/0406035].
4. For a review of solar neutrino experiments see A. B. McDonald, New J. Phys. **6**, 121 (2004) [astro-ph/0406253] and references therein. For the latest atmospheric, K2K and MINOS neutrino oscillation results see H. Gallagher, Nucl. Phys. B Proc. Suppl. **143**, 79 (2005); M. H. Ahn *et al.*, Phys. Rev. D **74**, 072003 (2006) [hep-ex/0606032] and D. G. Michael *et al.*, [MINOS Collaboration], Phys. Rev. Lett. **97**, 191801 (2006) [hep-ex/0607088].
5. M. Maltoni, T. Schwetz, M. A. Tortola and J. W. F. Valle, New J. Phys. **6**, 122 (2004) [hep-ph/0405172 v5].
6. J. Schechter and J. W. F. Valle, Phys. Rev. D **24**, 1883 (1981), Err. Phys. Rev. D **25**, 283 (1982).
7. E. K. Akhmedov, Phys. Lett. B **213**, 64 (1988); C. S. Lim and W. J. Marciano, Phys. Rev. D **37**, 1368 (1988).
8. L. Wolfenstein, Phys. Rev. D **17** 2369 (1978); J. W. F. Valle, Phys. Lett. B **199** 432 (1987).
9. J. W. F. Valle, Prog. Part. Nucl. Phys. **26**, 91 (1991).
10. G. Mangano *et al.*, Nucl. Phys. B **756**, 100 (2006) [hep-ph/0607267]; A. de Gouvea and J. Jenkins, Phys. Rev. D **74** 033004 (2006) [hep-ph/0603036]; J. Barranco, O. G. Miranda, C. A. Moura and J. W. F. Valle, Phys. Rev. D **73** 113001 (2006) [hep-ph/0512195].
11. S. Davidson, C. Pena-Garay, N. Rius and A. Santamaria, JHEP **0303** 011 (2003) [hep-ph/0302093].
12. Z. Berezhiani and A. Rossi, Phys. Lett. B **535** 207 (2002) [hep-ph/0111137].
13. M. Apollonio *et al.*, [CHOOZ Collaboration], Phys. Lett. B **466**, 415 (1999) [hep-ex/9907037].
14. A. Friedland, C. Lunardini and C. Pena-Garay, Phys. Lett. B **594**, 347 (2004) [hep-ph/0402266].
15. M. M. Guzzo, P. C. de Holanda and O. L. G. Peres, Phys. Lett. B **591**, 1 (2004) [hep-ph/0403134].
16. Z. Berezhiani, R. S. Raghavan and A. Rossi, Nucl. Phys. B **638**, 62 (2002) [hep-ph/0111138].
17. UNO collaboration Homepage: http://superk.physics.sunysb.edu/nngroup/uno/main.html.
18. N. Fornengo, M. Maltoni, R. T. Bayo and J. W. F. Valle, Phys. Rev. D **65** (2002) 013010.
19. A. Friedland, C. Lunardini and M. Maltoni, Phys. Rev. D **70**, 111301 (2004) [hep-ph/0408264].
20. A. Friedland and C. Lunardini, Phys. Rev. D **72** (2005) 053009 [hep-ph/0506143].

# Cosmology and neutrino masses

Sergio Pastor

*Instituto de Física Corpuscular (CSIC-Universitat de València)*
*Apdo. correos 22085, E-46071 Valencia, Spain*

**Abstract.** Neutrinos can play an important role in the evolution of the Universe, modifying some of the cosmological observables. We describe how the precision of present cosmological data can be used to learn about neutrino properties, in particular their mass. We show how the analysis of current cosmological observations provides an upper bound on the sum of neutrino masses, with improved sensitivity from future cosmological measurements.

**Keywords:** Early Universe; Neutrinos
**PACS:** 14.60.Pq, 95.35.+d, 98.80.Es

## INTRODUCTION

Neutrino cosmology is one of the best examples of the very close ties that have developed between nuclear physics, particle physics, astrophysics and cosmology. In this contribution I present the most interesting aspects, focusing on the information that the analysis of cosmological data can provide on neutrino masses. For a more detailed discussion of neutrinos in cosmology, see [1, 2, 3].

## THE COSMIC NEUTRINO BACKGROUND

The existence of the cosmic neutrino background (CNB) is a generic feature of the standard hot big-bang model. Although not detected yet, the presence of the CNB is indirectly established through its influence on cosmological observables. Here we review its evolution and main properties.

At large temperatures frequent weak interactions kept cosmic neutrinos of any flavour ($\nu_e, \nu_\mu, \nu_\tau$) in equilibrium. While coupled to the rest of the primeval plasma (relativistic $e^\pm$ and photons), neutrinos had the momentum spectrum

$$f_{\rm eq}(p,T) = [\exp((p-\mu_\nu)/T)+1]^{-1}\,, \tag{1}$$

i.e. a Fermi-Dirac spectrum with temperature $T$. We have included a neutrino chemical potential $\mu_\nu$ that would exist in the presence of a neutrino asymmetry, but it can be shown [4] that its contribution can be safely ignored.

As the Universe cools, the weak interaction rate falls below the expansion rate and neutrinos decouple from the rest of the plasma, at a decoupling temperature $T_{\rm dec} \simeq$ 1 MeV. After decoupling the spectrum in Eq. (1) is preserved, since both neutrino momenta and temperature redshift identically with the expansion. Since active neutrino masses are not much larger than 1 eV, they were ultra-relativistic at $T_{\rm dec}$ and Eq. (1) does not depend on neutrino masses even after decoupling.

CP917, *Particles and Fields,* edited by H. Castilla Valdez, J. C. D'Olivo, and M. A. Perez

Shortly after the temperature drops below the electron mass, favouring the annihilations of $e^{\pm}$ that transfer their entropy into photons but not into the decoupled neutrinos. This leads to $T_\gamma/T_\nu = (11/4)^{1/3} \simeq 1.40102$, the ratio between the temperatures of relic photons and neutrinos. It turns out that a detailed calculation of neutrino decoupling leads to small non-thermal distortions in the neutrino spectra and a slightly smaller increase of the comoving photon temperature (see [1] for a list of works). The most recent analysis [5] includes the effect of flavour neutrino oscillations, and finds that these distortions lead to a contribution of relativistic relic neutrinos to the total energy density of $N_{\rm eff} \simeq 3.046$ (see Eq. (2)).

Any quantity related to relic neutrinos can be calculated after decoupling with the spectrum in Eq. (1) and $T_\nu$. For instance, the number density per flavour $n_\nu = (3/11)\,n_\gamma = (6\zeta(3)/11\pi^2)\,T_\gamma^3$ is fixed by the temperature, leading to a present value of 113 neutrinos and antineutrinos of each flavour per cm$^3$. Instead, the energy density for massive neutrinos should in principle be calculated numerically, with two well-defined analytical limits: $\rho_\nu = (7\pi^2/120)(4/11)^{4/3}\,T_\gamma^4\ (m_\nu \ll T_\nu)$ and $\rho_\nu = m_\nu n_\nu\ (m_\nu \gg T_\nu)$.

## EXTRA RADIATION AND THE EFFECTIVE NUMBER OF NEUTRINOS

Together with photons, neutrinos fix the expansion rate when the Universe is dominated by radiation. Their contribution to the total radiation content can be parametrized in terms of the effective number of neutrinos $N_{\rm eff}$,

$$\rho_{\rm r} = \rho_\gamma + \rho_\nu = \left[1 + \frac{7}{8}\left(\frac{4}{11}\right)^{4/3} N_{\rm eff}\right] \rho_\gamma\,, \tag{2}$$

where we normalized to the photon energy density because its value today is known from the measurement of the temperature of the cosmic microwave background (CMB). This equation is valid for complete neutrino decoupling and relativistic neutrinos.

We know that there are three light neutrinos sensitive to weak interactions (flavour or active neutrinos) from the analysis of the invisible $Z$-boson width at LEP, $N_\nu = 2.984 \pm 0.008$ [6], and an accurate analysis of neutrino decoupling shows that they contribute as $N_{\rm eff} \simeq 3.046$. Any departure of $N_{\rm eff}$ from this last value would be due to non-standard neutrino features or to the contribution of other relativistic relics [1, 7].

The contribution of $N_{\rm eff}$ to the expansion rate during Big-Bang Nucleosynthesis (BBN) fixes the produced abundances of light elements, in particular that of $^4$He (for a recent review, see [8]). On the other hand, BBN is the last period of the Universe sensitive to neutrino flavour, since electron neutrinos and antineutrinos play a direct role in the $n \leftrightarrow p$ processes. Thus the comparison of theoretical predictions and experimental data on the primordial abundances leads to constraints on the value of $N_{\rm eff}$ and/or on non-standard features on the $\nu_e/\bar\nu_e$ spectra. In addition, a non-standard value of $N_{\rm eff}$ would modify the transition epoch from radiation to matter domination, which has consequences on cosmological observables such as the power spectrum of CMB anisotropies, leading to independent bounds on the radiation content.

Here, as an example, we describe the results of a very recent analysis [9] (see the references therein for a list of recent works), that considers both BBN and CMB/LSS data to calculate the allowed regions on the plane defined by $N_{\text{eff}}$ and the baryon contribution to the present energy density. For the BBN analysis they include the $^4$He and D abundances, leading to the allowed range $N_{\text{eff}} = 3.1^{+1.4}_{-1.2}$ (95% C.L.). They also consider the most recent cosmological data on CMB temperature anisotropies and polarization, galaxy clustering from SDSS and 2dF and luminosity distances of type Ia Supernovae. The allowed range in this case is $N_{\text{eff}} = 5.2^{+2.7}_{-2.2}$, while adding data on the Lyman-$\alpha$ absorption clouds and the Baryonic Acoustic Oscillations (BAO) from SDSS, is $N_{\text{eff}} = 4.6^{+1.6}_{-1.5}$. These ranges are in reasonable agreement with the standard prediction of $N_{\text{eff}} \simeq 3.046$, and show that an allowed region for $N_{\text{eff}}$ common at early (BBN) and more recent epochs exists, although some tension remains when adding Lyman-$\alpha$ and BAO data. However, the reader should be cautious in the interpretation of this tension as an indication for relativistic degrees of freedom beyond flavour neutrinos.

## MASSIVE NEUTRINOS IN COSMOLOGY

Nowadays there exist compelling evidences for non-zero neutrino masses from the experimental data on flavour neutrino oscillations [10, 11]. Oscillation experiments measure two differences of squared neutrino masses ($3\sigma$ ranges from [10])

$$|\Delta m^2_{31}| = |m^2_3 - m^2_1| = (2.6 \pm 0.6) \times 10^{-3}\ \text{eV}^2 \quad \Delta m^2_{21} = m^2_2 - m^2_1 = (7.9^{+1.0}_{-0.8}) \times 10^{-5}\ \text{eV}^2$$

Unfortunately oscillation experiments are insensitive to the absolute scale of neutrino masses, and two schemes are still possible, known as normal (NH) and inverted (IH) hierarchies, characterized by the sign of $\Delta m^2_{31}$. For small values of the lightest neutrino mass $m_0$, i.e. $m_1$ ($m_3$) for NH (IH), the mass states follow a hierarchical scenario, while for masses much larger than the differences all neutrinos share in practice the same mass and then we say that they are degenerate. In general, the relation between the individual masses and their sum is found numerically.

Cosmology is at first order sensitive to the sum of neutrino masses but blind to mixing angles or phases, while laboratory experiments such as neutrinoless double beta decay [12] and tritium beta decay [13] are sensitive to combinations of masses and mixing parameters called $m_{\beta\beta}$ and $m_\beta$. Since the current limits from tritium beta decay are $m_\beta \lesssim 2.2$ eV (95% CL) from the Troitsk and Mainz experiments, the sum of neutrino masses is restricted to the range $0.06(0.1) \lesssim \sum_i m_i/\text{eV} \lesssim 6$. The allowed regions in the parameter space defined by any pair of parameters $(\sum_i m_i, m_{\beta\beta}, m_\beta)$ can be found in [11].

Do neutrino oscillations have an effect on any cosmological epoch? In the standard picture all flavour neutrinos were produced with the same energy spectrum, so no effect is expected from oscillations among these three states (up to small spectral distortions, see [5]). But there are two cases where neutrino oscillations could have cosmological consequences: flavour oscillations with non-zero relic neutrino asymmetries and active-sterile neutrino oscillations (see e.g. Sec. 5 in [14]).

A priori, massive neutrinos are excellent candidates for being the Dark Matter, the dominant non-baryonic component of the matter density in the Universe. We know that they exist and their energy density in units of the critical value is

$$\Omega_\nu = \rho_\nu/\rho_c^0 = M_\nu/(93.14 h^2 \text{ eV}) . \quad (3)$$

where $h \equiv H_0/(100 \text{ km s}^{-1} \text{ Mpc}^{-1})$ is the present value of the Hubble parameter. The neutrino density fraction is usually defined with respect to the total matter density, $f_\nu \equiv \Omega_\nu/\Omega_m$. The total mass $M_\nu \equiv \sum_i m_i$ includes all masses of the neutrino states which are non-relativistic today, at least two because both $(\Delta m_{31}^2)^{1/2} \simeq 0.05$ eV and $(\Delta m_{21}^2)^{1/2} \simeq 0.009$ eV are larger than $T_\nu \simeq 1.7 \times 10^{-4}$ eV. The range of present values of $\Omega_\nu$ compatible with oscillation data and the approximate bounds from beta decay experiments is $0.0013\,(0.0022) \lesssim \Omega_\nu \lesssim 0.13$ (for $h \approx 0.7$).

In general, a cosmological upper bound on $\Omega_\nu$ has been used since the 1970s to constrain the possible values of neutrino masses. If we demand that neutrinos should not be heavy enough to overclose the Universe ($\Omega_\nu < 1$), we obtain an upper bound $M_\nu \lesssim 45$ eV. Moreover, since from present analyses of cosmological data we know that the contribution of matter is $\Omega_m \simeq 0.3$, the neutrino masses should obey the stronger bound $M_\nu \lesssim 15$ eV, which is roughly only a factor 2 worse than that from tritium decay.

Dark matter particles with a large velocity dispersion such as that of neutrinos are called hot dark matter (HDM), and they affect the evolution of cosmological perturbations in a particular way: the density contrasts are erased on wavelengths smaller than a mass-dependent free-streaming scale (see [15] for a historical review). If HDM dominates, this suppression contradicts various observations and this is why the attention then turned to cold dark matter (CDM) candidates, i.e. particles which were fully non-relativistic at the epoch when the universe became matter-dominated. Still in the mid-1990s it appeared that adding a small contribution of HDM fitted better the observational data on density fluctuations at small scales than a pure CDM model. However, within the presently favoured ΛCDM model dominated at late times by a cosmological constant (or some form of dark energy) there is no need for a significant HDM contribution. Instead, we use the available cosmological data to find how large neutrino masses can be.

An interesting case is that of a sterile neutrino with a mass in the keV range slightly mixed with the flavour neutrinos, so that it could be produced by active-sterile oscillations but not fully thermalized, playing the role of dark matter instead of the usual CDM. Due to their large thermal velocity (slightly smaller than that of active neutrinos), they would behave as Warm Dark Matter and erase small-scale cosmological structures. Their mass can be bounded from below using Lyman-$\alpha$ forest data, and from above using X-ray observations. The viability of this scenario is currently under careful examination (see e.g. [16] for a recent discussion and a list of references).

## EFFECTS OF NEUTRINO MASSES ON COSMOLOGY

Here we describe very briefly the effects caused by neutrino masses on the main cosmological observables. For a more detailed discussion, see [3].

In general, neutrino masses of the order of 1 eV (about $f_\nu \leq 0.1$) were still relativistic at the time of recombination between electrons and nucleons, i.e. after photon decoupling. Thus they only affect the shape of the power spectrum of CMB temperature anisotropies through the modified background evolution[1], related to the fact that these neutrinos account today as matter but to radiation at that time [3]. This changes some characteristic times and scales in the cosmological evolution, and affects mainly the amplitude of the first acoustic peak as well as the location of all peaks. Thus, it is possible to constrain neutrino masses using CMB data alone [3, 18], down to the level at which this background effect is masked by instrumental noise or cosmic variance, or by parameter degeneracies in extended models (such as varying $N_{\rm eff}$).

The current Large Scale Structure (LSS) of the Universe is probed by the matter power spectrum, observed with various techniques. It is defined as the two-point correlation function of non-relativistic matter fluctuations in Fourier space $P(k,z) = \langle|\delta_{\rm m}(k,z)|^2\rangle$, where the matter perturbation $\delta_{\rm m} = \sum_i \bar{\rho}_i\,\delta_i / \sum_i \bar{\rho}_i$, includes all components (CDM, baryons, non-relativistic neutrinos, ...) and represents indifferently the energy or mass power spectrum of non-relativistic matter. The shape of $P(k)$ is affected by the free-streaming caused by small neutrino masses and thus it is the key observable for constraining $m_\nu$. The physical effect of free-streaming is to damp small-scale neutrino density fluctuations: neutrinos cannot be confined into (or kept outside of) regions smaller than the free-streaming length. Instead, on scales much larger, the neutrino velocity effectively vanishes and after the non-relativistic transition the neutrino perturbations behave like CDM: modes with $k < k_{\rm nr} = 0.018\,\Omega_{\rm m}^{1/2}(m/{\rm eV})^{1/2}\,h{\rm Mpc}^{-1}$ [3], are never affected by free-streaming and evolve like in a pure ΛCDM model. There exist additional effects of massive neutrinos related to the background evolution and a gravitational back-reaction effect that also damp the metric perturbations on those scales (see Sec. 4.5 of [3]). The combined effect of neutrino masses is an attenuation of small-scale perturbations for $k > k_{\rm nr}$, as shown in Fig. 13 of [3]. For small values of $f_\nu$ this effect is approximated in the large $k$ limit by the well-known linear expression $P(k)^{f_\nu}/P(k)^{f_\nu=0} \simeq 1 - 8 f_\nu$ [19].

Is it possible to mimic the effect of massive neutrinos on $P(k)$ with some combination of other cosmological parameters? Ideally, if we could measure $P(k)$ on a large interval of $k$ values, the effect of neutrino masses would be non-degenerate, because of its very characteristic step-like shape. The problem is that usually $P(k)$ can only be accurately measured in the intermediate region where the mass effect is neither null nor maximal. In this region, neutrino masses affect the slope of the matter power spectrum in a way which can be easily confused with the effect of other cosmological parameters. Therefore, we need to combine LSS data with other cosmological data, in particular the CMB anisotropy spectrum, which could lift most of the degeneracies.

## CURRENT BOUNDS ON NEUTRINO MASSES

Let us summarize the bounds on the total neutrino mass obtained from current cosmological data. Note that the upper bounds on $M_\nu$ are all based on the Bayesian inference method, and are given at 95% C.L. after marginalization over all free cosmological pa-

[1] If neutrinos were heavier than a few eV, they would already be non-relativistic at photon decoupling and would have more complicated consequences for the CMB [17].

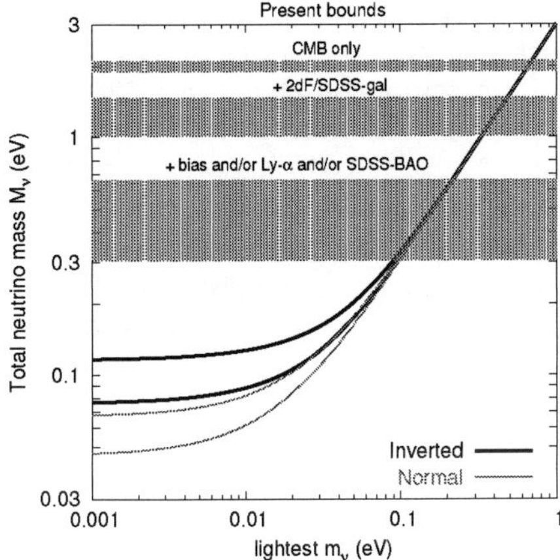


**FIGURE 1.** Current upper bounds (95%CL) from cosmological data on $M_\nu$, compared to the values in agreement at a $3\sigma$ level with oscillation data.

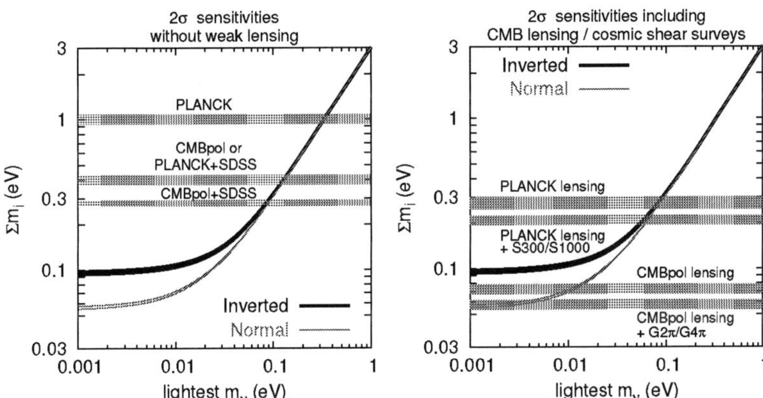


**FIGURE 2.** Forecast $2\sigma$ sensitivities to $M_\nu$ from future cosmological experiments compared to the values from oscillation data (assuming a future 5% determination). Left: future CMB experiments (without lensing extraction), alone and combined with the SDSS galaxy redshift survey. Right: future CMB experiments including lensing information, alone and combined with future cosmic shear surveys. Here CMBpol refers to a hypothetical CMB satellite roughly corresponding to the INFLATION PROBE mission.

rameters. For a discussion on this statistical method and the references for experimental data or parameter analyses, see [3].

One can see in Fig. 1 that a single cosmological bound on neutrino masses does not exist: the published results were grouped in bands according to the included set of data, while their thickness roughly describe the spread of values obtained from similar cosmological data[2]. One finds that $M_\nu$ should be less than 2-3 eV (CMB data only,

[2] Note that most of the analyses considered a minimal ΛCDM model plus neutrino masses, but some of them included also other cosmological parameters. Other cosmological data, such as the HST measure-

mostly from WMAP) or 0.9-1.7 eV (CMB & 2dF/SDSS galaxy clustering data, i.e. the shape of the matter power spectrum), which is probably a quite conservative upper bound. Thus current cosmological data probe the region of three degenerate neutrino masses, reaching the lowest values when data from Lyman-$\alpha$ and/or BAO-SDSS are included or the bias is fixed. In such cases the contribution of a total neutrino mass of the order $0.3-0.6$ eV could be already disfavoured. In [11] these cosmological bounds are compared with those coming from laboratory experiments.

We emphasize that the impressive cosmological bounds on $M_\nu$ in Fig. 1 may change whenever a new parameter degeneracy with the neutrino masses arises. For instance, if extra radiation exists the bound on $M_\nu$ gets less stringent [20], an interesting result for the 4-neutrino mass schemes that also incorporate the LSND results [21], which would require a fourth sterile neutrino with $\mathcal{O}$(eV) mass (see [22] for a recent cosmological analysis). Other possible parameter degeneracy exists between $M_\nu$ and the parameter $w$, that characterizes the equation of state of the dark energy $X$ ($p_X = w\rho_X$), but it can be broken adding BAO data [23]. Finally, the cosmological implications of neutrino masses could be very different for non-standard CNB spectrum or evolution [3].

## FUTURE SENSITIVITIES ON NEUTRINO MASSES

In the near future we will have more precise data on cosmological observables from various experimental techniques: CMB anisotropies measured with ground-based experiments or satellites such as PLANCK, galaxy redshift surveys, galaxy cluster surveys, etc. In particular, it has been recently emphasized the potentiality for measuring small neutrino masses of weak lensing experiments, which will look for the lensing effect caused by the large scale structure of the neighboring universe, either on the CMB signal [24] or on the apparent shape of galaxies (measured by cosmic shear surveys, see e.g. [25]). If the characteristics of these future experiments are known with some precision, it is possible to assume a "fiducial model", i.e. a cosmological model that would yield the best fit to future data, and to estimate the error on a particular parameter that will be obtained after marginalizing the hypothetical likelihood distribution over all the other free parameters (see Sec. 6 of [3] for further details). Technically, the simplest way to forecast this error is to compute a Fisher matrix, a technique has been widely used in the literature, now complemented by Monte Carlo methods.

Fig. 2 is a graphical summary of the forecast sensitivities to neutrino masses of different cosmological data. We see that there are very good prospects for testing neutrino masses in the degenerate and quasi-degenerate mass regions above 0.2 eV or so. A detection at a significant level of the minimal value of the total neutrino mass in the IH scheme will demand the combination of CMB lensing and cosmic shear surveys, whose more ambitious projects will provide a $2\sigma$ sensitivity to the minimal value in the case of NH (of order 0.05 eV). The combination of CMB data with future galaxy cluster surveys [26], derived from the same weak lensing observations, as well as X-ray and Sunyaev–Zel'dovich surveys, should yield a similar sensitivity.

---

ment of $h$ or type Ia supernovae luminosity, are sometimes added.

## CONCLUSIONS

Neutrinos, despite the weakness of their interactions, can play an important role in Cosmology as reviewed in this contribution. In addition, cosmological data can provide information on the properties of these elusive particles, such as the effective number of neutrinos and the total neutrino mass, complementary to terrestrial experiments. The analysis of current cosmological data leads to bounds on the degenerate mass region, and in the near future we could even hope to test the minimal values of neutrino masses guaranteed by flavour neutrino oscillations, using data from new cosmological experiments. Therefore, we expect that neutrino cosmology will remain an active research field in the next years.

## ACKNOWLEDGMENTS

I thank the organizers of the VI SILAFAE for their invitation and hospitality. This work was supported by the European Network of Theoretical Astroparticle Physics ILIAS/N6 under contract number RII3-CT-2004-506222 and the Spanish grants FPA2005-01269 and GV/05/017 of Generalitat Valenciana, as well as by a Ramón y Cajal contract of MEC.

## REFERENCES

1. A.D. Dolgov, *Phys. Rep.* **370**, 333 (2002).
2. S. Hannestad, *Ann. Rev. Nucl. Part. Sci.* **56**, 137 (2006).
3. J. Lesgourgues and S. Pastor, *Phys. Rep.* **429**, 307 (2006).
4. A. Dolgov et al, *Nucl. Phys. B* **632**, 363 (2002).
5. G. Mangano et al, *Nucl. Phys. B* **729**, 221 (2005).
6. W.M. Yao et al, *J. Phys. G* **33**, 1 (2006).
7. S. Sarkar, *Rep. Prog. Phys.* **59**, 1493 (1996).
8. G. Steigman, *Int. J. Mod. Phys. E* **15**, 1 (2006).
9. G. Mangano et al, astro-ph/0612150.
10. M. Maltoni et al, *New J. Phys.* **6**, 122 (2004).
11. G.L. Fogli et al, *Prog. Part. Nucl. Phys.* **57**, 742 (2006).
12. S.R. Elliott and P. Vogel, *Ann. Rev. Nucl. Part. Sci.* **52**, 115 (2002).
13. K. Eitel, *Nucl. Phys. Proc. Suppl.* **143**, 197 (2005).
14. S. Pastor, lectures at 61st SUSSP (Taylor & Francis Academic Pub., 2007), in press.
15. J.R. Primack, astro-ph/0112336.
16. M. Viel et al, *Phys. Rev. Lett.* **97**, 071301 (2006).
17. S.Dodelson et al, *Astrophys. J.* **467**, 10 (1996).
18. K. Ichikawa et al, *Phys. Rev. D* **71**, 043001 (2005).
19. W. Hu et al, *Phys. Rev. Lett.* **80**, 5255 (1998).
20. P. Crotty et al, *Phys. Rev. D* **69**, 123007 (2004).
21. A. Aguilar et al [LSND Collaboration], *Phys. Rev. D* **64**, 112007 (2001).
22. S. Dodelson et al, *Phys. Rev. Lett.* **97**, 041301 (2006).
23. A. Goobar et al, *J. Cosmol. Astrop. Phys.* **0606**, 019 (2006).
24. J. Lesgourgues et al, *Phys. Rev. D* **73**, 045021 (2006).
25. S. Hannestad et al, *J. Cosmol. Astrop. Phys.* **0606**, 025 (2006).
26. S. Wang et al, *Phys. Rev. Lett.* **95**, 011302 (2005).

# High energy neutrinos from optically thick sources

M. Kachelrieß* and R. Tomàs†

*Institutt for fysikk, NTNU Trondheim, N–7491 Trondheim, Norway
†AHEP Group, Institut de Física Corpuscular - C.S.I.C/Universitat de València
Edifici Instituts d'Investigació, Apt. 22085, E-46071 València, Spain

**Abstract.** We calculate the yield and flavor content of high energy neutrinos produced in astrophysical sources with negligible magnetic fields varying their interaction depth. We take into account the scattering of secondaries on background photons as well as the direct production of neutrinos in decays of charm mesons. If multiple scattering of nucleons becomes important, the neutrino spectra from meson and muon decays are strongly modified with respect to transparent sources. Characteristic for neutrino sources containing photons as scattering targets is a strong energy-dependence of the ratio $R^0$ of $\nu_\mu$ and $\nu_e$ fluxes at the sources, ranging from $R^0 = \phi_\mu/\phi_e \sim 0$ below threshold to $R^0 \sim 4$ close to the energy where the decay length of charged pions and kaons equals their interaction length on target photons. Above this energy, the neutrino flux is strongly suppressed and depends mainly on charm production.

**PACS:** 95.85.Ry, 98.70.Sa, 14.60Lm, 14.60.Pq

Experimental high energy neutrino physics has become one of the most active areas of astroparticle physics, offering among others the prospect of identifying the sources of ultra-high energy cosmic rays [1]. High energy neutrinos from astrophysical sources are the decay products of secondary mesons produced by scattered high energy protons on background protons or photons. In the case of sources transparent to hadronic interactions one can use the observed ultra-high energy cosmic ray (UHECR) flux to set upper limits on possible neutrino fluxes [2]. Dropping the assumption of transparent sources and hidding the acceleration region by sufficient material absorbing ultra-high energy cosmic rays allows one to avoid the cosmic ray limits. One might therefore speculate that large neutrino fluxes at high and ultra-high energies might be produced in opaque sources.

In this contribution we report on a calculation of the flux of high energy neutrinos produced as secondaries in astrophysical sources. We have considered sources with interaction depth ranging from nearly transparent to opaque. We have put emphasis on sources with such high densities that magnetic fields can be neglected and multiple scatterings are important [3].

We have idealized a hidden neutrino source as an homogeneous slab filled with photons in which high energy protons are injected with an energy spectrum $dN/dE \propto E^{-2.2}$ and $E_{\max} = 10^{24}$ eV. Each source is fully characterized by its length $L$ and a thermal photon distribution $n(\varepsilon)$. This phenomenological description has allowed us to perform a systematic analysis of different sources according to their thickness, parametrized by the "interaction depth" $\tau = L/l_{\rm int} = L\sigma n_\gamma$, from nearly transparent to opaque. For the discussion of our numerical results it is useful to introduce the critical

CP917, *Particles and Fields,* edited by H. Castilla Valdez, J. C. D'Olivo, and M. A. Perez

dimensionless energy $x_{cr} = E_{cr}\omega/m_p^2$, where $E_{cr}$ is defined as the energy at which the decay length $\ell_{1/2}$ of a meson is equal to its interaction length $\ell_{int}$. Thus, the "interaction depth" $\tau$ determines the fraction of nucleons that scatter on photons, while the parameter $x \equiv E\omega/m_p^2$ controls if mesons mainly decay ($x \ll x_{cr}$) or scatter ($x \gg x_{cr}$).

In Fig. 1, we show the interaction length of protons, pions and kaons in a thermal bath of photons with temperature $T = 10^4$ K together with their decay length. The fate of a hadron is characterized by two important energies: The threshold energy at $x_{th} \sim 0.1$, below which photo-meson reactions are exponentially suppressed, and the critical energy $x_{cr}$, above which most mesons will scatter before decaying. If the decay length and the interaction length cross inside the source the effect of multiple scattering has to be taken into account. Main consequence is a strong suppression of the high energy neutrino flux produced in meson decays for $x \gtrsim x_{cr}$. The shorter life-time of heavier mesons leads to a shift of their critical energy, i.e. $x_{cr}^K > x_{cr}^\pi$. In the case of opaque sources the range with unsuppressed neutrino fluxes, $0.1 \sim x_{th} \lesssim x \lesssim x_{cr} \propto 1/T^2$, shrinks for increasing $T$ and becomes a narrow peak around $x_{th}$ for $T \gtrsim 5 \times 10^5$ K.

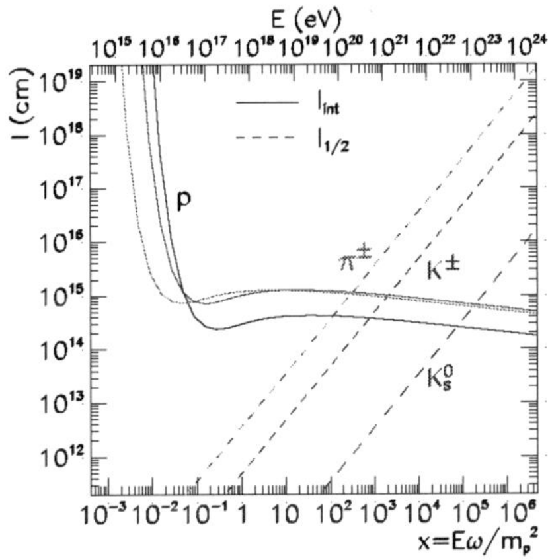


**FIGURE 1.** Interaction length (solid lines) for proton (black), charged pions (red) and charged kaons (blue) with thermal photons at a temperature $T = 10^4$ K, together with the decay length (dashed lines) for $\pi^\pm$, $K^\pm$ and $K_S^0$.

In Fig. 2, we show the neutrino yields, defined as $Y_\nu(E) = \phi_\nu(E)/(\tau\phi_p(E))$, from meson decays for a source with $T = 10^4$ K and different interaction depths. For $\tau \lesssim 1$, multiple scattering is negligible, $\phi_\nu \propto \tau$ and the neutrino yields expressed as functions of $x$ are thus independent from $\tau$. Moreover, the neutrino yield from charm meson decays is clearly subdominant. Increasing the size of the source, multiple scattering cannot be neglected anymore and three different $x$ regions can be distinguished. For $x < x_{cr}^\pi$, most pions decay before scattering and, as in the case of transparent sources, the neutrino yield is dominated by the contribution of charged pions. In the intermediate $x$ range, $x_{cr}^\pi \lesssim x \lesssim x_{cr}^K$, multiple pion scattering on photons becomes effective and therefore neutrinos from kaon decays start to provide the most important contribution to the total neutrino yield. At even higher energies, $x \gtrsim x_{cr}^K$, both pion and kaon neutrinos become strongly suppressed as these mesons most often scatter before decaying. Hence at high energies, neutrinos from decays of charm mesons represent the main component of the total neutrino yield.

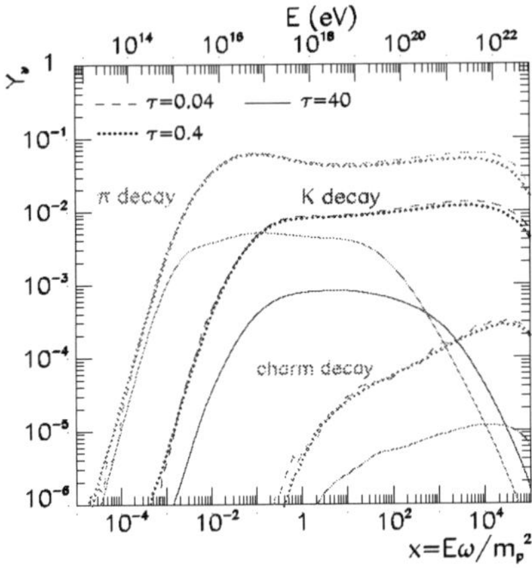


**FIGURE 2.** Neutrino yield $Y_\nu$ as function of $x = E\omega/m_p^2$ (bottom axis) and $E$ (upper axis) from pion (magenta), kaon (blue), and charm decays (red) for $\tau = 0.04, 0.4$ and 40 for $T = 10^4$ K.

In Fig. 3 we show the effect of the interaction depth in the (unnormalized) proton and neutrino fluxes for a source at $T = 10^5$ K. In the case of thin sources, $\tau \lesssim$ a few, we observe that the final proton flux is only slightly distorted relative to the initial flux, and therefore provides a non-negligible contribution to the observed UHECRs. The neutrino flux in these sources is basically proportional to the depth. For opaque sources the final proton flux becomes strongly suppressed above threshold, $E_p^{\text{th}} \approx m_p m_\pi/(2\varepsilon_\gamma)$. For such high depths most pions and kaons scatter before decaying and as a consequence the flux of neutrinos becomes also suppressed at energies higher than $E_{\text{cr}}$. The only effect of further increasing $\tau$ is an increase of the low energy neutrinos, as the high number of scatterings leads to more low energy mesons. Finally, we want to illustrate how the high-energy suppression of neutrino fluxes in opaque sources depends on the temperature of the thermal photon distribution. Figure 4 shows (unnormalized) fluxes for an opaque source, $\tau = 40$, at $T = 10^4$ K in the top and $T = 10^7$ K in the bottom. As previously mentioned, the final proton flux is strongly suppressed above threshold, whereas the neutrino flux rises already at lower energies, because they carry away only a fraction of the proton energy. The energy range where the neutrino flux is maximal extends approximately from $E_p^{\text{th}}$ to $E_{\text{cr}}$, which for the source at $T = 10^4$ K, goes roughly from $10^{16}$ eV to $10^{20}$ eV. This range is strongly reduced for increasing $T$ as $E_{\text{cr}} \propto 1/n \propto 1/T^3$, whereas $E_{\text{th}} \propto 1/T$. Therefore, critical and threshold energies merge and only a a narrow energy range around $E \approx 10^{12}$ eV with unsuppressed neutrino flux remains for $T = 10^7$ K, cf. the lower panel of Fig. 4.

As summary we show in Fig. 5 the basic conditions of density and size that must be fulfilled by an opaque source to lead to large high energy neutrino fluxes. The first requirement to evade the UHECR limit on the high energy neutrino fluxes is opacity, that means the dominance of multiple scattering at high energies. This condition amounts to $\tau \gtrsim 1$ or $Ln_\gamma \gtrsim 1/\sigma$, see Fig. 5. The second requirement has to do with the energy

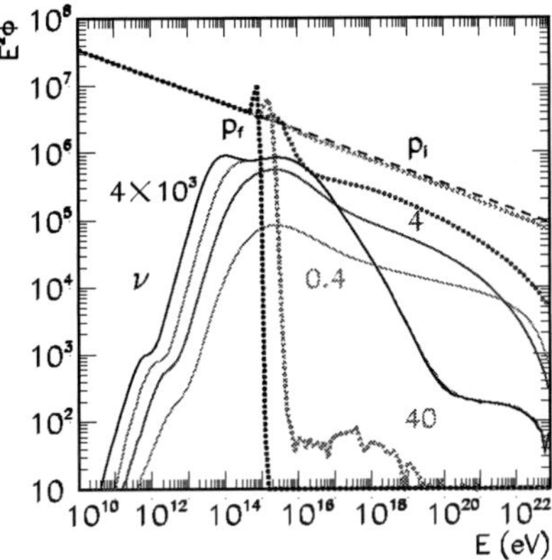


**FIGURE 3.** Unnormalized fluxes of inital (dashed) and final (dotted) protons, as well as the total neutrino (solid), for a source at $T = 10^5$ K and with interaction depths $\tau = 0.4$ (magenta), 4 (blue), 40 (red), and $4 \times 10^3$ (black).

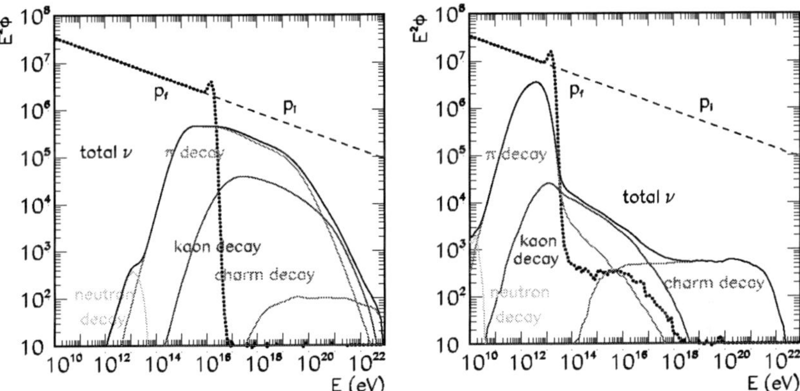


**FIGURE 4.** Unnormalized fluxes of initial (dashed) and final (dotted) protons, as well as the total neutrino (solid), for a source with interaction depth $\tau = 40$ at $T = 10^4$ K (left) and $T = 10^7$ K (right). The different contributions to the total neutrino flux are also shown.

range where the neutrino flux can be large. The limits of this range are provided by $E_{\rm th}$ and $E_{\rm cr}$: below the threshold mesons are not efficiently produced, and above the critical energy they scatter before decay leading to a significant suppression of the neutrino flux. In Fig. 5 one can notice how a source with photon density $n_\gamma \gtrsim$ few $\times\, 10^{16}$ g/cm$^3$ (few $\times\, 10^{11}$ g/cm$^3$) can not provide large neutrinos fluxes at energies larger than $10^{16}$ eV ($10^{21}$ eV). Consequently the best candidates to generate large high energy neutrino fluxes are very big opaque sources with low densities.

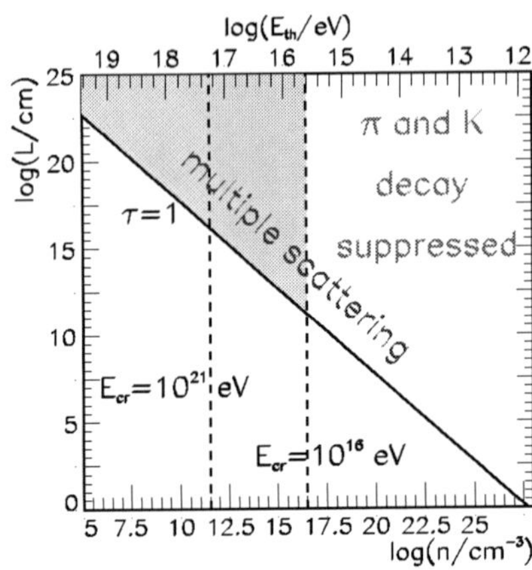


**FIGURE 5.** Contours of constant optical depth $\tau = 1$ and critical energies $E_{cr} = 10^{16}$ and $E_{cr} = 10^{21}$, as function of the density $n$ (bottom axis) and threshold energy $E_{th}$ (top axis), and size of the source, $L$. The regions where multiple scattering is relevant and the $\pi$ and $K$ neutrinos are suppressed are also shown. The shaded areas correspond to the parameter space leading to a large neutrino flux at $E = 10^{16}$ eV (cyan and yellow) and $E = 10^{21}$ eV (yellow).

Both neutrino telescopes and extensive air shower experiments can not only detect high energy neutrinos but also have some flavor discrimination possibilities [4, 5, 6]. The main observable for the neutrino flavor composition is the ratio of track to shower events in a neutrino telescope, $R_\mu = \phi_\mu/(\phi_e+\phi_\tau)$. In Fig. 6, we show the total flavor ratio at the source $R^0 \equiv Y_{\nu_\mu}/Y_{\nu_e}$ expected for different sources. In the case of a transparent source the value of $R^0$ is small at low $x$ because almost all neutrinos are produced in the decay of neutrons. Once the contribution from charged pion decay becomes dominating, $R^0$ approaches the standard value two. In contrast, in the case of an opaque source, $\tau = 40$, $R^0$ is strongly energy-dependent. In particular, there is a bump in the range $100 \lesssim x \lesssim 10^4$ and $1 \lesssim x \lesssim 100$ for the case $T = 10^4$ K and $10^5$ K, respectively, related to the range where kaon neutrinos dominate. The value of $R^0$ at the bump is determined by the interplay of the charged and neutral kaon neutrinos. Finally, in both cases the asymptotic value is $R^0 \sim 1$, which corresponds to the ratio expected when the decay of charm mesons dominate. The neutrino spectra at the source are modulated by oscillation [4, 7]. Therefore the expected flavor ratio $R_\mu$ at the Earth will be different from the original one $R^0$: it significantly depends on the neutrino mixing parameters, and especially on $\vartheta_{23}$. Therefore the observation of a strong energy dependence of $R_\mu$ will not only be a hint on the kind of source but also allow obtaining complementary information on the neutrino mixing angles.

## ACKNOWLEDGMENTS

R. T. is greatly grateful to local organizers and specially O. Miranda for the invitation, local support, as well as the hospitality shown during the Conference. This work has

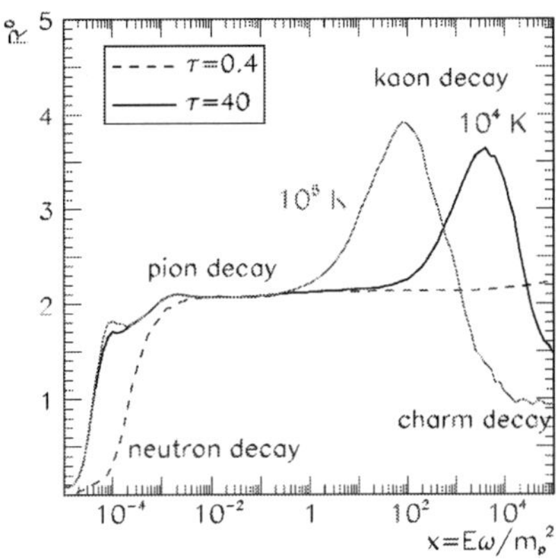


**FIGURE 6.** Total flavor ratio for a source at $T = 10^4$ K and $\tau = 0.4$ (blue dashed), $T = 10^4$ K and $\tau = 40$ (blue solid), and $T = 10^5$ K and $\tau = 40$ (solid solid). The main reactions giving rise to the neutrino signal are also shown.

been partially supported by the Juan de la Cierva programme, an ERG from the European Comission, and by the Spanish grant FPA2005-01269.

## REFERENCES


1. V. Beresinsky, "Ultra high energy neutrino astronomy," Nucl. Phys. Proc. Suppl. **151**, 260 (2006) ; P. Lipari, "Perspectives of High Energy Neutrino Astronomy," astro-ph/0605535.
2. E. Waxman and J. N. Bahcall, "High energy neutrinos from astrophysical sources: An upper bound," Phys. Rev. D **59**, 023002 (1999). K. Mannheim, R. J. Protheroe and J. P. Rachen, "On the cosmic ray bound for models of extragalactic neutrino production," Phys. Rev. D **63**, 023003 (2001).
3. M. Kachelrieß and R. Tomàs, "High energy neutrino yields from astrophysical sources. I: Weakly magnetized sources," Phys. Rev. D **74**, 063009 (2006).
4. J. G. Learned and S. Pakvasa, "Detecting tau-neutrino oscillations at PeV energies," Astropart. Phys. **3**, 267 (1995).
5. H. Athar, G. Parente and E. Zas, "Prospects for observations of high energy cosmic tau-neutrinos," Phys. Rev. D **62**, 093010 (2000).
6. J. F. Beacom *et al.*, "Measuring flavor ratios of high energy astrophysical neutrinos," Phys. Rev. D **68**, 093005 (2003) [Erratum-ibid. D **72**, 019901 (2005)].
7. H. Athar, M. Jezabek and O. Yasuda, "Effects of neutrino mixing on high-energy cosmic neutrino flux," Phys. Rev. D **62** 103007 (2000).

# The Flavor of Gravity

R. Delbourgo

*University of Tasmania, Private Bag 37, GPO Hobart, Australia 7001*

**Abstract.** I show how the observed particle spectrum in its various flavors and generations can be accommodated within a scheme involving five complex anticommuting Lorentz scalar coordinates $\zeta$, carrying "property'. A general relativistic extension of the scheme shows that gravity lies in the $x-x$ sector, gauge fields in the $x-\zeta$ sector and Higgs field in the $\zeta-\zeta$ sector.

**Keywords:** anticommuting coordinates, gravity, flavor
**PACS:** 04.50.+h,11.30.Hv,11.90.+t,12.10.-g

## FLAVOR OR PROPERTY

In a paper bearing the same title [1] and in a subsequent article concerned with flavor mixing [2], a scheme was presented where flavor was built up from a set of five complex anticommuting $\zeta$, each of which carried a specific property, such as chromicity, leptonicity and neutrinicity. In this talk I would like to summarize the salient features of the scheme and stress its principal predictions and where it differs from the standard model of particle physics.

First I'll quickly run over the advantages of having extra fermionic rather than bosonic degrees of freedom which are attached to spacetime. Next I will show how flavor can be built up from polynomials over the $\zeta$ coordinates and show how the quarks and leptons *in their generations* can be naturally accommodated. In addition the scheme allows for spinor sextets and curious $D$-type quarks which are electroweak singlets, as we shall see. Next I will outline the nature of the Higgs multiplets and how spinor masses are obtained. Finally in the last part I will describe a general relativistic version of the extended spacetime and show how gauge fields lie in the $\zeta-x$ sector of the supermetric.

## C-NUMBERS AND A-NUMBERS

| Bose-Einstein | Fermi-Dirac |
|---|---|
| $xy=+yx$ | $\xi\eta=-\eta\xi$ |
| $[a,a^\dagger]=1$ | $\{a,a^\dagger\}=1$ |
| $(e^{\alpha+\beta E}-1)^{-1}$ | $(e^{\alpha+\beta E}+1)^{-1}$ |
| $\int e^{-c^*Dc}d^2c\propto[\det D]^{-1}$ | $\int e^{-c^*Dc}d^2c\propto[\det D]^{+1}$ |
| O($2n$) $\sim$ Sp(-$2n$) | Sp($2n$) $\sim$ O(-$2n$) |

Anticommuting operators were formulated in the 1920s but anticommuting numbers (a-nos.) have been used in physics only since about 1970, when BRST and SUSY came to

CP917, *Particles and Fields,* edited by H. Castilla Valdez, J. C. D'Olivo, and M. A. Perez

the fore. (However they were invented by Grassmann to simplify differential geometry much earlier.) Commuting or c-nos, connected with BE statistics, are common but a-nos. have only been widely used since 1970 and are associated with FD statistics. The above table summarizes the main sign factors that are encountered.

BRST uses Lorentz scalar and vector a-numbers (pairs for ghosts and $\overline{\text{ghosts}}$) but SUSY uses Lorentz spinors. BRST is good for proving renormalizability and gauge invariance of gauge models, but despite SUSY's great allure, its applicability has turned out to be problematic: nature shows no signs of supersymmetric partner particles or states (photinos, squarks, gravitino,..)! The main (theoretical) attractions of SUSY are

- a consistent nontrivial higher group $\supset$ the Poincare group: $\{Q_\alpha, Q_\beta\} = (\gamma.PC)_{\alpha\beta}$,
- it unifies bosons and fermions,
- cancellation of $\infty$s between boson and fermion loops, improves renormalizability,
- it allows a supersymmetric generalization of gravity (SUGRA),
- its extension lets 'internal symmetries' be incorporated through generators $Q^n_\alpha$,
- it can be generalized to string/brane theory.

## NEGATIVE DIMENSIONS AND SUPERFIELDS

Since $(\int e^{-c^*Dc}d^2c)^{n_B}(\int e^{-c^*Dc}d^2c)^{n_F} \propto D^{n_F-n_B}$ we can construe c-coordinates as adding to dimension and a-coordinates as subtracting dimension. (This is also confirmed [3] by group theory associated with particular representations of $O(2n)$ and $Sp(2n)$.) Hence matching c-coordinates with a-coordinates (and likewise BE fields with FD fields) we might get *zero* total dimensions = # dimensions of universe before the BIG BANG. Therefore to the four $x^m$ of space-time add four $\zeta^\mu$, or equivalently add five complex $\zeta^\mu$, but only take half the states to get the statistics correct. Associate these *Lorentz scalar* a-nos. with 'property' or 'internal symmetry'; this is where we differ crucially from standard SUSY – there is spin proliferation. Make the following charge $Q$ and fermion number $F$ assignments so as to obtain a sensible fundamental particle spectrum:

$$Q(\zeta^{0,1,2,3,4}) = (0,1/3,1/3,1/3,-1); F(\zeta^{0,1,2,3,4}) = (1,-1/3,-1/3,-1/3,1).$$

and obtain other properties as composites of these. An Sp(10) or SU(5) group [4] is natural in this context. The appropriate representations of those groups arise simply through superfield expansions. Since the product of two a-nos. is a (nilpotent) c-no., a Bose superfield $\Phi$ should be a Taylor series in even powers of $\zeta, \bar{\zeta}$ and a spinor superfield $\Psi_\alpha$ a series in odd powers of $\zeta, \bar{\zeta}$ — up to the 5th:

$$\Phi(x,\zeta,\bar{\zeta}) = \sum_{even\, r+\bar{r}} (\bar{\zeta})^{\bar{r}} \phi_{(\bar{r}),(r)} (\zeta)^r, \quad \Psi_\alpha(x,\zeta,\bar{\zeta}) = \sum_{odd\, r+\bar{r}} (\bar{\zeta})^{\bar{r}} \psi_{\alpha(\bar{r}),(r)} (\zeta)^r. \qquad (1)$$

Crudely on coordinate $\zeta$ we can regard label 0 as *neutrinicity*, labels $i$ =1 - 3 as (down) *chromicity*, label 4 as *charged leptonicity*. Thus the products $\zeta^1\zeta^2\zeta^3, \bar{\zeta}_i\zeta^i$ are color neutral.

## CONJUGATION AND SUPERDUALITY

The expansions (1) produce too many states, $\psi_\alpha$ and $\phi$, viz. 512, so they need to be cut down. A primary way is to impose self-conjugation whereby $\psi_{(r),(\bar{r})} = \psi^{(c)}_{(\bar{r}),(r)}$, corresponding to reflection along the main diagonal. Secondarily impose (anti) self-duality corresponding to reflection about the cross diagonal, specifically $\psi_{(r),(\bar{r})} = -\psi_{(5-\bar{r}),(5-r)}$. Note that the dual $\times$ of a field term has exactly the same $Q$ and $F$ as the field. For example, $(\bar{\zeta}_\alpha \zeta^\mu \zeta^\nu)^\times = \frac{1}{3!}\varepsilon^{\rho\sigma\tau\mu\nu}\bar{\zeta}_\rho\bar{\zeta}_\sigma\bar{\zeta}_\tau.\frac{1}{4!}\varepsilon_{\alpha\beta\gamma\delta\varepsilon}\zeta^\beta\zeta^\gamma\zeta^\delta\zeta^\varepsilon$. In particular $\bar{\zeta}_0\bar{\zeta}_4\zeta^1\zeta^2\zeta^3$ and $\bar{\zeta}_4\zeta^0\zeta^1\zeta^2\zeta^3$ are self-dual, so imposing antiduality eliminates these unwanted states, since they respectively have $F = 3$ & $Q = -2$. Focussing on the source field $\Psi$, the resulting square contains the following varieties of up ($U$), down ($D$), charged lepton ($L$) and neutrinos ($N$), where the subscript distinguishes between repetitions. In the table below, $\times$ are duals, * are conjugates:

| $r\backslash\bar{r}$ | 0 | 1 | 2 | 3 | 4 | 5 |
|---|---|---|---|---|---|---|
| 0 | | $L_1, N_1, D_5^c$ | | $L_5^c, D_1, U_1$ | | |
| 1 | * | | $L_{2,3}, N_{2,3}, D^c_{3,6,7}, U_3^c$ | | $L_6^c, D_2, U_2$ | |
| 2 | | * | | $L_4, N_4, D^c_{4,8}, U_4^c$ | | $\times$ |
| 3 | * | | * | | $\times$ | |
| 4 | | * | | * | | $\times$ |
| 5 | * | | * | | * | |

Note that the fermions come in 4s, 6s and 8s comfortably containing the known three generations. However observe that whereas $U, D$ in the first and second generation are bona fide weak isospin doublets the third and fourth family $U, D$ are accompanied by another exotic color triplet quark, called $X$ say, having charge $Q = -4/3$ and make up a weak isospin triplet. [This is a departure from the standard model!] Specifically the weak isospin generators are:

$$T_+ = \zeta^0\partial_4 - \bar{\zeta}_4\bar{\partial}^0, \quad T_- = \zeta^4\partial_0 - \bar{\zeta}_0\bar{\partial}^4;$$

$$2T_3 = [T_+, T_-] = \zeta^0\partial_0 - \zeta^4\partial_4 + \bar{\zeta}_4\bar{\partial}^4 - \bar{\zeta}_0\bar{\partial}^0,$$

so we encounter doublets like $(N_1, L_1), (U_1, D_1) \sim (\zeta^0, \zeta^4), (\zeta^0, \zeta^4)\zeta^j\zeta^k$, weak isosinglets like $L_5, D_6 \sim \bar{\zeta}_1\bar{\zeta}_2\bar{\zeta}_3, \bar{\zeta}_i(\bar{\zeta}_0\zeta^0 + \bar{\zeta}_4\zeta^4)$, and triplets like $(U_3, D_3, X_3) \sim (-\bar{\zeta}_4\zeta^0, [\bar{\zeta}_0\zeta^0 - \bar{\zeta}_4\zeta^4]/\sqrt{2}, \bar{\zeta}_0\zeta^4)$.

## EXOTIC PARTICLES & GENERATIONS

The $U$-states arise from combinations like $\zeta^i\zeta^j\zeta^0$ lying in the $\overline{10}$-fold SU(5) combination $\zeta^\lambda\zeta^\mu\zeta^\nu$ and as $\bar{\zeta}_k\bar{\zeta}_4\zeta^0$. (Note that $\zeta^0\zeta^4\zeta^k$ has the exotic value $F = 5/3$ and cannot be identified with $U^c$.) The $D$-states occur similarly, as do the $N$'s and $L$'s. Observe that $L_{5,6} \sim \bar{\zeta}_3\bar{\zeta}_2\bar{\zeta}_1$ and $D_{5,6,7,8} \sim \bar{\zeta}_k$ surprisingly are weak isosinglets, which again differs from the standard model.

'Pentaquarks' such as $\Theta^+ \sim uudd\bar{s}$ & $\Xi^{--} \sim ddss\bar{u}$ were recently discovered(?) with quite narrow widths. Models have been advanced to describe these new resonances as well as tetraquark mesons, but who is to say unequivocally that they are not composites of ordinary quarks and the fourth $U$-quark or other $D$-family quarks which my scheme indicates, not to mention color sextet quarks? Further, I get a fourth neutrino, which could be essentially sterile and might help explain the mystery of $\nu$ masses. There is a great deal of richness in the magic square states which provide fertile ground for research.

## MASS MATRICES AND MIXING

More pressing is the mass matrix which affects quarks as well as leptons. If we assume it is due to a Higgs $\Phi$ field's expectation values and we assume that $\Phi$ is antiself-dual too, there are nine colorless possibilities having $F = Q = 0$:

- one $\phi_{(0)(0)} = \langle\phi\rangle = \mathscr{M}$
- one $\phi_{(0)(4)} = \langle\phi_{1234}\rangle = \mathscr{H}$ complex
- three $\phi_{(1)(1)} = \langle\phi_0^0, \phi_4^4, \phi_i^i\rangle = \mathscr{A}, \mathscr{B}, \mathscr{C}$
- four $\phi_{(2)(2)} = \langle\phi_{04}^{04}, \phi_{0k}^{0k}, \phi_{4k}^{4k}, \phi_{ij}^{ij}\rangle = \mathscr{D}, \mathscr{E}, \mathscr{F}, \mathscr{G}$,

others being related by duality. Thus the mass matrix calculation is a difficult task. Lepton and quark mixings arise via the Lagrangian $\int d^5\zeta d^5\bar{\zeta}\; \bar{\Psi}\langle\Phi\rangle\Psi$. A preliminary investigation [2] of this problem suggests that is it is unlikely to be solved by an anti-selfdual $\Phi$ so the problem is twice as difficult! One important point is that $U$ and $D$ mixing matrices are to be diagonalised via unitary transformations $V(U)$ & $V(D)$. However since the coupling of the weak bosons is *different* for isodoublets $U_{1,2}$ $D_{1,2}$ and isotriplets $U_{3,4}$ $D_{3,4}$ (and zero for the weak isosinglets $D_{5-8}$), we cannot just evaluate $V^{-1}(U)V(D)$ for the "unitary" CKM matrix now. Rather the weak interactions connecting $U$ & $D$ mass-diagonalised quarks will not be quite unitary (because of the different coupling factors):

$$L_{\text{weak}}/g_w = W^+(\bar{U}_1D_1 + \bar{U}_2D_2) + \sqrt{2}W^+(\bar{U}_3D_3 + \bar{U}_4D_4 + \bar{D}_3X_3 + \bar{D}_4X_4)$$
$$+W^-(\bar{D}_1U_1 + \bar{D}_2U_2) + \sqrt{2}W^-(\bar{D}_3U_3 + \bar{D}_4U_4 + \bar{X}_3D_3 + \bar{X}_4D_4) + W^3 \text{ terms}$$

This represents another departure from the standard picture: nonunitarity of the CKM matrix is thus a test of the scheme.

Also CP violation is an intrinsic feature of the formalism because $\mathscr{H}$ is naturally complex, unlike the other expectation values $\mathscr{A}, \mathscr{B}....\mathscr{M}$. There are two such $\mathscr{H}$ doublets when $\Phi$ has no particular duality.

## GENERALIZED RELATIVITY

Where are the gauge fields? Maybe one can mimic SUSY and supergauge the massless free action for $\Psi$, but without added complication of spin. But there is a more compelling way, which has the benefit of incorporating gravity! Construct a fermionic version of Kaluza-Klein (KK) theory, this time without needing to handle infinite modes which arise from compressing normal bosonic coordinates [5]. These points are relevant:

- One must introduce a fundamental length in the extended $X$, as property $\zeta$ has no dimensions; maybe this is the gravity scale $\kappa = \sqrt{8\pi G_N}$?
- Gravity (plus gauge field products) fall within the $x-x$ sector, gauge fields in $x-\zeta$ and the Higgs scalars must form a matrix in $\zeta-\zeta$,
- Gauge invariance is connected with no. of $\zeta$ so SU(5) or perhaps a subgroup,
- There is no place for a gravitino as spin is absent ($\zeta$ are Lorentz scalar),
- There are necessarily a small *finite* number of modes.

The real metric is $ds^2 = dx^m dx^n G_{nm} + dx^m d\zeta^\nu G_{\nu m} + dx^n d\bar{\zeta}^{\bar{\mu}} G_{\bar{\mu}n} + d\bar{\zeta}^{\bar{\mu}} d\zeta^\nu G_{\bar{\mu}\nu}$ ($\bar{\zeta}^{\bar{\mu}} \equiv \bar{\zeta}_\mu$) where the tangent space limit corresponds to Minkowskian $G_{ab} \to I_{ab} = \eta_{ab}, G_{\bar{\alpha}\beta} \to I_{\bar{\alpha}\beta} = \Lambda^2 \delta_{\alpha\beta}$, multiplied at least by $(\bar{\zeta}\zeta)^5$ to arrange correct property integration. Proceeding to curved space the components should contain the force fields, leading one to a 'superbein'

$$\bar{E}_M^A = \begin{pmatrix} e_m{}^a & i\Lambda(A_m)_\mu^\alpha \zeta^\mu \\ 0 & \Lambda\delta_\mu^\alpha \end{pmatrix},$$

and the metric "tensor" obtained as

$$ds^2 = dx^m dx^n g_{nm} + 2\Lambda^2 [d\bar{\zeta}^{\bar{\mu}} - i dx^m \bar{\zeta}^{\bar{\kappa}} (A_m)_{\bar{\kappa}}^{\bar{\mu}}] \delta_{\bar{\mu}\nu} [d\zeta^\nu + i dx^n (A_n)_\lambda^\nu \zeta^\lambda]; \; g_{mn} = e_m^a e_n^b \eta_{ab}.$$

Gauge symmetry corresponds to the special change $\zeta^\mu \to \zeta'^\mu = [\exp(i\Theta(x))]_\nu^\mu \zeta^\nu$ with $x' = x$. Given the standard transformation law

$$G_{\zeta m}(X) = \frac{\partial X'^R}{\partial x^m} \frac{\partial X'^S}{\partial \zeta} G'_{SR}(X')(-1)^{[R]} = \frac{\partial \zeta'}{\partial \zeta} G'_{\zeta m} - \frac{\partial \bar{\zeta}'}{\partial x^m} \frac{\partial \zeta'}{\partial \zeta} G'_{\zeta\bar{\zeta}},$$

this translates into the usual gauge variation (a matrix in property space),

$$A_m(x) = \exp(-i\Theta(x))[A'_m(x) - i\partial_m] \exp(i\Theta(x)).$$

The result is consistent with other components of the metric tensor but does not fix what (sub)group is to be gauged in property space although one most certainly expects to handle the nonabelian color group and the abelian electromagnetic group, so as to agree with physics. $\mathrm{SU}(2)_L$ emerges if we take the magic square to be left-handed and weak isospin to act only on $\zeta$ coordinates (not $\bar{\zeta}$).

## NOTATIONAL NICETIES

This has been detailed elsewhere [1] and only the main points will be repeated now. We have to be **exceedingly** careful with the order of quantities and of labels. For derivatives the rule is $dF(X) = dX^M(\partial F/\partial X^M) \equiv dX^M \partial_M F$, **not** with the $dX$ on the right, and for products of functions: $d(FG..) = dF\,G + F\,dG + \ldots$ Coordinate transformations read $dX'^M = dX^N(\partial X'^M/\partial X^N)$, *in that particular order.* Since $ds^2 = dX^N dX^M G_{MN}$. The symmetry property of the metric is $G_{MN} = (-1)^{[M][N]} G_{NM}$. Let $G^{LM} G_{MN} = \delta^L_N$ for the inverse metric, so $G^{MN} = (-1)^{[M]+[N]+[M][N]} G^{NM}$.

Changing coordinate system from $X$ to $X'$, we have to be very careful with signs and orders of products, things we normally ignore; the correct transformation law is

$$G_{NM}(X) = \left(\frac{\partial X'^R}{\partial X^M}\right)\left(\frac{\partial X^S}{\partial X^N}\right) G'_{SR}(X')\,(-1)^{[N]([R]+[M])}$$

Transformation laws for contravariant and covariant vectors read:

$$V'^M(X') = V^R(X)\left(\frac{\partial X'^M}{\partial X^R}\right) \quad \text{and} \quad A'_M(X') = \left(\frac{\partial X^R}{\partial X'^M}\right) A_R(X),$$

in the order stated. Thus the invariant contraction is $V'^M(X')A'_M(X') = V^R(X)A_R(X) = (-1)^{[R]}A_R(X)V^R(X)$. The inverse metric $G^{MN}$ can be used to raise and lower indices as well as forming invariants, so for instance $V_R \equiv V^S G_{SR}$ and $V'^R V'^S G'_{SR} = V^M V^N G_{NM}$.

The next issue is covariant differentiation; insist that $A_{M;N}$ should transform like $T_{MN}$,

$$\text{viz. } T'_{MN}(X') = (-1)^{([S]+[N])[R]}\left(\frac{\partial X^R}{\partial X'^M}\right)\left(\frac{\partial X^S}{\partial X'^N}\right) T_{RS}(X).$$

With some work we find, $A_{M;N} = (-1)^{[M][N]} A_{M,N} - A^L \Gamma_{\{MN,L\}}$, where the connection is given by

$$\Gamma_{\{MN,L\}} \equiv [(-1)^{([L]+[M])[N]} G_{LM,N} + (-1)^{[M][L]} G_{LN,M} - G_{MN,L}]/2 = (-1)^{[M][N]}\Gamma_{\{NM,L\}}.$$

Another useful formula is ${\Gamma_{MN}}^K \equiv (-1)^{[L]([M]+[N])}\Gamma_{\{MN,L\}} G^{LK} = (-1)^{[M][N]}{\Gamma_{NM}}^K$, whereupon $A_{M;N} = (-1)^{[M][N]} A_{M,N} - {\Gamma_{MN}}^L A_L$. Similarly one can show that for double index tensors the correct differentiation rule is

$$T_{LM;N} \equiv (-1)^{[N]([L]+[M])} T_{LM,N} - (-1)^{[M][N]}{\Gamma_{LN}}^K T_{KM} - (-1)^{[L]([M]+[N]+[K])}{\Gamma_{MN}}^K T_{LK}.$$

As a nice check the covariant derivative of the metric properly vanishes:

$$G_{LM;N} \equiv (-1)^{[N]([L]+[M])} G_{LM,N} - (-1)^{[L][M]}\Gamma_{\{LN,M\}} - \Gamma_{\{MN,L\}} \equiv 0.$$

Moving on, $A_{K;L;M} - (-1)^{[L][M]} A_{K;M;L} \equiv (-1)^{[K]([L]+[M])} {R^J}_{KLM} A_J$, defines the Riemann curvature tensor

$$\begin{aligned} {R^J}_{KLM} \;\equiv\;& (-1)^{[K][M]}({\Gamma_{KM}}^J)_{,L} - (-1)^{[L]([K]+[M])}({\Gamma_{KL}}^J)_{,M} \\ & + (-1)^{[M]([K]+[L])+[K][L]}{\Gamma_{KM}}^N{\Gamma_{NL}}^J - (-1)^{[K]([M]+[L])}{\Gamma_{KL}}^N{\Gamma_{NM}}^J. \end{aligned}$$

Evidently, ${R^J}_{KLM} = -(-1)^{[L][M]}{R^J}_{KML}$ and, less obviously, the cyclical relation takes the form

$$(-1)^{[K][L]}{R^J}_{KLM} + (-1)^{[L][M]}{R^J}_{LMK} + (-1)^{[M][K]}{R^J}_{MKL} = 0.$$

The fully covariant Riemann tensor is $R_{JKLM} \equiv (-1)^{([J]+[K])[L]} R^N{}_{KLM} G_{NJ}$ featuring:

$$
\begin{aligned}
R_{JKLM} &= -(-1)^{[L][M]} R_{JKML} = -(-1)^{[J][K]} R_{KJLM}, \\
0 &= (-1)^{[J][L]} R_{JKLM} + (-1)^{[J][M]} R_{JLMK} + (-1)^{[J][K]} R_{JMKL} \\
R_{JKLM} &= (-1)^{([J]+[K])([L]+[M])} R_{LMJK}.
\end{aligned}
$$

Finally the Ricci tensor and scalar curvature ($R \equiv G^{MK} R_{KM}$) arise from:

$$
R_{KM} \equiv (-1)^{[J]+[K][L]+[J]([K]+[M])} G^{LJ} R_{JKLM} = (-1)^{[L]([K]+[L]+[M])} R^L{}_{KLM} = (-1)^{[K][M]} R_{MK},
$$

## CURVATURES OF SPACE-TIME-PROPERTY

Write everything in terms of the real coordinates $\xi, \eta$ rather than complex $\zeta = (\xi + i\eta)/\sqrt{2}$ to avoid confusion. Consider just one extra pair and the following two examples.

(1) Decoupled property and space-time, but both curved:

$$
\begin{aligned}
ds^2 &= dx^m dx^n G_{nm}(x,\xi,\eta) + 2i d\xi d\eta G_{\eta\xi}(x,\xi,\eta) \\
&\equiv dx^m dx^n g_{nm}(x)(1 + if\xi\eta) + 2i\Lambda^2 d\xi d\eta (1 + ig\xi\eta)
\end{aligned}
$$

from which we can read off the metric components ($G^{LM} G_{MN} \equiv \delta^L_N$)

$$
\begin{pmatrix} G_{mn} & G_{m\xi} & G_{m\eta} \\ G_{\xi n} & G_{\xi\xi} & G_{\xi\eta} \\ G_{\eta n} & G_{\eta\xi} & G_{\eta\eta} \end{pmatrix} = \begin{pmatrix} g_{mn}(1+if\xi\eta) & 0 & 0 \\ 0 & 0 & -i\Lambda^2(1+ig\xi\eta) \\ 0 & i\Lambda^2(1+ig\xi\eta) & 0 \end{pmatrix},
$$

$$
\begin{pmatrix} G^{lm} & G^{l\xi} & G^{l\eta} \\ G^{\xi m} & G^{\xi\xi} & G^{\xi\eta} \\ G^{\eta m} & G^{\eta\xi} & G^{\eta\eta} \end{pmatrix} = \begin{pmatrix} g^{lm}(1-if\xi\eta) & 0 & 0 \\ 0 & 0 & -i(1-ig\xi\eta)/\Lambda^2 \\ 0 & i(1-ig\xi\eta)/\Lambda^2 & 0 \end{pmatrix}.
$$

The non-zero connections in the property sector are

$$
\Gamma_{\xi\eta}{}^{\xi} = -\Gamma_{\eta\xi}{}^{\xi} = ig\xi, \quad \Gamma_{\xi\eta}{}^{\eta} = -\Gamma_{\eta\xi}{}^{\eta} = ig\eta, \text{ so}
$$

$$
\begin{aligned}
R^{\eta}{}_{\xi\eta\eta} &= -2ig(1+ig\xi\eta) = -R^{\xi}{}_{\eta\xi\xi} \\
R^{\xi}{}_{\xi\xi\eta} &= -ig(1+ig\xi\eta) = -R^{\eta}{}_{\eta\eta\xi} \\
\text{so } R_{\xi\eta} &= -R_{\eta\xi} = 3ig(1+ig\xi\eta).
\end{aligned}
$$

Consequently the total curvature is given by

$$
R = G^{mn} R_{nm} + 2G^{\eta\xi} R_{\xi\eta} = R^{(g)}(1 - if\xi\eta) - 6g/\Lambda^2.
$$

Since $\sqrt{-G_{..}} = -i\Lambda^2\sqrt{-g_{..}}(1+2if\xi\eta)(1+ig\xi\eta)$, we obtain an action

$$
\equiv \frac{1}{2\Lambda^4} \int R\sqrt{G_{..}}\, d^4x\, d\eta\, d\xi = \frac{1}{2\kappa^2} \int d^4x \sqrt{-g_{..}} \left[ R^{(g)} + \lambda \right] \tag{2}
$$

where $\kappa^2 \equiv 8\pi G_N = \Lambda^2/(f+g)$, $R^{(g)}$ is the standard gravitational curvature and $\lambda = -6g(2f+g)/\Lambda^2(f+g)$ corresponds to a cosmological term.

(2) Our second example leaves property space flat (in the $\eta,\xi$ sector) but introduces a U(1) gauge field $A$, governed by the metric

$$\begin{pmatrix} G_{mn} & G_{m\xi} & G_{m\eta} \\ G_{\xi n} & G_{\xi\xi} & G_{\xi\eta} \\ G_{\eta n} & G_{\eta\xi} & G_{\eta\eta} \end{pmatrix} = \begin{pmatrix} g_{mn}(1+if\xi\eta)+2i\Lambda^2\xi A_m A_n \eta & i\Lambda^2 A_m\xi & i\Lambda^2 A_m\eta \\ i\Lambda^2 A_n\xi & 0 & -i\Lambda^2 \\ i\Lambda^2 A_n\eta & i\Lambda^2 & 0 \end{pmatrix}.$$

Simplify the analysis somewhat by going to flat (Minkowski) space first as there are then fewer connections. After some work and evaluation of Christoffel symbols, we obtain ($F_{mn} \equiv A_{m,n} - A_{n,m}$ and $R_{\xi\eta} = -\Lambda^4 F_{kl}F^{lk}\xi\eta$.)

$$R_{km} = R^l{}_{klm} - R^\xi{}_{k\xi m} - R^\eta{}_{k\eta m} = -i\Lambda^2(A_{k,l}+A_{l,k})F^l{}_m\xi\eta/2 + \text{total der.}$$

$$R_{k\xi} = R^l{}_{kl\xi} + R^\xi{}_{k\xi\xi} + R^\eta{}_{k\eta\xi} = i\Lambda^2[F^l{}_{k,l}\xi/2 + A^l F_{k,l}\eta] + \text{total der.}$$

$$R_{k\eta} = R^l{}_{kl\eta} + R^\xi{}_{k\xi\eta} + R^\eta{}_{k\eta\eta} = i\Lambda^2[F^l{}_{k,l}\eta/2 - A^l F_{k,l}\xi] + \text{total der.}$$

Then covariantize by including the gravitational component $g_{mn}(1+if\xi\eta)$ so

$$R = G^{mn}R_{nm} + 2G^{m\xi}R_{\xi m} + 2G^{m\eta}R_{\eta m} + 2G^{\eta\xi}R_{\xi\eta} \to R^{(g)} - 3i\Lambda^2 g^{km}g^{ln}F_{kl}F_{nm}\xi\eta/2$$

Finally rescale $A$ to identify the answer as electromagnetism + gravitation:

$$\int R\sqrt{-G..}d^4x d\eta d\xi/4\Lambda^4 = \int d^4x\sqrt{-g..}\left[R^{(g)}/2\kappa^2 - F^{kl}F_{kl}/4\right], \qquad (3)$$

where $\kappa^2 = \Lambda^2/f = 8\pi G_N$. It is a nice feature of the formalism that the gauge field Lagrangian arises from space-property terms — like the standard KK model (from the tie-up between ordinary space-time and the fifth dimension).

## ACKNOWLEDGMENTS

I would like to thank Professor Bashir for inviting me to Mexico and for his hospitality.

## REFERENCES

1. R. Delbourgo, *J. Phys.* **A39**, 5175-5185 (2000).
2. R. Delbourgo, *J. Phys.* **A39**, 14735-14744 (2006).
3. A. McKane, *Phys. Lett.* **A76**, 22 (1980); P. Cvitanovic and A.D. Kennedy, *Phys. Scripta* **26**, 5 (1982); I.G. Halliday and R.M. Ricotta, *Phys. Lett.* **B193**, 241 (1987); G.V. Dunne, *J. Phys.* **A22**, 1719 (1989).
4. P.D. Jarvis and M. White, *Phys. Rev.* **D43**, 4121 (1991); R. Delbourgo, P.D. Jarvis, R.C. Warner, *Aust. J. Phys.* **44**, 135 (1991).
5. R. Delbourgo, S. Twisk and R. Zhang, *Mod. Phys. Lett.* **A3**, 1073 (1988); P. Ellicott and D.J. Toms, *Class. Quant. Grav.* **6**, 1033 (1989);

# Quantum gravity and cosmological observations

Martin Bojowald

*Institute for Gravitational Physics and Geometry, The Pennsylvania State University, 104 Davey Lab, University Park, PA 16801, USA*

**Abstract.** Quantum gravity places entirely new challenges on the formulation of a consistent theory as well as on an extraction of potentially observable effects. Quantum corrections due to the gravitational field are commonly expected to be tiny because of the smallness of the Planck length. However, a consistent formulation now shows that key features of quantum gravity imply magnification effects on correction terms which are especially important in cosmology with its long stretches of evolution. After a review of the salient features of recent canonical quantizations of gravity and their implications for the quantum structure of space-time a new example for potentially observable effects is given.

**Keywords:** Quantum gravity, early universe, curvature perturbation
**PACS:** 98.80.Cq, 98.80.Qc, 04.60.Pp

## GRAVITY

The gravitational field is the only known fundamental force not yet quantized despite of more than six decades of research. Difficulties arise due to two key properties: Although gravity is the dominant player on cosmic scales, it is weak in usual regimes of particle physics. Strong quantum gravity effects, possibly accessible to observations, thus require large gravitational fields which are realized only in exotic situations such as the very early universe or black holes. But then, the classical field grows without bound, implying space-time singularities. Secondly, gravity is conceptually very different from other interactions due to the equivalence principle: gravity is a manifestation of space-time geometry. The full space-time metric $g_{\mu\nu}$ is the physical object to be quantized, not perturbations $h_{\mu\nu}$ on a background space-time such as Minkowski space $\eta_{\mu\nu}$.

Thus, any quantization of gravity able to describe these phenomena faithfully must be non-perturbative and background independent for most applications. A new framework is required which does not refer to causality or vacuum states and other concepts which are available only once a metric has already been specified. One has to make use of the quantum structure of space and time itself.

Although mathematically involved, this is now available in broad form due to research in the past 15 years. It is not yet uniquely formulated, but several characteristic properties have been revealed. All this is essential for non-singular quantum space-times, making sense even at the big bang, but also for quantum corrections in strong field regimes as they might be observable as remnants of the very early universe. The stage is thus provided for the first potentially observable effects of quantum gravity.

CP917, *Particles and Fields*, edited by H. Castilla Valdez, J. C. D'Olivo, and M. A. Perez

## POSSIBLE EFFECTS?

There is a dimensional argument which usually is taken as proof that quantum gravity effects are tiny, too tiny to be observable anytime soon. Given Planck's constant $\hbar$ and Newton's constant $G$, one can (in units where the speed of light is $c = 1$) define the Planck length $\ell_P = \sqrt{G\hbar} \approx 10^{-33}$cm. Its value is tiny compared to any scales we can probe directly, or equivalently the Planck mass $M_P = \sqrt{\hbar/G} \approx 10^{19}$GeV is huge compared to the mass of any elementary object. Only negligible correction terms are then expected from dimensionless combinations of available length scales, e.g. of the order $\ell_P H \approx 10^{-60}$ in cosmology with the current Hubble length $H^{-1} = a/\dot{a}$. Indeed, correction terms in low energy effective actions, obtained by perturbative approaches on a background space-time, give only such negligible terms in equations of motion [1].

However, for quantum gravity the low energy effective action is too special unless one is interested only in scattering of gravitons. A low energy effective action is obtained by an expansion around the vacuum state of quantum field theory on a background. The concept of a vacuum itself changes in background independent approaches since a vacuum state, defined e.g. as the unique Poincaré invariant state of a quantum representation, refers to symmetries of a background space-time. (Or, as the ground state of the Hamiltonian a vacuum is uniquely defined only for time-independence, which requires a symmetry.) Moreover, the gravitational Hamiltonian as it arises from the action is always unbounded from below and thus lacks a ground state in the usual sense.

There is an additional expectation from quantum gravity, namely that space has a discrete structure on very small scales. One can think of this structure as an irregular lattice whose typical plaquette size $p$ is close to $\ell_P^2$. But unlike the Planck length, this is a geometrical parameter or field specifying the quantum gravity state and can thus be dynamical. This parameter brings in crucial information from quantum gravity, unlike $\ell_P$ which is determined simply by parameters of quantum mechanics and classical gravity.

In such a situation, there are three length parameters: the macroscopic scale $H^{-1}$, the fundamental scale $p$ and the dimensional Planck length $\ell_P$. Any dimensional argument must then fail since dimensionless combinations of three length parameters can have any value depending on which geometric mean $l_1^{x_1} l_2^{x_2} l_3^{x_3}$ with $x_1 + x_2 + x_3 = 0$ is relevant in a given situation. In other words, a large dimensionless parameter exists such as the number $N$ of lattice sites which may enter and magnify correction terms. The precise form of corrections can then only be determined by detailed calculations taking into account the discrete structure of space. (See also [2] for a critique of dimensional arguments in quantum gravity.)

## SPACE-TIME STRUCTURE

A quantum theory for space-time structure is required, which is not unlike atomic pictures in condensed matter physics. From the two basic properties of general relativity we obtain two important lessons for the formulation of quantum gravity:

- Do not use a split $g_{\mu\nu} = \eta_{\mu\nu} + h_{\mu\nu}$ of the space-time metric as a background $\eta_{\mu\nu}$ with a perturbation $h_{\mu\nu}$, as this does not allow one to describe a lattice structure. One rather has, at a fundamental level, metrics which are distributional and supported only on lattice links and vertices.

- Do not use low energy effective theory but suitable generalizations. Low energy theory does not apply due to the unboundedness of the Hamiltonian, and it would even be unclear what "low energy" refers to given that there are no observer independent concepts of energy in general relativity.

All these techniques are now available in the framework of loop quantum gravity [3, 4, 5]. In particular, cosmological effects are calculable with recent progress. The six main ingredients in this general scheme are described in what follows.

## 1. New variables

A canonical formulation of general relativity was originally developed in ADM variables [6] given by the spatial metric $q_{ab}$ and momenta related to extrinsic curvature $K_{ab}$ (or $\dot{q}_{ab}$). This refers to a chosen time coordinate, but the theory is independent of that choice if certain constraints are satisfied. These constraints are equivalent to Einstein's equation and implement general covariance. However, when trying to quantize a theory whose dynamical variable is the metric, it is difficult to turn tensor components such as $q_{ab}$ itself into operators. In generally covariant systems one has to take into account arbitrary, non-linear changes of coordinates $x'(x)$. A tensor such as $q_{ab}$ then transforms as $q_{a'b'} = (\partial x^a/\partial x'^{a'})(\partial x^b/\partial x'^{b'})q_{ab}$ which leads to coordinate dependent factors. A quantization of gravity needs to represent the field $q_{ab}$ and its momenta as operators on a Hilbert space, but the definition must be independent of spatial coordinates which are not defined on the Hilbert space. This is the key difference to quantum field theories defined on, say, Minkowski space. This background space-time allows only Poincaré transformations $x'(x)$ as coordinate changes, which are linear. The coordinate change of tensors is then a simple linear transformation by spatially constant matrices, which can easily be extended to operators.

This problem, as it turned out, can be circumvented by using a new set of variables [7, 8]: $q_{ab}$ is replaced by a co-triad $e_a^i$ (three co-vector fields such that $e_a^i e_b^i = q_{ab}$), which then defines the densitized triad $E_i^a = |\det e_b^j| e_i^a$. Similarly, one defines $K_a^i := e_i^b K_{ab}$ and then the connection $A_a^i = \Gamma_a^i - K_a^i$ with $\Gamma_a^i = -\varepsilon^{ijk} e_j^b(\partial_{[a} e_{b]}^k + \frac{1}{2} e_k^c e_a^l \partial_{[c} e_{b]}^l)$.

## 2. Basic objects

These are variables as in non-Abelian gauge theories, the Ashtekar–Barbero connection $A_a^i$ canonically conjugate to "electric fields" $E_i^a$: $\{A_a^i(x), E_j^b(y)\} = 8\pi G \delta_a^b \delta_j^i \delta(x,y)$. The gauge group here is the group of triad rotations $E_i^a \mapsto R_i^j E_j^a$, $R \in SO(3)$, which leave the metric unchanged. Moreover, for any curve $e$ and surface $S$ in space we define holonomies and fluxes [9]

$$h_e(A) = \mathscr{P}\exp \int_e A_a^i \tau_i \dot{e}^a \mathrm{d}t \quad , \quad F_S(E) = \int_S \mathrm{d}^2 y n_a E_i^a \tau_i \tag{1}$$

with the tangent vector $\dot{e}^a$ of $e$, the co-normal $n_a$ of $S$ and Pauli matrices $\tau_i$. These quantities have many advantages. (i) They do not have any indices and are thus scalar quantities. No complicated tensor transformations under non-linear coordinate transformations arise. (ii) A smearing is automatically included, resulting in well-defined Poisson brackets to become commutators, free of any delta functions. (iii) No reference is made to any metric other than that determined by $E^a_i$, and background independence of the classical theory is thereby respected.

## 3. Representation

One can thus construct a quantum representation of these smeared basic fields and then impose the necessary constraints as operators. A convenient way to do this is the connection representation, where holonomies are used as multiplication operators "creating" spin networks [10] $T_{g,j,C}(A) = \prod_{v\in g} C_v \cdot \prod_{e\in g} \rho_{j_e}(h_e(A))$ as an orthonormal basis of $h_e$-dependent functions, where $g$ is an oriented graph with labels $j$ as SU(2) representations $\rho_j$ on edges, and $C$ as gauge invariant contraction matrices in vertices.

## 4. Discrete geometry

Fluxes are conjugate to holonomies and thus become derivative operators

$$\hat{F}^i_S f_g = -8\pi i G\hbar \int_S \mathrm{d}^2 y n_a \frac{\delta}{\delta A^i_a(y)} f_g(h(A)) = -8\pi i \ell_{\mathrm{P}}^2 \sum_{e\in g} \int_S \mathrm{d}^2 y n_a \frac{\delta(h_e)^A_B}{\delta A^i_a(y)} \frac{\partial f_g(h)}{\partial (h_e)^A_B} \quad (2)$$

when acting on a state $f_g$ as a linear combination of spin networks. With $\int_S \mathrm{d}^2 y n_a \delta h_e/\delta A^i_a(y) = \frac{1}{2}\tau_i \int_S \mathrm{d}^2 y \int_e \mathrm{d}t n_a(y) \dot{e}^a \delta(e(t),y) h_e$ we have non-zero contributions only if $S$ intersects edges of $g$, and contributions are determined by su(2) derivatives $J_i = \mathrm{tr}(\tau_i h\partial/\partial h)$ acting on holonomies. As invariant derivatives on a compact group (identical to angular momentum operators), a discrete spectrum of fluxes and thus discrete spatial geometry results. This also extends to other geometrical operators such as the area operator, obtained as a quantization of $A(S) = \int_S \mathrm{d}^2 y \sqrt{E^a_i n_a E^b_i n_b}$, [11, 12]

$$\hat{A}(S) f_{g,j} = 4\pi \ell_{\mathrm{P}}^2 \sum_{p\in S\cap g} \sqrt{j_p(j_p+1)} f_{g,j}$$

(assuming no intersections in vertices of $g$). Similarly the volume operator has a discrete spectrum receiving contributions only from vertices in a region [11, 13].

The representation used so far is not only an explicitly known one, but is in fact, to a large degree, unique using the covariance under spatial diffeomorphisms [14, 15]. As it happens sometimes also in quantum field theories on a background, symmetry reduces the freedom in choosing a representation. The large symmetry group of diffeomorphisms present due to background independence selects a unique representation. This differs from usual Fock spaces for particle excitations, but Fock states can be reproduced as distributional states. There is thus a tight kinematical setting, although quantization ambiguities such as factor ordering will as usually arise for composite operators, in particular the constraints, specifying the dynamics.

## Example: Cubic lattice

Labels $j$ appearing on edges of graphs determine the geometry through flux values of the densitized triad. They are half-integer and thus imply a minimum non-zero flux given by $4\pi\ell_{\rm P}^2$. The states are all that is present in the quantum theory to determine geometry. The lattice thus *is space,* and no continuum limit is to be taken as one would do it in lattice gauge theories. From the labels we obtain lattice fluxes $p_{v,I} = F_{S_{v,I}} = 8\pi\ell_{\rm P}^2 j_{v,I}$ depending on the direction $I$ of an edge and its starting vertex $v$. This is the state dependent scale of geometry introduced before as the additional length scale provided by discrete quantum gravity. This scale can be inhomogeneous if labels differ much for different edges, and it is dynamical (as well as, possibly, the lattice structure such as the number of vertices). States, in this way, determine which physical scales are relevant. Recent developments from different directions within loop quantum gravity have now converged to such structures [16, 17, 18].

## 5. Gravitational dynamics

Dynamics of space-time is determined by the Hamiltonian constraint which, when solved, is supposed to show which special superpositions of lattices are allowed for generally covariant states. This constraint is implemented through lattice Hamiltonians which change the labels and possibly the graph when acting on a state. Thanks to spatial discreteness, those operators are well-defined even with matter contributions: there are no UV divergences [19, 20]. As even classical expressions are complicated non-polynomial functions of the basic fields, the operators are only barely tractable. Although they have been defined rigorously, they have quantization ambiguities (such as factor ordering and several other choices) and do not easily reveal interesting solutions. But they do display characteristic properties which follow from spatial discreteness and are common to all available constructions. They can be tested with suitable approximation schemes, which currently include symmetries [21, 22] or perturbations [18, 23].

More in detail, the classical expression of the constraint functional (to be imposed for all spatial functions $N(x)$) is

$$H[N] = \frac{1}{16\pi G}\int_\Sigma {\rm d}^3x N\left(\varepsilon_{ijk}F^i_{ab}\frac{E^a_jE^b_k}{\sqrt{|\det E|}} - 4(A^i_a-\Gamma^i_a)(A^j_b-\Gamma^j_b)\frac{E^{[a}_iE^{b]}_j}{\sqrt{|\det E|}}\right) \qquad (3)$$

which requires an inverse determinant of $E^a_i$. This is not available immediately due to the fact that fluxes, and also the volume operator, have discrete spectra containing zero and thus no inverse operators. But one can use the identity [19]

$$\left\{A_a^i, \int \sqrt{|\det E|} d^3x\right\} = 2\pi G \varepsilon^{ijk} \varepsilon_{abc} \frac{E_j^b E_k^c}{\sqrt{|\det E|}} \tag{4}$$

to obtain a well-defined quantization through a Poisson bracket, which then becomes a commutator of holonomies and volume. For the curvature components $F_{ab}^i$ we use $s_1^a s_2^b F_{ab}^i \tau_i = \Delta^{-1}(h_\alpha - 1) + O(\Delta)$ and write it in terms of a holonomy $h_\alpha$ around a square loop of coordinate size $\Delta$ and with tangent vectors $s_1^a$ and $s_2^a$ at $v$ [24]. Finally, an extrinsic curvature operator for $A_a^i - \Gamma_a^i$ in (3) can be derived as a double commutator.

## 6. Effective theory

In any quantum field theory, especially one with a highly complicated Hamiltonian, progress can be made only with suitable approximations to compute physical effects. One of the main tools in particle physics is the low energy effective action which allows powerful applications for instance in perturbation theory. The lattice Hamiltonians we have here are different from quantum field theory Hamiltonians on a background, and conceptually also from lattice gauge theory. For instance, they are unbounded from below already classically. No ground state is available to expand around as done in low energy effective actions. It is thus necessary to generalize effective theory which has been accomplished [25, 26]. Effective equations, in general terms, are obtained from an analysis of the coupled dynamics of $n$-point functions. (In quantum mechanics, these are spread and deformations of a wave packet back-reacting on the peak position.) Effective dynamics is given by the expectation value of the Hamiltonian in suitable semiclassical states, with a precise specification depending on the regime to be analyzed.

Applied to lattice Hamiltonians, we can already draw one important conclusion: local coefficients of the effective Hamiltonian appear as functions of $\ell_P^2/p_{I,v} \gg \ell_P^2 H^2$ (after, e.g., using (4) with local lattice building blocks). Quantum effects are thus much larger than they would be in low energy effective theory due to the discrete structure [27].

## Summary of Quantum effects

Typical properties of effective Hamiltonians as they are also known from effective actions are [26]: (i) Factors such as the quantized inverse metric determinant give rise to modified small-scale behavior of coefficients (possibly related to boundedness of classically diverging curvature expressions near singularities [28, 29]). (ii) Replacing local curvature and connection components by holonomies along extended loops implies non-locality or, when Taylor expanded in an effective Hamiltonian, higher order spatial derivatives. (iii) The coupling of $n$-point functions in general effective equations implies, as usually, new quantum degrees of freedom (related to higher time derivatives).

Properties (i) and (ii) are typical holonomy effects of the loop quantization which was forced upon us by background independence while (iii) is a genuine quantum effect.

Both (ii) and (iii) correspond to higher curvature terms, while (i) corrects geometrical factors purely from quantum geometry.

## APPLICATION: HOW BIG IS THE TYPICAL QUANTUM SCALE?

Lattice states as solutions to the Hamiltonian constraint are difficult to find even numerically. But orders of magnitude of corrections can be estimated based on two different roles played by the fundamental scale $p$. First, $p$ determines the number of lattice sites within a typical macroscopic scale which we take to be $H^{-1}$: for larger $p$, the lattice is coarser and discreteness corrections arise. *Continuum physics* requires $p \ll H^{-2}$. Secondly, the size of $p$ signals quantum effects since it is proportional to the quantum number of a state. For smaller $p$, a lower "excitation level" is realized and thus one has larger quantum effects. *Semiclassical physics* requires $p \gg \ell_{\rm P}^2$.

These are two opposite requirements, leaving an allowed range $\ell_{\rm P}^2 \ll p \ll H^{-2} = \frac{3}{8\pi G\rho}$ where we took the Hubble length as the macroscopic scale relevant for cosmology, and computed it in terms of energy density $\rho$ using the Friedmann equation. This range is wide in the late universe, where quantum corrections can be almost arbitrarily small. But it is more narrow in the early universe and during inflation where the typical energy scale implies $1 \gg \ell_{\rm P}^2/p \gg 10^{-6}$.

Direct observations of effects from $\ell_{\rm P}^2/p \approx 10^{-6}$ would not be observable soon, but magnifications can occur if they add up during long cosmic evolution. This is in fact realized [27] as it follows from cosmological perturbation equations derived from the effective Hamiltonian [30]: One of the metric modes, $\Phi$, satisfies $\ddot{\Phi} + (1 + \nu)\dot{\Phi}/\eta + \varepsilon\Phi/\eta^2 = 0$ on large scales as a differential equation in conformal time $\eta$ and with $\nu$ being related to the matter equation of state $w$. The quantum correction $\varepsilon$ changes the behavior crucially compared to the classical situation where $\varepsilon = 0$. Solutions for constant $\nu$ are $\Phi(\eta) = \eta^\lambda$ with $\lambda = -\frac{\nu}{2} \pm \frac{1}{2}\sqrt{\nu^2 - 4\varepsilon}$. For $\varepsilon = 0$, one mode is constant, one decays, which is known as conservation of curvature perturbations. Loop quantum gravity implies $\varepsilon < 0$, such that the constant mode becomes slightly growing and curvature perturbations are not exactly preserved.

Since $\varepsilon$ has a definite sign, which is a robust property in loop quantum gravity independently of quantization ambiguities [31, 32], small corrections indeed add up during evolution. During inflation, conformal time for the largest visible modes changes by a factor $e^{-60}$. Thus, a factor $e^{-60\varepsilon} \approx 1 + 10^2|\varepsilon|$ results, magnifying the correction to at least $10^{-4}$ for $\varepsilon$, as estimated above, of the size $10^{-6}$.

## SUMMARY

Effects from the basic scale of quantum gravity in phenomenological equations can thus be computed with recent advances, and they reveal sometimes surprising implications. Classical modes (or also gravitons) are not fundamental but collective excitations out of the microscopic discrete state. Basic excitations are rather the scale parameters $p_{v,I}$ which, when excited inhomogeneously, can give rise to the classical fields or, in some approximation, gravitons on large scales.

Systematic calculations are now possible, mainly due to advances in the understanding of effective theory, and can be applied to different regimes. While characteristic effects are robust under quantization ambiguities, ambiguity parameters might become relevant for possible observations with more precise data. Cosmological observations thus have a good chance of revealing quantum gravity effects in the near future and to guide further constructions to complete the theory.

## ACKNOWLEDGMENTS

Results reported here were obtained through work funded by NSF grant PHY0554771.

## REFERENCES


1. N. Kaloper, M. Kleban, A. Lawrence, and S. Shenker, *Phys. Rev. D* **66**, 123510 (2002), hep-th/0201158.
2. D. Meschini, gr-qc/0601097.
3. C. Rovelli, *Quantum Gravity*, Cambridge University Press, Cambridge, UK, 2004.
4. A. Ashtekar, and J. Lewandowski, *Class. Quantum Grav.* **21**, R53–R152 (2004), gr-qc/0404018.
5. T. Thiemann, gr-qc/0110034.
6. R. Arnowitt, S. Deser, and C. W. Misner, *The Dynamics of General Relativity*, Wiley, New York, 1962.
7. A. Ashtekar, *Phys. Rev. D* **36**, 1587–1602 (1987).
8. J. F. Barbero G., *Phys. Rev. D* **51**, 5507–5510 (1995), gr-qc/9410014.
9. C. Rovelli, and L. Smolin, *Nucl. Phys. B* **331**, 80–152 (1990).
10. C. Rovelli, and L. Smolin, *Phys. Rev. D* **52**, 5743–5759 (1995).
11. C. Rovelli, and L. Smolin, *Nucl. Phys. B* **442**, 593–619 (1995), erratum: *Nucl. Phys. B* **456**, 753 (1995), gr-qc/9411005.
12. A. Ashtekar, and J. Lewandowski, *Class. Quantum Grav.* **14**, A55–A82 (1997), gr-qc/9602046.
13. A. Ashtekar, and J. Lewandowski, *Adv. Theor. Math. Phys.* **1**, 388–429 (1997), gr-qc/9711031.
14. J. Lewandowski, A. Okołów, H. Sahlmann, and T. Thieman, *Commun. Math. Phys.* **267**, 703–733 (2006), gr-qc/0504147.
15. C. Fleischhack, math-ph/0407006 and *Phys. Rev. Lett.* **97**, 061302 (2006).
16. A. Ashtekar, T. Pawlowski, and P. Singh, *Phys. Rev. D* **74**, 084003 (2006), gr-qc/0607039.
17. K. Giesel, and T. Thiemann, gr-qc/0607099.
18. M. Bojowald, *Gen. Rel. Grav.* **38**, 1771–1795 (2006), gr-qc/0609034.
19. T. Thiemann, *Class. Quantum Grav.* **15**, 839–873 (1998), gr-qc/9606089.
20. T. Thiemann, *Class. Quantum Grav.* **15**, 1281–1314 (1998), gr-qc/9705019.
21. M. Bojowald, and H. A. Kastrup, *Class. Quantum Grav.* **17**, 3009–3043 (2000), hep-th/9907042.
22. M. Bojowald, *Living Rev. Relativity* **8**, 11 (2005), gr-qc/0601085.
23. M. Bojowald, H. Hernández, M. Kagan, and A. Skirzewski, gr-qc/0611112.
24. C. Rovelli, and L. Smolin, *Phys. Rev. Lett.* **72**, 446–449 (1994), gr-qc/9308002.
25. M. Bojowald, and A. Skirzewski, *Rev. Math. Phys.* **18**, 713–745 (2006), math-ph/0511043.
26. M. Bojowald, and A. Skirzewski, hep-th/0606232.
27. M. Bojowald, H. Hernández, M. Kagan, P. Singh, and A. Skirzewski, *Phys. Rev. Lett.* **98**, 031301 (2007), astro-ph/0611685.
28. M. Bojowald, *Phys. Rev. D* **64**, 084018 (2001), gr-qc/0105067.
29. M. Bojowald, *Phys. Rev. Lett.* **86**, 5227–5230 (2001), gr-qc/0102069.
30. M. Bojowald, H. Hernández, M. Kagan, P. Singh, and A. Skirzewski, *Phys. Rev. D* **74**, 123512 (2006), gr-qc/0609057.
31. M. Bojowald, *Class. Quantum Grav.* **19**, 5113–5130 (2002), gr-qc/0206053.
32. M. Bojowald, *Pramana* **63**, 765–776 (2004), gr-qc/0402053.

# Quantum-gravity phenomenology, Lorentz symmetry, and the SME

Ralf Lehnert

*Center for Theoretical Physics, Massachusetts Institute of Technology, Cambridge, MA 02139*

**Abstract.** Violations of spacetime symmetries have recently been identified as promising signatures for physics underlying the Standard Model. The present talk gives an overview over various topics in this field: The motivations for spacetime-symmetry research, including some mechanisms for Lorentz breaking, are reviewed. An effective field theory called the Standard-Model Extension (SME) for the description of the resulting low-energy effects is introduced, and some experimental tests of Lorentz and CPT invariance are discussed.

**Keywords:** Lorentz violation, quantum gravity, Planck-scale physics, Standard-Model Extension
**PACS:** 11.30.Cp, 11.30.Er, 12.60.-i

## INTRODUCTION

Perhaps the most challenging open question in present-day fundamental physics concerns a unified quantum theory of all interactions including gravity. To date, a tremendous amount of theoretical work has been devoted to various approaches addressing this question. However, experimental efforts in this line of research are hampered, primarily because of the expected Planck suppression of the corresponding effects at presently attainable energies. A possible avenue to circumvent this issue is to scan the predictions of a given candidate underlying theory for effects that could be present already at lower energy scales. For example, one can search for novel particles, such as those required by supersymmetry, or large extra dimension.

Another promising approach is to ask which type of effect *can* be measured with Planck precision, and then determine whether such effects are indeed allowed in theoretical approaches to a more fundamental theory. In this context, tests of invariance properties appear to be an excellent candidate: symmetries allow *exact* theoretical predictions, and they are typically amenable ultrahigh-precision experiments. From a quantum-gravity perspective, spacetime symmetries could be particularly promising: gravity governs the dynamics of space and time, so that a quantum theory of gravitation is likely to affect the structure of spacetime. For example, typical underlying models can involve more than four dimensions, operator-valued noncommuting coordinates, or a certain discreteness of space.

The present talk further explores the idea that spacetime symmetries—and in particular Lorentz invariance—may be violated by small amounts as a result of more fundamental physics. We begin by studying the interplay between various spacetime symmetries. Then, two sample mechanisms for Lorentz and CPT breakdown in Lorentz symmetric approaches to underlying physics are reviewed. The last section recounts the basic philosophy behind the construction of the Standard-Model Extension (SME), which

CP917, *Particles and Fields*, edited by H. Castilla Valdez, J. C. D'Olivo, and M. A. Perez

provides the modern effective-field-theory framework for describing Lorentz and CPT violation. This section also lists some tests of Lorentz and CPT symmetry.

## THE INTERPLAY BETWEEN VARIOUS SPACETIME-SYMMETRY VIOLATIONS

Spacetime transformations that, to the best of our knowledge, are associated with a symmetry of nature are closely intertwined in the Poincaré group. Consider the case in which symmetry is lost under one (or more) Poincaré transformations. The question arises as to whether the remaining transformations still determine an invariance, or whether the violation of one set of spacetime symmetries typically leads to the breakdown of other invariances contained in the Poincaré group. The purpose of the present section is to explore this question.

Let us first consider the case of a discrete transformation—the combination of charge conjugation (C), parity inversion (P), and time reversal (T). The famous CPT theorem [1] states roughly that, under a few mild assumptions, CPT invariance is a consequence of conventional quantum mechanics and Lorentz symmetry. Thus, when CPT invariance is broken, one or more ingredients of the CPT theorem can no longer hold true. The questions arises which one of these ingredients should be abandoned. Both the Lorentz and the CPT transformations involve spacetime, which suggests that CPT breaking implies Lorentz violation. This result has been proved in the framework of axiomatic field theory by Greenberg [2]. This "anti-CPT theorem" essentially asserts that in local, unitary, relativistic point-particle quantum field theories a breakdown of CPT symmetry always comes with Lorentz violation. Note, however, that the converse of this result, i.e., Lorentz violation implies CPT breakdown, is false in general. We see that as a corollary, CPT tests are at the same time Lorentz tests. We finally note that other types of CPT breaking would require further deviations from conventional physics, such as unconventional quantum mechanics [3].

We next look at a situation in which translational invariance is violated. (A mechanism for this effect is studied in a subsequent section of this talk.) In this case, the energy–momentum tensor $\theta^{\mu\nu}$, which generates translations, is usually no longer conserved. To see that this also affects Lorentz symmetry, we consider the angular-momentum tensor $J^{\mu\nu}$, given by

$$J^{\mu\nu} = \int d^3x \left(\theta^{0\mu}x^{\nu} - \theta^{0\nu}x^{\mu}\right). \qquad (1)$$

In the Poincaré-symmetric case, this tensor is the generator of Lorentz transformations. Note that energy–momentum tensor $\theta^{\mu\nu}$ appears in the definition of $J^{\mu\nu}$. Since $\theta^{\mu\nu}$ is taken as not conserved in the present case, $J^{\mu\nu}$ will usually depend on time. Then, the expression (1) no longer determines the conventional time-independent Lorentz-transformation generators. In fact, such generators will typically no longer exist, and then Lorentz invariance ceases to be assured. In other words, translation-symmetry breaking is in such cases associated with Lorentz violation.

## EXAMPLES OF MECHANISMS FOR SPACETIME-SYMMETRY VIOLATION

The preceding section has shown that the breakdown of certain subsets of spacetime symmetries can result in the violation of additional spacetime invariances. But the question remains how spacetime symmetries can be broken in a Poincaré-invariant underlying theory in the first place. Various mechanisms for symmetry violations have been devised in a number of approaches to fundamental physics, such as strings [4], spacetime foam [5], noncommutative geometry [6], nontrivial spacetime topology [7], and cosmologically varying scalars [8]. The present section gives a somewhat more detailed description of two of the above mechanisms—spontaneous Lorentz and CPT breaking in string theory and Lorentz and CPT violation through spacetime-dependent scalars.

**Spontaneous Lorentz and CPT violation.** Spontaneous breakdown of Lorentz and CPT symmetry can occur in the context of the field theory of the open bosonic string [4]. The mechanism of spontaneous symmetry violation is quite attractive from a theoretical viewpoint because the dynamics remains invariant under the symmetry in question. It is the ground-state solution of the system that fails to exhibit all invariances of the Hamiltonian, and is therefore said to break the corresponding symmetries. Instances of spontaneous symmetry violation can readily be identified in solid-state physics, in the physics of elastic media, and in elementary-particle theory.

In what follows, we will discuss three sample physical systems. The features of these examples will enable us to gain further intuition about spontaneous Lorentz and CPT breakdown in a step-by-step fashion. An illustration that supports these three examples is given in Fig. 1.

Our first system is classical electrodynamics. Within this context, the energy density $V(\vec{E},\vec{B})$ of any pattern of electric and magnetic fields $\vec{E}$ and $\vec{B}$, respectively, is determined by

$$V(\vec{E},\vec{B}) = \tfrac{1}{2}\left(\vec{E}^2 + \vec{B}^2\right), \tag{2}$$

where we have employed natural units. Given any solution of the classical Maxwell equations, Eq. (2) yields the energy stored in the electromagnetic fields. Any field configuration involving a nonzero field is associated with a strictly positive energy. Only when both $\vec{E}$ and $\vec{B}$ are zero everywhere, we have vanishing field energy. The lowest-energy configuration of a system is usually identified with its ground state or vacuum. It is thus apparent that in classical electrodynamics the vacuum contains no fields and the Maxwell vacuum is empty.

The second sample system we examine is a Higgs-type field $\phi$. Unlike the electromagnetic fields considered above, the Higgs field is a scalar. Scalar fields of this type are believed to occur in Nature; they are, in fact, part of the Standard Model of particle physics. As before, we consider the energy density of $\phi$. In the absence of spacetime dependence, i.e., $\partial_\mu \phi = 0$, the energy density $V(\phi)$ obeys

$$V(\phi) = (\phi^2 - \lambda^2)^2, \tag{3}$$

where $\lambda$ is a constant. A possible spacetime dependence of $\phi$ would add (positive definite) kinetic-type energy to this expression, which is uninteresting in the present context. We therefore see that the lowest possible energy state of $\phi$ possesses zero energy. As opposed to the electrodynamics case, this state is attained for *finite* field values $\phi = \pm\lambda$. It is thus apparent that in the presence of such Higgs-type fields the vacuum is filled with a condensate $\phi_{vac} \equiv \langle\phi\rangle = \pm\lambda$. In quantum theory, the condensate $\langle\phi\rangle$ is referred to as the vacuum expectation value (VEV) of $\phi$. One of the physical effects associated with the VEV of the Standard-Model Higgs is to cause mass terms for many elementary particles. It is important to note that $\langle\phi\rangle$ is a scalar, so it does *not* select a Lorentz-violating spacetime direction.

The third sample system we consider concerns a (hypothetical) vector field $\vec{C}$. Such a vector field clearly lacks observational evidence. However, such a field may be contained in candidate fundamental theories, such as strings. We focus here on the non-relativistic physics, which is sufficient to illustrate rotation violation; a relativistic generalization can easily be obtained. Paralleling the previous Higgs-field case, we take the potential-energy density for $\vec{C}$ to be given by

$$V(\vec{C}) = (\vec{C}^2 - \lambda^2)^2. \tag{4}$$

Spacetime dependence of $\vec{C}$ would contribute positive-definite kinetic-energy contributions. Equation (4) shows that $E = 0$ is the lowest energy for the system. As for the Higgs field, this lowest-energy state requires $\vec{C}$ to be nonvanishing: $\vec{C}_{vac} \equiv \langle\vec{C}\rangle = \vec{\lambda}$. Here, $\vec{\lambda}$ is any spacetime-constant vector that obeys $\vec{\lambda}^2 = \lambda^2$. As before, the vacuum fills with the VEV of the field $\vec{C}$ and is therefore not empty. Since $\langle\vec{C}\rangle = \vec{\lambda}$ is a constant vector, the vacuum contains an intrinsic direction, which violates rotational invariance and therefore also Lorentz symmetry.

**Cosmologically varying scalars.** As noted at the beginning of this section, a scalar varying on cosmological scales leads to Lorentz violation. This can be established with our result from the previous section that the breakdown of translational invariance (here via the spacetime dependence of the scalar) typically leads to the loss of Lorentz symmnetry. This effect is independent of the mechanism driving the variation of the scalar. The remainder of this section gives a more detailed discussion of this result with the goal of providing further intuition.

We begin by establishing the effect at the Lagrangian level. To this end, consider two scalar fields $\phi$ and $\Phi$ and a varying coupling $\xi(x)$. It is the spacetime dependence of $\xi(x)$ that will lead to Lorentz violation; $\phi$ and $\Phi$ are sample dynamical variables that could in principle be replaced by vector or spinor fields. Let the Lagrangian $\mathscr{L}$ of the system contain a kinetic-type term of the form $\xi(x)\,\partial^\mu\phi\,\partial_\mu\Phi$. A suitable integration by parts at the level of the action will produce a boundary term that can be dropped while leaving unaffected the equations of motion. The resulting Lagrangian $\mathscr{L}'$ will then contain a term of the form

$$\mathscr{L}' \supset -K^\mu\phi\,\partial_\mu\Phi. \tag{5}$$

Here, the vector $K^\mu \equiv \partial^\mu\xi$ is an external nondynamical 4-vector. Such a quantity breaks Lorentz symmetry because it selects a preferred direction in spacetime. If $\xi$ varies on

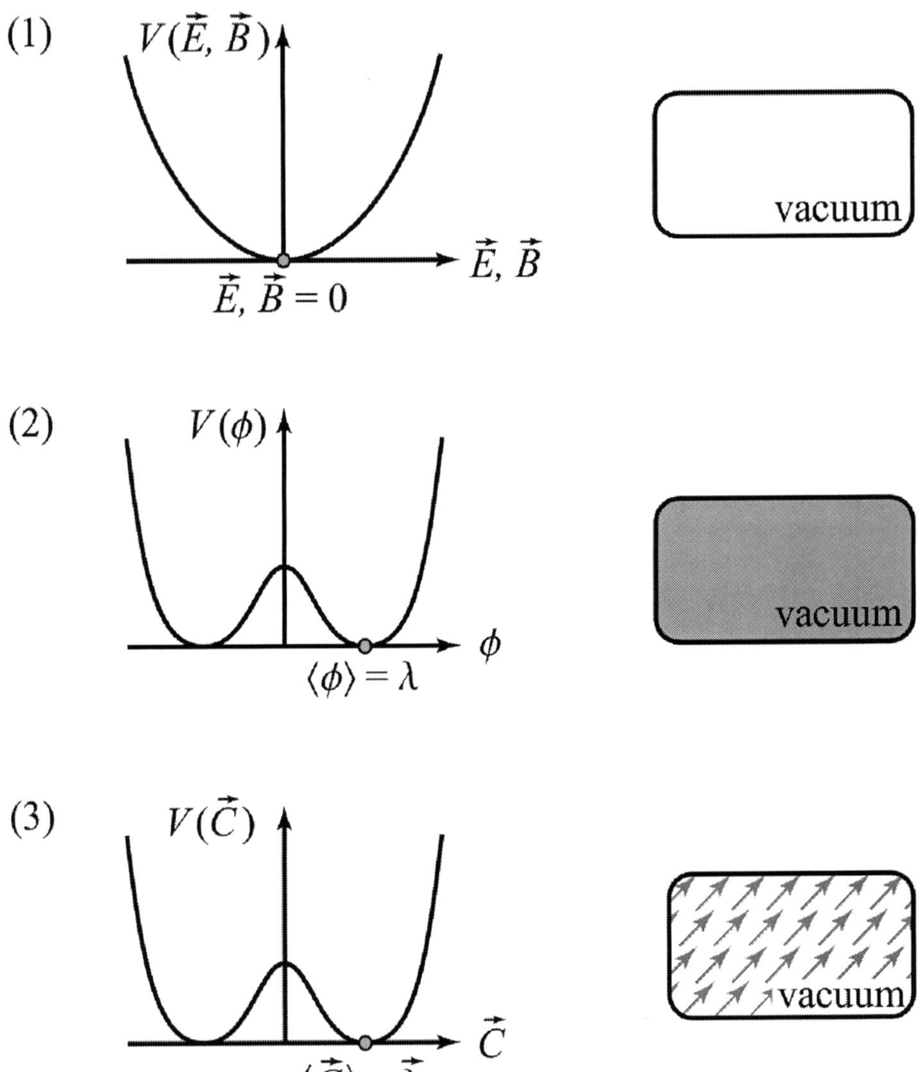


**FIGURE 1.** Spontaneous symmetry breaking. In Maxwell's classical electromagnetism (1), the state with the lowest energy is attained for $\vec{E} = 0$ and $\vec{B} = 0$. The classical Maxwell vacuum is therefore empty. The Higgs-type field (2) possesses interactions with an energy density $V(\phi)$ that triggers a non-zero value of $\phi$ in the ground state. The vacuum contains a scalar condensate, which is shown in gray. Lorentz and CPT symmetry are still intact. (However, other, internal symmetries may be broken.) Vector fields occurring, for instance, in string field theory (3) can have energy densities analogous to those of the Higgs field. Such interactions would lead to a nonvanishing field value in the lowest-energy configuration. The resulting VEV of such a vector field causes a special direction in the vacuum, which breaks Lorentz and possibly CPT invariance.

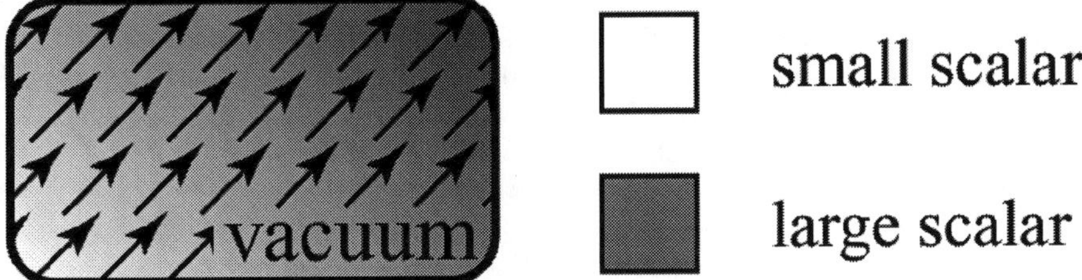


**FIGURE 2.** Lorentz breakdown via spacetime-dependent scalars. The shade of gray displays the size of the scalar. Lighter areas correspond to smaller values and darker areas to larger values. The black arrows symbolize the gradient of the scalar. This gradient selects a preferred direction in spacetime, so that Lorentz symmetry is violated.

cosmological scales, $K^\mu$ is essentially a constant with respect to local physics, such as solar-system physics.

To gain further understanding of the Lorentz breakdown resulting from a varying scalar consider the following intuitive picture. The variation of the scalar implies that there is some region in which its 4-gradient is nonzero. Such a 4-gradient is associated with a preferred direction in spacetime. This is illustrated in Fig. 2. For instance, consider a particle that possesses interactions with this background scalar. When the motion of such a particle is along the gradient of the scalar, its propagation features may be different from the situation in which this particle moves perpendicular to the gradient. Since physically inequivalent directions correspond to anisotropies, rotational invariance, and thus Lorentz symmetry, must be violated.

## THE STANDARD-MODEL EXTENSION

Once Lorentz and CPT violation is identified as a possible signature for physics underlying the Standard Model (and possibly arising at the Planck scale), it is desirable to have at ones disposal a test framework for the description of the resulting low-energy effects. Perhaps the most important use of such a test framework is the identification of suitable Lorentz tests. Another important advantage of a test model is provided by the fact that it can be used to compare different Lorentz and CPT tests if the model is broad enough. The consistency of the low-energy framework may also put some theoretical constraints on possible high-energy underlying models.

A test model for a certain high-energy underlying theory can usually be obtained by taking the low-energy limit of that theory. In the present case of Lorentz and CPT violation we do not follow this approach for two reasons. First, no complete and realistic fundamental theory is currently known. It is therefore desirable to have test framework for Lorentz and CPT symmetry that is relatively independent of the details of candidate underlying models. This permits a comprehensive search for Lorentz and CPT breakdown. Second, in some approaches to quantum gravity the physical vacuum state is currently unknown. For those models, the standard determination of a unique low-energy limit therefore fails. In light of these facts, we must rather proceed by constructing a test framework of sufficient generality to include the largest class of possible Lorentz and CPT violations consistent with certain more fundamental principles.

In what follows, we review the construction of the flat-spacetime limit of the Standard-Model Extension (SME), which is the modern framework for the description of the low-energy effects of Lorentz- and CPT breakdown [9].[1] The basic idea is to start with the Standard-Model Lagrangian $\mathscr{L}_{\mathrm{SM}}$ and add terms that violate Lorentz and possibly also CPT symmetry:

$$\mathscr{L}_{\mathrm{SME}} = \mathscr{L}_{\mathrm{SM}} + \delta\mathscr{L} \,. \tag{6}$$

Here, the SME Lagrangian is denoted by $\mathscr{L}_{\mathrm{SME}}$ and Lorentz- and CPT-breaking corrections are collected in $\delta\mathscr{L}$. As discussed above, we need to construct $\delta\mathscr{L}$ by hand. From the arguments in the previous sections we know that the symmetry-violating effects appear as background vectors or tensors in the vacuum. To be observable, they must couple to conventional fields. Because we insist on the fundamental principle of coordinate independence, this coupling must be a covariant contraction, so that $\delta\mathscr{L}$ transforms as a scalar under changes of the (inertial) coordinate system. For example,

$$\delta\mathscr{L} \supset b^{\mu}\,\overline{\psi}\gamma_5\gamma_{\mu}\psi \,. \tag{7}$$

Here, $b^{\mu}$ is a Lorentz- and CPT-violating background assumed to be generated in some underlying theory. It is a free coefficient, which can be searched for in suitable tests. Clearly, $b^{\mu}$ must be extremely small on observational grounds. The quantity $\overline{\psi}\gamma_5\gamma_{\mu}\psi$ denotes the usual chiral current of a Standard-Model fermion. Note that all Lorentz indices are properly contracted, so that $b^{\mu}\,\overline{\psi}\gamma_5\gamma_{\mu}\psi$ is a coordinate scalar.

Clearly, this construction yields infinitely many contributions to $\delta\mathscr{L}$, most of which would be expected to be subleading. For phenomenological purposes, it therefore seems practical and justified to consider only a subset of contributions to $\delta\mathscr{L}$ that satisfies certain additional requirements. For example, power-counting renormalizability, translational invariance, and the usual gauge symmetries are commonly imposed. This "minimal SME" has provided the basis for a number of investigations of Lorentz and CPT breakdown involving mesons [11], baryons [12, 13, 14], electrons [15, 16, 17], photons [18], muons [19], and the Higgs sector [20]. An analysis involving the gravity sector has recently also been performed [21]. We note that neutrino-oscillation experiments offer discovery potential [9, 22].

## ACKNOWLEDGMENTS

The author would like to thank the organizers for the invitation to this stimulating meeting and for financial support. This work was also supported by the U.S. Department of Energy under cooperative research agreement No. DE-FG02-05ER41360 and by the European Commission under Grant No. MOIF-CT-2005-008687.

---

[1] A similar methodology can also be applied in curved-spacetime situations involving gravity [21].

# REFERENCES


1. See, e.g., R.G. Sachs, *The Physics of Time Reversal*, University of Chicago Press, Chicago, 1987.
2. O.W. Greenberg, Phys. Rev. Lett. **89**, 231602 (2002).
3. J. Bernabeu, N.E. Mavromatos, and S. Sarkar, Phys. Rev. D **74**, 045014 (2006).
4. V.A. Kostelecký and S. Samuel, Phys. Rev. D **39**, 683 (1989); Phys. Rev. Lett. **63**, 224 (1989); **66**, 1811 (1991); V.A. Kostelecký and R. Potting, Nucl. Phys. B **359**, 545 (1991); Phys. Lett. B **381**, 89 (1996); Phys. Rev. D **63**, 046007 (2001); V.A. Kostelecký *et al.*, Phys. Rev. Lett. **84**, 4541 (2000).
5. G. Amelino-Camelia *et al.*, Nature (London) **393**, 763 (1998); D. Sudarsky *et al.*, Phys. Rev. D **68**, 024010 (2003); J. Alfaro *et al.*, Phys. Rev. D **70**, 084002 (2004); M. Bojowald *et al.*, Phys. Rev. D **71**, 084012 (2005).
6. I. Mocioiu *et al.*, Phys. Lett. B **489**, 390 (2000); S.M. Carroll *et al.*, Phys. Rev. Lett. **87**, 141601 (2001); A. Anisimov *et al.*, Phys. Rev. D **65**, 085032 (2002).
7. F.R. Klinkhamer, Nucl. Phys. B **578**, 277 (2000).
8. V.A. Kostelecký *et al.*, Phys. Rev. D **68**, 123511 (2003); O. Bertolami *et al.*, Phys. Rev. D **69**, 083513 (2004).
9. D. Colladay and V.A. Kostelecký, Phys. Rev. D **55**, 6760 (1997); **58**, 116002 (1998); V.A. Kostelecký and R. Lehnert, Phys. Rev. D **63**, 065008 (2001).
21. V.A. Kostelecký, Phys. Rev. D **69**, 105009 (2004); R. Bluhm and V.A. Kostelecký, Phys. Rev. D **71**, 065008 (2005).
11. KTeV Collaboration, H. Nguyen, hep-ex/0112046; FOCUS Collaboration, J.M. Link *et al.*, Phys. Lett. B **556**, 7 (2003); OPAL Collaboration, R. Ackerstaff *et al.*, Z. Phys. C **76**, 401 (1997); DELPHI Collaboration, M. Feindt *et al.*, preprint DELPHI 97-98 CONF 80 (1997); BELLE Collaboration, K. Abe *et al.*, Phys. Rev. Lett. **86**, 3228 (2001); BaBar Collaboration, B. Aubert *et al.*, Phys. Rev. Lett. **92**, 181801 (2004); BaBar Collaboration, B. Aubert *et al.*, hep-ex/0607103; V.A. Kostelecký and R. Potting, Phys. Rev. D **51**, 3923 (1995); D. Colladay and V.A. Kostelecký, Phys. Lett. B **344**, 259 (1995); Phys. Rev. D **52**, 6224 (1995); V.A. Kostelecký and R. Van Kooten, Phys. Rev. D **54**, 5585 (1996); O. Bertolami *et al.*, Phys. Lett. B **395**, 178 (1997); N. Isgur *et al.*, Phys. Lett. B **515**, 333 (2001); V.A. Kostelecký, Phys. Rev. Lett. **80**, 1818 (1998); Phys. Rev. D **61**, 016002 (2000).
12. D. Bear *et al.*, Phys. Rev. Lett. **85**, 5038 (2000); D.F. Phillips *et al.*, Phys. Rev. A **62**, 063405 (2000); Phys. Rev. D **63**, 111101 (2001); M.A. Humphrey *et al.*, Phys. Rev. A **68**, 063807 (2003); V.A. Kostelecký and C.D. Lane, Phys. Rev. D **60**, 116010 (1999); J. Math. Phys. **40**, 6245 (1999).
13. R. Bluhm *et al.*, Phys. Rev. Lett. **88**, 090801 (2002); R. Lehnert, Phys. Rev. D **68**, 085003 (2003).
14. F. Canè *et al.*, Phys. Rev. Lett. **93**, 230801 (2004); P. Wolf *et al.*, Phys. Rev. Lett. **96**, 060801 (2006).
15. H. Dehmelt *et al.*, Phys. Rev. Lett. **83**, 4694 (1999); R. Mittleman *et al.*, Phys. Rev. Lett. **83**, 2116 (1999); G. Gabrielse *et al.*, Phys. Rev. Lett. **82**, 3198 (1999); R. Bluhm *et al.*, Phys. Rev. Lett. **82**, 2254 (1999); Phys. Rev. Lett. **79**, 1432 (1997); Phys. Rev. D **57**, 3932 (1998); C.D. Lane, Phys. Rev. D **72**, 016005 (2005).
16. L.-S. Hou *et al.*, Phys. Rev. Lett. **90**, 201101 (2003); R. Bluhm and V.A. Kostelecký, Phys. Rev. Lett. **84**, 1381 (2000); B.R. Heckel *et al.*, Phys. Rev. Lett. **97**, 021603 (2006).
17. H. Müller *et al.*, Phys. Rev. D **68**, 116006 (2003); R. Lehnert, J. Math. Phys. **45**, 3399 (2004); B. Altschul, Phys. Rev. Lett. **96**, 201101 (2006); R. Lehnert, Phys. Rev. D **74**, 125001 (2006).
18. S.M. Carroll *et al.*, Phys. Rev. D **41**, 1231 (1990); V.A. Kostelecký and M. Mewes, Phys. Rev. Lett. **87**, 251304 (2001); Phys. Rev. D **66**, 056005 (2002); astro-ph/0702379; J. Lipa *et al.*, Phys. Rev. Lett. **90**, 060403 (2003); Q. Bailey and V.A. Kostelecký, Phys. Rev. D **70**, 076006 (2004); R. Lehnert and R. Potting, Phys. Rev. Lett. **93**, 110402 (2004); Phys. Rev. D **70**, 125010 (2004); B. Feng *et al.*, Phys. Rev. Lett. **96**, 221302 (2006); V.A. Kostelecký and M. Mewes, Phys. Rev. Lett. **97**, 140401 (2006); C.D. Carone, M. Sher, and M. Vanderhaeghen, Phys. Rev. D **74**, 077901 (2006); B. Altschul, Phys. Rev. Lett. **98**, 041603 (2007); A.J. Hariton and R. Lehnert, Phys. Lett. A (in press), hep-th/0612167.
19. V.W. Hughes *et al.*, Phys. Rev. Lett. **87**, 111804 (2001); R. Bluhm *et al.*, Phys. Rev. Lett. **84**, 1098 (2000); E.O. Iltan, JHEP **0306**, 016 (2003).
20. D.L. Anderson *et al.*, Phys. Rev. D **70**, 016001 (2004).
21. Q.G. Bailey and V.A. Kostelecký, Phys. Rev. D **74**, 045001 (2006).
22. V. Barger *et al.*, Phys. Rev. Lett. **85**, 5055 (2000); J.N. Bahcall *et al.*, Phys. Lett. B **534**, 114 (2002); V.A. Kostelecký and M. Mewes, Phys. Rev. D **69**, 016005 (2004); Phys. Rev. D **70**, 031902 (2004); Phys. Rev. D **70**, 076002 (2004); T. Katori *et al.*, Phys. Rev. D **74**, 105009 (2006).

# Symmetries and Renormalization of Noncommutative Field Theory

Richard J. Szabo

*Department of Mathematics and Maxwell Institute for Mathematical Sciences, Heriot-Watt University, Colin Maclaurin Building, Riccarton, Edinburgh EH14 4AS, U.K.*

**Abstract.** An overview of recent developments in the renormalization and in the implementation of spacetime symmetries of noncommutative field theory is presented, and argued to be intimately related.

**Keywords:** Noncommutative field theory, Renormalization, Duality, Twisted symmetries
**PACS:** 11.10.Gh, 11.10.Nx, 11.30.Ly

## INTRODUCTION

### Noncommutative spaces

Noncommutative spaces are typically defined in physics by promoting coordinates $x^i$ of spacetime to hermitean operators $\hat{x}^i$ satisfying commutation relations of the form $[\hat{x}^i,\hat{x}^j] = \mathrm{i}\,\theta^{ij}$, where $\theta^{ij}$ is an antisymmetric tensor which can be position dependent. Such spaces are believed to be important for the understanding of quantum gravity. Recent realizations of noncommutativity in such contexts have been described for $\kappa$-Minkowski space [1] and in doubly special relativity [2], among many others.

A much simpler system in which we can see noncommutative spaces arising is the Landau problem [3], which deals with the motion of charged particles confined to a plane and subjected to a perpendicularly applied uniform magnetic field $B$. The lagrangian is $\mathscr{L}_m = \frac{m}{2}\,\dot{\mathbf{x}}^2 - e\,\dot{\mathbf{x}}\cdot\mathbf{A}$, which in a symmetric gauge for the vector potential $\mathbf{A}$ reduces in the strong field limit $B \gg m$ to $\mathscr{L}_0 = -\frac{eB}{2}\,(\dot{x}\,y - \dot{y}\,x)$. This limiting lagrangian is of first order in time derivatives, implying that upon canonical quantization the coordinates describe a noncommutative space $[\hat{x},\hat{y}] = \frac{\mathrm{i}}{eB}$. This model will play a prominent role in much of our subsequent discussion.

### Noncommutative field theory

The huge interest and large amount of activity came with the seminal paper [4] wherein a very natural and precise physical realization of noncommutative spaces was illustrated, in complete analogy with the Landau problem above. It was shown that string theory, in the presence of D-branes and background "magnetic" fields, reduces in a particular low-energy limit to field theory on a noncommutative space [5, 6]. A "toy" model for this scenario, and the one this overview will focus on, is the Moyal

CP917, *Particles and Fields*, edited by H. Castilla Valdez, J. C. D'Olivo, and M. A. Perez

space for which the tensor $\theta^{ij}$ is constant and nondegenerate. The coordinate operators then satisfy Heisenberg commutation relations. In particular, one has the coordinate uncertainty relations $\Delta x^i \Delta x^j \geq \frac{1}{2}|\theta^{ij}|$. Fields on Moyal spaces are multiplied using the associative, noncommutative *star-product*, replacing the usual pointwise multiplication $\phi \cdot \psi$ with

$$\phi \star \psi := \phi \, \exp\left(\tfrac{\mathrm{i}}{2} \overleftarrow{\partial}_i \theta^{ij} \overrightarrow{\partial}_j\right) \psi \, . \tag{1}$$

Whatever our motivation may be, one of the most interesting and profound prospects of noncommutative field theory is that it has the possibility of providing diffeomorphism invariant field theories, and hence models that we might wish to call noncommutative gravity [7]. However, in order to provide viable models of quantum gravity, one is first faced with the technical task of determining whether or not they make sense. The two main issues are the precise implementation of spacetime symmetries at both the classical and quantum level, and the renormalizability of the quantum field theory. This overview will focus on the very interesting recent understanding of these two technical issues, and argue that the two problems are in fact intimately related to one another.

## Violations of Lorentz invariance

The constant tensor $\theta^{ij}$ gives a prefered directionality in space. In string theory, the resulting loss of Lorentz invariance is due to the expectation value of a background supergravity field. As a consequence, noncommutative field theory is *not* invariant under rotations or boosts of localized field configurations within a *fixed* observer inertial frame of reference, *i.e.* under particle Lorentz transformations. This observation has been exploited to construct Lorentz-violating extensions of the standard model in [8].

Some possible resolutions to this symmetry breaking are provided by:

1. Varying $\theta^{ij}$, so that one essentially integrates over all possible backgrounds to manifestly reinstate general covariance. This has been beautifully implemented in [9] within the framework of $C^*$-algebras, but it requires dealing with a host of different noncommutative spaces not of the Moyal type which makes extracting physical quantities rather difficult.
2. Dimensional reductions of noncommutative gauge theory in higher dimensions which induce teleparallel theories of gravity in lower dimensions [7, 10]. However, the formalism is somewhat limited in the types of gauge theories of gravity one can obtain in this way, and moreover the teleparallelism is embedded in a complex way.
3. Exploiting well-known quantum group techniques to reinterpret noncommutative field theory as a twist deformed quantum field theory [11, 12]. This realization has been a focus of intense activity in recent years and will be the route we take in the second part of this overview.

## UV/IR mixing

In momentum space, the replacement of the pointwise products of fields with the star-product (1) amounts to multiplying the usual convolution products $\tilde{\phi}(k)\,\tilde{\psi}(q)$ of Fourier transforms by a momentum dependent phase factor $\mathrm{e}^{\,\mathrm{i}k\times q}$, where $k\times q := \frac{1}{2}k_i\,\theta^{ij}\,q_j$. In a scalar field theory with polynomial interactions, a typical interaction vertex with $n$ incoming momenta $k_1,\dots,k_n$ will thus contribute

$$\exp\Big(\mathrm{i}\sum_{I<J} k_I\times k_J\Big)\,. \tag{2}$$

This is effective at energies $E$ with $E\sqrt{\theta}\ll 1$. *Planar diagrams* are those Feynman ribbon graphs which can be drawn without crossing any lines [13, 14], and their values coincide with those of the ordinary $\theta=0$ field theory up to possible phase factors which depend only on external momenta. *Non-planar diagrams*, on the other hand, contain crossings and additional phase factors depending on internal momenta such as (2), whose net result is that an ultraviolet cutoff $\Lambda$ on the graph induces an effective infrared cutoff $\Lambda_0 = 1/\theta\,\Lambda$ [15].

This entangling of momentum scales ruins Wilsonian renormalization, and two resolutions to this problem are to modify standard noncommutative field theory to either:

1. Duality covariant noncommutative field theory [16]–[18]. This is described in the next section.
2. Twisted noncommutative field theory [19]. This is explained in the second part.

# RENORMALIZATION

## Scalar field theory

Consider interacting charged scalar fields in four dimensions with the euclidean action

$$S=\int \mathrm{d}^4x\,\Big[\phi^\dagger\,(\partial_i^2+m^2)\phi+\tfrac{\lambda}{2}\,\phi^\dagger\star\phi\star\phi^\dagger\star\phi\Big] \tag{3}$$

where we have used $\int \mathrm{d}^4x\;\phi\star\psi=\int \mathrm{d}^4x\;\phi\;\psi$ for any pair of fields, so that the free field theory is unaffected by the noncommutative deformation. The quantum field theory is given by the Green's functions which are defined as the $2n$-point correlation functions $G_n(x_I,y_I)=\langle\phi^\dagger(x_1)\,\phi(y_1)\cdots\phi^\dagger(x_n)\,\phi(y_n)\rangle$. As an example, consider the contribution of the one-loop tadpole graph to the two-point function in the case of real scalar fields $\phi=\phi^\dagger$.[1] If the external legs carry momentum $p$, then the planar tadpole diagram gives the momentum space Feynman integral $(\lambda/6)\int \mathrm{d}^4k/(2\pi)^4\;(k^2+m^2)^{-1}$ which exhibits the standard ultraviolet divergence of scalar $\phi^4$-theory in four dimensions. On the other

[1] The case of complex scalar fields is much more subtle, as shown in [20].

hand, the non-planar tadpole diagram gives

$$\frac{\lambda}{12}\int\frac{\mathrm{d}^4k}{(2\pi)^4}\,\frac{\mathrm{e}^{2\mathrm{i}k\times p}}{k^2+m^2}=\frac{\lambda}{48\pi^2}\sqrt{\frac{m^2}{(\theta\,p)^2}}\,K_1\Big(\sqrt{m^2\,(\theta\,p)^2}\,\Big) \tag{4}$$

which diverges as $(\theta\,p)^{-2}$ as $p\to 0$. The model is thus *not* renormalizable, because the quantum field theory is not covariant under "UV/IR duality" [17] as we now explain.

## UV/IR duality

To resolve the problems associated with UV/IR mixing, we introduce a covariant version of noncommutative field theory in which the ultraviolet and infrared regimes are indistinguishable [16]. It is defined by modifying the action (3) to

$$S=\int\mathrm{d}^4x\,\sqrt{g}\,\Big[\phi^\dagger\,(g^{ij}D_iD_j+m^2)\phi+\tfrac{\lambda}{2}\,(\phi^\dagger\star\phi)^2\Big]\ , \tag{5}$$

where $g_{ij}$ is a constant metric tensor and $D_i=\partial_i-\frac{\mathrm{i}}{2}B_{ij}x^j$ are gauge covariant derivatives in a magnetic background characterized by another nondegenerate, constant antisymmetric tensor $B_{ij}$. The new derivatives obey the commutation relations $[D_i,D_j]=-2\mathrm{i}B_{ij}$, so that the modification of the kinetic term in (5) can be thought of as inducing a "noncommutative momentum space". The quantum field theory now has a duality under Fourier transformation of the fields, given by the covariant transformation rules

$$\begin{aligned} S[\phi]\big|_{\lambda,g,B,\theta} &= |\det B|\,S[\tilde\phi]\big|_{\tilde\lambda,\tilde g,\tilde B,\tilde\theta}\ ,\\ \tilde G_n(\theta^{-1}x_I,\theta^{-1}y_I)\big|_{\lambda,g,B,\theta} &= |\det B|^{n/2}\,G_n(x_I,y_I)\big|_{\tilde\lambda,\tilde g,\tilde B,\tilde\theta} \end{aligned} \tag{6}$$

where the dual parameters are defined by $\tilde\lambda=\lambda/\sqrt{\det\theta}$, $\tilde g=-B^{-1}gB^{-1}$, $\tilde B=B^{-1}$ and $\tilde\theta=\theta^{-1}$.

## Renormalization of the duality-covariant field theory

The key to the renormalization of the duality-covariant model is the use of a "matrix basis" for the expansion of fields given by [17, 21]

$$\phi(x)=\sum_{\ell,n\in\mathbb{N}_0^2}\phi^\dagger_{\ell n}\,\varphi_{\ell n}(x)\ , \tag{7}$$

where $\varphi_{\ell n}$ are the "Landau wavefunctions" which at $B=\theta^{-1}$ are the eigenfunctions of the kinetic operator in (5) with $-D_i^2\varphi_{\ell n}=\mathrm{Pf}(\theta)^{-1}\,(\ell_1+\ell_2-1)\,\varphi_{\ell n}$. They form a complete orthornormal system of $L^2$-fields which multiply together like matrix units $|\ell_1,\ell_2\rangle\langle n_1,n_2|$ under the star-product as $\varphi_{\ell n}\star\varphi_{\ell' n'}=\mathrm{Pf}(4\pi\,\theta)^{-1}\,\delta_{n\ell'}\,\varphi_{\ell n'}$. The noncommutative field theory (5) then becomes an infinite-dimensional complex matrix model

$$S=\mathrm{Tr}\Big(\phi^\dagger\mathscr{G}\phi+\tfrac{\lambda}{2}\,\mathrm{Pf}(4\pi\,\theta)\,(\phi^\dagger\,\phi)^2\Big)\ . \tag{8}$$

One of the beautiful calculations performed in [17] is that of the propagator $\mathscr{G}^{-1}$ as the formal inverse of the infinite matrix $\mathscr{G} = (\mathscr{G}_{\ell n})$, which uses hypergeometric Meixner $q$-polynomials. The natural ultraviolet cutoff is now on the matrix dimension as $\ell_1 + \ell_2, n_1 + n_2 \leq N$. One then slices the propagator in the renormalization group with sharp bounds on the matrix indices $\ell, n$. By using the Wilson-Polchinski renormalization group equations, one proves in this way that the duality-covariant field theory is renormalizable to all orders [17].

### Beyond perturbation theory

The beta-functions of the couplings $B$ and $\lambda$ in the duality-covariant field theory have been computed to all orders in [22, 23]. They are of the usual sign for any magnetic background $B \neq \theta^{-1}$. At the special point $B = \theta^{-1}$ a number of remarkable things happen:

- $\beta_B = \beta_\lambda = 0$, hence the renormalized coupling flows to a finite bare coupling and the field theory is asymptotically safe. This allows in principle a nonperturbative construction of the quantum field theory.
- The noncommutative quantum field theory is completely duality *invariant*.
- The matrix $\mathscr{G}$ in (8) is diagonal with entries the Landau energies, and the field theory becomes an exactly solvable large $N$ matrix model with a huge unitary symmetry $\phi \mapsto U \phi U^{-1}$, $U \in U(N)$ reflecting the degeneracy of Landau levels. In terms of fields this symmetry corresponds to the transformations $\phi(x) \mapsto (U \star \phi \star U^\dagger)(x)$ with $U \star U^\dagger = U^\dagger \star U = 1$, which generate the infinite unitary group $U(\infty)$ representing "deformed" canonical transformations of the spacetime [24].

Some of these considerations have been generalized to noncommutative $\phi^3$-theory by mapping it onto the Kontsevich matrix model [25], and to the noncommutative Gross-Neveu model [26]. However, there are difficulties associated with the *nonperturbative* renormalizability of the self-dual model [21], due to the overly large $U(\infty)$ symmetry which appears to kill all non-trivial dynamics,[2] and we must search for an alternative way to implement the symplectomorphism symmetry.

## TWISTED SYMMETRIES

### Twist deformations

Suppose that $X$ is an infinitesimal symmetry transformation of fields, denoted $\phi \mapsto X \triangleright \phi$. Then the action of $X$ on tensor products of fields is implemented by a *coproduct* $\Delta$ with $\phi \otimes \psi \mapsto \Delta(X) \triangleright (\phi \otimes \psi)$. The *primitive* coproduct $\Delta = \Delta_0$, with $\Delta_0(X) =$

---

[2] Of course in the real case there is no $U(\infty)$ symmetry which kills the dynamics. In [27] numerical evidence is provided for the renormalizability of a somewhat different class of noncommutative scalar field theories.

$X \otimes 1 + 1 \otimes X$, is covariant with respect to the pointwise product $m_0(\phi \otimes \psi) = \phi \cdot \psi$ in the sense that

$$X \triangleright m_0(\phi \otimes \psi) = m_0 \, \Delta_0(X) \triangleright (\phi \otimes \psi) \; . \tag{9}$$

For example, for the translation generator $X = P_i = -\mathrm{i}\,\partial_i$ the covariance (9) is just the usual Leibniz rule $P_i(\phi \cdot \psi) = (P_i\phi) \cdot \psi + \phi \cdot (P_i\psi)$. While $\Delta_0$ is *not* covariant with respect to the star-product, the *twist deformed* coproduct $\Delta = \Delta_\theta$ *is* [11, 12],[28]–[31]. It is defined by rewriting the star-product (1) as $\phi \star \psi = m_\theta(\phi \otimes \psi) := m_0(\mathscr{F}\phi \otimes \psi)$, where

$$\mathscr{F} = \exp\Big(-\tfrac{\mathrm{i}}{2}\,\theta^{ij}\,P_i \otimes P_j\Big) \tag{10}$$

is an abelian Drinfeld twist, and forming the similarity transformation

$$\Delta_\theta(X) = \mathscr{F}^{-1}\,\Delta_0(X)\,\mathscr{F} = \mathscr{F}^{-1}\,(X \otimes 1 + 1 \otimes X)\mathscr{F} \; . \tag{11}$$

## Twisted spacetime symmetries

Using (11) one can straightforwardly work out twisted Poincaré transformations generated by the usual linear and angular momentum operators $P_i = -\mathrm{i}\,\partial_i$ and $M_{ij} = -\mathrm{i}\,(x_i\,\partial_j - x_j\,\partial_i)$. One finds $\Delta_\theta(P_i) = \Delta_0(P_i)$, reflecting the fact that noncommutative field theory is translationally invariant, whereas

$$\Delta_\theta(M_{ij}) = \Delta_0(M_{ij}) + \tfrac{\mathrm{i}}{2}\,\theta^{kl}\,(\eta_{ik}\,P_j - \eta_{jk}\,P_i) \otimes P_l + \tfrac{\mathrm{i}}{2}\,\theta^{kl}\,P_k \otimes (\eta_{il}\,P_j - \eta_{jl}\,P_i) \; , \tag{12}$$

reflecting that it is not invariant under boosts or rotations. However, from (12) one finds that noncommutative field theory is invariant under *twisted* particle transformations, because

$$M_{kl} \triangleright [x^i, x^j]_\star = m_\theta\,\Delta_\theta(M_{kl}) \triangleright (x^i \otimes x^j - x^j \otimes x^i) = 0 \tag{13}$$

is equivalent to $M_{kl} \triangleright \theta^{ij} = 0$. This symmetry can be generalized to linear affine transformations $x \mapsto L\,x + a$, $\theta \mapsto L\,\theta\,L^\top$ using covariance of the Moyal star-product [32]. One can in fact generalize the symmetry to twisted diffeomorphisms, generated by arbitrary smooth vector fields $X = X^i(x)\,\partial_i$ [31]. One generically has $\Delta_\theta(X) \neq \Delta_0(X)$, but one can construct a "twisted" tensor calculus such that star-products of tensor fields transform as tensors. However, only unimodular $U(\infty)$ twisted transformations, with $\partial_i X^i = 0$, preserve action functionals of noncommutative field theories [33].

## Twisted noncommutative quantum field theory

Given a set of one-particle wavefunctions $\phi(x)$, two-particle wavefunctions are constructed as $(\phi \otimes \psi)(x_1, x_2)$. The flip map $\sigma_0(\phi \otimes \psi) = \psi \otimes \phi$ is only superselected when $\theta = 0$, as then $\sigma_0\,\Delta_0 = \Delta_0\,\sigma_0$. For $\theta \neq 0$, we use instead the "twisted" flip operator

$\sigma_\theta = \mathscr{F}^{-1}\,\sigma_0\,\mathscr{F} = \mathscr{F}^{-2}\,\sigma_0$ with $\sigma_\theta^2 = 1\otimes 1$ and $\sigma_\theta\,\Delta_\theta = \Delta_\theta\,\sigma_\theta$ ($\mathscr{F}^{-2}$ is the corresponding R-matrix). If $c(p)$ are the usual (bosonic or fermionic) oscillators of the undeformed quantum field theory, then this modifies them to the twisted oscillators

$$a(p) = c(p)\;\exp\left(\tfrac{\mathrm{i}}{2}\,\theta^{ij}\,p_i\,P_j\right) \tag{14}$$

with the commutation relations $a(p)\,a(q) = \pm\;\mathrm{e}^{2\mathrm{i}\,p\times q}\,a(q)\,a(p)$. In particular, on free quantum fields the creation parts obey

$$\phi^{(+)}(x_1)\,\phi^{(+)}(x_2) = \pm\;\exp\left(\mathrm{i}\,\theta^{ij}\,\frac{\partial}{\partial x_2^i}\,\frac{\partial}{\partial x_1^j}\right)\;\phi^{(+)}(x_2)\,\phi^{(+)}(x_1)\;. \tag{15}$$

This modifies the ordinary Feynman path integral measure $\prod_x \mathrm{d}\phi(x)$ and defines a *braided quantum field theory* with covariant Wick expansions [34]. The two main consequences of these arguments are:

1. For a spin 0 field with interaction hamiltonian density of the form $\mathscr{H}_{\mathrm{I}}(x) = \phi(x)\star\cdots\star\phi(x)$, the corresponding S-matrix is independent of $\theta$ and hence there is *no* UV/IR mixing in twisted quantum field theory [19, 30].
2. Even *free* twisted noncommutative field theory is *non-local*, because one has $\langle q|[\phi(x),\phi(y)]|p\rangle \neq 0$ for $q\neq p$ and space-like separations [35].

## Twisted Noether symmetries

Twisted diffeomorphisms do not appear to be *bonafide* physical symmetries, because they do not act solely on fields. They modify the usual Leibniz rule (represented by the primitive coproduct $\Delta_0$) through transformation of the star-product as

$$\delta_X(\phi\star\psi) := m_\theta\,\Delta_\theta(X)\triangleright(\phi\otimes\psi) = (\delta_X\phi)\star\psi + \phi\star(\delta_X\psi) + \phi(\delta_X\star)\psi\;. \tag{16}$$

The extra variation seems to present an obstacle to application of the *standard* Noether procedure and of the Ward identities in the quantum field theory. One resolution, proposed in [36] (see also [33]), is to use a proper covariant noncommutative differential calculus to perform the Noether analysis relating fields and conserved charges. However, the physical origins of these symmetries are unclear, and in particular string theory is unable to account for twisted diffeomorphisms [37]. The brane-induced low-energy dynamics of closed string theory in the presence of a $B$-field is *much* richer than any noncommutative theory of gravity based solely on star-products.

## ACKNOWLEDGMENTS

This work was supported in part by the EU-RTN Network Grant MRTN-CT-2004-005104. The author gratefully thanks David Vergara and the other organisors of the Silafae conference for the hospitality in a beautiful and stimulating atmosphere.

## REFERENCES


1. M. Dimitrijevic *et al.*, *Eur. Phys. J. C* **31**, 129–138 (2003).
2. J. Kowalski-Glikman and S. Nowak, *Phys. Lett. B* **539**, 126–132 (2002).
3. R. J. Szabo, *Int. J. Mod. Phys. A* **19**, 1837–1862 (2004).
4. N. Seiberg and E. Witten, *JHEP* **9909**, 032 (1999).
5. M. R. Douglas and N. A. Nekrasov, *Rev. Mod. Phys.* **73**, 977–1029 (2001).
6. R. J. Szabo, *Phys. Rept.* **378**, 207–299 (2003).
7. R. J. Szabo, *Class. Quant. Grav.* **23**, R199–R242 (2006).
8. S. M. Carroll *et al.*, *Phys. Rev. Lett.* **87**, 141601 (2001).
9. S. Doplicher, K. Fredenhagen and J. E. Roberts, *Commun. Math. Phys.* **172**, 187–220 (1995).
10. E. Langmann and R. J. Szabo, *Phys. Rev. D* **64**, 104019 (2001).
11. M. Chaichian, P. P. Kulish, K. Nishijima and A. Tureanu, *Phys. Lett. B* **604**, 98–102 (2004).
12. J. Wess, "Deformed Coordinate Spaces: Derivatives," in *Mathematical, Theoretical and Phenomenological Challenges Beyond the Standard Model*, Vrnjacka Banja, 2003, pp. 122–128.
13. T. Filk, *Phys. Lett. B* **376**, 53–58 (1996).
14. N. Ishibashi, S. Iso, H. Kawai and Y. Kitazawa, *Nucl. Phys. B* **573**, 573–593 (2000).
15. S. Minwalla, M. Van Raamsdonk and N. Seiberg, *JHEP* **0002**, 020 (2000).
16. E. Langmann and R. J. Szabo, *Phys. Lett. B* **533**, 168–177 (2002).
17. H. Grosse and R. Wulkenhaar, *Commun. Math. Phys.* **256**, 305–374 (2005).
18. V. Rivasseau, F. Vignes-Tourneret and R. Wulkenhaar, *Commun. Math. Phys.* **262**, 565–594 (2006).
19. A. P. Balachandran, A. Pinzul and B. A. Qureshi, *Phys. Lett. B* **634**, 434–436 (2006).
20. I. Y. Aref'eva, D. M. Belov and A. S. Koshelev, A note on UV/IR for noncommutative complex scalar field, Preprint arXiv:hep-th/0001215.
21. E. Langmann, R. J. Szabo and K. Zarembo, *JHEP* **0401**, 017 (2004).
22. H. Grosse and R. Wulkenhaar, *Eur. Phys. J. C* **35**, 277–282 (2004).
23. M. Disertori, R. Gurau, J. Magnen and V. Rivasseau, Vanishing of beta-function of noncommutative $\phi^4$ theory to all orders, Preprint arXiv:hep-th/0612251.
24. F. Lizzi, R. J. Szabo and A. Zampini, *JHEP* **0108**, 032 (2001).
25. H. Grosse and H. Steinacker, *JHEP* **0608**, 008 (2006).
26. F. Vignes-Tourneret, Renormalization of the orientable noncommutative Gross-Neveu model, Preprint arXiv:math-ph/0606069.
27. W. Bietenholz, F. Hofheinz and J. Nishimura, *JHEP* **0406**, 042 (2004).
28. V. G. Drinfeld, *Alg. Anal.* **1N6**, 114–148 (1989).
29. N. Reshetikhin, *Lett. Math. Phys.* **20**, 331–335 (1990).
30. R. Oeckl, *Nucl. Phys. B* **581**, 559–574 (2000).
31. P. Aschieri *et al.*, *Class. Quant. Grav.* **22**, 3511–3532 (2005).
32. J. M. Gracia-Bondia, F. Lizzi, F. R. Ruiz and P. Vitale, *Phys. Rev. D* **74**, 025014 (2006).
33. M. Riccardi and R. J. Szabo, Wilson loops and area-preserving diffeomorphisms in twisted noncommutative gauge theory, Preprint arXiv:hep-th/0701273.
34. R. Oeckl, *Commun. Math. Phys.* **217**, 451–473 (2001).
35. A. P. Balachandran *et al.*, Statistics and UV/IR mixing with twisted Poincaré invariance, Preprint arXiv:hep-th/0608179.
36. A. Agostini *et al.*, Generalizing the Noether theorem for Hopf algebra spacetime symmetries, Preprint arXiv:hep-th/0607221.
37. L. Alvarez-Gaume, F. Meyer and M. A. Vazquez-Mozo, *Nucl. Phys. B* **753**, 92–117 (2006).

# Towards a dual description of QCD

Martin Kruczenski

*Department of Physics, Purdue University,*
*525 Northwestern Avenue, W. Lafayette, IN 47907-2036.*
*E-mail: markru@purdue.edu*

**Abstract.** Recently, the AdS/CFT correspondence showed it was possible to understand certain strongly coupled gauge theories in terms of string theory. Applying those ideas to QCD is not straight-forward and can proceed in two ways: deforming known examples of AdS/CFT to include theories closer to QCD or understanding the mechanism for emergence of strings in the known examples and apply those methods to QCD. In this talk I describe some recent work done in these two approaches and in which I was involved. One is the computation of the meson spectrum for certain strongly coupled gauge theories and the other is a direct relation between operators and strings in a particular sector of $\mathcal{N} = 4$ SYM.

**Keywords:** String theory, QCD, AdS/CFT
**PACS:** 11.25.Tq, 11.15.-q, 11.15.Pg, 11.25.-w

## INTRODUCTION

String theory studies a quantum and relativistic theory whose fundamental entities are extended one dimensional objects called strings[1]. This generalizes quantum field theory which studies point particles and allows to describe an infinite number of particles of different masses in a unified way as different modes of oscillation of a string. The original motivation was as a phenomenological model to understand the large number of mesonic and baryonic resonances that were found in hadron collisions. A simple way to see why this model arose, is to consider a rotating string and compute the dependence of its energy $E$ with its angular momentum $J$. The result: $E \sim \sqrt{J}$ fits well with certain mesons as shown in fig.1 although we should emphasize that the agreement is not perfect and not all mesons fit so well in these straight lines called Regge trajectories. Although a somewhat promising model, when strings were thought as fundamental, two main problems arose. First the theory contained a tachyon, meaning that it was unstable. This problem was later solved [1] by including fermions and constructing the theory of superstrings. The second problem was that the theory was consistent only in 26 dimensions (or 10 dimensions for the superstrings). As a result of these difficulties and the proposal of the quark model, the idea of strings as the fundamental object underlying strong interactions was abandoned.

In the quark model, mesons are understood as bound states of quarks ($q$) and anti-quarks ($\bar{q}$). In spite of the phenomenological success of such model, the original goal

[1] An account of the early period of string theory that we describe here can be found for example in the classic book by M.Green, J. Schwarz and E. Witten [1] including references to the original works.

CP917, *Particles and Fields,* edited by H. Castilla Valdez, J. C. D'Olivo, and M. A. Perez

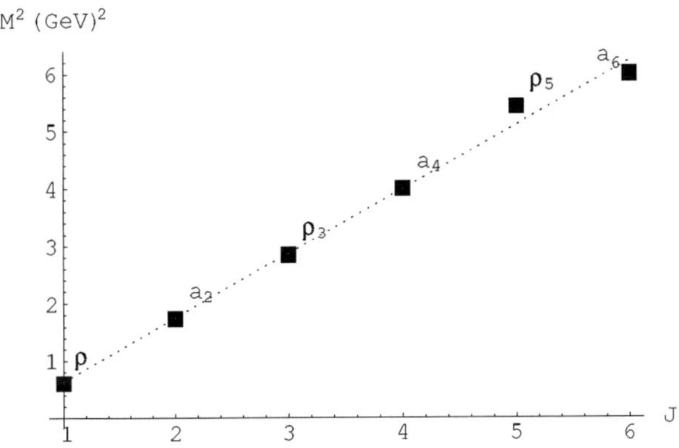


**FIGURE 1.** Example of Regge trajectory. Data taken from [2]. (The last two mesons $\rho_5$ and $a_6$ need confirmation according to the same source).

of computing the mesons masses was not fully achieved since solving the $q\bar{q}$ bound state problem turned out to be extremely difficult because, at the energies of interest, the theory is strongly coupled. For that reason the string model remained as a useful phenomenological model. The basic idea was to think that the color-electric flux lines between quark an anti-quark condensed into a tube or vortex which could be understood as a string. The model can be improved in various ways, for example by introducing explicitly quark masses but always thought as an effective description of the theory. In fact 't Hooft proposed a way to make this idea of an effective description of a gauge theory in terms of strings more precise by generalizing the color $SU(3)$ symmetry of QCD to $SU(N)$. In the limit $N \to \infty$, $g_{YM} \to 0$ (where $g_{YM}$ is the coupling constant of the gauge theory) keeping $g_{YM}^2 N = \lambda$ fixed, the gauge theory resembles a string theory. What happens is that in such limit the Feynman diagrams for a given process contribute differently according to their topology. At lowest order (order $N^0$) only planar diagrams contribute, namely those that can be drawn on a plane without the propagators crossing each other. Then those that can be drawn on the surface of a torus (order $N^{-2}$) then on a surface with two handles (order $N^{-4}$) etc. This is the same classification as the diagrams of string theory where the lowest order comes from diagrams where the world-sheet has the topology of a sphere then a torus, a surface with two handles etc. Although this looks like a promising way to derive a string model directly from large-N QCD, already the planar approximation implies summing up an infinite number of planar diagrams and is not well understood.

For that reason, string theory evolved independently of the strong interactions motivated instead by the fact that superstrings provided a consistent theory of quantum gravity. As mentioned, superstrings are consistent only in ten dimension but by compactifying six of them also semi-realistic models that unified all interactions can be constructed.

Recently the situation changed when Maldacena proposed the AdS/CFT correspondence [3] which relates a particular gauge theory, $\mathcal{N} = 4$ SYM with type IIB string theory which was one of the superstring theories derived as a possibly way to quantize

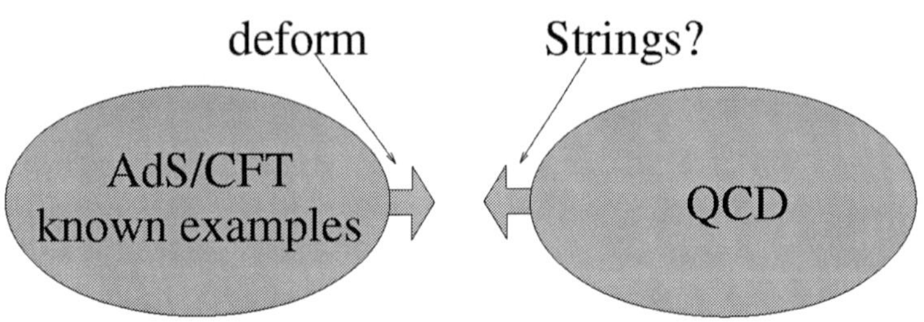


**FIGURE 2.** To apply AdS/CFT ideas to QCD we have to connect the known examples with QCD by deforming them or by deriving directly a string description of QCD.

gravity. As mentioned, it has to live in ten dimensions which in this case are not flat. The string theory lives in $AdS_5 \times S^5$ space which has the same symmetries as the gauge theory. The string coupling constant $g_s$ is given by $g_s = g_{YM}^2$ and is very small in the 't Hooft limit. The radius of the sphere $R$ and that of $AdS$ space (which is also $R$) is related to the 't Hooft coupling by $\frac{R}{l_s} = \lambda^{\frac{1}{4}}$. If $\lambda$ is large the strings can be treated classically except at very low energies where they can be quantized as in flat space. If $\lambda$ is small the strings behave quantum mechanically and at the moment is not known how to properly describe the strings in such regime. On the other hand for small $\lambda$ the field theory calculations provide a perturbative expansion in $\lambda$. In this sense the classical strings and the perturbative gauge theory are dual descriptions of the same object, valid in different regimes. A way to compare both is to match operators in the field theory with strings and compare the conformal dimension of the operators with the energy of the corresponding string. The AdS/CFT correspondence establishes that they should agree. Checking such agreement however is not easy since, as we said, the respective calculations are in principle valid for different values of $\lambda$. On the other hand we can also use this idea to compute field theory quantities at large $\lambda$, something beyond the reach of perturbation theory.

If we want to use these ideas in more realistic settings such as QCD two main paths can be followed. One is to deform the theory so that it becomes closer to QCD. Other is to understand more deeply how the string description emerges from the field theory and then use that to study QCD and see if a string description appears also in that case. We summarize this point in fig.2. In the rest of this talk we provide an example of each of these approaches. First we discuss a deformation by which we include quarks in the set up and compute the meson masses. Then we describe a precise map between strings and operators that appear in a particular sector of the correspondence.

## MESONS IN ADS/CFT

In the original version of the correspondence, all fields are in the adjoint representation and therefore the theory has no analog to quarks. In [4] Karch and Katz made explicit the way in which quarks should be incorporated. A D-brane is inserted in $AdS_5$ space. Strings extending from the D-brane all the way to the horizon are identified with quarks or anti-quarks depending on their orientation. A bound state of a quark and an anti-quark is simply a string starting and ending on the D-brane. If such a string rotates it describes a meson of high spin. Therefore we see that, in AdS/CFT, mesons are rotating strings but the string rotates in five dimensions! One consequence is that the spectrum is not necessarily of the Regge type. In fact in this case we study the problem [5] and found that the spectrum is of the Coulomb type. This is in agreement with the fact that, in this theory, the quark anti-quark potential decreases as $1/L$ [6] with the distance $L$ (as opposed to the linearly rising potential characteristic of confining theories). This results follow from studying classical strings. However when the angular momentum is small we have to take into account quantum effects, in particular the fact that the angular momentum is quantized. The cases where $J = 0, \frac{1}{2}, 1$ are special and those mesons turn out to be much lighter than the higher spin ones, although none of them is massless (for non-zero quark mass). In fact, such spectrum can be computed exactly in the large-N approximation [5]. The mass gap is found to be of order $M_0 = \frac{m_q}{\sqrt{\lambda}}$ where $m_q$ is the quark mass and, since we are working for $\lambda \gg 1$ the mass gap is small compared with the natural scale $m_q$.

In a following paper [7] we studied a confining case. In order to do that we found convenient to add $D6$ branes to Witten's confining background [8]. Other possibilities of doing the same are described in [9, 10]. Let us briefly summarize the main properties of the confining case [7]:

- The spectrum of low spin states can be computed, albeit numerically and not exactly as before.
- For zero quark mass ($m_q = 0$) there is a massless meson that we call $\Phi$ ($M_\phi = 0$ which can be understood as a Goldstone boson of chiral symmetry breaking. Notice that chiral symmetry is a symmetry of the quantum theory in the large N limit since the action of an instanton is divergent in that limit.
- A non vanishing chiral condensate $\langle \bar{q}q \rangle$ appears and its value can be computed.
- For $m_q \neq 0$ the meson $\Phi$ gets a mass equal to $M_\phi^2 = -\frac{m_q}{f_\Phi^2} \langle \bar{q}q \rangle$, which is the so called Gell-Mann-Oakes-Renner relation. Here $f_\Phi$ is the $\Phi$-meson decay constant.
- Large spin mesons can be described by rotating strings and reproduce the model of two masses rotating around each other while tied by a string [11].

We see from our example and also [9, 10] that even in the case of non-supersymmetric confining theories we are able to compute the mesons spectrum and reproduce many qualitative features of QCD. However the theory we described has many other particles that just gluons and quarks. The natural question is if those extra particles can be decoupled to get pure QCD. The answer is that they can be decoupled but in a regime where we lose control over the calculations that lead to the meson spectrum and therefore further work would be required to make contact with real QCD.

## SPIN CHAINS AND STRINGS

Now, based on [12, 13, 14, 15], we consider the other possible approach which is to understand better the emergence of strings from the gauge theory in the known example where AdS/CFT works. If that succeeds then we can try to apply the same procedures to QCD but, here, we are just going to show, in a very precise manner, how strings emerge in a particular sector of the $\mathcal{N} = 4$ theory. The $\mathcal{N} = 4$ SYM theory has a gluon $A_\mu$, four fermions $\psi^{A=1\ldots4}$ and six scalars $\phi^{a=1\ldots6}$, all transforming in the adjoint representation of the gauge group $SU(N)$. From those six (real) scalars we are going to consider four and make the following complex combinations:

$$X = \Phi^1 + i\Phi^2, \quad Y = \Phi^3 + i\Phi^4. \tag{1}$$

With these fields we form the following gauge invariant operators:

$$\mathcal{O} = \mathrm{Tr}(XXXYY\ldots XYYX), \tag{2}$$

where to be definite we can consider $J_1$ fields $X$ and $J_2$ $Y$'s. Since they are $N \times N$ matrices they do not commute. Thus, even if we fix the total number, there are many operators because we can put the $X$'s and the $Y$'s in different order. Now we have to remind ourselves that $\mathcal{N} = 4$ SYM is a conformal theory so an important property of the operators is their conformal dimension. The conformal dimension is a function of the coupling and can be computed in perturbation theory by evaluating the correlation functions between operators. In the case of operators like $\mathcal{O}$ the conformal dimension in the free theory is $J = J_1 + J_2$ and we are going to be interested in computing its first correction in the 't Hooft coupling. Although a one-loop calculation seems straightforward, the problem that arises is that all the operators with the same number of $X$'s and $Y$'s mix with each other and therefore, for large $J$ we have to diagonalize a huge matrix. How to do that was shown by Minahan and Zarembo [16] who related the problem of diagonalizing such matrix to the one of diagonalizing the Hamiltonian of a known physical system. They proposed to map the operators to configurations of a spin $\frac{1}{2}$ chain by replacing each field $X$ by a spin up and each field $Y$ by a spin down:

$$\mathcal{O} = \mathrm{Tr}(XXXYY\ldots XYYX) \quad \Leftrightarrow \quad |\uparrow\uparrow\uparrow\downarrow\downarrow\ldots\uparrow\downarrow\downarrow\uparrow\rangle. \tag{3}$$

Linear combinations of operators can be mapped into generic states of the spin chain and the matrix of anomalous dimensions can be mapped into a Hamiltonian acting on the spin chain. At 1-loop and in the large N limit, the operator turns out to be [16]:

$$H = \frac{\lambda}{4\pi^2} \sum_{j=1}^{J} \left( \frac{1}{4} - \vec{S}_j \cdot \vec{S}_{j+1} \right), \tag{4}$$

which is the standard ferromagnetic Heisenberg Hamiltonian. Let us analyze this result. The factor of $\lambda$, the 't Hooft coupling comes from the fact that this is a one loop calculation. The operators $\vec{S}_j$ are the standard spin operators given by the Pauli matrices acting on site $j$ of the chain. The whole operator $H$ is a convenient way to express the mixing under renormalization between operators of type $\mathcal{O}$. More precisely, an eigenstate of $H$ is a linear combination of states of the type shown in (3) and, by using the relation depicted in such equation, equivalent to an operator linear combination of those of type $\mathcal{O}$. Such linear combination has a definite conformal dimension at one loop equal to the corresponding eigenvalue of $H$. Summarizing, diagonalizing $H$ allows us to compute the one loop anomalous dimensions that we wanted. For example the ground state is with all spins parallel and the first excited states are spin waves where one spin is turned over. These spin waves where originally identified with string excitations by Berenstein, Maldacena and Nastase [17]. Here we want to go beyond that and consider more generic states. What actually happens is that if one turns over more spins, the low energy states are still described by spin waves which interact according to a well known action called the ferromagnetic sigma model. The variables in the effective action are two angles $(\theta, \phi)$ which determine, in polar coordinates, the direction in which the spin at each site points. Moreover, in the long wave-length limit, we can consider the spin chain to be continuous and parametrize it with a spacial coordinate $\sigma$. The effective action then reads:

$$S_{\text{eff.}} = J\left\{-\frac{1}{2}\int d\sigma d\tau \cos\theta\, \partial_\tau\phi - \frac{\lambda}{32\pi J^2}\int d\sigma d\tau \left[(\partial_\sigma\theta)^2 + \sin^2\theta\,(\partial_\sigma\phi)^2\right]\right\}, \quad (5)$$

where we rescaled $\sigma$ so that it runs from 0 to $2\pi$ instead of 0 to $J$. In the limit $J \to \infty$ keeping $\lambda/J^2$ fixed, the dynamics becomes classical and computing the anomalous dimensions becomes equivalent to solving the classical equations of motion of the action (5). Note, however, that the limit only makes sense if we take $\lambda$ to infinity contradicting the fact that we are doing just a one loop calculation. It has been argued in a related context [18] that higher order corrections are absent and therefore we expect no corrections to the $\lambda$ dependence from higher loops. Thus, what we have found, is that to compute the anomalous dimensions of operators of the type $\mathcal{O}$ we need to solve the classical equations of a two dimensional sigma model. However, solving the classical equations of motion of a two dimensional sigma model is precisely what one does to compute the motion of a string!. Therefore we can say that, by analyzing the field theory we have concluded that a good way to solve a particular problem, namely the computation of anomalous dimensions of operators of type $\mathcal{O}$ is by studying the classical equations of motion of a string in a particular background. Moreover, it turns out that the action is precisely the same action that AdS/CFT predicts the dual string should have. In that way we can claim that we have shown exactly how strings appear from the field theory and obtained a precise map between operators and strings. The down side is that we were able to do so only for a very particular theory and a very restricted set of operators.

## ACKNOWLEDGMENTS

The work regarding mesons in AdS/CFT that we describe here was done in collaboration with D. Mateos, R. Myers and D. Winters. The work regarding spin chains and strings was clarified in work with A. Tseytlin and A. Ryzhov. I am grateful to A. Guijosa for hospitality at UNAM, Mexico and for various discussion while there. I am also very grateful to the organizers of VI-Silafae/XII-MSPF, Puerto Vallarta, for the opportunity to speak in such a stimulating and interesting conference.

# REFERENCES


1. M. B. Green, J. H. Schwarz and E. Witten, "Superstring Theory", Cambridge University Press 1995.
2. W.-M. Yao et al., J. Phys. G 33, 1 (2006) (http://pdg.lbl.gov/index.html)
3. J. Maldacena, "The large *N* limit of superconformal field theories and supergravity," Adv. Theor. Math. Phys. **2**, 231 (1998) [Int. J. Theor. Phys. **38**, 1113 (1998)], hep-th/9711200,
   S. S. Gubser, I. R. Klebanov and A. M. Polyakov, "Gauge theory correlators from non-critical string theory," Phys. Lett. B **428**, 105 (1998) [arXiv:hep-th/9802109],
   E. Witten, "Anti-de Sitter space and holography," Adv. Theor. Math. Phys. **2**, 253 (1998) [arXiv:hep-th/9802150],
   O. Aharony, S. S. Gubser, J. M. Maldacena, H. Ooguri and Y. Oz, "Large N field theories, string theory and gravity," Phys. Rept. **323**, 183 (2000) [arXiv:hep-th/9905111].
4. A. Karch and E. Katz, JHEP **0206**, 043 (2002) [arXiv:hep-th/0205236].
5. M. Kruczenski, D. Mateos, R. C. Myers and D. J. Winters, JHEP **0307**, 049 (2003) [arXiv:hep-th/0304032].
6. J. M. Maldacena, Phys. Rev. Lett. **80**, 4859 (1998) [arXiv:hep-th/9803002],
   S. J. Rey and J. T. Yee, Eur. Phys. J. C **22**, 379 (2001) [arXiv:hep-th/9803001].
7. M. Kruczenski, D. Mateos, R. C. Myers and D. J. Winters, JHEP **0405**, 041 (2004) [arXiv:hep-th/0311270].
8. E. Witten, Adv. Theor. Math. Phys. **2**, 505 (1998) [arXiv:hep-th/9803131].
9. J. Babington, J. Erdmenger, N. J. Evans, Z. Guralnik and I. Kirsch, Phys. Rev. D **69**, 066007 (2004) [arXiv:hep-th/0306018].
10. T. Sakai and S. Sugimoto, Prog. Theor. Phys. **113**, 843 (2005) [arXiv:hep-th/0412141].
11. M. Kruczenski, L. A. P. Zayas, J. Sonnenschein and D. Vaman, JHEP **0506**, 046 (2005) [arXiv:hep-th/0410035].
12. M. Kruczenski, Phys. Rev. Lett. **93**, 161602 (2004) [arXiv:hep-th/0311203].
13. M. Kruczenski, A. V. Ryzhov and A. A. Tseytlin, Nucl. Phys. B **692**, 3 (2004) [arXiv:hep-th/0403120].
14. M. Kruczenski and A. A. Tseytlin, JHEP **0409**, 038 (2004) [arXiv:hep-th/0406189].
15. R. Hernandez and E. Lopez, JHEP **0404**, 052 (2004) [arXiv:hep-th/0403139].
16. J. A. Minahan and K. Zarembo, JHEP **0303**, 013 (2003) [arXiv:hep-th/0212208].
17. D. Berenstein, J. M. Maldacena and H. Nastase, JHEP **0204**, 013 (2002) [arXiv:hep-th/0202021].
18. A. Santambrogio and D. Zanon, Phys. Lett. B **545**, 425 (2002) [arXiv:hep-th/0206079].

# QUANTUM GRAVITY AND MAXIMUM ATTAINABLE VELOCITIES IN THE STANDARD MODEL

Jorge Alfaro

*Facultad de Física, Pontificia Universidad Católica de Chile
Casilla 306, Santiago 22, Chile.*

**Abstract.**
A main difficulty in the quantization of the gravitational field is the lack of experiments that discriminate among the theories proposed to quantize gravity. Recently we showed that the Standard Model(SM) itself contains tiny Lorentz invariance violation(LIV) terms coming from QG. All terms depend on one arbitrary parameter $\alpha$ that set the scale of QG effects. In this talk we review the LIV for mesons nucleons and leptons and apply it to study several effects, including the GZK anomaly.

**Keywords:** Lorentz invariance violation, Quantum gravity
**PACS:** 04.60.Pp, 11.15.-q

## INTRODUCTION

Recently several proposal have been advanced to select theories and predict new phenomena associated to the Quantum gravitational field [1, 2, 3]. Most of the new phenomenology is associated to some sort of Lorentz invariance violations(LIV's)[4, 5, 6]. More recently [7], this approach has been subjected to severe criticism.

In previous papers [8], we asserted that the main effect of QG is to deform the measure of integration of Feynman graphs at large four momenta by a tiny LIV. The classical lagrangian is unchanged. In a similar manner, we can say that QG deforms the metric of space-time, introducing a tiny LIV proportional to (d-4)$\alpha$, d being the dimension of space time in Dimensional Regularization and $\alpha$ is the only arbitrary parameter in the model. Such small LIV could be due to quantum fluctuations of the metric of space-time produced by QG:virtual black holes as suggested in[1], D-branes as in [10], compactification of extra-dimensions or spin-foam anisotropies [11]. A precise derivation of $\alpha$ will have to wait for additional progress in the available theories of QG[1] An intriguing possibility may be provided by the anisotropy between spatial and temporal directions found necessary to recover our universe at macroscopic scales in a recent numerical simulation of Quantum Gravity[9].

It is possible to have modified dispersion relations without a preferred frame(DSR)[12]. Notice, however, that in our case the classical lagrangian is invariant under usual linear

[1] Such derivation must explain why the LIV parameter is so small. Progress in this direction is in [3]. There $\alpha$ appears as $(l_P/L)^2$, where $l_P$ is Planck's lenght and $L$ is defined by the semiclassical gravitational state in Loop Quantum Gravity. If $L \sim 10^{11} l_P$, an $\alpha$ of the right order is obtained

CP917, *Particles and Fields,* edited by H. Castilla Valdez, J. C. D'Olivo, and M. A. Perez

Lorentz transformations but not under DSR. So our LIV is more similar to radiative breaking of usual Lorentz symmetry than to DSR. Moreover the regulator R defined below and the deformed metric (3) are given in a particular inertial frame, where spatial rotational symmetry is preserved. That is why, in this paper we are ascribing to the point of view of [5] which is widely used in the literature. The preferred frame is the one where the Cosmic Background Radiation is isotropic.

Within the Standard Model, such LIV implies several remarkable effects, which are wholly determined up to one arbitrary parameter ($\alpha$).The main effects are:

The maximal attainable velocity for particles is not the speed of light, but depends on the specific couplings of the particles within the Standard Model. Noticeably, this LIV of the dispersion relations is the only acceptable, according to the very stringent bounds coming from the Ultra High Energy Cosmic Rays (UHECR) spectrum[13]. Moreover, the specific interactions between particles in the SM, determine different maximum attainable velocities for each particle, a necessary requirement to explain the Greisen[14],Zatsepin and Kuz'min[15](GZK) anomaly[5, 13, 16]. Since the Auger[17] experiment is expected to produce results in the near future, powerful tests of Lorentz invariance using the spectrum of UHECR will be available.

Also birrefringence occurs for charged leptons, but not for gauge bosons. In particular, photons and neutrinos have different maximum attainable velocities. This could be tested in the next generation of neutrino detectors such as NUBE[12, 13].

This paper is organized as follows: In chapter 2, we present the LIV cutoff regulator; section 3 contains the effect of the regulator on One Particle Irreducible Green functions(1PI); section 4 defines LIV dimensional regularization ; Explicit one loop computations are contained in section 5; the LIV for mesons and baryons is found in chapter 6; Reactions thresholds and bounds on $\alpha$ are derived in section 7; Section 8 contains our conclusions.

## CUTOFF REGULATOR

To see what are the implications of the asymmetry in the measure for renormalizable theories, we will represent the Lorentz asymmetry of the measure by the replacement

$$\int d^d k -> \int d^d k R(\frac{k^2+\alpha k_0^2}{\Lambda^2})$$

Here R is an arbitrary function, $\Lambda$ is a cutoff with mass dimensions, that will go to infinity at the end of the calculation. We normalize $R(0) = 1$ to recover the original integral. $R(\infty) = 0$ to regulate the integral. $\alpha$ is a real parameter. Notice that we are assuming that rotational invariance in space is preserved. More general possibilities such as violation of rotational symmetry in space can be easily incorporated in our formalism.

This regulator has the property that for logarithmically divergent integrals, the divergent term is Lorentz invariant whereas when the cutoff goes to infinity a finite LIV part proportional to $\alpha$ remains.

## ONE LOOP ANALYSIS

We study explicitly the case of bosons. A similar analysis apply for fermions and gauge bosons[8]. Let D be the naive degree of divergence of a One Particle Irreducible (1PI) graph. The change in the measure induces modifications to the primitively log divergent integrals(D=0) In this case, the correction amounts to a finite LIV. The finite part of 1PI Green functions will not be affected. Therefore, Standard Model predictions are intact, except for the maximum attainable velocity for particles[5] and interaction vertices, which receive a finite wholly determined contribution from Quantum Gravity.

Let us analyze the primitivily divergent 1PI graphs for bosons first. **Self energy:** $\chi(p) = \chi(0) + A^{\mu\nu} p_\mu p_\nu + convergent, A^{\mu\nu} = \frac{1}{2}\partial_\mu \partial_\nu \chi(0)$. We have:

$$A^{\mu\nu} = c_2 \eta^{\mu\nu} + a^{\mu\nu}$$

$c_2$ is the log divergent wave function renormalization counterterm; $a^{\mu\nu}$ is a finite LIV. The on-shell condition is:

$$p^2 - m^2 - a^{\mu\nu} p_\mu p_\nu = 0$$

If spatial rotational invariance is preserved, the nonzero components of the matrix $a$ are:

$$a^{00} = a_0; \;\; a^{ii} = -a_1$$

So the maximum attainable velocity for this particle will be:

$$c_m = \sqrt{\frac{1-a_1}{1-a_0}} \sim 1 - (a_1 - a_0)/2 \tag{1}$$

By doing explicit computations for all particles in the SM, we get definite predictions for the LIV, assuming a particular regulator $R$. However, the dependence on $R$ amounts to a multiplicative factor. So ratios of LIV's are uniquely determined. Explicit computations are simplified by using Dimensional Regularization as explained below.

## LIV DIMENSIONAL REGULARIZATION

We generalize dimensional regularization to a d dimensional space with an arbitrary constant metric $g_{\mu\nu}$. We work with a positive definite metric first and then Wick rotate. We will illustrate the procedure with an example. Here $g = det(g_{\mu\nu})$ and $\Delta > 0$.

$$\frac{1}{\sqrt{g}} \int \frac{d^d k}{(2\pi)^d} \frac{k_\mu k_\nu}{(k^2+\Delta)^n} =$$

$$\frac{1}{\sqrt{g}\Gamma(n)} \int_0^\infty dt t^{n-1} \int \frac{d^d k}{(2\pi)^d} k_\mu k_\nu e^{-t(g^{\alpha\beta} k_\alpha k_\beta + \Delta)} =$$

$$\frac{1}{(4\pi)^{d/2}} \frac{g_{\mu\nu}}{2} \frac{\Gamma(n-1-d/2)}{\Gamma(n)} \frac{1}{\Delta^{n-1-d/2}} \tag{2}$$

In the same manner, after Wick rotation, we obtain Appendix A4 of [12]. These definitions preserve gauge invariance, because the integration measure is invariant under shifts. To get a LIV measure, we assume that

$$g^{\mu\nu} = \eta^{\mu\nu} + (4\pi)^2\alpha\eta^{\mu 0}\eta^{\nu 0}Res_{\varepsilon=0} \tag{3}$$

where $\varepsilon = 2 - \frac{d}{2}$ and $Res_{\varepsilon=0}$ is the residue of the pole at $\varepsilon = 0$. A formerly divergent integral will have a pole at $\varepsilon = 0$, so when we take the physical limit, $\varepsilon -> 0$, the answer will contain a LIV term.

That is, LIV dimensional regularization consists in: 1)Calculating the d-dimensional integrals using a general metric $g_{\mu\nu}$. 2) Gamma matrix algebra is generalized to a general metric $g_{\mu\nu}$. 3) At the end of the calculation, replace $g^{\mu\nu} = \eta^{\mu\nu} + (4\pi)^2\alpha\eta^{\mu 0}\eta^{\nu 0}Res_{\varepsilon=0}$ and then take the limit $\varepsilon -> 0$.

To define the counterterms, we used the minimal substraction scheme(MSS); that is we substract the poles in $\varepsilon$ from the 1PI graphs.

LIV Dimensional Regularization reinforces our claim that these tiny LIV's originates in Quantum Gravity. In fact the sole change of the metric of space time is a correction of order $\varepsilon$ to the Minkowsky metric and this is the source of the effects studied above. Quantum Gravity is the strongest candidate to produce such effects because the gravitational field is precisely the metric of space-time and tiny LIV modifications to the flat Minkowsky metric may be produced by quantum fluctuations.

## LIV in the Standard Model

We follow [20, 21, 8] and use LIV Dimensional Regularization.

**Photons**

The LIV photon self-energy in the SM is:

$$L\Pi^{\mu\nu}(q) = -\frac{23}{3}e^2\alpha q_\alpha q_\beta$$
$$(\eta^{\alpha\beta}\delta_0^\mu\delta_0^\nu + \eta^{\mu\nu}\delta_0^\alpha\delta_0^\beta - \eta^{\nu\beta}\delta_0^\mu\delta_0^\alpha - \eta^{\mu\alpha}\delta_0^\nu\delta_0^\beta) \tag{4}$$

It follows that the maximal attainable velocity is

$$c_\gamma = 1 - \frac{23}{6}e^2\alpha \tag{5}$$

We have included coupling to quarks and charged leptons as well as 3 generations and color.

**Fermions**

In the SM, the fermion self-energy is given by:

$$(4\pi^2)\Sigma(q) = -\frac{1}{\varepsilon}\not{q}\sum_{graphs}(|c_V + c_A|^2 P_L + (|c_V - c_A|^2 P_R) + finite \tag{6}$$

where the fermion-gauge boson vertex is:

$$i\gamma^k(c_V - c_A\gamma^5) \tag{7}$$

and $P_L(P_R)$ are the L(R) helicity projectors.

Therefore

$$L\Sigma(q) = \frac{\alpha}{2} q_0 \gamma^0 \sum_{graphs} (|c_V + c_A|^2 P_L + (|c_V - c_A|^2 P_R) \tag{8}$$

We apply this last result to neutrinos and charged leptons below.

**Neutrinos:** The maximal attainable velocity is

$$c_\nu = 1 - (3 + tan^2\theta_w)\frac{g^2\alpha}{8} \tag{9}$$

In this scenario, we predict that neutrinos [18] emitted simultaneously with photons in gamma ray bursts will not arrive simultaneously to Earth . The time delay during a flight from a source situated at a distance $D$ will be of the order of $(5 \times 10^{-23})D/c \sim 5 \times 10^{-6}$ s, assuming $D = 10^{10}$ light-years. No dependence of the time delay on the energy of high energy photons or neutrinos should be observed(contrast with [1]). Photons will arrive earlier since $\alpha < 0$(See below). These predictions could be tested in the next generation of neutrino detectors [19].

Using $R_\xi$-gauges we have checked that the LIV is gauge invariant. The gauge parameter affects the Lorentz invariant part only.

**Electron self-energy in the Weinberg-Salam model. Birrefringence:**

Define: $e_L = \frac{1-\gamma^5}{2}e$, $e_R = \frac{1+\gamma^5}{2}e$, where $e$ is the electron field. We get

$$c_L = 1 - (\frac{g^2}{cos^2\theta_w}(sin^2\theta_w - 1/2)^2 + e^2 + g^2/2)\frac{\alpha}{2}; \tag{10}$$

$$c_R = 1 - (e^2 + \frac{g^2 sin^4\theta_w}{cos^2\theta_w})\frac{\alpha}{2} \tag{11}$$

The difference in maximal speed for the left and right helicities is $\sim (5 \times 10^{-24})$.

## MESONS AND BARYONS

In order to calculate the maximal attainable velocity of hadrons, we will use effective lagrangians.

We use the results of [22, 23] for the wave function renormalization of pions and nucleons in the chiral lagrangian and Heavy Baryon Chiral Perturbation Theory. They get:

$$Z_\pi^{-1} = 1 - \frac{4m_\pi^2}{3(4\pi)^2F^2}\frac{1}{\varepsilon} + finite \tag{12}$$

$$Z_N^{-1} = 1 - \frac{9g_A^2 m_\pi^2}{4(4\pi)^2F^2}\frac{1}{\varepsilon} + finite \tag{13}$$

Here, $m_\pi$ is the renormalized pion mass, $F$ is the renormalized decay constant of pions and $g_A$ is the axial vector coupling constant, in the chiral limit.

Using the LIV metric, we can read off the maximal attainable velocities for pions and nucleons:

$$c_\pi = 1 + \frac{2m_\pi^2 \alpha}{3F^2}$$
$$c_N = 1 + \frac{9m_\pi^2 g_A^2 \alpha}{8F^2} \qquad (14)$$

## REACTION THRESHOLDS AND BOUNDS

Knowing the LIV for nucleons, pions, photons and electrons, we proceed to study the reactions involved in the GZK cutoff. We follow the discussion in [5, 13, 8].

**Photo-Pion Production** $\gamma + p \to p + \pi$

Let us begin with the photo-pion production $\gamma + p \to p + \pi$. Considering the corrections provided in the dispersion relation (14) for pions and nucleons, we note that, for the photo-pion production to proceed, the following condition must be satisfied

$$2\,\delta c E_\pi^2 + 4E_\pi \omega \geq \frac{m_\pi^2(2m_p + m_\pi)}{m_p + m_\pi}. \qquad (15)$$

where $E_\pi$ is the produced pion energy and $\delta c = c_p - c_\pi$.

**Pair Creation** $\gamma + p \to p + e^+ + e^-$

Pair creation, $\gamma + p \to p + e^+ + e^-$, is greatly abundant in the sector previous to the GZK limit. When the dispersion relations for fermions are considered for both protons and electrons, it is possible to find

$$\delta c \frac{m_e}{m_p} E^2 + E\omega \geq m_e(m_p + m_e), \qquad (16)$$

where $E$ is the incident proton energy and $\delta c = c_p - c_e$.

Combining the two reactions and the standard values, $m_\pi = 139Mev, g_A = 1.26, F = 92.4Mev$, we get an upper and lower bound on $\alpha$

$$2.2 \times 10^{-21} > -\alpha > 1.3 \times 10^{-24} \qquad (17)$$

First of all, we notice that $\alpha < 0$, in order to suppress the photopion production, thus removing the GZK cutoff. This implies that photons are the fastest particles and they arrive before neutrinos coming from the same source of GRB. Moreover, photons become unstable. They decay in a electron positron pair above an energy $E_0$[5]. See below.

Since $c_{photon} > c_{proton}$, the strong bound of [27] is avoided: Proton is stable under Cerenkov radiation in vacuum.

If no GZK anomaly is confirmed in future experimental observations, then we should state a stronger bound for the difference $c_\pi - c_p$. Using the same assumptions to set the restriction (17) when the primordial proton reference energy is $E_{\rm ref} = 2 \times 10^{20}$ eV, it is possible to find

$$|c_\pi - c_p| < 2.3 \times 10^{-23}. \qquad (18)$$

In terms of $\alpha$, this last bound may be read as

$$|\alpha| < 9.1 \times 10^{-24}, \tag{19}$$

**Photon unstability**

It has been pointed out in [27, 5] that if $c_{photon} > c_{electron}$ then the process $\gamma \rightarrow e^{+} + e^{-}$ is allowed above an energy $E_0$:

$$E_0 = m_e \sqrt{\frac{2}{\delta c}} \tag{20}$$

where $\delta c = c_\gamma - c_e$.

In our case, we have:

$$\delta c_L = -\alpha(\frac{23}{6}e^2 - (\frac{g^2}{cos^2\theta_w}(sin^2\theta_w - 1/2)^2 + e^2 + g^2/2)/2) \tag{21}$$

$$\delta c_R = -\alpha(\frac{23}{6}e^2 - (e^2 + \frac{g^2 sin^4\theta_w}{cos^2\theta_w})/2) \tag{22}$$

Therefore, with

$$\begin{aligned} EL_0 &= 2.3 \times 10^8 Gev \\ ER_0 &= 1.9 \times 10^8 Gev \end{aligned} \tag{23}$$

So, we should not detect photons with energies above $2.3 \times 10^8 Gev$

**Neutral pion Stability**

Following [5] we study the main decay process of neutral pion $\pi_0 \rightarrow \gamma + \gamma$ .This becomes possible if $c_\gamma > c_\pi$ and above an energy

$$E_\pi = \frac{m_\pi}{\sqrt{2(c_\gamma - c_\pi)}} \tag{24}$$

Using the bound $c_\gamma - c_\pi < 10^{-22}$ obtained in [28], we get

$$|\alpha| < 5.4 \times 10^{-23} \tag{25}$$

In our numerical estimates we have chosen $\alpha = -5 \times 10^{-23}$.

We get $E_\pi = 10^{19} eV$. Therefore we expect that neutral pions above this energy are stable, so they could be a primary component of UHECR. Photons will be unstable above this energy by the same mechanism. Notice however that photons are unstable at a lower energy due to electron-positron pair creation (23).

## CONCLUSIONS

In this paper we have computed the LIV induced by Quantum Gravity on Baryons and Mesons, using the Chiral Lagrangian approach. This permitted to fix that $\alpha < 0$, in order to explain the GZK anomaly. Studying several available processes, we found bounds on $\alpha$:

From pair creation and absence of photopion creation: $2.2 \times 10^{-21} > -\alpha > 1.3 \times 10^{-24}$.

From pion stability and the most stringent experimental bound found in [28]: $|\alpha| < 5.4 \times 10^{-23}$.

Then, several predictions are obtained:Photons are unstable above an energy $2.3 \times 10^8 Gev$.

Neutral pions are stable above an energy $E_\pi = 10^{19} eV$; so they could be a primary component of UHECR, thus evading the GZK cutoff.

Moreover, in time of flight experiments, photons will arrive before neutrinos, assuming that they were emitted simultaneously at the source. No energy dependence of the time delay should be observed. The time delay during a flight from a source situated at a distance $D$ will be of the order of $(5 \times 10^{-23})D/c \sim 5 \times 10^{-6}$ s, assuming $D = 10^{10}$ light-years.

## ACKNOWLEDGMENTS

The author wants to thank the organizers of the VI Silafae for the pleasent atmosphere they created in Puerto Vallarta. He also wants to thank Prof. Roberto Percacci for pointing out the results of reference [9]. The work of JA is partially supported by Fondecyt # 1060646.

## REFERENCES

1. Amelino-Camelia, G. et al., Nature 393, 763 (1998).
2. Gambini, R. and Pullin, J. , Phys. Rev. D 59, 124021 (1999).
3. Alfaro, J. Morales-Técotl, H.A. and Urrutia, L.F. , Phys. Rev. Lett. 84, 2318 (2000);Phys. Rev. D 65, 103509(2002).
4. Colladay, D. and Kostelecky, V.A., Phys. Rev. D58, 116002(1998).
5. Coleman, S.and Glashow, S.L., Phys. Rev. D 59, 116008 (1999).
6. Bertolami, O., Colladay, Kostelecky, V.A. and Potting, R., Phys. Lett. B395(1997)178; Bertolami, O. , Threshold Effects and Lorentz Symmetry, hep-ph 0301191.
7. Collins, J. et al , Phys.Rev.Lett.93, 191301(2004).
8. Alfaro, J.,Phys.Rev.Lett.94:221302,2005;Phys.Rev.D72:024027,2005.
9. J. Ambjorn, J. Jurkiewicz and R. Loll, Phys. Rev. D72, 064014(2005).
10. J. Ellis, N.E. Mavromatos, D.V. Nanopoulos, G. Volkov, Gen.Rel.Grav. 32 (2000) 1777-1798.
11. For a comprehensive review on the loop quantum gravity framework see for example: Rovelli, C. , Quantum Gravity, Cambridge University Press, Cambridge (UK) 2004.
12. G. Amelino-Camelia, Int. J. Mod. Phys. **D 11** (2002) 35-60, Phys. Lett. **B 510** (2001) 255-263; N. Bruno, G. Amelino-Camelia, and J. Kowalski-Glikman, Phys. Lett. **B 522** (2001) 133; G. Amelino-Camelia, Nature **418** (2002) 34; G. Amelino-Camelia Int. J. Mod. Phys. **D 12** (2003) 1211;J. Magueijo and L. Smolin, Phys. Rev. Lett. **88** (2002) 190403; J. Magueijo and L. Smolin, Phys. Rev. **D 67** 044017 (2003).
13. Alfaro,J. and Palma,G., Phys. Rev. D 67,083003(2003);Phys. Rev. D **65**, 103516 (2002).
14. Greisen,K. Phys.Rev.Lett.16,748 (1966).
15. Zatsepin,G.T. and Kuz'min, V.A.,JETP Lett.4:78(1966), Pisma Zh.Eksp.Teor.Fiz.4:114(1966).
16. T.Kifune, Astroph. J. Lett. **518**, 21 (1999); G.Amelino-Camelia and T.Piran, Phys. Lett. B **497**, 265 (2001); J.Ellis, N.E.Mavromatos and D.V.Nanopoulos, Phys. Rev. D **63**, 124025 (2001); G.Amelino-Camelia and T.Piran, Phys. Rev. D **64**, 036005 (2001); G.Amelino-Camelia, Phys. Lett. B **528**, 181 (2002).

17. See http://www.auger.org/auger.html.
18. Waxman,E. and Bahcall, J.,Phys.Rev.Letts.78, 2292(1997).
19. Roy,M., Crawford,H.J. and Trattner, A., astro-ph/9903231.
20. Peskin, M. and Schroeder D., An introduction to Quantum Field Theory, Addison-Wesley Publishing Company, New York 1997.
21. Pokorski, S., Gauge Field Theories, Cambridge Monographs on Mathematical Physics, Cambridge University Press 2000.
22. G. Ecker and M. Mojzis, Phys. Lett. B 410(1997)266.
23. H. Fearing, R. Lewis, N. Mobed and S. Scherer, Phys. Rev. D 56 (1997) 1783.
24. V. Berezinsky, A.Z. Gazizov and S.I. Grigorieva, hep-ph/0107306; hep-ph/0204357.
25. F.W. Stecker and S.L. Glashow, Astropart.Phys. **16**, 97 (2001).
26. M. Takeda *et al.*, Phys. Rev. Lett. **81**, 1163 (1998). For an update see M. Takeda *et al.*, Astrophys. J. **522**, 225 (1999).
27. S. Coleman and S.L. Glashow, Phys. Lett. B 405, 249(1997).
28. E. Antonov et al.,JETP Letters 73(2001)446.

# Spatial confinement and thermal deconfinement in the Gross-Neveu model

J. M. C. Malbouisson*, F. C. Khanna†, A. P. C. Malbouisson** and A. E. Santana‡

*Instituto de Física, Universidade Federal da Bahia, 40210-340, Salvador, BA, Brazil
†Theoretical Physics Institute, University of Alberta, Edmonton, Alberta T6G 2J1, Canada
**Centro Brasileiro de Pesquisas Físicas/MCT, 22290-180, Rio de Janeiro, RJ, Brazil
‡Instituto de Física, Universidade de Brasília, 70910-900, Brasília, DF, Brazil

**Abstract.** We discuss the occurrence of spatial confinement and thermal deconfinement in the massive, $D$-dimensional, Gross-Neveu model with compactified spatial dimensions.

**Keywords:** Gross-Neveu model, Spatial confinement, Thermal deconfinement
**PACS:** 11.10.Kk, 11.10.Wx

## 1. INTRODUCTION

Hadronic matter shows very impressive quantum features. The strong interaction between its constituents (quarks and gluons) obligate them to live, at low temperatures, spatially confined within distances of the order of 1 fm and gather themselves in multi-particle colorless states. In usual scenarios of the early evolution of the universe, as it cooled down, hadrons appeared in a confining phase transition (with quarks and gluons condensing into material “droplets") that occurred at a temperature estimated of the order of 200 MeV. Also, as probed by deep inelastic scattering processes, the interaction between quarks becomes very weak at very high energies (that is, very short distances), a regime in which they are asymptotically free.

The established theory of the strong interaction in the context of the standard model, quantum chomodynamics (QCD), should then be able to predict such facts and also to account for the binding nuclear forces. However, due to the intricate (non-Abelian) structure of QCD, no analytical results exist taking into account simultaneously all those aspects of hadronic matter. Lattice calculations have been implemented to simulate the behavior of the theory in the confining region, both at zero and finite temperatures, providing (among others results) an estimate of the confining temperature.

The difficulty of handling analytically QCD has stimulated, over the last decades, the use of phenomenological approaches and the study of effective, simplified, theories to get clues of the behavior of hadronic systems. The simplest effective model which may be conceived to describe quark interactions, is a direct four-fermion coupling where the gluon fields are integrated out and all color degrees of freedom are ignored, in a way similar to the Fermi treatment of the weak interaction; this corresponds to the Gross-Neveu model [1], considered in space-time dimension $D = 4$.

In this communication, we discuss the occurrence of spatial confinement and thermal deconfinement in the $N$-component, massive, Gross-Neveu model, with all spatial

CP917, *Particles and Fields,* edited by H. Castilla Valdez, J. C. D'Olivo, and M. A. Perez

coordinates compactified through a generalization of the Matsubara formalism, in the large-$N$ limit. The present work generalizes, for arbitrary space-time dimension $D$, previous ones where the Gross-Neveu model has been considered in $D=3$ with one [2, 3] or two [4] spatial dimensions compactified. The results obtained have similar structure for all dimensions which permits to consider $D=4$, even the Gross-Neveu model not being perturbatively renormalizable.

## 2. COMPACTIFIED MASSIVE GROSS-NEVEU MODEL

We consider the Wick-ordered, $N$-component, massive Gross-Neveu (GN) model in a $D$-dimensional Euclidean space, described by the Lagrangian density (in natural units)

$$\mathscr{L} =: \bar{\psi}(x)(i\not\nabla + m)\psi(x) : + \frac{u}{2}(: \bar{\psi}(x)\psi(x) :)^2, \tag{1}$$

where $m$ is the mass, $u$ is the coupling constant and $x$ is a point of $\mathbf{R}^D$. The quantity $\psi(x)$ is a spin $\frac{1}{2}$ field having $N$ (flavor) components, $\psi^a(x)$, $a = 1,2,...,N$, summation over flavor and spin indices is understood, and we shall take the large-$N$ limit ($N \to \infty$).

Our goal is to determine the renormalized large-$N$ (*effective*) coupling constant when $d$ ($\leq D$) Euclidean coordinates, say $x_1,\ldots,x_d$, are compactified. The compactification is engendered via a generalized Matsubara prescription, which corresponds to considering the system in a topology $S^{1_1} \times \cdots S^{1_d} \times R^{D-d}$. In other words, the coordinates $x_i$ are restricted to segments of length $L_i$ ($i = 1,2,....d$), with the field $\psi(x)$ satisfying anti-periodic boundary conditions. For spatial coordinates, these conditions are equivalent to bag model ones (no outgoing currents) [5, 6] while, if one of the coordinates is the imaginary time ($x_d$, say), this corresponds to considering finite temperature ($L_d = \beta = 1/T$). Accordingly, Feynman rules should change to the Matsubara replacements

$$\int \frac{dk_i}{2\pi} \to \frac{1}{L_i}\sum_{n_i=-\infty}^{+\infty}, \qquad k_i \to \nu_i = \frac{2\pi(n_i+\frac{1}{2})}{L_i}, \qquad i = 1,2,\ldots,d. \tag{2}$$

The large-$N$ effective coupling constant between the fermions will be defined in terms of the four-point function at zero external momenta. At leading order in $\frac{1}{N}$, summing chains of one-loop (bubble) diagrams, the $\{L_i\}$-dependent four-point function is

$$\Gamma^{(4)}_{Dd}(0;\{L_i\},u) = \frac{u}{1+Nu\Sigma_{Dd}(\{L_i\})}, \tag{3}$$

with the $\{L_i\}$-dependent Feynman one-loop diagram given by

$$\Sigma_{Dd}(\{L_i\}) = \frac{1}{L_1\cdots L_d}\sum_{\{n_i\}=-\infty}^{\infty}\int \frac{d^{D-d}k}{(2\pi)^{D-d}}\left[\frac{m^2-\mathbf{k}^2-\sum_{i=1}^d \nu_i^2}{(\mathbf{k}^2+\sum_{i=1}^d \nu_i^2+m^2)^2}\right], \tag{4}$$

where $\nu_i = 2\pi(n_i+\frac{1}{2})/L_i$ ($i = 1,\ldots,d$) are the Matsubara frequencies and $\mathbf{k}$ stands for a continuous $(D-d)$-dimensional vector in momentum space.

To define a renormalized coupling constant, we have to handle the ultraviolet divergences of $\Sigma_{Dd}(\{L_i\})$. To simplify the regularization procedure, we introduce the dimensionless quantities $b_i = (mL_i)^{-2}$ ($i = 1,\ldots,d$) and $q_j = k_j/2\pi m$ ($j = d+1,\ldots,D$), and write the one-loop diagram as

$$\begin{aligned}\Sigma_{Dd}(\{b_i\}) &= \Sigma_{Dd}(s;\{b_i\})|_{s=2} \\ &= \frac{m^{D-2}}{4\pi^2}\sqrt{b_1\cdots b_d}\left\{\frac{1}{2\pi^2}U_{Dd}(s;\{b_i\}) - U_{Dd}(s-1;\{b_i\})\right\}\Bigg|_{s=2}, \end{aligned}\tag{5}$$

where

$$U_{Dd}(\mu;\{b_i\}) = \sum_{\{n_i\}=-\infty}^{\infty}\int\frac{d^{D-d}q}{\left[\mathbf{q}^2+\sum_{j=1}^{d}b_j(n_j+\frac{1}{2})^2+(2\pi)^{-2}\right]^{\mu}}. \tag{6}$$

This shows explicitly that $\Sigma_{Dd}$ has dimension of mass$^{D-2}$, which is the inverse of the mass dimension of the coupling constant.

We calculate the function $U_{Dd}(\mu;\{b_i\})$ by employing concurrently dimensional and analytical regularizations. First, the integral over $\mathbf{q} = (q_{d+1},\ldots,q_D)$ in Eq. (6) is performed using well-known dimensional regularization formulas; the resulting expression can then be written in terms of multiple ($d$-dimensional) Epstein-Hurwitz *zeta*-functions, by transforming the summations over half-integers into sums over integers; and, finally, using the analytic extension of the Epstein-Hurwitz *zeta*-function to the whole complex plane [7, 8, 9], we obtain, after a long calculation,

$$U_{Dd}(\mu;\{b_i\}) = \frac{2^{2\mu-D}\pi^{2\mu-\frac{D}{2}}}{\Gamma(\mu)}\frac{1}{\sqrt{b_1\cdots b_d}}\left[\Gamma\left(\mu-\frac{D}{2}\right)+2^{\frac{D}{2}}W_{Dd}(\mu;\{b_i\})\right], \tag{7}$$

with $W_{Dd}(\mu;\{b_i\})$ given by

$$W_{Dd}(\mu;\{b_i\}) = 2^{1-\mu}\sum_{j=1}^{d}2^{2j}\sum_{\{\rho_j\}}\sum_{\{c_{\rho_k}=1,4\}}\left(\prod_{k=1}^{j}\frac{(-1)^{c_{\rho_k}-1}}{\sqrt{c_{\rho_k}}}\right)F_{Dj}(\mu;c_{\rho_1}b_{\rho_1},\ldots,c_{\rho_j}b_{\rho_j}), \tag{8}$$

where $\{\rho_j\}$ stands for the set of all combinations of the indices $\{1,2,\ldots,d\}$ with $j$ elements and the functions $F_{Dj}(\mu;a_1,\ldots,a_j)$ ($j = 1,\ldots,d$) are defined by

$$F_{Dj}(\mu;a_1,\ldots,a_j) = \sum_{n_1,\ldots,n_j=1}^{\infty}\left(2\sqrt{\frac{n_1^2}{a_1}+\cdots+\frac{n_j^2}{a_j}}\right)^{\mu-\frac{D}{2}}K_{\mu-\frac{D}{2}}\left(2\sqrt{\frac{n_1^2}{a_1}+\cdots+\frac{n_j^2}{a_j}}\right), \tag{9}$$

with $K_\alpha(z)$ denoting the Bessel function of the third kind. Substituting Eq. (7) into Eq. (5), we obtain an analytic extension of $\Sigma_{Dd}(s;\{b_i\})$ for complex values of $s$:

$$\begin{aligned}\Sigma_{Dd}(s;\{b_i\}) &= \frac{m^{D-2}\pi^{\frac{D}{2}}}{(2\pi)^{D-2(s-2)}\Gamma(s)}\left\{(s-1-D)\Gamma\left(s-1-\frac{D}{2}\right)\right. \\ &\left. + 2^{\frac{D}{2}}\left[2W_{Dd}(s;\{b_i\})-(s-1)W_{Dd}(s-1;\{b_i\})\right]\right\}. \end{aligned}\tag{10}$$

The first term in this expression for $\Sigma_{Dd}(s;\{b_i\})$, involving the $\Gamma$-functions, does not depend on the compactification lengths $\{L_i\}$ and diverges as $s \to 2$ for even dimensions $D \geq 2$. In the modified minimal subtraction scheme we use, this polar term is suppressed to give the renormalized single bubble function, $\Sigma^R_{Dd}(\{b_i\})$; such a subtraction corresponds to a finite renormalization in the case of odd dimensions. In all cases, then, the $\{L_i\}$-dependent, renormalized, one-loop diagram arises from the regular part of the analytic extension of the Epstein-Hurwitz *zeta*-functions:

$$\Sigma^R_{Dd}(\{b_i\}) = \frac{m^{D-2}}{(2\pi)^{\frac{D}{2}}} \left[2W_{Dd}(2;\{b_i\}) - W_{Dd}(1;\{b_i\})\right]. \tag{11}$$

We proceed analyzing the behavior of the large-$N$ coupling constant in various cases.

## 3. LARGE-$N$ RENORMALIZED COUPLING CONSTANT

We define the coupling constant, interpreted as measuring the strength of the interaction between the fermions, in terms of the four-point function at fixed external momenta, taking $\mathbf{p} = 0$. Using Eqs. (3) and (11), and taking the limit $N \to \infty$ and $u \to 0$ (with $Nu = \lambda$ fixed), we find the large-$N$ renormalized ($\{L_i\}$-dependent) coupling constant as

$$g_{Dd}(\{b_i\},\lambda) = \lim_{N\to\infty}\left[N\Gamma^{(4)R}_{Dd}(0,\{b_i\},u)\right] = \frac{\lambda}{1+\lambda\,\Sigma^R_{Dd}(\{b_i\})}. \tag{12}$$

From the dependence of $\Sigma^R_{Dd}$ on $\{b_i\}$, dictated by the functions $K_{\mu-\frac{D}{2}}$ in Eq. (9), we conclude that, if all $b_i$ tend to zero (i.e., if $\{L_i \to \infty\}$), $\Sigma^R_{Dd} \to 0$ and, as expected,

$$\lim_{\{L_i\to\infty\}} g_{Dd}(\{b_i\},\lambda) = \lambda; \tag{13}$$

that is, when all compactification lengths tend to $\infty$, the renormalized coupling constant reduces to the effective fixed coupling constant in free space at zero temperature, $\lambda$. On the other hand, if any of the $b_i$ goes to $\infty$ ($L_i \to 0$), the renormalized single bubble diagram $\Sigma^R_{Dd}$ diverges, implying the vanishing of the renormalized coupling constant $g_{Dd}$, irrespective of the value of $\lambda$. This suggests that the system presents an ultraviolet asymptotic-freedom type of behavior for short distances and/or for high temperatures.

Interesting features appear if $\Sigma^R_{Dd}$ acquires negative values; in such a case, depending on the value of $\lambda$, the renormalized coupling constant may diverge for finite values of the lengths $L_i$. We now investigate such a possibility, considering the the fully compactified system at zero ($d = D-1$) and finite temperatures, $x_D$ being the time coordinate.

### 3.1. Compactified GN model at $T = 0$

We consider initially the case $D = 2$, with only the spatial coordinate compactified, and fix $b_1 = b = (mL)^{-2}$; Eq. (11) then becomes

$$\Sigma^R_{21}(L) = 2E_1(2mL) - E_1(mL), \tag{14}$$

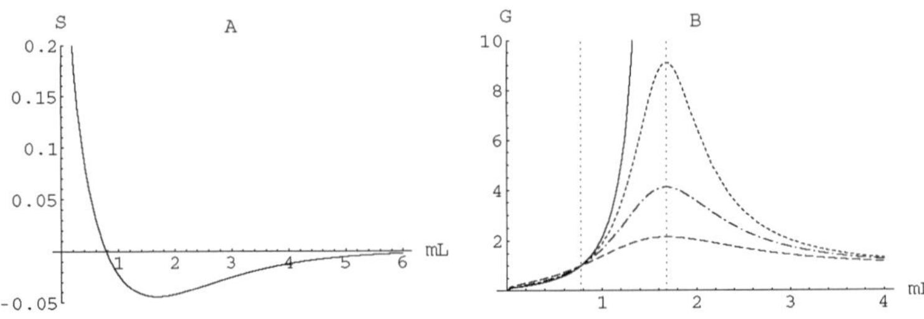


**FIGURE 1.** A - Plot of $S = \Sigma^R_{21}(L)$ as a function of *mL*. B - Relative effective coupling constant, $G = g_{21}(L,\lambda)/\lambda$, as a function of *mL* for some values of $\lambda$: 20.0 (dotted line), 17.0 (dotted-dashed line), 12.0 (dashed line) and 22.5 (full line). The dotted vertical lines, passing by $L^{(2)}_{\rm min} \simeq 0.78m^{-1}$ and $L^{(2)}_{\rm max} \simeq 1.68m^{-1}$, are plotted as a visual guide.

where the function $E_1(x)$ is given by

$$E_1(x) = \frac{1}{\pi}\sum_{n=1}^{\infty}\left[(xn)K_1(xn) - K_0(xn)\right]. \tag{15}$$

We can calculate $\Sigma^R_{21}(L)$ numerically with high precision, by truncating the series defining $E_1(y)$ at $n = N$, for a large enough $N$; the behavior of $\Sigma^R_{21}(L)$ is shown in Fig. 1-A. We find that $\Sigma^R_{21}(L)$ diverges ($\to +\infty$) when $L \to 0$ and tends to 0, through negative values, as $L \to \infty$. We find that $\Sigma^R_{21}(L)$ vanishes for a specific value of $L$, which we denote by $L^{(2)}_{\rm min}$, being negative for all $L > L^{(2)}_{\rm min}$; it also assumes a minimum (negative) value for a value of $L$ we denote by $L^{(2)}_{\rm max}$, for reasons that will be clarified later. Numerically, we find: $L^{(2)}_{\rm min} \simeq 0.78m^{-1}$; $L^{(2)}_{\rm max} \simeq 1.68m^{-1}$; and $\Sigma^{R\,\rm min}_{21} \simeq -0.0445$.

The fact that $\Sigma^R_{21}(L)$ is negative for $L > L^{(2)}_{\rm min}$ has a remarkable influence on the behavior of the renormalized coupling constant. For $D = 2$, Eq. (12) becomes

$$g_{21}(L,\lambda) = \frac{\lambda}{1+\lambda\,\Sigma^R_{21}(L)}. \tag{16}$$

We first note that, independent of the value of $\lambda$, $g_{21}(L,\lambda)$ approaches 0 as $L \to 0$; therefore, the system presents a kind of asymptotic-freedom behavior for short distances. On the other hand, starting from a low value of $L$ (within the region of asymptotic freedom) and increasing the size of the system, $g_{21}$ will present a divergence at a finite value of $L$ ($L^{(2)}_c$), if the value of the fixed coupling constant ($\lambda$) is high enough. In fact, this will happen for all values of $\lambda$ above the "critical value" $\lambda^{(2)}_c = (-\Sigma^{R\,\rm min}_{21})^{-1} \simeq 22.5$. We interpret this result by stating that, in the strong-coupling regime ($\lambda \geq \lambda^{(2)}_c$) the system gets spatially confined in a segment of length $L^{(2)}_c(\lambda)$. The behavior of $g_{21}(L,\lambda)$ as a function of *mL* is illustrated in Fig. 1-B, for some values of $\lambda$.

From the definition of $\lambda^{(2)}_c$, we find that, for $\lambda = \lambda^{(2)}_c$, the divergence of $g_{21}(L,\lambda)$ is reached as $L$ approaches the value that makes $\Sigma^R_{21}$ minimal, which we denoted by

$L_{\max}^{(2)}$. On the other hand, since $g_{21}(L,\lambda \to \infty) = \Sigma_{21}^{R}(L)$, $L_c^{(2)}(\lambda)$ tends to $L_{\min}^{(2)}$, the zero of $\Sigma_{21}^{R}$, as $\lambda \to \infty$. In other words, the confining length $L_c^{(2)}(\lambda)$ pertains to the interval $\left(L_{\min}^{(2)}, L_{\max}^{(2)}\right]$. For a given value of $\lambda$, the confining length $L_c^{(2)}(\lambda)$ can be found numerically by determining the smallest solution of the equation $1+\lambda\Sigma_{21}^{R}(L)=0$. These results are presented in Fig. 2-A, where we plot $mL_c^{(2)}(\lambda)$ as a function of $l=\lambda/\lambda_c^{(2)}$.

Similar results are obtained for higher dimensions, in particular for $D=3$ and $D=4$ when all the spatial coordinates are compactified to the same length $L$; plots of $\Sigma_{32}^{R}(L)$ and $\Sigma_{43}^{R}(L)$, and those of $g_{32}(L,\lambda)$ and $g_{43}(L,\lambda)$, have the same shapes as that in Fig. 1 but with rather distinct scales. Also, although the values of $\lambda_c^{(3)}$ and $\lambda_c^{(4)}$ are very different from $\lambda_c^{(2)}$, the minimum and maximum values of $L_c^{(3)}$ and $L_c^{(4)}$ are close to the corresponding values for $D=2$, the behavior of $L_c^{(D)}(\lambda)$ being like that in Fig. 2-A.

### 3.2. Compactified GN model at $T \neq 0$

We now discuss the effect of raising the temperature on the renormalized effective coupling constant for the Gross-Neveu model with all spatial dimensions compactified. Finite temperature is introduced through the compactification of the time coordinate, with the compactification length given by $L_D = \beta = 1/T$. Although in an Euclidean theory time and space coordinates are treated on the same footing, the interpretation of their compactifications are rather distinct. On general grounds, we expect that the dependence of $\Sigma_D^R$ and $g_D$ on $\beta$ should follow similar patterns as that for the dependence with $L$. In fact, as $\beta \to 0$ (that is, $T \to \infty$), $\Sigma_D^R \to \infty$ implying that $g_D \to 0$, independently of the value of $\lambda$; thus, we have asymptotic-freedom behavior for very high temperatures. Therefore, we expect that, starting from the compactified model at $T=0$ with $\lambda \geq \lambda_c^{(D)}$, raising the temperature will lead to the suppression of the divergence of $g_D$ and the consequent deconfinement of the system.

Consider the fully compactified GN model in $D=2$, with the time coordinate ($x_2$) compactified in a "length" $L_2=\beta=1/T$; the $(L,\beta)$- dependent bubble diagram, with $L$ and $\beta$ measured in units of $m^{-1}$, is then given by

$$\begin{aligned}\Sigma_{22}^{R}(L,\beta) &= 2E_1(2L)-E_1(L)+2E_1(2\beta)-E_1(\beta) \\ &\quad +2E_2(L,\beta)-4E_2(2L,\beta)-4E_2(L,2\beta)+8E_2(2L,2\beta), \quad (17)\end{aligned}$$

where the function $E_1(x)$ is given by Eq. (15) and the function $E_2(x,y)$ is defined by

$$\begin{aligned}E_2(x,y) &= \frac{1}{\pi}\sum_{n,l=1}^{\infty}\left\{\left(\sqrt{x^2n^2+y^2l^2}\right)K_1\left(\sqrt{x^2n^2+y^2l^2}\right)\right. \\ &\quad \left. -K_0\left(\sqrt{x^2n^2+y^2l^2}\right)\right\}. \quad (18)\end{aligned}$$

Firstly, notice that if $\beta \to \infty$, all terms depending on $\beta$ vanish and $\Sigma_{22}^{R}(L,\beta)$ reduces to the expression for $T=0$, $\Sigma_{21}^{R}(L)$. On the other hand, independently of the value of $\lambda$, if

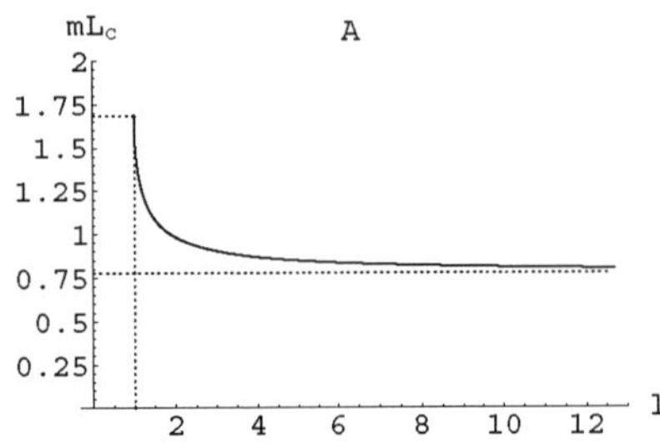


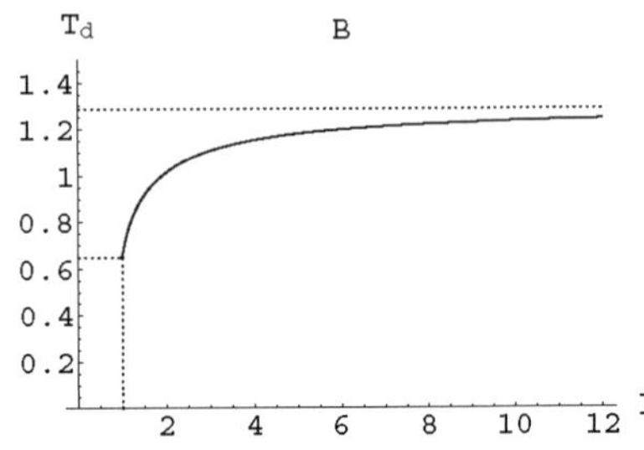


**FIGURE 2.** A - Zero-temperature confining length (in units of $m^{-1}$), as a function of $l = \lambda/\lambda_c^{(2)}$; the horizontal dashed lines correspond to the limiting values $mL_{\min}^{(2)} \simeq 0.78$ and $mL_{\max}^{(2)} \simeq 1.68$. B - Deconfining temperature $T_d^{(2)}(\lambda)$ (in units of $m$), as a function of $l = \lambda/\lambda_c^{(2)}$; the horizontal dashed lines correspond to the limiting values $T_{\min}^{(2)} \simeq 0.65m$ and $T_{\max}^{(2)} \simeq 1.29m$.

$\beta \to 0$, $\Sigma_{22}^R(L,\beta) \to \infty$ and the system becomes asymptotically free. Since the system is spatially confined in a length $L_c^{(2)}(\lambda)$, for $\lambda \geq \lambda_c^{(2)}$ at $T = 0$, we expect that by raising the temperature we reach a value for which the divergence in $g_2$ disappears and the system becomes spatially unconfined.

We can determine $T_d^{(2)}(\lambda)$ by analyzing the behavior of $g_{22}^{-1}(L,\beta,\lambda)$ as $\beta$ is decreased from infinite, for a given value of $\lambda \geq \lambda_c^{(2)}$; the deconfining temperature corresponds to the value of $\beta$, $\beta_d^{(2)}(\lambda) = 1/T_d^{(2)}(\lambda)$, for which the minimum of $g_{22}^{-1}$ (taken as a function of $L$) vanishes. We find that $T_d^{(2)}(\lambda) \in [T_{d\min}^{(2)}, T_{d\max}^{(2)})$, with $T_{d\min}^{(2)} = 0.65m$ and $T_{d\max}^{(2)} = 1.29m$; the dependence of the deconfining temperature on $\lambda$ is presented in Fig. 2-B. Although the involved expressions are more complicated, we find that the deconfining temperature for the cases $D = 3$ and $D = 4$ behave similarly to $T_d^{(2)}(\lambda)$, with values of the same order of magnitude.

## 4. FINAL REMARKS

We have analyzed the $N$-component D-dimensional massive Gross-Neveu model with compactified spatial dimensions, both at zero and finite temperatures. The large-$N$ renormalized coupling constant $g$, for $T = 0$, shows a kind of asymptotic freedom behavior, vanishing when the comapctification length tends to zero, irrespective of the value of the fixed coupling constant $\lambda$. In the strong coupling regime, where the fixed coupling constant is greater than some critical value, starting from small compactification lengths and increasing the size of the system, a divergence of the renormalized effective coupling constant appears at a given length $L_c(\lambda)$, signalizing that the system gets spatially confined. When the temperature is raised, a deconfining transition occurs at a temperature $T_d(\lambda)$, as the minimum value of the inverse of the renormalized effective coupling constant reaches zero. Such general aspects of the compactified GN model hold for arbitrary dimensions $D$ in a similar fashion to the case of $D = 2$, presented in detail here.

**TABLE 1.** Limit values of $L_c^{(D)}(\lambda)$ and $T_d^{(D)}(\lambda)$, with $m$ being the constituent quark mass ($\approx 350\,\mathrm{MeV}$).

| $D$ | $L_{\min}^{(D)}$ [fm] | $L_{\max}^{(D)}$ [fm] | $T_{\min}^{(D)}$ [MeV] | $T_{\max}^{(D)}$ [MeV] |
|---|---|---|---|---|
| 2 | 0.45 | 0.96 | 227 | 451 |
| 3 | 0.74 | 1.2 | 189 | 304 |
| 4 | 0.96 | 1.35 | $< 170$ | 245 |

It should be emphasized that these results are intrinsic to the model, not emerging from any adjustment. The limit values of $L_c^{(D)}(\lambda)$ and $T_d^{(D)}(\lambda)$ depend only on the fermion mass. Thus, to get an estimate of these values we have to fix the parameter $m$. To do so, we consider the Gross-Neveu model as an effective theory for the strong interaction (in which the gluon propagators have been shrunk, similarly to the Fermi treatment of the week force) and take $m$ to be the constituent quark mass, $m \approx 350\,\mathrm{MeV} \simeq 1.75\,\mathrm{fm}^{-1}$ [10]. With such a choice, we obtain the limit values of the confining length and the deconfining temperature presented in Table 1, for the cases where $D = 2,3,4$. These values should be compared with the experimentally measured proton charge diameter ($\approx 1.74\,\mathrm{fm}$) [11] and the estimated deconfining temperature ($\approx 200\,\mathrm{MeV}$) for hadronic matter [12].

## ACKNOWLEDGMENTS

This work was partially supported by CNPq/MCT (Brazil) and NSERC (Canada). J.M.C.M. thanks M. A. Pérez and H. Castilla for the invitation, and FAPESB/BA (Brazil) for financial aid, for his participation in the VI-SILAFAE.

## REFERENCES

1. D. J. Gross and A. Neveu, *Phys. Rev. D* **10**, 3235 (1974).
2. A. P. C. Malbouisson, J. M. C. Malbouisson, A. E. Santana and J. C. Silva, *Phys. Lett. B* **583**, 373 (2004).
3. F. C. Khanna, A. P. C. Malbouisson, J. M. C. Malbouisson, H. Queiroz, T. M. Rocha-Filho, A. E. Santana and J. C. Silva, *Phys. Lett. B* **624**, 316 (2005).
4. A. P. C. Malbouisson, F. C. Khanna, J. M. C. Malbouisson and A. E. Santana, *Braz. J. Phys.* **36**, 1165 (2006).
5. A. Chodos, R. L. Jaffe, K. Johnson, C. B. Thorn and V. F. Weisskopf, *Phys. Rev D* **9**, 3471 (1974).
6. C. A. Lutken and F. Ravndal, *J. Phys. A: Math. Gen.* **21**, L793 (1988).
7. E. Elizalde and A. Romeo, *J. Math. Phys.* **30**, 1133 (1989).
8. K. Kirsten, *J. Math. Phys.* **35**, 459 (1994).
9. A. P. C. Malbouisson, J. M. C. Malbouisson and A. E. Santana, *Nucl. Phys. B* **631**, 83 (2002).
10. Particle Data Group, *Phys. Lett. B* **592**, 1 (2004); see page 475.
11. S. G. Karshenboim, *Can. J. Phys.* **77**, 241 (1999).
12. A. Smilga, *Lectures on QCD*, World Scientific, Singapore, 2001, p. 279.

# QED in the worldline representation

Christian Schubert

*Instituto de Física y Matemáticas, Universidad Michoacana de San Nicolás de Hidalgo, Edificio C-3, Ciudad Universitaria, C.P. 58040 Morelia, Michoacan, Mexico, schubert@ifm.umich.mx*

**Abstract.** Simultaneously with inventing the modern relativistic formalism of quantum electrodynamics, Feynman presented also a first-quantized representation of QED in terms of worldline path integrals. Although this alternative formulation has been studied over the years by many authors, only during the last fifteen years it has acquired some popularity as a computational tool. I will shortly review here three very different techniques which have been developed during the last few years for the evaluation of worldline path integrals, namely (i) the "string-inspired formalism", based on the use of worldline Green functions, (ii) the numerical "worldline Monte Carlo formalism", and (iii) the semiclassical "worldline instanton" approach.

**Keywords:** Quantum electrodynamics, perturbation theory, worldline, string inspired formalism
**PACS:** 11.15.Bt,11.15.Kt,11.25.Db,12.20.Ds

## 1. FEYNMAN'S WORLDLINE REPRESENTATION OF QED

In 1950 Feynman presented, in an appendix to one of his groundbreaking papers on the modern, manifestly relativistic formalism of perturbative QED [1], also a first-quantized formulation of scalar QED, "for its own interest as an alternative to the formulation of second quantization". There he provides a simple rule for constructing the complete scalar QED S-matrix by representing virtual scalars and photons in terms of relativistic particle path integrals, and coupling them in all possible ways. Restricting ourselves, for simplicity, to the purely photonic part of the S-matrix (no external scalars), and moreover to the "quenched" contribution (only one virtual scalar), this "worldline representation" can be given most compactly in terms of the (quenched) effective action $\Gamma[A]$:

$$\Gamma_{\rm scalar}[A] = \int d^4x\,\mathscr{L}_{\rm scalar}[A] = \int_0^\infty \frac{dT}{T}\,\mathrm{e}^{-m^2T}\int_{x(T)=x(0)}\mathscr{D}x(\tau)\,e^{-S[x(\tau)]} \tag{1}$$

Here $T$ denotes the proper-time of the scalar particle in the loop, $m$ its mass, and $\int_{x(T)=x(0)}\mathscr{D}x(\tau)$ a path integral over all closed loops in spacetime with fixed periodicity in the proper-time. The worldline action $S[x(\tau)]$ has three parts,

$$S = S_0 + S_{\rm ext} + S_{\rm int} \tag{2}$$

(see fig. 1). Of these, the kinetic term $S_0$ describes the free propagation of the scalar, $S_{\rm ext}$ its interaction with the external field, and $S_{\rm int}$ the corrections due to internal photon exchanges in the loop. The connection to a standard Feynman diagrammatic description is made simply by expanding out the two interaction exponentials.

CP917, *Particles and Fields,* edited by H. Castilla Valdez, J. C. D'Olivo, and M. A. Perez

$$S_0 = \int_0^T d\tau \frac{\dot{x}^2}{4} \qquad \text{(free propagation)}$$

$$S_{\text{ext}} = ie \int_0^T \dot{x}^\mu A_\mu(x(\tau)) \qquad \text{(external photons)}$$

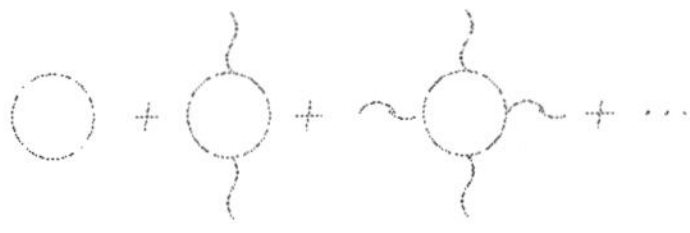

$$S_{\text{int}} = -\frac{e^2}{8\pi^2} \int_0^T d\tau_1 \int_0^T d\tau_2 \frac{\dot{x}(\tau_1) \cdot \dot{x}(\tau_2)}{(x(\tau_1) - x(\tau_2))^2} \qquad \text{(internal photons)}$$

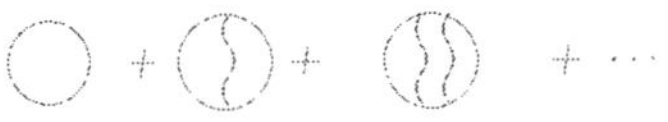

**FIGURE 1.** Perturbative expansion of the worldline path integral.

The generalization of this representation to include multiple scalar loops and open scalar lines is straightforward. It yields a first-quantized representation of the full effective action, and thus, by Fourier transformation, of the S-matrix.

While for scalar QED this representation is essentially unique, when generalizing it to spinor QED one has a number of choices. First, worldline representations of spin half particles can be derived either from standard first-order Dirac theory, or from its second-order formulation (see [2] and refs. therein), based on the identity

$$(\not{\partial} + e\not{A})^2 = -(\partial + ieA)^2 - \frac{i}{2} e \sigma^{\mu\nu} F_{\mu\nu} \tag{3}$$

Contrary to the situation in second-quantized field theory, in the worldline formalism the second-order approach is the more standard one, and I will restrict myself to it in this review (see [3, 4] for the first-order approach). Using (3) one arrives at a worldline representation for the fermion QED effective action $\Gamma_{\text{spinor}}[A]$ which, at the one-loop level, differs from the scalar one (1) only by the addition of a global factor of $-\frac{1}{2}$, and the insertion of a *spin factor* $S[x,A]$ under the path integral [5],

$$S[x,A] = \text{tr}_\Gamma \mathscr{P} \exp\left[\frac{i}{2} e \sigma^{\mu\nu} \int_0^T d\tau F_{\mu\nu}(x(\tau))\right] \tag{4}$$

A more modern way of writing the same spin factor is in terms of an additional Grassmann path integral [6, 7, 8, 9],

$$S[x,A] = \int \mathscr{D}\psi(\tau)\exp\left[-\int_0^T d\tau\left(\frac{1}{2}\psi\cdot\dot{\psi} - ie\psi^\mu F_{\mu\nu}\psi^\nu\right)\right] \quad (5)$$

Here the path integration is over the space of anticommuting functions antiperiodic in proper-time, $\psi^\mu(\tau_1)\psi^\nu(\tau_2) = -\psi^\nu(\tau_2)\psi^\mu(\tau_1)$, $\psi^\mu(T) = -\psi^\mu(0)$. The main advantage of introducing this second path integral is that, as it turns out, there is a "worldline" supersymmetry between the coordinate function $x(\tau)$ and the spin function $\psi(\tau)$ [8]. Although this supersymmetry is broken by the different periodicity conditions for $x$ and $\psi$, it still has a number of useful computational consequences. In particular, introducing a worldline superformalism allows one to combine the two path integrals, and to write down a formula for $\Gamma_{\text{spinor}}[A]$ which is completely analogous to (1), (2), including the internal photon corrections [10].

A third, even more subtle, way of implementing spin on the worldline is the "Polyakov spin factor", a purely geometric quantity depending only on the worldline itself [11, 12]. This aspect of spin has been studied by a number of authors, but usually using the first-order formalism; the form of the spin factor appropriate for the second-order formalism has been established only recently [13].

A vast amount of work has been done on these QED worldline representations and their generalizations to other background fields and couplings (see [14] for an extensive list of references). Nevertheless, most of the earlier work on this subject is concerned with formal aspects of relativistic particle Lagrangians, rather than with attempts at performing state-of-the-art calculations in quantum field theory (notable exceptions are [15, 16] and [17]). It is only during the last fifteen years that the usefulness of the first-quantized approach as an alternative to standard Feynman diagrammatic methods has been seriously investigated. This development was triggered by string theory, where first-quantized path integrals are the standard tool for perturbative calculations of scattering amplitudes, and by the fact that many amplitudes in quantum field theory can be represented as infinite string tension limits of the corresponding string amplitudes. Bern and Kosower [18] investigated this limit in detail for the case of the QCD $N$ gluon amplitudes, and found in this way a new set of rules for the construction of these amplitudes. Shortly later, Strassler's work [19] showed that the same type of "string-inspired" representation of photon/gluon amplitudes can also be obtained more directly by evaluating the corresponding first-quantized path integrals using appropriate worldline Green functions. This approach was then generalized to some cases of multiloop amplitudes [20, 10, 21, 22], as well as to QED amplitudes in a constant external field [23, 24]. Moreover, the success of this program led to the development of alternative techniques for the calculation of worldline path integrals. In this talk, I will restrict myself to the computational aspects of the worldline representation, and to the prototypical case, QED. I will discuss, in turn, three quite different methods which are now available for the computation of the scalar/spinor QED worldline path integrals, (i) the original "string-inspired" approach, (ii) the numerical "worldline Monte Carlo" method and (iii) the semiclassical "worldline instanton" technique.

## 2. THE "STRING-INSPIRED" APPROACH

The strategy in the "string-inspired approach" is simple. The path integral(s) will be manipulated into Gaussian form, after which they can be performed using worldline correlators with the appropriate periodicity properties. For the one-loop path integrals (1),(5) those are [25, 19]

$$
\begin{aligned}
\langle y^\mu(\tau_1) y^\nu(\tau_2)\rangle &= G_B(\tau_1,\tau_2)\,\delta^{\mu\nu}\\
G_B(\tau_1,\tau_2) &= |\tau_1-\tau_2| - \frac{1}{T}\Bigl(\tau_1-\tau_2\Bigr)^2\\
\langle \psi^\mu(\tau_1) \psi^\nu(\tau_2)\rangle &= G_F(\tau_1,\tau_2)\,\delta^{\mu\nu}\\
G_F(\tau_1,\tau_2) &= \mathrm{sign}(\tau_1-\tau_2)
\end{aligned}
\tag{6}
$$

Here $y^\mu(\tau)\equiv x^\mu(\tau)-x_0^\mu$, where $x_0^\mu$ denotes the loop center of mass, $x_0^\mu\equiv\frac{1}{T}\int_0^T d\tau x^\mu(\tau)$. There is a certain freedom in choosing the coordinate correlator $G_B$ (see, e.g., [14]); the one given above is usually the most convenient one, but some of the references given below use the alternative $G_B(\tau_1,\tau_2)=|\tau_1-\tau_2|-(\tau_1+\tau_2)+\frac{2}{T}\tau_1\tau_2$.
Now, to get gaussian form,

- Expand all the interaction exponentials $\mathrm{e}^{-S_{\mathrm{ext}}[x(\tau)]}$ etc.
- For the effective action: Taylor expand the external field at the loop center of mass,

$$
A^\mu(x(\tau)) = \mathrm{e}^{y(\tau)\cdot\partial}A^\mu(x_0)
\tag{7}
$$

- For the $N$ photon amplitudes : Expand the field in $N$ plane waves,

$$
A^\mu(x(\tau)) = \sum_{i=1}^{N}\varepsilon_i^\mu\,\mathrm{e}^{ik_i\cdot x(\tau)}
\tag{8}
$$

- Exponentiate the denominator of the photon insertion terms,

$$
\begin{aligned}
-\frac{e^2}{8\pi^2}\int_0^T d\tau_i\int_0^T d\tau_j\frac{\dot{x}(\tau_i)\cdot\dot{x}(\tau_j)}{(x(\tau_i)-x(\tau_j))^2} \quad &=\\
-\frac{e^2}{2}\int_0^\infty\frac{d\bar{T}}{(4\pi\bar{T})^2}\int_0^T d\tau_2\,\dot{x}(\tau_i)\cdot\dot{x}(\tau_j)\exp\left[-\frac{(x(\tau_i)-x(\tau_j))^2}{4\bar{T}}\right]&
\end{aligned}
\tag{9}
$$

For example, for the four-photon amplitude in scalar QED this procedure yields the following integral representation:

$$\Gamma[k_1,\varepsilon_1;\ldots;k_4,\varepsilon_4] = e^4\int_0^\infty \frac{dT}{T}[4\pi T]^{-\frac{D}{2}}e^{-m^2T}\prod_{i=1}^4\int_0^T d\tau_i\, Q_4\, \mathrm{e}^{G_{Bij}k_i\cdot k_j} \tag{10}$$

$$Q_4 = Q_4^4+Q_4^3+Q_4^2-Q_4^{22} \tag{11}$$

$$\begin{aligned}
Q_4^4 &= \dot G_{B12}\dot G_{B23}\dot G_{B34}\dot G_{B41}Z_4(1234)+2\text{ permutations}\\
Q_4^3 &= \dot G_{B12}\dot G_{B23}\dot G_{B31}Z_3(123)\dot G_{B4i}\varepsilon_4\cdot k_i+3\text{ perm.}\\
Q_4^2 &= \dot G_{B12}\dot G_{B21}Z_2(12)\Big\{\dot G_{B3i}\varepsilon_3\cdot k_i\dot G_{B4j}\varepsilon_4\cdot k_j+\frac{1}{2}\dot G_{B34}\varepsilon_3\cdot\varepsilon_4\Big[\dot G_{B3i}k_3\cdot k_i-\dot G_{B4i}k_4\cdot k_i\Big]\Big\}\\
&\quad +5\text{ perm.}\\
Q_4^{22} &= \dot G_{B12}\dot G_{B21}Z_2(12)\dot G_{B34}\dot G_{B43}Z_2(34)+2\text{ perm.}
\end{aligned} \tag{12}$$

$$\begin{aligned}
Z_2(ij) &\equiv \varepsilon_i\cdot k_j\varepsilon_j\cdot k_i-\varepsilon_i\cdot\varepsilon_j k_i\cdot k_j\\
Z_n(i_1i_2\ldots i_n) &\equiv \mathrm{tr}\prod_{j=1}^n\Big[k_{i_j}\otimes\varepsilon_{i_j}-\varepsilon_{i_j}\otimes k_{i_j}\Big]\quad (n\geq 3)
\end{aligned} \tag{13}$$

Here each of the integrals $\int_0^T d\tau_i$ represents one of the four photon legs moving around the scalar loop. $G_{Bij}$ stands for $G_B(\tau_i,\tau_j)$ and $\dot G_{Bij}$ for its derivative. Repeated indices are to be summed over $1,\ldots,N$. In writing down the integrand, a number of integrations by parts have already been performed which eliminated all second derivatives of the $G_{Bij}$'s that initially appear in the factor $Q_4$. Moreover, this factor has been decomposed into terms $Q_N^m$ according to "cycle content", indicated by the superscript, where a "cycle" is a factor of $\dot G_{B_{i_1i_2}}\dot G_{B_{i_2i_3}}\dot G_{B_{i_3i_4}}\cdots\dot G_{B_{i_ni_1}}$. Such a '$\tau$-cycle' always gets multiplied by a 'Lorentz-cycle' $Z_n(i_1i_2\ldots i_n)$, which is the trace of the products of the field strength tensors associated to the corresponding legs. The various $Q_N^m$'s are individually gauge invariant. This representation can be generalized to an arbitrary number of photons, and is unique if one requests manifest permutation symmetry in the photon legs [18, 19, 26]. The generalization to higher loop photon amplitudes can be achieved either using (9), or by explicit sewing of pairs of external photons, or more directly using generalized worldline Green functions adapted to the multiloop graph topologies [20, 21, 27].

In all cases the arising multiple integrals are equivalent to standard Feynman parameter integrals. For example, in (10) and its $N$-point generalizations, the prefactor $Q_N$ relates to a Feynman numerator and the exponential factor to the universal denominator of one-loop $N$-point integrals (see, e.g., [28]). However, the string-inspired representation has a number of interesting properties which are not usually manifest in standard Feynman parameter calculations:

1. The use of the cycle notation allows a more compact way of writing the $N$ photon amplitudes than usual.

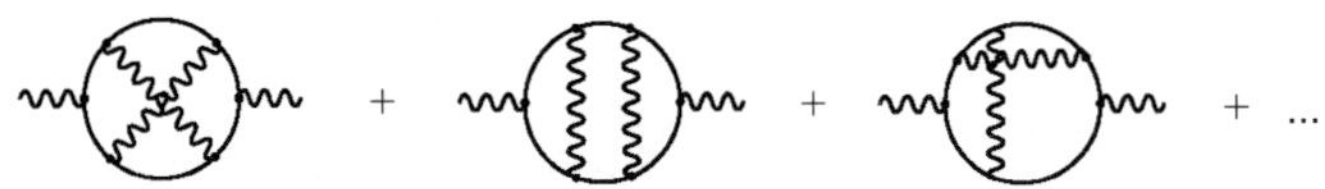

**FIGURE 2.** Diagrams contributing to the three loop QED photon propagator.

2. There is a simple "cycle replacement rule" which allows one to construct the integrand of the spinor loop $N$ photon amplitude from the scalar loop one [18, 19]. This implies that, in this formalism, the calculations of the same quantity in scalar and spinor QED are not independent; any spinor QED calculation yields the corresponding scalar QED result as a byproduct.
3. The integral (10) and its higher $N$ generalizations represent the whole amplitude, with no need to sum over permutations. This property is not very relevant at the one-loop level, but becomes interesting at higher loop orders. In the QED case, it generally allows one to combine into one integral all contributions from Feynman diagrams which can be identified by letting photon legs slide along scalar/electron loops or lines. As an example, we show in fig. 2 the "quenched" contributions to the three-loop photon propagator.
   This property is particularly interesting in view of the fact that it is precisely this type of sums of diagrams which in QED generally leads to extensive cancellations between diagrams, and to final results which are substantially simpler than intermediate ones (see [29] and refs. therein). And for the two-loop QED $\beta$ function indeed a way was found for calculating the corresponding integral in a way which avoided splitting up the multiple parameter integral into sectors with a fixed ordering of the photon legs, and which led to dramatic simplifications [10]. However, so far no generalization of the method used there to higher loop orders has been found.
4. The string-inspired method provides a particularly convenient way of implementing constant external fields in QED calculations. Photon amplitudes or effective lagrangians in a constant field are obtained from the corresponding vacuum quantities simply by substituting the vacuum Green functions $G_{B,F}(\tau_1,\tau_2)$ by appropriate field-dependent Green functions $\mathcal{G}_{B,F}(\tau_1,\tau_2;F)$, and by a change of the free worldline path integral determinants [23, 24] (see also [30]). In particular, the "cycle replacement rule" carries over to the constant field case.

The efficiency of the technique for QED in a constant field has, at the one-loop level, been demonstrated by recalculations of the photon splitting amplitude in a magnetic field [31], and of the one-loop vacuum polarization in a general constant field [32]. At the two-loop level it has been extensively applied to the QED effective Lagrangian in a constant field. This includes recalculations of the standard two-loop Euler-Heisenberg Lagrangians [24, 33, 34], closed-form expressions for the weak field expansion coefficients of the magnetic two-loop Euler-Heisenberg Lagrangian [35], and a generalization of these Lagrangians to the case of a self-dual Euclidean field [36]. I will show here only the last result, which is particularly nice: a Euclidean self-dual field fulfills

$F_{\mu\nu} = \frac{1}{2}\varepsilon_{\mu\nu\alpha\beta}F^{\alpha\beta}$, which implies that the square of the field strength tensor is proportional to the unit matrix, $F_{\mu\nu}F^{\nu\lambda} = -f^2\delta_\mu^\lambda$. Remarkably, this simplifies matters so much that all parameter integrals can be done in closed form, leading to the following explicit formulas for the two-loop scalar and spinor QED effective Lagrangians in a constant self-dual field [36],

$$\begin{aligned} \mathscr{L}^{(2)}_{\rm scalar}(\kappa) &= \alpha\frac{m^4}{(4\pi)^3}\frac{1}{\kappa^2}\left[\frac{3}{2}\xi^2(\kappa)-\xi'(\kappa)\right] \\ \mathscr{L}^{(2)}_{\rm spinor}(\kappa) &= -2\alpha\frac{m^4}{(4\pi)^3}\frac{1}{\kappa^2}\left[3\xi^2(\kappa)-\xi'(\kappa)\right] \end{aligned} \tag{14}$$

Here $\kappa = \frac{m^2}{2ef}$ and

$$\xi(x) \equiv -x\left(\frac{\Gamma'(x)}{\Gamma(x)} - \ln(x) + \frac{1}{2x}\right) \tag{15}$$

The simplicity of these results made it possible to use this self-dual case for studying the generic properties of the weak and strong field expansions of Euler-Heisenberg Lagrangians [37], as well as the corresponding maximally helicity violating components of the two-loop $N$-photon amplitudes in the low energy limit [36].

Apart from the effective action in a constant field, the string-inspired method has also been extensively applied to the calculation of the full one-loop QED effective action in an arbitrary background field, using the derivative expansion [38, 39].

Moreover, at the one-loop level the method has been generalized to the finite temperature case, both to obtain representations of the $N$ - photon amplitudes [40, 41] and for the effective action in a general background [42].

Finally, as should be clear from the above, work on the string-inspired technique in QED so far has been concerned mainly with purely photonic amplitudes. The extension to amplitudes involving also external scalars [22] or fermions [43, 44] is possible without difficulties, although its practical usefulness is difficult to judge at present due to a lack of state-of-the-art applications.

## 3. THE "WORLDLINE MONTE CARLO" APPROACH

During the last few years it was found that Feynman's worldline representation (1) and its various generalizations are also very amenable to a direct numerical evaluation using standard Monte Carlo techniques [45, 46, 47]. At the one-loop level, this approach has been shown to work well for QED effective actions in quite general background fields. This includes singular fields such as a magnetic field of the step-function type [45]. It also extends to the imaginary part of the effective action [48], to be discussed in section 4 below). First steps towards a multiloop extension have been taken in [49], although, as with all numerical approaches to quantum field theory, implementing the full renormalization program in such a formalism poses a formidable challenge.

The Monte Carlo approach appears to hold particular promise for the calculation of Casimir energies for arbitrary geometries [47, 50, 51]. Although this has so far been done only for scalar fields, not for the QED case, it shall be discussed here since it is interesting to see how Dirichlet boundary conditions can be implemented at the level of the path integral (see the recent [52]) for a treatment of boundary conditions in the "string-inspired" approach). For a (massless) scalar field in a background potential $V(x)$, the gauge coupling term in Feynman's path integral (1) has to be replaced by $-\int_0^T d\tau V(x(\tau))$. Dirichlet boundary conditions on an (infinitely thin) surface $\Sigma$ can be implemented by choosing

$$V(\mathbf{x}) \;=\; \lambda \int_\Sigma d\sigma\, \delta^3(\mathbf{x}-\mathbf{x}_\sigma) \tag{16}$$

with $\lambda \to \infty$. Considering the case of two disjoint surfaces $\Sigma = \Sigma_1 \bigcup \Sigma_2$, it can then easily be shown [47] that the Casimir interaction energy between the surfaces is given by [1]

$$E_{\text{Casimir}} \;=\; -\frac{1}{2}\frac{1}{\sqrt{4\pi}}\int_0^\infty \frac{dT}{T^{\frac{3}{2}}}\int d^3x_0 \int \mathscr{D}\mathbf{x}(\tau)\, e^{-\frac{1}{4}\int_0^T d\tau \dot{\mathbf{x}}^2}\, \Theta_\Sigma[\mathbf{x}(\tau)] \tag{17}$$

where $\Theta_\Sigma = 1$ if the path $\mathbf{x}(\tau)$ intersects both $\Sigma_1$ and $\Sigma_2$, and $\Theta_\Sigma = 0$ otherwise. The trivial time component of the path integral has been integrated out. $\mathbf{x_0}$ denotes the loop center of mass of the remaining integral over spatial loops; the energy density at a point $\mathbf{x_0}$ is obtained by restricting the path integral to worldloops with that point as their common center of mass.

Figure 3 shows the Casimir energy density obtained by a numerical evaluation of (17) for the case of two infinite perpendicular plates at separation $a$ [50] [2]

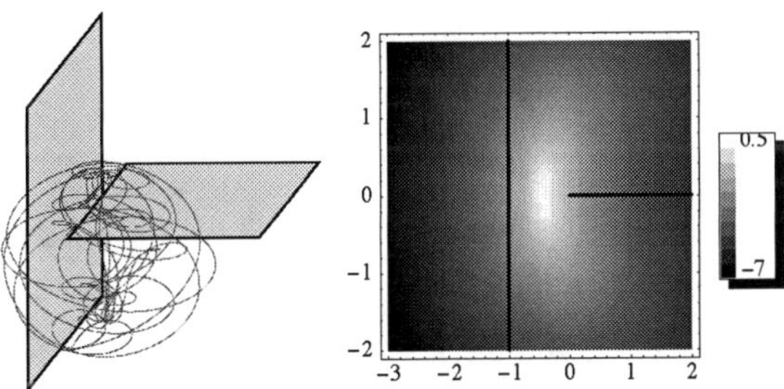


**FIGURE 3.** Left panel: sketch of the perpendicular-plates configuration with (an artist's view of) a typical worldline that intersects both plates. Right panel: Density plot of the effective action density $\mathscr{L}$ for the perpendicular plates case; the plot shows $\ln(2(4\pi a^2)^2|\mathscr{L}|)$.

---

[1] When comparing with [47] note that they use a different normalization of the free path integral.

[2] Fig. 3 is reprinted from [50] with the permission of the authors and of the American Institute of Physics.

A particularly nice example is the case of a sphere of radius $R$ and an infinite plate at a distance $a$ [51]. Here an exact result for the Casimir interaction energy is known as a function of $a/R$ for $a/R \gtrsim 0.1$ [53]. Fig. 4 shows the result of the worldline evaluation for this case. As can be seen from fig. 5, the result is in very good agreement with [53] for the whole parameter range [3].

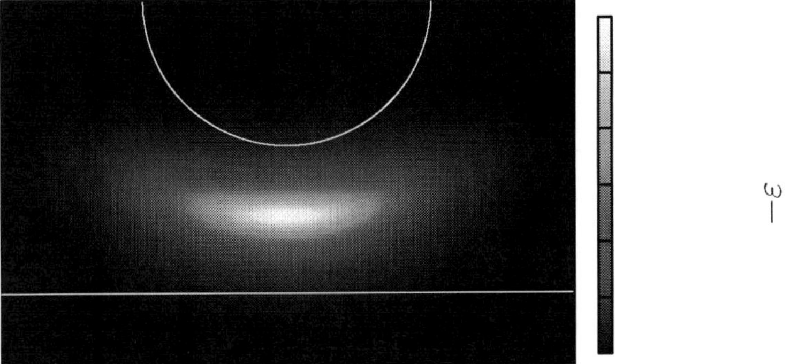

**FIGURE 4.** Contour plot of the negative Casimir interaction energy density for a sphere of radius $R$ above an infinite plate; the sphere-plate separation has been chosen as $a = R$. The plot results from a pointwise evaluation of eq. (17) using worldlines with a common center of mass.

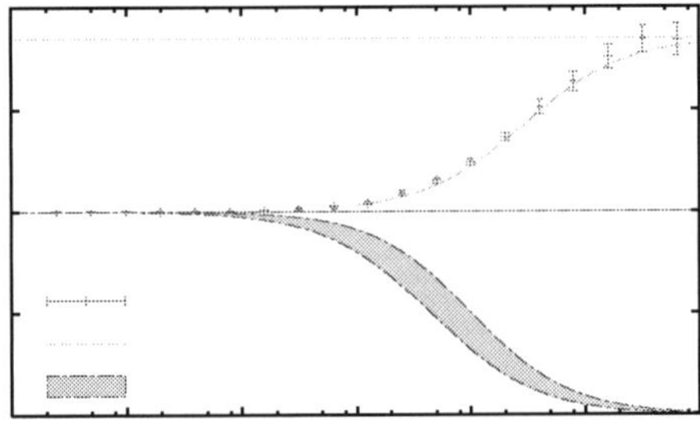

**FIGURE 5.** Casimir interaction energy of a sphere with radius $R$ and an infinite plate vs. the curvature parameter $a/R$. The energy is normalized with respect to the zeroth-order proximity force approximation (PFA) [51]. The numerical worldline result is compared to the exact result of [53] and the PFA estimate.

---

[3] Figs. 4 and 5 are reprinted from [51] with the permission of the authors and of the American Physical Society (copyright by APS, http://link.aps.org/abstract/PRL/v96/e220401) .

# 4. THE "WORLDLINE INSTANTON" APPROACH

We come now to a third approach which is more specialized than the two previous ones, since it applies only to the imaginary part of amplitudes or the effective action. Although the basic idea was presented by Affleck et al. in 1982 [17], it seems not to have been followed up on until very recently [54, 55, 56].

The work of Affleck et al. concerned the imaginary part of the scalar QED effective Lagrangian in a constant electric field. This quantity has been of much interest ever since Schwinger showed [57] that its existence implies the possibility of electron-positron pair creation in vacuum by the electric field. For small production rates this rate is simply given by twice the imaginary part itself, $P_{\text{production}} \approx 2\text{Im}\mathscr{L}[E]$. Moreover, Schwinger was able to explicitly calculate the imaginary part in terms of a sum of exponentials,

$$\text{Im}\mathscr{L}_{\text{spinor}}(E) = \frac{m^4}{8\pi^3}\beta^2 \sum_{n=1}^{\infty} \frac{1}{n^2} \exp\left[-\frac{\pi n}{\beta}\right] \tag{18}$$

($\beta = eE/m^2$). In this sum the $n$th term relates to the coherent production of $n$ pairs by the field. The appearance of $E$ in the denominator of the exponents indicates that the pair creation effect is nonperturbative in nature. It also suggests an interpretation as a tunnel effect where a virtual electron-positron pair separates out and extracts a sufficient amount of energy from the field to turn real.

The corresponding formula for scalar QED differs only by a global factor and signs (due to the difference in statistics),

$$\text{Im}\mathscr{L}_{\text{scalar}}(E) = -\frac{m^4}{16\pi^3}\beta^2 \sum_{n=1}^{\infty} \frac{(-1)^n}{n^2} \exp\left[-\frac{\pi n}{\beta}\right] \tag{19}$$

The production rates are exponentially small for

$$E \ll E_{\text{crit}} = \frac{m^2}{e} = 1.3 \times 10^{18}\,\text{V/m} \tag{20}$$

Until a few years ago producing an electric field close to this critical field strength $E_{\text{crit}}$, and with a sufficient spatial extension, appeared far out of the reach of laboratory experiments. However, due to recent advances in laser technology it seems now conceivable that pair production could be seen in laser fields in the near future. The optical laser POLARIS, under construction at the Jena high-intensity laser facility, is projected to reach a maximal field strength of $E_{\max} \approx 2 \times 10^{14}\,\text{V/m}$ [58], while the European X-ray free electron laser (XFEL), under construction at DESY, is expected to come even closer, $E_{\max} \approx 1.2 \times 10^{16}\,\text{V/m}$ [59].

For realistic laser experiments it is usually far from clear whether the constant field approximation is justified; it would be preferable to have generalizations of Schwinger's formula (18) to inhomogeneous and time-dependent fields. This is a subject which has been pursued by many authors, beginning with Keldysh's seminal work in 1965

[60], and many results have been obtainedover the years (see, e.g. [60, 61, 62, 63, 64, 65, 66, 67, 68, 69]). However, considering this large volume of work there is a surprising lack of variety in the calculation methods used; excepting a few special field configurations, virtually all results of a more general nature have been obtained by WKB, or some variant of it. The formalism developed below is, while similar in spirit to WKB, technically quite different.

Let us start with retracing Affleck et al.'s [17] recalculation of the Schwinger formula for scalar QED, (19). At one loop, Feynman's representation (1) reads

$$\Gamma_{\text{scalar}}[A] = \int_0^\infty \frac{dT}{T} \, \mathrm{e}^{-m^2 T} \int \mathscr{D}x \, \mathrm{e}^{-\int_0^T d\tau \left( \frac{\dot{x}^2}{4} + ieA\cdot\dot{x} \right)} \tag{21}$$

Rescaling $\tau = Tu$, this becomes

$$\Gamma_{\text{scalar}}[A] = \int_0^\infty \frac{dT}{T} \, \mathrm{e}^{-m^2 T} \int \mathscr{D}x \, \mathrm{e}^{-\left( \frac{1}{T} \int_0^1 du \dot{x}^2 + ie \int_0^1 du A\cdot\dot{x} \right)} \tag{22}$$

The $T$ integral has a stationary point at

$$T_c = \frac{\sqrt{\int du \dot{x}^2}}{m} \tag{23}$$

This allows us to calculate its imaginary part using a stationary phase approximation, yielding

$$\mathrm{Im}\Gamma_{\text{scalar}} = \frac{1}{m} \sqrt{\frac{2\pi}{T_c}} \, \mathrm{Im} \int \mathscr{D}x \, \mathrm{e}^{-\left( m\sqrt{\int du \dot{x}^2} + ie \int_0^1 du A\cdot\dot{x} \right)} \tag{24}$$

with a new worldline action,

$$S = m\sqrt{\int du \dot{x}^2} + ie \int_0^1 du A \cdot \dot{x} \tag{25}$$

We would like to calculate the remaining path integral using a stationary phase approximation, too. The action (25) is stationary if

$$m \frac{\ddot{x}_\mu}{\sqrt{\int du \dot{x}^2}} = ieF_{\mu\nu}\dot{x}_\nu \tag{26}$$

Contracting with $\dot{x}^\mu$ yields $\dot{x}^2 = \text{constant} \equiv a^2$, and thus

$$m\ddot{x}_\mu = ieaF_{\mu\nu}\dot{x}_\nu \tag{27}$$

Thus the extremal action trajectory $x^{\rm cl}(u)$, the "worldline instanton", will be a periodic solution of the Lorentz force equation, with a parameter $a$ to be determined by the periodicity condition. Once the instanton is found, its worldline action immediately provides a semiclassical approximation for the imaginary part of the effective Lagrangian. Closer inspection shows that this approximation also corresponds to a weak field approximation [54]:

$$\mathrm{Im}\mathscr{L}_{\rm scalar}(E) \overset{\mathrm{E}\to 0}{\sim} \mathrm{e}^{-S[x^{\rm cl}]} \tag{28}$$

For a the case of a constant electric field, $\vec{E} = (0,0,E) = const.$, it is easy to see that the periodicity condition is solved by

$$a_n = \frac{m}{eE} 2n\pi, \qquad n \in \mathbf{Z}^+ \tag{29}$$

The worldline instantons are simply circles in the $t-z$ plane, with a winding number $n$:

$$\begin{aligned} x_n^{\rm cl}(u) &= \frac{m}{eE}\Big(x_1, x_2, \cos(2n\pi u), \sin(2n\pi u)\Big) \\ S[x_n^{\rm cl}] &= n\pi\frac{m^2}{eE} \end{aligned} \tag{30}$$

The evaluation of the worldline action on the $n$th instanton thus yields just the $n$th exponent in Schwinger's formula (19). Moreover, Affleck et al. were able to compute also the prefactor, which here involves the determinant of fluctuations around the instanton path (see below). Thus, this approach provides a very simple and elegant rederivation of Schwinger's result for scalar QED.

We will now sketch how to generalize this approach to general electric fields, as well as to the spinor QED case; see [54, 55, 56] for the details. Starting all over from Feynman's representation of the one-loop effective action in scalar QED (1), for the general case we prefer not to eliminate the $T$ integral, but rather to first seek a stationary phase approximation for the path integral. The stationarity condition is again the Lorentz force equation,

$$\ddot{x}_\mu = 2ieF_{\mu\nu}(x)\dot{x}_\nu \tag{31}$$

We fix a point on the loop:

$$\int_{x(T)=x(0)} \mathscr{D}x(\tau)\, e^{-S[x(\tau)]} = \int d^4x^{(0)} \int_{x(T)=x(0)=x^{(0)}} \mathscr{D}x(\tau)\, e^{-S[x(\tau)]} \tag{32}$$

Let us assume that we have found a worldline instanton $x^{\rm cl}(\tau)$, a classical solution with $x(T) = x(0) = x^{(0)}$. We expand around $x^{\rm cl}$:

$$x_\mu(\tau) = x^{\mathrm{cl}}_\mu(\tau) + \eta_\mu(\tau), \qquad \eta_\mu(0) = \eta_\mu(T) = 0. \tag{33}$$

Obtain the operator of quadratic fluctuations (Hess matrix) $\Lambda_{\mu\nu}$

$$\Lambda_{\mu\nu} = -\frac{1}{2}\delta_{\mu\nu}\frac{d^2}{d\tau^2} - \frac{d}{d\tau}Q_{\nu\mu} + Q_{\mu\nu}\frac{d}{d\tau} + R_{\mu\nu}, \tag{34}$$

where

$$Q_{\mu\nu} = \frac{\partial^2 L}{\partial x_\mu \partial \dot{x}_\nu}, \qquad R_{\mu\nu} = \frac{\partial^2 L}{\partial x_\mu \partial x_\nu} \tag{35}$$

Find the zero modes $\eta^{(\lambda)}_\nu(\tau)$ of $\Lambda_{\mu\nu}$,

$$\Lambda_{\mu\nu}\eta^{(\lambda)}_\nu = 0 \tag{36}$$

with initial value conditions

$$\eta^{(\lambda)}_\nu(0) = 0, \qquad \dot{\eta}^{(\lambda)}_\nu(0) = \delta_{\nu\lambda}, \qquad (\mu,\nu = 1,2,3,4) \tag{37}$$

Evaluate them at $\tau = T$. Then, the final result for the semiclassical approximation becomes [55]

$$\int_{x(T)=x(0)=x^{(0)}} \mathscr{D}x(\tau)\, e^{-S[x(\tau)]} = \frac{\mathrm{e}^{i\theta}\,\mathrm{e}^{-S[x^{\mathrm{cl}}](T)}}{(4\pi T)^2} \sqrt{\frac{\left|\det\left[\eta^{(\lambda)}_{\mu,\mathrm{free}}(T)\right]\right|}{\left|\det\left[\eta^{(\lambda)}_{\mu}(T)\right]\right|}} \tag{38}$$

Note that the calculation of this quite general fluctuation determinant has been reduced to the one of the $4\times 4$ matrix of zero modes. This remarkable simplification relies on a theorem by Levit and Smilansky [70]. Finally, at the very end one does $\int dT$ using the stationary phase method again.

For a number of classical special cases of "planar" non-constant fields, such as the single-pulse time dependent field [63, 64], the single-bump space dependent field [62], the sinusoidal time-dependence [61, 64], and the sinusoidal space-dependence, the worldline instantons can be found explicitly in terms of special functions [54], leading to simple explicit formulas for the semiclassical exponent. For example, for the single-pulse field

$$E(t) = E\,\mathrm{sech}^2(\omega t) \tag{39}$$

the stationary action is [54]

$$S_{\text{pulse}} = n\frac{m^2\pi}{eE}\left(\frac{2}{1+\sqrt{1+\gamma^2}}\right) \tag{40}$$

where $n$ is the winding number and $\gamma \equiv \frac{m\omega}{eE}$ the "adiabaticity parameter" [60]. For the analogous single-bump field

$$E(x_3) = E\,\text{sech}^2(kx_3) \tag{41}$$

one finds

$$S_{\text{bump}} = n\frac{m^2\pi}{eE}\left(\frac{2}{1+\sqrt{1-\tilde{\gamma}^2}}\right) \tag{42}$$

where $\tilde{\gamma} = mk/eE$. Note that $S_{\text{pulse}}$ decreases with $\gamma$, so that the pair production rate increases, while it is the other way round for $S_{\text{bump}}$. We believe that this is just an instance of a quite general fact: *Inhomogeneity in time tends to shrink the size of the wordline instantons, leading to an increase in the pair production rate; spatial inhomogeneity increases the instanton size and decreases the pair production rate.*

The single-bump field also provides an excellent opportunity to test the validity of the semiclassical approximation, since for this case an exact integral representation has been obtained by Nikishov [62], suitable for numerical evaluation, and moreover there is a worldline Monte Carlo result [48]. As shown in figure 6, all three results are in close agreement over the whole range of the inhomogeneity parameter $\tilde{\gamma}$. Note that the imaginary part vanishes for $\tilde{\gamma} > 1$, which in the instanton approach simply means that instanton solutions cease to exist.

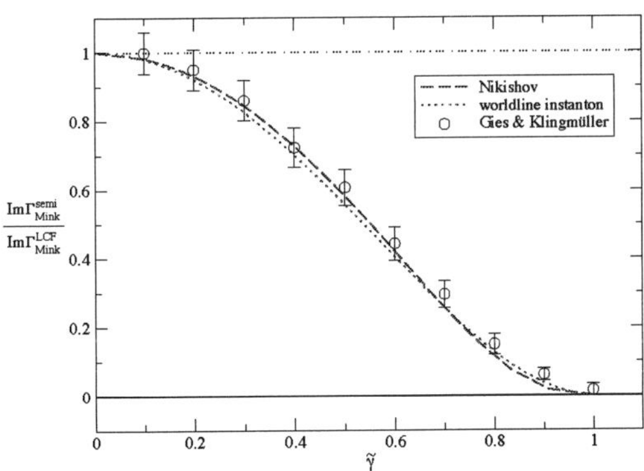


**FIGURE 6.** Plot of $\text{Im}\Gamma^{\text{semi}}/\text{Im}\Gamma^{\text{LCF}}$ for a space-dependent field $E(x) = E\,\text{sech}^2(kx)$.

Here the amplitude has been normalized by the weak field limit of the "locally constant field approximation" [55].

Although the method works very well for these classical cases, these are rather special configurations, and known to be amenable also to WKB methods. The existence of closed-form expressions for the worldline instanton can be expected only for a very restricted classes of fields. Much more interesting is the fact that the worldline instanton equations (27) and the zero mode equations (36) are ordinary differential equations, which leads us to expect that they can be solved numerically for more general classes of fields than have been treated by WKB. Dunne and Wang [56] have very recently applied this numerical approach to a class of electric fields which depend nontrivially on two spatial coordinates, parametrized by

$$A_4(\vec{x}) = -i\frac{E}{k}f(\vec{x}) \tag{43}$$

($\gamma = mk/eE$). Fig. 7 [4] shows their results for two examples,

$$\begin{aligned} f(\vec{x}) &= \frac{k(x_1+x_2)}{1+k^2(x_1^2+x_2^2)} \\ f(\vec{x}) &= k(x_1+x_2)\,\mathrm{e}^{-k^2(x_1^2+x_2^2)} \end{aligned} \tag{44}$$

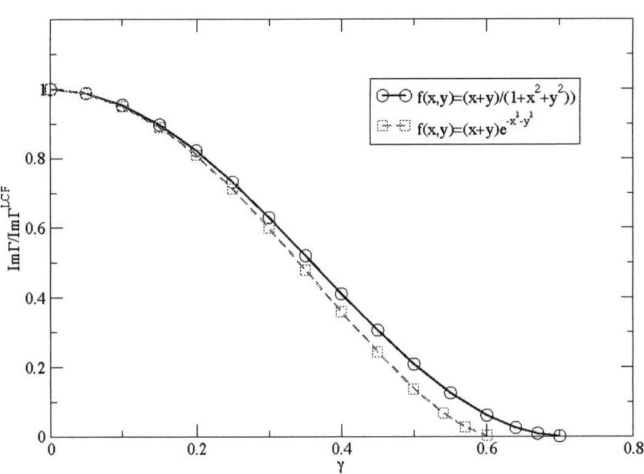


**FIGURE 7.** Pair creation rates for two nonplanar cases.

Note that again $\mathrm{Im}\Gamma$ vanishes for sufficiently large inhomogeneities.

Finally, making the transition from scalar to spinor QED in the instanton approach is straightforward if we use Feynman's original implementation of spin in the worldline path integral. Up to the global normalization, it amounts to multiplying with the spin factor (4) evaluated on the instanton trajectory, $S[x^{\mathrm{cl}}, A]$. For the classical cases discussed

[4] Reproduced from [56] with the permission of the authors.

above, and more generally for all "planar" fields the path ordering in the spin factor has no effect, and one obtains simply

$$S[x^{\rm cl},A] \;=\; 4\cos\left[eT\int_0^1 du\,E(x(u))\right] = 4(-1)^n \qquad (45)$$

with $n$ the winding number. Thus for planar fields one finds the same simple relation between ${\rm Im}\mathscr{L}_{\rm scalar}(E)$ and ${\rm Im}\mathscr{L}_{\rm spinor}(E)$ as in the constant $E$ case, eqs.(18),(19)! It is an interesting open question whether this property might even extend to general electric fields.

## 5. SUMMARY

The purpose of this short review was to show that the worldline approach in its various versions is turning into an efficient alternative to second-quantized methods for an increasing range of problems in QED. Unfortunately, despite of the restriction to QED it was not possible here to give due credit to all relevant work. Among other things, it was not possible here to discuss the worldline variational approach of [71, 72], nor the nonperturbative propagator calculations of [73].

Let me conclude with summarizing the main advantages which one can hope to achieve using the worldline formalism: (i) Compact parameter integral representations for arbitrary QED multiloop amplitudes (ii) Easy implementation of constant external fields (iii) Reliable numerical (Monte Carlo) results for one loop effective actions and Casimir energies (iv) Reliable numerical results for pair creation rates in arbitrary electric fields.

## REFERENCES

1. R. P. Feynman, *Phys. Rev.* **80**, 440 (1950).
2. A. Morgan, *Phys. Lett. B* **351**, 249 (1995).
3. A. A. Migdal, *Nucl. Phys. B* **265**, 594 (1986).
4. C. D. Fosco, J. Sánchez-Guillén, and R. A.Vázquez, *Phys. Rev. D* **69**, 105022 (2004); *Phys. Rev. D* **73**, 045010 (2006).
5. R. P. Feynman, *Phys. Rev.* **84**, 108 (1951).
6. E. S. Fradkin, *Nucl. Phys. B* **76**, 588 (1966).
7. F. A. Berezin and M. S. Marinov, *Ann. Phys.* **104**, 336 (1977).
8. L. Brink, S. Deser, B. Zumino, P. Di Vecchia, and P. S. Howe, *Phys. Lett. B* **64**, 435 (1976).
9. L. Brink, P. Di Vecchia, and P. Howe, *Nucl. Phys. B* **118**, 76 (1977).
10. M.G. Schmidt, and C. Schubert, *Phys. Rev. D* **53**, 2150 (1996).
11. A. Strominger, *Phys. Lett. B* **101**, 271 (1981).
12. A. M. Polyakov, *Mod. Phys. Lett. A* **3**, 335 (1988).
13. H. Gies, and J. Hämmerling, *Phys. Rev. D* **72**, 065018 (2005).
14. C. Schubert, *Phys. Rep.* **355**, 73 (2001).
15. M. B. Halpern, and W. Siegel, *Phys. Rev. D* **16**, 2486 (1977).
16. M.B. Halpern, A. Jevicki, and P. Senjanoivc, *Phys. Rev. D* **16**, 2476 (1977).
17. I.K. Affleck, O. Alvarez, and N.S. Manton, *Nucl. Phys. B* **197**, 509 (1982).
18. Z. Bern, and D.A. Kosower, *Nucl. Phys. B* **379**, 451 (1992).
19. M.J. Strassler, *Nucl. Phys. B* **385**, 145 (1992).
20. M.G. Schmidt, and C. Schubert, *Phys. Lett. B* **331**, 69 (1994).

21. K. Roland, and H.-T. Sato, *Nucl. Phys. B* **480**, 99 (1996); *Nucl. Phys. B* **515**, 488 (1998).
22. K. Daikouji, M. Shino, and Y. Sumino, *Phys. Rev. D* **53**, 4598 (1996).
23. R. Shaisultanov, *Phys. Lett. B* **378**, 354 (1996).
24. M. Reuter, M.G. Schmidt, and C. Schubert, *Ann. Phys. (N.Y.)* **259**, 313 (1997).
25. A. M. Polyakov, *Gauge fields and strings*, Harwood Academic Publishers, 1987.
26. C. Schubert, *Eur. Phys. J. C* **5**, 693 (1998).
27. P. Dai, and W. Siegel, YITP-SB-06-29, hep-th/0608062.
28. C. Itzykson, and J. Zuber, *Quantum field theory*, McGraw-Hill, 1985.
29. D.J. Broadhurst, R. Delbourgo, and D. Kreimer, *Phys. Lett. B* **366**, 421 (1996).
30. D.G.C. McKeon, and T.N. Sherry, *Mod. Phys. Lett. A* **9**, 2167 (1994).
31. S.L. Adler, and C. Schubert, *Phys. Rev. Lett.* **77**, 1695 (1996).
32. C. Schubert, *Nucl. Phys. B* **585**, 407 (2000).
33. D. Fliegner, M. Reuter, M.G. Schmidt, and C. Schubert, *Theor. Math. Phys.* **113**, 1442 (1997).
34. B. Körs, and M.G. Schmidt, *Eur. Phys, J. C* **6**, 175 (1999).
35. G.V. Dunne, A. Huet, D. Rivera, and C. Schubert, *JHEP* **11**, 013 (2006).
36. G.V. Dunne, and C. Schubert, *JHEP* **0208**, 0053 (2002).
37. G.V. Dunne, and C. Schubert, *JHEP* **0206**, 0042 (2002).
38. M.G. Schmidt, and C. Schubert, *Phys. Lett. B* **318**, 438 (1993).
39. V.P. Gusynin, and I.A. Shovkovy, *Can. J. Phys.* **74**, 282 (1996); *J. Math. Phys.* **40**, 5406 (1999); I.A. Shovkovy, *AMS/IP Stud. Adv. Math.* **13**, 467 (1999).
40. D.G.C. McKeon, and A. Rebhan, *Phys. Rev. D* **47**, 5487 (1993); *Phys. Rev. D* **49**, 1047 (1994).
41. H.-T. Sato, *J. Math. Phys.* **40**, 6407 (1999).
42. I.A. Shovkovy, *Phys. Lett. B* **441**, 313 (1998).
43. D.G.C. McKeon, and A. Rebhan, *Phys. Rev. D* **48**, 2891 (1993).
44. A.I. Karanikas, and C.N. Ktorides, *JHEP* **9911**, 033 (1999); *Phys. Lett. B* **500**, 75 (2001).
45. H. Gies, and K. Langfeld, *Nucl. Phys. B* **613**, 353 (2001); *Int. J. Mod. Phys. A* **17**, 966 (2002).
46. M.G. Schmidt, and I.O. Stamatescu, hep-lat/0201002; *Mod. Phys. Lett. A* **18**, 1499 (2003).
47. H. Gies, K. Langfeld, and L. Moyaerts, *JHEP* **0306**, 018 (2003).
48. H. Gies, and K. Klingmüller, *Phys. Rev. D* **72**, 065001 (2005).
49. H. Gies, J. Sánchez-Guillen, and R.A. Vázquéz, *JHEP* **0508**, 067 (2005).
50. H. Gies, and K. Klingmüller, *J. Phys. A* **39**, 6415 (2006).
51. H. Gies, and K. Klingmüller, *Phys. Rev. Lett.* **96**, 220401 (2006).
52. F. Bastianelli, O. Corradini, and E. Latini, hep-th/0701055.
53. A. Bulgac, P. Magierski, and A. Wirzba, *Phys. Rev. D* **73**, 025007 (2006).
54. G.V. Dunne, and C. Schubert, *Phys. Rev. D* **72**, 105004 (2005).
55. G.V. Dunne, H. Gies, C.Schubert, and Q.-h. Wang, *Phys. Rev. D* **73**, 065028 (2006).
56. G.V. Dunne, and Q.-h. Wang, hep-th/0608020.
57. J. Schwinger, *Phys. Rev.* **82**, 664 (1951).
58. T. Heinzl et al., hep-ph/0601076.
59. TESLA technical design report.
60. L.V. Keldysh, *JETP* **20**, 1307 (1965).
61. E. Brezin, and C. Itzykson, *Phys. Rev. D* **2**, 1191 (1970).
62. A.I. Nikishov, *Nucl. Phys. B* **21**, 346 (1970).
63. N.B. Narozhnyi, and A.I. Nikishov, *Yad. Fiz.* **11**, 1072 (1970).
64. V.S. Popov, *JETP* **34**, 709 (1971); V.S. Popov, and M.S. Marinov, *Yad. Fiz.* **16**, 809 (1972).
65. M. Stone, *Phys. Rev. D* **14**, 3568 (1976).
66. W. Greiner, B. Müller, and J. Rafelski, *Quantum Electrodynamics of Strong Fields*, Springer, 1985.
67. A.B. Balantekin, J.E. Seger, and S.H. Fricke, *IJMP A* **6**, 695 (1991).
68. G.V. Dunne, and T. Hall, *Phys. Rev. D* **58**, 105022 (1998).
69. S.P. Kim, and D. Page, *Phys. Rev. D* **73**, 065020 (2006); hep-th/0701047.
70. S. Levit, and U. Smilansky, *Ann. Phys.* **103**, 198 (1977).
71. R. Rosenfelder, and A.W. Schreiber, *Phys. Rev. D* **53**, 3337 (1996); *Eur. Phys. J. C* **37**, 161 (2004).
72. A. Alexandrou, R. Rosenfelder, and A.W. Schreiber, *Phys. Rev. D* **62**, 085009 (2000).
73. C. Savkli, J. Tjon, and F. Gross, *Phys. Rev. C* **60**, 055210 (1999); Erratum-ibid. **61**, 069901 (2000); *Phys. Rev. C* **62**, 116006 (2000).

# Detection of GRBs in Auger and the LAGO project

X. Bertou for the Pierre Auger Observatory and the LAGO Project

*Centro Atómico Bariloche, S. C. de Bariloche, 8400 Río Negro, Argentina*

**Abstract.** The detection of a high energy tail of photons from GRBs should be possible with the single particle counting technique. A group of detectors would detect a higher rate of events on a short time scale, of the order of the second. This technique has already been used with plastic scintillators. Use of water Cherenkov detectors gives a better sensitivity with respect to scintillators as secondary photons can be detected through their conversion in the water volume. This increases significantly the detection efficiency as the ratio photons to electrons in GRB photon showers is about 9 to 1 in the shower tail. The Pierre Auger Observatory, with 1600 water Cherenkov detectors, is a good candidate to check the efficiency of the technique. The Large Aperture GRB Observatory (LAGO) uses similar detectors at higher altitude (Puebla 4550 m a.s.l., Chacaltaya 5250 m a.s.l.), and will bring better sensitivity to the high energy part of the GRB photon spectrum with much less surface to be covered.

**Keywords:** Gamma Ray Burst, single particle technique, high altitude sites
**PACS:** 98.70.Rz

## INTRODUCTION

Since their discovery at the end of the 60's[1], the Gamma Ray Bursts (GRB) have been of high interest to the astrophysicists. A GRB is characterised by a sudden emission of gamma rays during a very short period of time (between 0.1 and 100 second). The luminosity reached during this flare is typically between $10^{51}$ and $10^{55}$ ergs, should the emission be isotropic. The astrophysical source of these bursts is still not clear but candidates would be coalescence of compact objects (neutron stars), and mechanisms based on internal shocks of a relativistic wind in a compact source give good agreement between the theory and the observations (see [2] for a recent review).

A first large data set of GRB was provided by the BATSE instrument on board the Compton Gamma Rays Observatory (1991-2000). More GRBs where then detected by BEPPO-SAX (1997-2002). Current GRBs are registered by HETE, INTEGRAL and Swift. In the last 5 years afterglows were observed allowing a much better understanding of the GRB phenomena. Most observations have however been done below 1 GeV, and the presence of a high energy component in the GRB spectrum is still a mystery.

GLAST will be the next generation of GRB satellite experiment and should be launched in the fall of 2007. Its sensibility should allow to get individual GRB spectra up to 300 GeV. In the meanwhile, the only way to get to the high energy emission of GRB is to work on the ground.

A classic technique to use is called "single particle technique". When high energy photons from a GRB reach the atmosphere, they produce a cascade of secondary particles that one can detect. The energies are still quite low to produce a shower detectable

CP917, *Particles and Fields,* edited by H. Castilla Valdez, J. C. D'Olivo, and M. A. Perez

at ground level (even at high altitudes). However, as we expect a lot of these photons to arrive during the burst, in a short period of time (typically, one second), some of them could produce a few hits in a ground based detector, on a time scale of one second. One would therefore see an increase of the background rate on all the detectors. This technique has already been applied in INCA[3] in Bolivia and ARGO[4] in Tibet. A general study of this technique can be found in [5]. Up to now, it has only been applied to arrays of scintillators. We will present here a study using Water Cherenkov Detectors (WCD).

Currently under construction in Argentina, the Pierre Auger Observatory will consist of 1600 WCD of 12 tons of water (10 $m^2$ of surface) overlooked by 3 PMTs, spread over a total surface of 3000 $km^2$. It is aimed at the understanding of the highest energy cosmic rays detected up to now, above $10^{20}$ eV. It has started stable data taking in January 2004 with 150 detectors, and has now more than 1200 WCD in operation after three years of deployment. Even if it is located at relatively low altitude (1400 m a.s.l), its large size makes it a good GRB detector.

A similar sensitivity can be achieved by instrumenting only a few tens of square meters of WCD but at higher altitude. This is the purpose of the LAGO project, and two sites are being instrumented. The Sierra Negra site, near Puebla, México, is located 4550 m a.s.l. and has 20 $m^2$ of WCD in operation as of January 2007. The Chacaltaya site, near La Paz, Bolivia, is located at 5250 m a.s.l. and will host 12 $m^2$ of WCD in the first half of 2007. With optimised software, these observatories should reach a better sensitivity than the Pierre Auger site.

This paper will detail the results of simulations determining the sensitivity of WCD to GRBs, present the capabilities of the Pierre Auger Observatory and describe the current status of the LAGO project.

## RESPONSE OF WCD TO GRB

### GRB photons cascades in the atmosphere

In order to determine the response of ground based WCD to secondary particles in cascades produced by high energy photons from GRBs, 1.4 billions showers with primary energies of 10 MeV to 10 TeV were generated, for zenith angles from 0 to 30 degrees. Ground was set at 1400 m, the altitude of the southern site of Auger. No thinning was applied (meaning all the particles for each shower were followed until absorbed in the atmosphere or on the ground). Simulations were performed using the Corsika[6] shower generator.

The number of particles at ground level increases as a function of energy as a power law of index $\approx 1.7$. About 90% of the particles reaching the ground are secondary photons. This is a very important point which gives strength to WCD with respect to scintillators or RPCs, since WCD usually can detect photons by pair-conversion in their water volume, while scintillators and RPCs used in cosmic ray experiments are too thin to provoke such a conversion.

The dependence of this number of particle as a function of zenith angle is a direct consequence of the absorption in the atmosphere, and is reduced by a factor 10 when going from vertical to 30°.

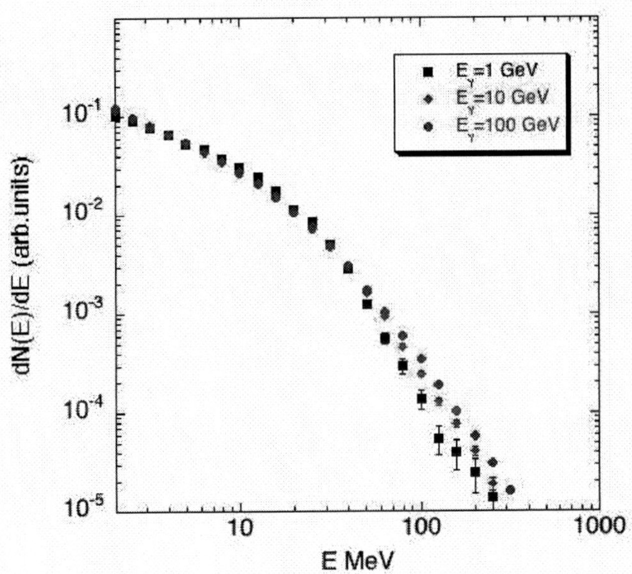


**FIGURE 1.** Energy spectrum of secondary photons at ground (1400 m a.s.l.) for three different primary photon energy, normalised. Except for the high energy tail, the spectra are very similar.

The energy spectrum of secondaries (see figure 1) is found to be almost insensitive to the primary energy, and exhibits a break at about 20 MeV. The high energy secondary tail, above 50 MeV, does however depend on the primary energy.

## WCD response to secondaries

The response of a WCD to incoming secondaries was simulated using the GEANT4 based WCD simulation of the Pierre Auger Observatory. Secondary photons and electrons in the 1 MeV - 1 GeV energy range were simulated with incident angle between 0 and 60°. Different trigger strategies were studied in the case of the Pierre Auger Observatory, and the 4 ADC threshold on a single PMT ($\approx$ 20 MeV) was determined to be the optimum given the electronic noise and low gain of the PMTs (these parameters are governed by the UHECR detection of Auger and can not be modified for GRB detection purposes).

Simulations of different WCD geometries and threshold were performed for the LAGO project, and the optimal water height was found to be 1.4 m, while the efficiency is found to increase with the detector size, for a unique central PMT. Given the evolution of the cost of WCD with size, and the difficulty of deployment of large WCD, the optimal size was determined to be 4 $m^2$ per PMT.

As the detection of a GRB will be claimed when the signal to square root of noise reaches a specific level, the optimal threshold was found to be the single photo-electron.

While this cannot be achieved for Auger, this is the design goal of LAGO.

In order to reduce the background, one can apply a high threshold cut to remove muons. This is a very efficient cut for the Pierre Auger Observatory, as muons represent a strong contribution to the background at low altitudes. This cut is however not so efficient at higher altitude, where the dominant background is low energy photons and electrons.

## SENSITIVITY OF THE PIERRE AUGER OBSERVATORY

One can convolve the response of the Auger WCD to the secondaries produced by a flux of high energy photons from a GRB to determine the probability of detecting such a burst in Auger. If one considers a GRB emitting a luminosity $L$ (erg.s$^{-1}$) in an energy range $[E_{min}, E_{max}]$ at a redshift $z$, the flux at Earth and the implied detection condition will be given by :

$$\Phi = \frac{L}{4\pi d_L^2(z)} = \int_{\frac{E_{min}}{1+z}}^{\frac{E_{max}}{1+z}} En(E)\,dE \Longrightarrow$$
$$\int_{\frac{E_{min}}{1+z}}^{\frac{E_{max}}{1+z}} P_{detec}(E)n(E)\,dE \geq \frac{s}{S_\theta A}\sqrt{\frac{B}{N\Delta t}} \qquad (1)$$

where $d_L$ is the luminosity distance of the source and $n(E)$ is the differential spectrum of photons (s$^{-1}$.cm$^{-2}$.erg$^{-1}$), $P_{detec}(E)$ is the mean number of triggering particles from a shower induced by a photon of energy $E$, $s$ is the desired statistical significance (in the following we will take $s$=7), $A$ is the area of a tank, $N$ the number of tanks, $B$ the background counting rate (for the trigger strategy we choose the background is around 1.8 kHz/tank), $\Delta t$ is the duration of the particle flux, $S_\theta$ is a geometric factor depending on the zenith angle of the gamma.

The high energy component of GRBs is very badly known and one can use as a first approximation extrapolation of BATSE and EGRET measurements, using a power law with an index of 2 to 3 (the mean observed value being around 2.2), and a cut-off energy between 100 MeV and 10 TeV. The minimum fluence that can be detected for a one second burst is given in figure 2. Assuming typical fluences of $10^{-6}$ to $5\times10^{-4}$ erg.cm$^{-2}$, one expects possible detection of vertical bursts for a cut-off energy above 100 GeV.

These sensitivity curves, when compared to similar ones obtained for scintillators[5], shows that the Pierre Auger Observatory is 7 to 8 times more sensitive for a similar area. The 16 000 m$^2$ of Auger are as sensitive as 1000 m$^2$ of scintillators at about 4700 m a.s.l., being almost as efficient as the ARGO YBJ observatory (for the single particle mode technique only, as the Pierre Auger Observatory has no way of reconstructing low energy showers).

It is worthwhile to note that should the signal detected by MILAGRITO for GRB 970417a be true[7], Auger would also have detected it (if in its field of view, below 30° of zenith angle).

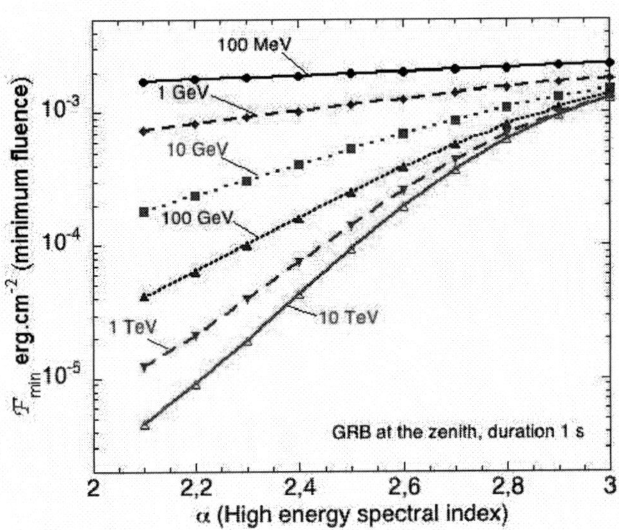


**FIGURE 2.** Minimum fluence between 10 MeV and $E_{max}$ for a vertical GRB of 1 s of duration to be observable by the Pierre Auger Observatory, as a function of its spectral index. Typical spectral index is ≈2.2, and fluences expected are about $10^{-6}$ to $5 \times 10^{-4}$ erg.cm$^{-2}$.

# THE LARGE APERTURE GRB OBSERVATORY

## Gaining in sensitivity

In order to improve the sensitivity of the Pierre Auger Observatory without having to instrument a huge surface, the LAGO project has deployed WCD at high altitude. Simulations show that the spectrum of secondaries at 5250 m a.s.l. (Chacaltaya) is not much different to the one at 1400 m a.s.l. (Auger) in its shape, with a normalisation factor of about 100. As the background is roughly a factor 10 higher[5], the signal to square root of noise gets a factor $100/\sqrt{10} = 32$ higher. This corresponds to a gain in surface of 1000. A total of 20 m$^2$ of WCD at Chacaltaya would therefore be as efficient as the whole Pierre Auger Observatory with respect of detecting GRBs with the single particle technique.

In order to achieve an even better sensitivity, one can alter the design parameters of Auger which were fixed for UHECR physics. By raising the PMT gain, one can reach the single photo-electron sensitivity and greatly increase the detection efficiency at low secondary energy. A proper determination of the resulting sensitivity needs a good understanding of the background at high altitude and is still under investigation.

Another way to improve the sensitivity is to get a finer time resolution. In Auger, the minimum time scale on which a GRB can be observed is one second (the counting rates are exported to the central data acquisition system every second). By reducing the time resolution of the LAGO data acquisition, one could detect very short bursts which would otherwise be covered by the noise if one had to integrate a whole second.

## Technical Design

The WCD used in the LAGO project are either commercial rotomolted plastic tanks (Sierra Negra site) or fibreglass ones (Chacaltaya site), with 4 $m^2$ of surface. Their inside is covered by commercial reflective material (Tyvek®), and they are filled up to a level of 1.4 m with pure water. A central PMT observes the water volume. Some tests are still under way whether one can improve the detection efficiency at low energy by using a wavelength shifter (Amino-G).

In order to reduce the costs of the project, the electronics and some of the PMTs used in the LAGO sites were obtained from the engineering phase of the Pierre Auger Observatory. The PMTs are operated at a higher gain in order to achieve the single photo-electron, and the triggering logic, hosted in a FPGA, has been changed in order to count particles at three different thresholds, as well as counting events below baseline, characteristics of high frequency noise produced by radio and lightning storms. This provides an important veto to reject false bursts. The time resolution has been set to 5 ms, and the counting rates of up to 6 detectors are read by a single Auger electronics. Timing is extracted from a GPS. Data is transmitted to an acquisition PC via a 115 200 bauds serial link, where it is stored for further analysis.

## Status

The Sierra Negra site consists of 4 large 4 $m^2$ detectors and 2 small 1 $m^2$ ones, all 6 connected to a single Auger prototype board. Data is being collected on a stable basis since January 2007, and first data analysis is expected for July 2007.

The Chacaltaya site is still not in operation but has a 1 $m^2$ prototype in operation at La Paz, at 3700 m a.s.l.. A first 4 $m^2$ detector is expected to be in operation for July 2007.

A possible new site in the northern Argentina in under investigation, as well as a site in Mérida, Venezuela, on top of the Pico Espejo (4765 m a.s.l.). A first detector could be in operation at these sites at the end of 2007.

## CONCLUSION

The Pierre Auger Observatory, despite its low altitude, has been shown to have a sensitivity to the high energy end of the GRB spectra similar to the one of higher altitude observatory, due to its sensitivity to secondary photons. It is however limited by the low gain at which its PMTs are operated, and its acquisition mode, allowing only to get counting rates once per second. It is nevertheless one of the best candidates to detect a GRB with the single particle technique.

The LAGO project is designing and operating various sites which present a similar sensitivity with a much smaller instrumented area by going to high altitude. When fully instrumented, the Sierra Negra and Chacaltaya sites will reach the full Pierre Auger sensitivity with only 40 and 20 $m^2$ of detector. In order to provide a significant gain with respect of Auger, modifications were introduced to increase the PMT gain in order to reach lower energy sensitivity, and the acquisition records rates every 5 ms in order to be able to detect very short bursts.

It is likely that at the moment GLAST is launched (fall 2007), the analysis of the data recorded will be ready and it will be known if a ground based observatory using the single particle technique can provide additional information to the one provided by satellites.

## ACKNOWLEDGMENTS

The LAGO project is very thankful to the Pierre Auger collaboration for the lending of the engineering equipment without which no progress could have been done.

The author wishes to thank the organisers for such a pleasant and efficient meeting.

## REFERENCES

1. Kleidesabel et. al, 1973, ApJ 182, 85
2. P. Mészáros, 2006, Rept.Prog.Phys. 69, 2259
3. Cabrera et. al, 1999, A&AS 138, 599
4. Bacci et. al, 1999, A&AS 138, 597
5. Verneto, 2000, APh 13, 75
6. Heck, http://www-ik.fzk.de/corsika/
7. R. Atkins et. al, 2000, ApJ 533, 119

# Recent Results from Milagro

Jordan A. Goodman for the Milagro Collaboration

*Department of Physics, University of Maryland, College Park, MD 20742-4111, USA*

**Abstract.** The diffuse gamma radiation arising from the interaction of cosmic ray particles with matter and radiation in the Galaxy is one of the few probes available to study the origin of the cosmic rays. Milagro is a water Cherenkov detector that continuously views the entire overhead sky. The large field-of-view combined with the long observation time makes Milagro the most sensitive instrument available for the study of large, low surface brightness sources such as the diffuse gamma radiation arising from interactions of cosmic radiation with interstellar matter. In this paper we report our results on diffuse emission from the galactic plane and a detailed study of the Cygnus region where we have discovered a new TeV gamma-ray source.

**Keywords:** Milagro; Galactic Sources; TeV Gamma-ray
**PACS:** 98.70.Rz

## INTRODUCTION

Milagro is the first large area, continuously operating, water Cherenkov detector used for gamma-ray astronomy. Milagro consists of a 24 million liter water reservoir instrumented with 723 photomultiplier tubes (PMTs) surrounded by an array of 175 water tanks. The top layer of 450 PMTs is beneath 1.3 meters of water and is used to trigger the detector and to reconstruct the direction of the primary gamma ray (or cosmic ray). The bottom layer of 273 PMTs is beneath 6 meters of water and is used to measure the penetrating component of air showers induced by hadronic cosmic rays. The PMT spacing in the reservoir is 2.7 meters and the area enclosed is 4800 $m^2$. The surrounding array of water tanks is dispersed over 34,000 $m^2$. Each tank is cylindrical with a 1.6 meter radius and a depth of 1 meter. The tanks are instrumented with a single PMT located at the top of the tank looking down into the water volume. Milagro began physics operation in 2000 taking data with the central water reservoir. In 2004, construction of the outrigger array was completed. Before the installation of the outriggers the small size of the water reservoir limited the sensitivity of Milagro such that the Crab Nebula was observed at $\sim 4\sigma$ in one year of operation. The completion of the outriggers enabled a large increase in the sensitivity of the instrument, enabling us to detect the Crab Nebula at over 8 standard deviations in a single year of observation. This factor of $\sim 2$ increase in sensitivity (as shown in figure 1) has dramatically changed our view of the high-energy sky. It also means that the data currently being taken now with Milagro is substantially more important than our original data and we are not simply increasing our sensitivity by the square root of time over a 6-year observational period.

To date, our most important observational results with Milagro have been: the first detection of TeV gamma rays from the Galactic plane [1], the mapping of the diffuse Galactic gamma-ray emission at TeV energies, including the detection of the Cygnus Region at high-significance (over $10\sigma$) [2] , the discovery of a new (slightly extended)

CP917, *Particles and Fields*, edited by H. Castilla Valdez, J. C. D'Olivo, and M. A. Perez

source of TeV gamma rays (MGRO J2019+37) embedded in the Cygnus Region, and the possible detection of a gamma-ray burst with our prototype instrument Milagrito [4]. In addition, we have detected TeV gamma rays from the active galaxies Mrk 501 [5], Mrk 421 [6] plus the Crab Nebula [7], set stringent upper limits on the prompt TeV emission from several gamma ray bursts [8], and performed the most sensitive survey of the northern hemisphere at TeV energies [6]

## SURVEY OF THE NORTHERN HEMISPHERE

The sensitivity of Milagro has been dramatically improved with the addition of the outrigger array. The effect of this gain in sensitivity is best demonstrated by comparing the results of our published sky survey [1] which used approximately 1000 days of data (top of figure 1) and our current data set of just over 2000 days of data (bottom of figure 1). While both the Crab Nebula and the active galaxy Mrk421 are visible in the top panel, the improvement since the outriggers is dramatic. The significance of the Crab Nebula has increased from $6\sigma$ to $14\sigma$ and the Galactic plane is now clearly visible, even in this broad map. Our current sensitivity is $8\sigma$ per year for a 1 Crab flux. (It should be noted that for a source with a spectrum harder than the Crab, our sensitivity improves).

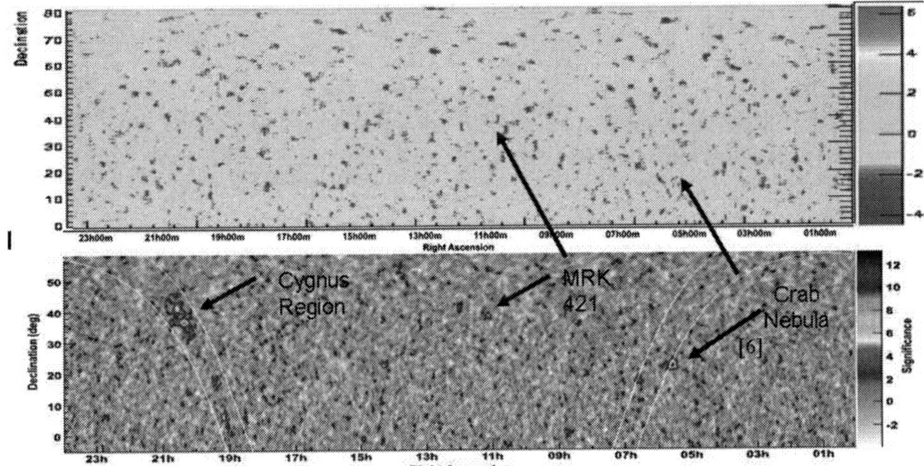


**FIGURE 1.** The northern sky seen in TeV gamma rays using Milagro data. The top panel shows the sky as observed before the outriggers and the bottom panel after the completion of the outrigger array.

## THE CYGNUS REGION AND THE DISCOVERY OF A TEV GAMMA-RAY SOURCE

In 2005, we published the first detection of diffuse TeV gamma-ray emission from the inner Galaxy [1]. The flux of the Milagro detection is not consistent with expectations from cosmic ray interactions if the local cosmic ray flux is indicative of the flux in

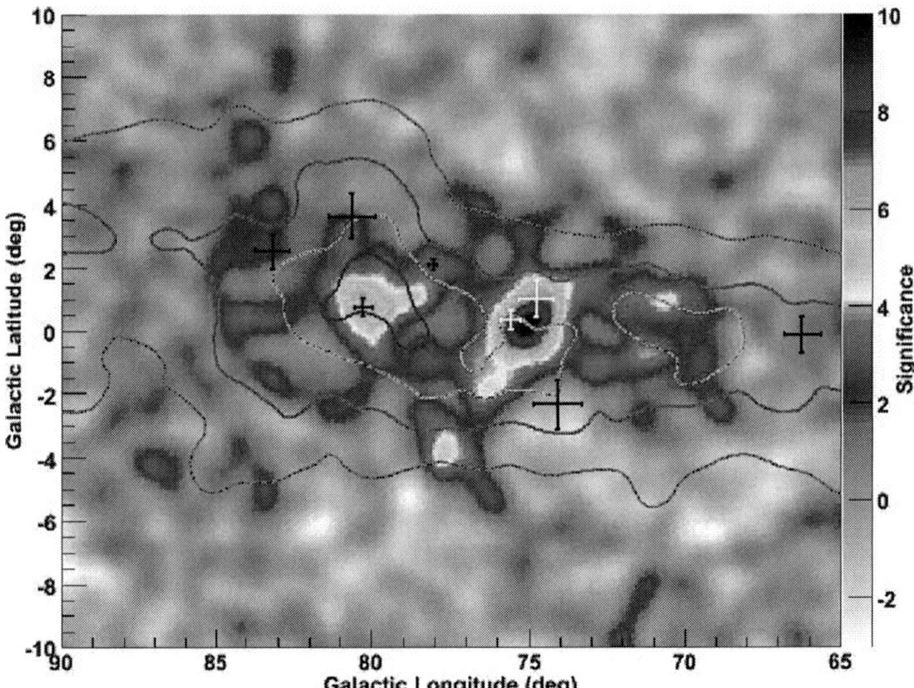


**FIGURE 2.** The Cygnus Region of the Galaxy as seen in TeV gamma rays. Superimposed on the image are contours showing the matter density in the region. The probability that the two distributions are not correlated is 1.5x10-6. The crosses show the location of the EGRET sources and their corresponding location errors.

the rest of the galaxy. There are several possible explanations for this excess: the local cosmic ray flux is unusually low, the local spectral index is soft relative to the rest of the Galaxy, and the existence of unresolved point sources. With the more recent data, taken since the completion of the outriggers, we have refined the analysis to investigate the Cygnus Region of the Galaxy in more detail. The Cygnus Region of the Galaxy is a natural laboratory for the study of cosmic ray origins. It contains a large column density of interstellar gas that should lead to strong emission of diffuse gamma rays and is also the home of potential cosmic-ray acceleration sites (Wolf-Rayet stars [9], OB associations [10], and supernova remnants [11]). Figure 2 shows a detailed view of the Cygnus Region in TeV gamma rays. Superimposed on the figure are contour lines indicating the matter density in the region and the location (and location errors) of the EGRET sources (all unidentified) in the region. There is definitive evidence for a new source of TeV gamma rays (MGRO J2019+37). MGRO J2019+37 is detected at over 10 standard deviations and its location is consistent with the location of two EGRET sources one of which has been tentatively identified with a pulsar wind nebula [12]. Though it is not evident from this figure, this source is most likely extended with a width of $0.32\pm0.12$ degrees. The best-fit location of this source is

R.A.=304.83° ±0.14°stat ±0.3°sys and Dec.= 36.83° ±0.08°stat ±0.25°sys. Assuming a differential source spectrum of $E^{-2.6}$, the Milagro flux measurement at the median energy of 12 TeV is given by $E^2$ dN/dE = (3.49 ± 0.47stat ± 1.05sys) x $10^{-12}$ TeV $cm^{-2}$ $s^{-1}$. A change in the assumed source spectral index from -2.6 to -2.7 changes the integral flux above 12 TeV by less than 7%.

The next brightest TeV region is just to the left of MGRO J2019+37 in Figure 2, at Galactic latitude of 80°, and is also coincident with an EGRET source (3EG J2033+4118) and the HEGRA source[13] TeV J2032+413. The HEGRA source was detected between 1 and 10 TeV with a differential photon spectral index of -1.9± 0.1stat± 0.3sys, which when extrapolated to 12 TeV gives $E^2$ dN/dE = (7.95± 2.7 stat) x $10^{-13}$ TeV $cm^{-2}$ $sec^{-1}$ . The Milagro flux in a 3x3 square degree region centered on the HEGRA source at 12 TeV is (2.41± 0.48stat± 072sys) x $10^{-12}$ $cm^{-2}$ $s^{-1}$ assuming a differential photon source spectrum of $E^{-2.6}$. Thus, the Milagro flux exceeds the HEGRA flux as is expected due to the additional contribution of the diffuse flux in this region. In fact, this region contains the largest matter density as can be seen from the contour lines of Figure 2. Further analysis is required to distinguish what fraction of the gamma rays observed by Milagro is from the HEGRA source or from the diffuse interactions.

The flux of diffuse emission from the Cygnus Region is measured after subtraction of the contribution from MGRO 2019+37. Using the GALPROP [14] program we estimate the expected flux of diffuse gamma rays. This estimate has contributions from a pion component arising from the interactions of cosmic rays with matter in the region and from gamma rays produced by inverse Compton interactions of cosmic-ray electrons with infrared radiation in the region. We find that the TeV gamma ray flux is about a factor of 5 larger than that predicted by this standard model of cosmic ray production in the Galaxy. One plausible explanation of this observation is that the Cygnus region contains cosmic-ray accelerators thereby increasing the cosmic-ray density in that region relative to the model predictions of GALPROP.

## REFERENCES


1. Atkins, R. et al. 2005 Phys Rev Lett 95, 251103.
2. Abdo A. et al. 2007, to appear AP J. 658: March 20, 2007
3. Smith, A. et al. 2006, Review, Rapporteur, and Highlight Papers of 29th ICRC in Pune IndiaÂą, 10, 227.
4. Atkins, R. et al. 2000, ApJ Lett, 533, 119 and Atkins, R. et al. 2003, ApJ, 583, 824.
5. Atkins, R. et al. 1999, ApJ Lett, 525, L25.
6. Atkins, R. et al. 2004, ApJ, 608, 680.
7. Atkins, R. et al. 2003, ApJ, 595, 803.
8. Atkins, R. et al. 2005, ApJ, 630, 996 and Atkins, R. et al., 2004 ApJ Lett, 604, 25.
9. Van der Hucht, K. A., 2001, New Astron. Rev. 45, 135.
10. Bochkarev, N. G. and Sitnik, T. G., 1985, Astrophys. Space Sci. 108, 237.
11. Breen, D. A., 2004, Bull. Astron. Soc. India 32, 325.
12. Roberts, M. S. E., et al., 2002, ApJ. 577, L19.
13. Aharonian, F. et al., 2005, A&A, 431, 197.
14. Strong, A. W., Moskalenko, I.V., and Reimer, O., 2004, ApJ, 613, 962.

# HAWC: A next generation all-sky gamma-ray telescope

Jordan A. Goodman for the HAWC Collaboration

*Department of Physics, University of Maryland, College Park, MD 20742-4111, USA*

**Abstract.** HAWC will have unprecedented sensitivity for a ground-based particle detector array. It will be capable of observing the Crab at the $5\sigma$ level with each transit while simultaneously observing the entire northern sky with $\sim$15 times the sensitivity of the Milagro detector. The design and performance of the HAWC water Cherenkov gamma-ray observatory is presented.

**Keywords:** TeV Gamma-Rays
**PACS:** 95.55.Ka

## MOTIVATION AND DESIGN OF HAWC

The HAWC observatory combines the Milagro[1] water Cherenkov air-shower detection technique with a very high altitude site. Re-deploying the existing Milagro photomultiplier tubes (PMTs) and electronics in a different configuration at an altitude above 4000m will lead to a sensitivity increase of a factor of $\sim$15 over Milagro. This dramatic improvement is due to three things: the increased altitude, the increased physical area, and the optical isolation of the PMTs; which results in a lower threshold, better angular resolution and significantly improved gamma-hadron separation. As a result of these improvements, HAWC will detect a $\sim 5\sigma$ signal from the Crab Nebula in a single 4-hr transit (compared to $\sim$5 months for Milagro). Unlike imaging atmospheric Cherenkov telescopes (IACT), HAWC is capable of continuous operation and has a field of view of nearly 2 sr. The total exposure (time $\times$ solid angle) of HAWC is 2000-4000 times higher than that of a modern IACT.

This sensitivity will enable very high energy gamma ray studies that are unattainable with the current suite of instruments. HAWC will monitor (for $\sim$4 hours every day) $\sim$1/3 of the entire sky. Over a 1 year observation period HAWC will perform an unbiased sky survey with a detection threshold of $\sim$50 mCrab, enabling the monitoring of known sources, the discovery of new sources of known types, and perhaps most importantly the discovery of new classes of TeV gamma ray sources. The sensitivity of HAWC to extended sources surpasses that of IACTs (in survey mode) for sources larger than 0.25°, as IACTs rely on their outstanding angular resolution for their high instantaneous sensitivity. HESS measurements clearly point to the existence of such diffuse objects within our Galaxy; since they observe many diffuse galactic sources clustered around their sensitivity limit, they may only be seeing "the tip of the iceberg". HAWC's large area and exposure will permit the study of energy spectra in the range from 10-100 TeV where other instruments are photon starved. With HAWC's large FOV, it will serve as an excellent monitor for mulit-wavelenth studies of transient sources with other instruments

CP917, *Particles and Fields*, edited by H. Castilla Valdez, J. C. D'Olivo, and M. A. Perez

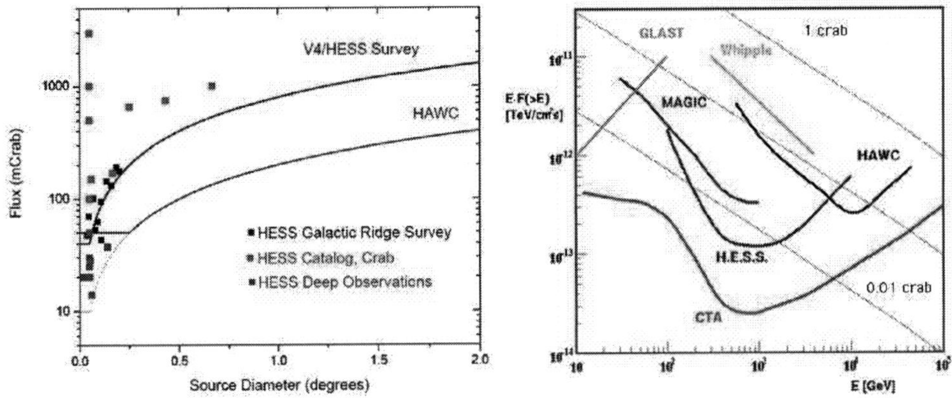


**FIGURE 1.** Sensistivity of HAWC and HESS to diffuse sources (left) and a comparison of HAWC 1 yr sensitivity to IACTs 50hrs on source

including X-ray, gamma-ray and particularly the neutrino telescopes such as Ice Cube that are under construction. See Figure 1

The HAWC concept builds upon experience with the Milagro detector. Milagro is the first large, uniformly instrumented, air shower array using water Cherenkov technology. The water Cherenkov technique is an excelent method for EAS detection because of its superior detection efficiency, calorimetric capability, and low cost. In addition, calorimetry of showers is possible in a water detector, so HAWC can readily distinguish low energy electrons and gamma rays from muons, hadrons and high energy EM particles. We can use this capability to distinguish, with unprecedented discrimating power, between gamma-ray (signal) and hadron (background) induced showers.

The HAWC detector consists of a 150m x 150m x 5m deep reservoir containing 125 Ml of filtered water. Milagro's 900 photomultiplier tubes will be secured on a 30x30 grid with 5m spacing. Curtains are deployed to optically isolate each PMT. The top of the PMTs will be located 4m below the surface of the water. Because of the curtains each sensor only detects light produced within its cell. This dramatically reduces the PMT noise rate, the trigger rate from muons and background rejection capabilities of HAWC. A test of this system has been performed in Milagro over a 4x4 array of PMTs. We observed that the singles rates in the isolated PMTs dropped by a factor of three (in agreement with Monte Carlo simulations) and that when a PMT is struck it is much more likely to have information useful to the event reconstruction algorithms.

The depth of the HAWC detector was selected as a compromise between timing resolution and gamma-hadron separation. Milagro has two layers of PMTs: a shallow layer at $\sim$1.5m depth for shower plane reconstruction, and a deep layer at $\sim$6m depth for gamma/hadron separation. In HAWC, the functionality of these two layers are combined into a single layer. For gamma-hadron separation, it is important to be able to distinguish between large and small energy depositions, so the PMT needs to be sufficiently deep that EM particles have interacted prior to reaching the depth of the PMT and the

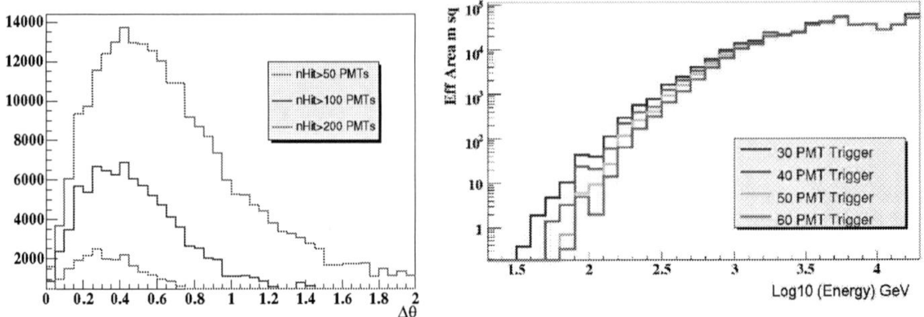


**FIGURE 2.** Angular Resolution as a function of event size (left) and effective Area of HAWC for showers with zenith < 30° passing the $\gamma$/h cut.

produced Cherenkov light has diffused. Monte Carlo studies indicate that a depth of ~4m or greater is sufficient for effective gamma/hadron separation and that greater depth is not advantageous. In general, the depth of the PMTs should be slightly smaller than their spacing making it possible for a PMT to view the entire surface of its cell (given the 41° Cherenkov angle in water). Given that the depth is selected to be 4m, a separation of 5m was found to be optimal.

## HAWC PERFORMANCE

The simulation for HAWC is an extension of the Milagro simulation software package. CORSIKA[2] is used to simulate gamma ray and hadron induced atmospheric showers. A custom detector simulation using GEANT[3] is used to propagate the secondary shower particles through the HAWC detector. The simulation has been thoroughly tested through comparison with Milagro data. Gamma ray and background rates are scaled from measured values in Milagro by comparing the predictions of the HAWC and Milagro simulations. By doing this, we not only remove potential systematic errors internal to the simulation from the air shower modeling, optical model, and detection efficiency, but also remove systematic errors in the measurement of gamma-ray fluxes and hadronic backgrounds provided by other experiments.

The angular resolution depends on the event size. Figure 2 shows the angular resolution of the HAWC detector for three different ranges of event size and the effective area for the HAWC detector. For triggered events with a median energy of ~3 TeV the angular resolution of HAWC will be $\sim 0.25°$.

Hadronic showers are identified through the pattern of energy deposition in the detector. While gamma-ray induced showers have compact cores with smoothly falling density, hadronic showers typically deposit large amounts of energy in distinct clumps far from the shower core. This is mostly due to the presence of hadrons and muons in hadronic showers. As a simple $\gamma$/hadron discriminator, we have extended the compactness parameter, C, developed for Milagro. Here C is defined as the total number of

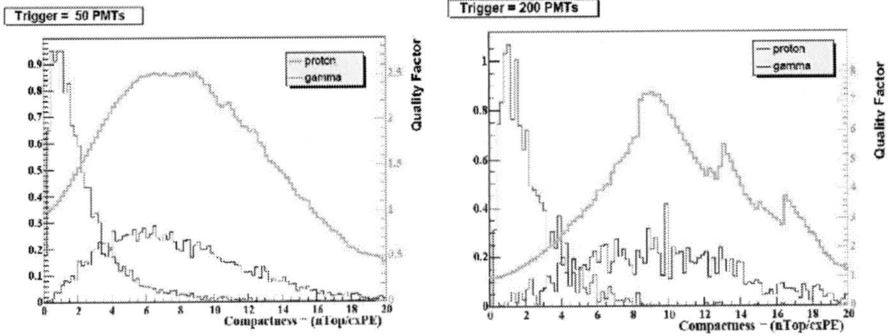


**FIGURE 3.** Gamma/hadron separation parameter distribution plotted for gamma and proton events. The Q value is plotted in green showing the increase in sensitivity achieved.

PMTs hit with amplitudes greater than 2 PEs divided by the largest pulse amplitude that is more than 30m from the reconstructed core position. Gamma-ray induced showers, with only small hits far from the core, have large values of C. Hadron induced showers with large hits far from the core have low values for C. Figure 3 shows Compactness distribution for gamma ray and proton triggers for small (nHit>50) and large (nHit>200) events. The same figure shows the Q (increase in sensitivity) as a function of the cut level. The efficiency for retaining gamma ray induced events increases with increasing energy, while it drops for proton-induced events. Therefore, the background rejection capability of HAWC improves with increasing energy.

Because the sensitivity of the detector is strongly dependent on the zenith angle of the source being studied, we compute the sensitivity by estimation of the number of signal and background events collected during a single transit of the source from horizon to horizon. The transit declination of the source is 22° and the detector is assumed to be located at a latitude of 37° (the location of Milagro), so the minimum zenith angle for the source is 15°. For a source transiting through the detector zenith the sensitivity is $\sim$25% higher and for a source transiting only within 25° of zenith, the sensitivity would be $\sim$25% lower. So the results given here represent a typical sensitivity for a sky survey of $\pm$25° declination. For a detector altitude assumed to be 4100m a.s.l., the altitude of the sites under consideration, HAWC will detect $\sim$50 gamma-ray events per transit with a threshold of 1 TeV and a significance $5\sigma$. Because the background rejection capability and angular resoultion of HAWC improves with energy the sensitivity of HAWC is better for sources with harder spectra. For hard spectrum sources without a high energy cut off, the exposure of HAWC will make it possible to detect and measure spectra of even dim sources well beyond 50 TeV.

## REFERENCES

1. Goodman, Jordan A., These Proceedings.
2. http://www-ik.fzk.de/corsika/
3. http://wwwinfo.cern.ch/asd/geant4/geant4.html

# Pierre Auger Enhancements: Transition from Galactic to Extragalactic Cosmic Ray Sources

A. Etchegoyen*, D. Melo,†, A. D. Supanitsky,† and M.C. Medina**

*Laboratorio Tandar - Comisión Nacional de Energía Atómica and Universidad Tecnológica Nacional, Regional Buenos Aires
†Laboratorio Tandar - Comisión Nacional de Energía Atómica
**Laboratorio Tandar - Comisión Nacional de Energía Atómica and CONICET.
Buenos Aires - Argentina

**Abstract.** The Pierre Auger Collaboration has decided to include detector enhancements in order to have unitary detection efficiencies down to $10^{17}$ eV in cosmic rays detection. These enhancements consist in high elevation telescopes and an infill area with both surface detectors and underground muon counters thus allowing a detailed study of the spectrum region where the cosmic rays sources are assumed to change from galactic to extragalactic origins.

**Keywords:** High energy cosmic rays, galactic to extragalactic sources, Pierre Auger Project
**PACS:** 98.70.Sa, 95.85.Ry, 95.55.Vj, 95.55.Cs

## INTRODUCTION

The Pierre Auger Project [1] studies the highest energies cosmic rays arriving from outer space at the earth surface with a very low flux. The questions to be elucidated are what are the origin, energy, production mechanism, and chemical composition of these cosmic rays. Auger aims at building two Observatories situated in both hemispheres and in 2000 the construction of the austral observatory started in the Malargüe, Province of Mendoza, Argentina.

Prior to Auger several other experiments were performed since the first detection of a cosmic ray shower with ascribed energy in excess of 100 EeV (1.0 EeV = $10^{18}$ eV) at Volcano Ranch, New Mexico, in 1962. They would generally deploy surface detectors over a given area with a chosen geometry in order to reconstruct the shower lateral distribution at earth surface. A different approach, the fluorescence telescope detection technique was developed and implemented by Fly's Eye/HiRes. They reconstruct the longitudinal shower profile by detecting the fluorescence light produced by excitation of atmospheric nitrogen by shower electrons. Surface detectors have a 100% duty cycle and make use of simulations in order to evaluate the primary energy, while the fluorescence technique has a 10 - 15% duty cycle (clear, dark nights) and does not resort to simulations for the energy estimation, although it relies on both the atmospheric fluorescence yield and on a continuous evaluation of the atmospheric light attenuation length. The aperture is well defined for the surface array at higher energies and it is calculated with Monte Carlo simulations for the telescopes. They are clearly two systems that complement each other.

Auger two distinctive features are its exceptional size and its hybrid nature. It spans

CP917, *Particles and Fields*, edited by H. Castilla Valdez, J. C. D'Olivo, and M. A. Perez

over an area of $3,000 km^2$ and it was originally designed to be constituted by 24 fluorescence detector (FD) telescopes, deployed in 4 buildings hosting 6 telescopes each, and 1600 surface detectors (SD) [1] disposed in a 1500 m equilateral triangle grid. The FD's are placed on the periphery of the SD array overlooking it and each one has an acceptance of $\sim 30^\circ \times 30^\circ$ in azimuth and elevation. Therefore Auger provides a large number of events with less systematic detection uncertainties.

Auger has now decided to include detector enhancements in order to both have unitary detection efficiencies down to 0.1 EeV and to improve primary type identification. These enhancements consist in 3 high elevation telescopes (HEAT, High Elevation Auger Telescopes, [2]) and an 85 detector graded infill area with both surface detectors and buried muon counters (AMIGA, Auger Muons and Infill for the Ground Array, [3]) and therefore Auger will total 27 telescopes, 1666 surface detectors and 85 muon counters. The scientific purpose of these enhancements is to allow a detailed study of the spectrum region where the cosmic rays sources are assumed to change from galactic to extragalactic origins.

The spectrum shows three features: the "knee" at $3-5\times 10^{15} eV$ where the spectral index changes from -2.7 to -3.1, the "second knee", with a further steepening of the spectrum, and the "ankle"(see [4] for a recent spectrum compilation of the second knee-ankle region). Above the knee the primary chemical composition becomes heavier which is consistent with supernova remnants failing to accelerate lighter nuclei any further.

The second knee was observed at $\sim$ 0.4 EeV by AKENO [5], Fly's Eye stereo [6], and at a higher energy by Yakutsk [7] and its physical interpretation is still quite uncertain. It might be due to attaining the maximum cosmic rays energies within the galaxy, i.e. primaries from the heaviest stable isotopes [8] since when this limit is reached the flux will necessarily decay with a larger spectral index. Even more it has been suggested [10, 11] that it could be the transition region from galactic to extragalactic primaries, i.e. on top of a diminishing galactic component a predominant contribution from extragalactic proton primaries.

The ankle was observed by Fly's Eye [6, 12, 13], Haverah Park [14] and HiRes [15] at $\sim$ 3 EeV. Also by Yakutsk [7] and AGASA [16] but at a higher energy, $\sim$ 10 EeV . There are two main physical interpretations of the ankle depending on where the transition from galactic to extragalactic sources takes place. If it occurs at the second knee, then the ankle (or the dip in the spectrum) is reported to arise from $e^-e^+$ pair production from extragalactic protons collisions with the microwave cosmic background radiation [10, 11]. On the other hand, if it occurs at the ankle, then the ankle would arise from the different spectrum indexes of the galactic and extragalactic components [17, 18, 19].

As such it is of uppermost relevance to study the 0.1 - 10 EeV energy range in order to cast light on the second knee and ankle. The two main requirements are good energy resolution in order to settle the current discrepancies among different experiments and primary type identification (statistical discrimination over this energy range will suffice) since as mentioned above the transition from galactic (heavy elements) to extragalactic (light elements) sources is directly linked to primary composition.

The two shower parameters relevant to composition are the atmospheric depth at shower maximum, $X_{max}$, and the shower muon content being other parameters noisy combination of them. Composition is very poorly understood in this energy range where varieties of mixed compositions are reported ranging from proton to iron dominated primaries (see [20] and references within). Still, composition can only be assessed within a given hadronic interaction model and therefore comparisons are only to be performed under those premises. Moreover, they would need a recalculation if a hadronic model is reformulated. Much more robust results are attained from the variation rate of either $X_{max}$ (called elongation rate) or muon content as a function of energy [21] by which composition changes may be assessed fairly independent of the assumed hadronic model.

## HEAT AND AMIGA

The current Auger layout has unitary trigger efficiencies at 0.5 and 3 EeV for hybrid [2] and SD [22] event reconstructions, respectively. Trigger efficiencies depend on mass composition and therefore such analyses will be biased if not performed in the full efficiency region. For instance, the SD array was estimated [4] to be $\sim$ 40% more efficient in detecting iron than proton initiated showers at 1.0 EeV. Higher elevation angle telescopes and closely spaced SD's will solve this bias and at the same time yield a much better shower reconstruction at lower energies and provide an overlap spectrum region for $E \leq 1.0 EeV$ with KASCADE-Grande [8]. Note that since the integrated cosmic ray flux rapidly decays as $\sim E^{-2}$, a smaller detector acceptance (i.e. both a reduced number of telescopes and infill area), is needed as shown in Fig. 1.

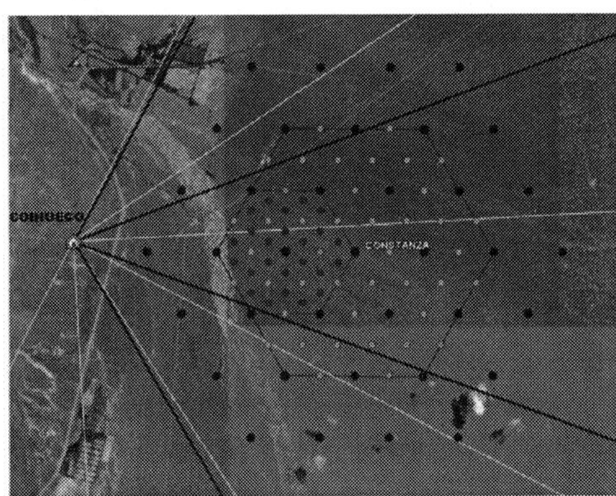


**FIGURE 1.** Layout of Auger enhancements [9]. FD telescopes including HEAT are placed at the Coihueco hill top and AMIGA graded infill are bounded by the two hexagons covering areas of 5.9 and 23.5 km$^2$. White and black lines show the six original and three enhanced telescopes FOV, respectively. Grey, light grey, and black dots indicate SD's plus buried muon counters, placed 433, 750, and 1500 m apart, respectively.

There will be three HEAT telescopes similar to the already deployed telescopes, which will be placed at Cerro Coihueco (alongside the remaining six telescopes) in three shelters with steel frame enclosures. The shelters will have two positions: horizontally resting on the floor for installation and maintenance and with the enclosure tilted in 28° for normal operation, permitting a seamless elevation acceptance up to $\sim$ 58° [2].

A telescope comprises a circular diaphragm with an annular Schmidt lens corrector ring, an UV filter, a segmented spherical mirror which focusses the fluorescence light onto a 22 rows $\times$ 20 columns hexagonal photomultiplier tube (PMT) camera, each PMT subtending $\sim 1.5^{\circ}$ in both elevation and azimuth [1]. Fig. 2 displays the simulated signals of a 1 EeV proton shower on the enhanced Auger telescope system. The extended FOV permits the inclusion of the shower maximum $X_{max}$ without which a shower can not be properly reconstructed.

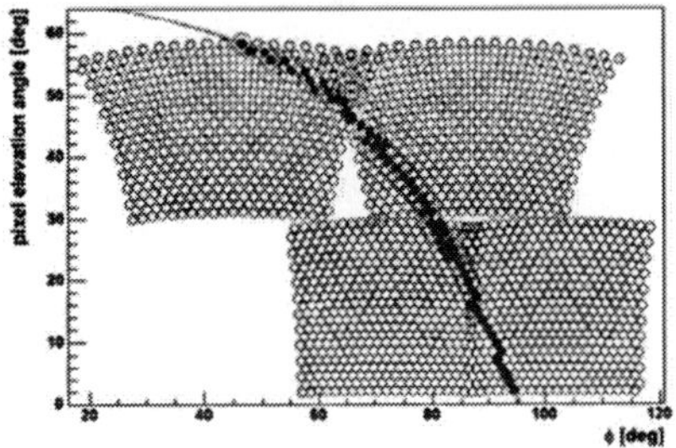


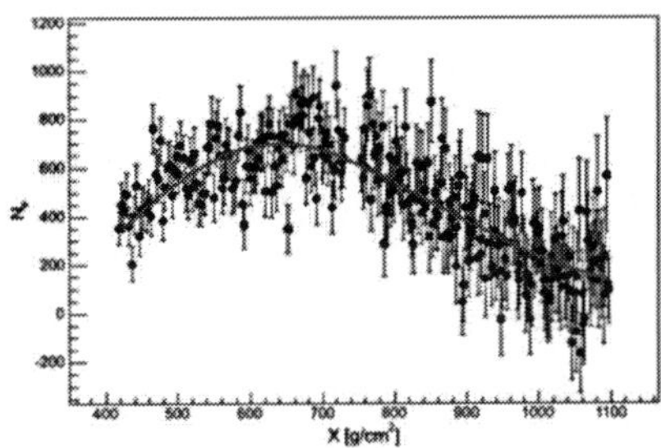


**FIGURE 2.** A 1.0 EeV simulated shower (*left*) illuminating 2 upper and 2 lower PMT cameras and (*right*) the reconstructed shower longitudinal profile.

AMIGA will be centered at SD named Constanza, $\sim 6.0$ km away from Coihueco and it consists in identical SD's to the existing ones [1] plus nearby $\sim 30m^2$ muon counters buried $\sim 2.5$ - 3.0 m underground alongside the electronics box. SD's are plastic tanks filled with 12 tons of hyper pure water in which shower secondaries produce Cherenkov light which is collected by three symmetrically placed 9-inch diameter PMT's on the top lit. The scintillator modules were developed following the MINOS design [23] and they are not supposed to reconstruct the shower parameters (which will be performed by the FD and SD systems) but just to count muons. The current baseline design calls for 400 cm long $\times$ 4.1 cm wide $\times$ 1.0 cm high strips of extruded polystyrene doped with fluors and co-extruded with a $TiO_2$ reflecting coating with a groove in where a wavelength shifter (WLS) fiber is glued and covered with a reflective foil (see Fig. 3). Each module will consist of 64 strips with the fibers ending in an optical connector matched to a 64 multianode PMT from the Hamamatsu H7546B series ($2mm \times 2mm$ pixel size).

**FIGURE 3.** Assembly at Argonne National Laboratory [24] of a 64 $200cm \times 4.1cm$ prototype (*left*) side view, and (*right*) front view displaying the 64 pixel optical connector.

The counter has two main design parameters, the strip length and the counter size. The strip length was chosen in order to have an electronic signal at the PMT anode for a muon impinging on the far end. The segmentation is fixed once the length is chosen since each strip is 4.1 cm wide. The counter size is set by the 64 pixel PMT so each module would have 64 strips and a counter could comprise a given number of modules. After simulations a 3 module counter was chosen, i.e. 192 segments. With this segmentation a pile-up uncertainty can be estimated for the closest counter to the core which is at a mean distance of 230 m for a 750 m grid assuming a 25 ns electronics: for 90 muons (as expected for a $\theta = 30°$ 1.0 EeV iron shower at 200 m from the core arriving in the first 25 ns bin) the uncertainty is just increased by 15% on top of the poisson fluctuations [25]. Note that single counter uncertainties will be diminished when a fit to a Muon Lateral Distribution Function (MLDF) is performed, as it will be now explained.

A final assessment of the muon counter performance was performed by fitting a MLDF and extracting a representative muon density from it at a given core distance which typically is taken as 600 m, $N_\mu(600)$ [26]. Recently KASCADE has proposed the following MLDF [27]:

$$\rho_\mu(r) = N_\mu \left(\frac{r}{r_0}\right)^{-\alpha} \left(1+\frac{r}{r_0}\right)^{-\beta} \left(1+\left(\frac{r}{10\, r_0}\right)^2\right)^{-\gamma} \quad (1)$$

As shown in [25] fixing $r_0 = 320m$, $\alpha = 0.75$, and $\gamma = 3$ and leaving $N_\mu$ and $\beta$ as free parameters give a satisfactory MLDF fit: it is seen that though iron initiated showers have smaller uncertainties since they have more muons than proton showers, the overall uncertainty ranges between 10- 15 %, which compares satisfactory with the 25 % shower energy uncertainty which in turns directly translates to the muon number uncertainty since the latter depends nearly proportional to shower energy. $N_\mu(600)$ values extracted in this fashion are used in the next section.

## Shower Reconstruction with Enhancements

Auger enhancements will provide a seamless energy region for E $\geq$ 0.1 EeV with unique experimental data that will allow to reconstruct shower parameters performing both thorough quality and composition analyses. In this work we use the QGSJET02 hadronic model interaction [28] for muon simulations and Gaisser-Hillas profiles with parameters extracted from QGSJET01 [29] for longitudinal profile simulations.

### *Reconstruction Quality*

Not only trigger efficiencies need be independent of composition, but also does the working data sample obtained after applying quality cuts. This issue is of particular interest for fluorescence and hybrid data samples since cuts on the composition dependent $X_{max}$ parameter are always applied. Simulations were performed which showed that $X_{max}$ sometimes lies outside the FOV even for the enhanced telescope and particularly for iron initiated showers and therefore imposing $X_{max}$ to lie inside the FOV might bias towards proton primaries, by up to $\sim 8\%/15\%$ at $10^{17.5}$ eV in the current simulations [30] after performing hybrid/monocular reconstructions, respectively with all quality cuts applied and showers with isotropic arrival directions with $\theta_{max} = 60°$ . There is a geometrical problem arising from showers coming from behind the telescope and forced to impinge within the close-by AMIGA area which may be tackled by imposing further cuts, for example on the arrival direction. Simulations with $\theta_{max} = 45°$ gave just a $\sim 3\%/5\%$ difference in efficiency with primary type.

Data quality may also be tested by operating HEAT on its horizontal position since the same shower will be directly measured by two telescopes [2] and by using the larger number of trigger tanks of AMIGA on high energy showers [3], as discussed below.

The shower impact area on the ground diminishes with the primary energy and therefore the SD spacing needs to accordingly decrease in order to have enough triggered stations as to allow a good quality shower reconstruction. This is clearly seen in the Surface Lateral Distribution Function (SLDF) of Fig. 4 where 4, 19, and 34 stations triggered for 1500, 750, and 433 m grids for a $10^{18.5}$ eV, $\theta = 30°$ iron initiated shower and QGSJET01 interaction.

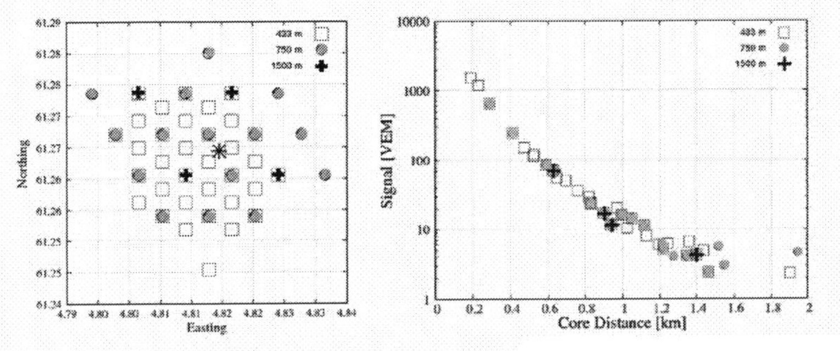


**FIGURE 4.** Simulated $10^{18.5}$ eV, $\theta = 30°$ iron initiated shower on 433, 750, and 1500 m grids with triggered tanks in open squares, grey circles, and black crosses, respectively; *(left)* plan view with asterisk at core impact position. Peripheral grey circles without a corresponding open square signal tanks outside the 433 m array; and *(right)* SLDF.

Actually, the advantage of a denser infill array is twofold: shower detection becomes fully efficient at lower energies as above mentioned but also reconstruction uncertainties for the larger array can be experimentally checked.

In [4] it is shown that the Auger SD system with 433, 750 and 1500 m grids becomes fully efficient at 0.01, 0.3, and 3.0 EeV, respectively. Inspection of Fig. 4 shows for an event on the verge of the fully efficiency regime for the 1500 m array that reconstructions with the different arrays will lead to experimentally check the Auger 4-tank reconstruction.

A further test to be performed is on a possible underestimate of hadronic interaction models of the shower muon contents [31] (note that at the Auger energy region there is no experimental data from collider accelerators). AMIGA muon counters might help in both solving this discrepancy and in calibrating indirect muon parameters which could then be used for the whole SD array.

## *Composition*

Although a large variety of composition estimates have been performed [20] by using different reconstructed shower parameters, they will only be a manifestation of the basic shower characteristics which are its muon contents and its atmospheric depth at shower maximum [32]. The three main primary types which might initiate high energy showers are usually classified in three groups, gammas, lighter, and heavy primaries. Gammas are more easily identified since they develop deeper in the atmosphere [33] and in particular they have much less muon contents [34], and therefore gamma-hadron identification is expected to be significantly easier than hadron identification. A simulation on the latter is shown in Fig. 5, where each simulated event is reconstructed for $X_{max}$ and $N_\mu(600)$, approximately including both detector and a 25 % energy uncertainties.

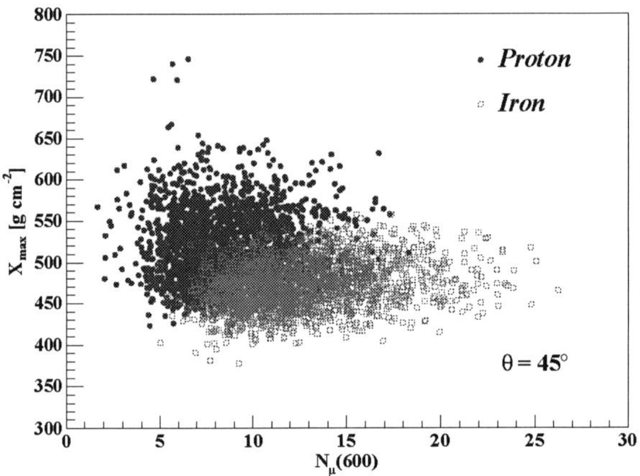


**FIGURE 5.** Simulated Depth-of-Maximum vs. Muon Number at 600 m from shower core for iron (grey) and proton (black) reconstructed showers (with a 25% energy uncertainty) with energies in the $[0.75, 1.25]$EeV interval. The events corresponds to showers initiated by proton and iron primaries with $\theta = 45°$ that follow a power law energy spectrum of spectral index $\gamma = -2.7$ (extracted from [25]).

A composition analysis performed on the simulated data of Fig. 5 showed a $\sim 90\%$ shower-to-shower primary identification [25]. Still, and as already mentioned it is more robust to assess a composition change over an energy range than to directly assign a composition. We performed an expected typical analysis for composition changes which is shown in Fig. 6 (see [21] for a pioneer work on multicomponent measurements).

Fig. 6 includes uncertainties arising from shower to shower fluctuations, detector simulations, and longitudinal and lateral profiles reconstructions. The elongation rate plot also includes energy uncertainties and a spectrum convolution. Statistical uncertainties are plotted and they are mostly within the point symbols. The elongation rate graph shows a $\sim 10g/cm^2$ systematic error for this type of reconstruction procedure while muon contents do not show such a problem at all. Both analyses show that the slopes are clearly reproduced after reconstruction and as such a composition variation over this energy range should be simultaneously identified and corroborated given a strong and convincing indication on where the transition from galactic to extragalactic sources takes

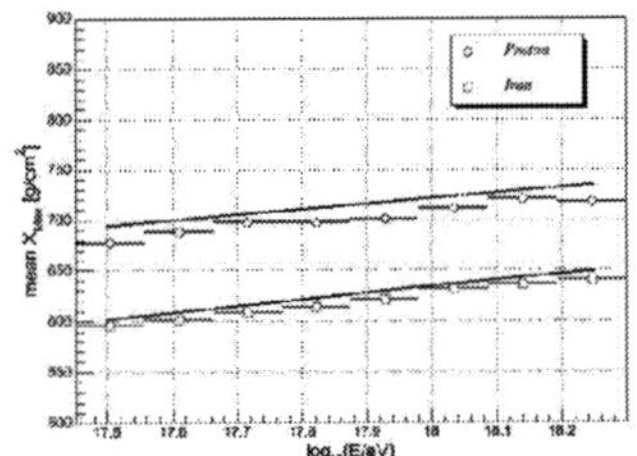

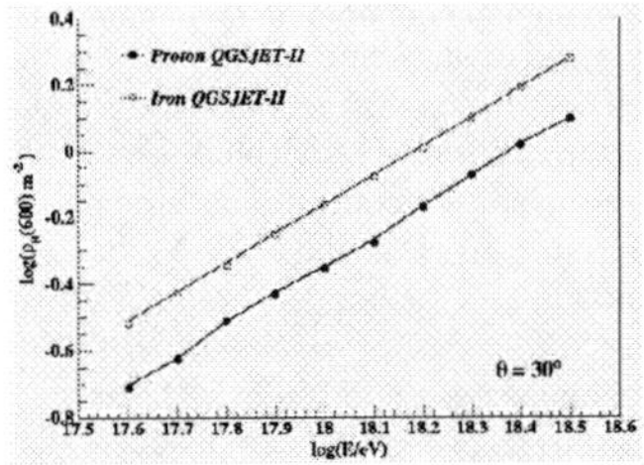


**FIGURE 6.** Simulated (*left*) elongation rate [30], and (*right*) shower muon contents [25]. Lines and dots are the input simulation and reconstructed values, respectively.

place.

## CONCLUSIONS

The extension of Auger to full efficiency down to 0.1 EeV with a direct measure of both $X_{max}$ and shower muon contents will open an unique experimental tool to study the cosmic ray spectrum in the second-knee and ankle region, where the transition from galactic to extragalactic sources is assumed to occur. Data with unprecedented precision will be available and in particular there will be a triple hybrid data set (fluorescence, muon, and tank detections) on which careful primary energy and composition analyses could be performed.

## REFERENCES


1. Auger Collaboration, J. Abraham et al. *Nucl. Inst. & Meth.*, **A523**, 50-95, (2004).
2. Auger HEAT Collaboration (corresponding author: H.O. Klages) "Enhancement of the Pierre Auger Observatorory by Additional Telescopes with Elevated Field of View: HEAT = High Elevation Auger Telescopes", *Design Report*, unpublished, 2006.
3. Auger AMIGA Collaboration (corresponding author: A. Etchegoyen) "AMIGA, Auger Muons and Infill for the Ground Array", *Design Report*, unpublished, 2006.
4. M.C. Medina et. al., *Nucl. Inst. and Meth.* **A** 566, 302-311 (2006), and astro-ph0607115.
5. Nagano M. et al., *J. Phys. G* **18**, 423,(1992).
6. Bird D. J. et al., *Phys. Rev. Lett.* **71**, 3401, (1993).
7. A.V. Glushkov et al., *Proc. 28th Int. Cosmic Ray Conf.* (Tukuba, Japan), 389, (2003).
8. A. Haungs et al. (KASCADE-Grande Coll.), *astro-ph*/0508286.
9. M. Gómez Berisso and M.C. Medina, personal communication.
10. V.S. Berenzinsky et al. *Phys. Lett. B* **612**, 147, (2005) and *astro-ph*/0502550.
11. V.S. Berenzinsky, *astro-ph*/0509069.
12. D.J. Bird et al., *Astrohys. J* **441**, 144, (1995).
13. T. Abu-Zayyad et al., *Astrophys. J* **557**, 686, (2001).
14. Ave M. et al., *Proc. 27th Int. Cosmic Ray Conf.* (Hamburg, Germany) **1**, 381, (2001) and *astro-ph*/0112253.
15. R.U. Abbasi et al. (HiRes Coll.), *Phys. Lett.* **B619**, 246,(2005).
16. M. Takeda et al., *Astropart. Phys.* **19**, 447, (2003).
17. A.M. Hillas, *Nucl. Phys. B (Proc. Suppl)* **136**, 139, (2004).
18. T. Wibig y A.W. Wolfendale, *J. Phys. G* **31**, 255, (2005).
19. D. Allard et al., *astro-ph*/0505566.

20. L. Anchordoqui et al., *hep-ph*/0407020.
21. T. Abuy-Zayyad et al., *Phys. Rev. Letts.* **84**, 4276-4279 (2000).
22. D. Allard et. al., *Proc. 29th Int. Cosmic Ray Conf.* (Pune, India) (2005).
23. MINOS Collaboration "The MINOS Detectors Technical Design Report", Version 1.0, October 1998.
24. S. Kuhlmann, personal communication.
25. A.D. Supanitsky, Ph.D thesis, Universidad de Nacional Buenos Aires.
26. N. Hayashida et al., *J. Phys. G: Nucl Part. Phys.* **21**, 1101-1119, (1995).
27. J. Buren et al., *Proc. 29th Int. Cosmic Ray Conf.*(Pune, India) **6**, 301, (2005).
28. S. Ostapchenko, *hep-ph/0501093 V1*, (2005).
29. N.N. Kalmykov, S.S. Ostapchenko, and A.I. Pavlov, *Nucl. Phys. B (Proc. Suppl)* **52B**, 17, (1997).
30. D. Melo, Ph.D thesis, Universidad Nacional San Martín.
31. P. Billoir and O. Blanch Bigas, personal communication.
32. A.S. Chou et al., *Proc. 29th Int. Cosmic Ray Conf.* (Pune, India), usa-chou-A-abs1-he14-poster (2005).
33. Auger Collaboration, J. Abraham et al. *Astropart. Phys*, accepted and *astro-ph*/0606619.
34. K. Shinozaki et al., *Proc. 21th Int. Cosmic Ray Conf.*, 346, (2001).

# Color Glass Condensate and Glasma

Larry McLerran

*Physics Department and Riken Brookhaven Center PO Box 5000, Brookhaven National Laboratory, Upton, NY 11973 USA*

**Abstract.** The physics of the scattering of very high energy strongly interacting particles is controlled by a new, universal form of matter, the Color Glass Condensate. This matter is the dominant contribution to the low x part of a hadron wavefunction. In collisions, this mater almost instantaneously turns into a Glasma. The Glasma initially has strong longitudinal color electric and magnetic fields, with topological charge. These fields melt into gluons. Due to instabilities, quantum noise is converted into classical turbulence, which may be responsible for the early thermalization seen in heavy ion collisions at RHIC.

## THE HIGH ENERGY LIMIT

The high energy limit of QCD is the limit where the energy of collisions goes to infinity, but the typical momentum transfer is finite. This momentum transfer can be much larger than $\Lambda_{QCD}$, but it is to remain fixed. This is not the short distance limit, where both momentum transfer and energy go to infinity. The short distance limit is understood using weak coupling perturbation theory. The high energy limit is that of non-perturbative phenomena such as Pomerons, Reggeons, unitarization etc. etc. One of the purposes of this lecture is to convince the reader that this non-perturbative limit of QCD is also a weak coupling limit

The Bjorken x variable can be understood as the ratio of the energy of the constituent of a hadron to that of the hadron itself in the reference frame where the hadron has large energy. The typical minimal value of x is

$$x_{min} \sim \Lambda_{QCD}/E_{hadron} \qquad (1)$$

The minimal value decreases as the hadron energy increases.

A hadron wavefunction has many different components. This is illustrated in Fig. 1.

A nucleon has a Fock space component with three quarks and no gluons, with an extra gluon and with many extra gluons. The components which control low energy scattering are those with three quarks and a few gluons. For high energy scattering processes, the typical matrix elements are those with three quarks, many quark anti-quark pairs, and even more gluons.

## WHAT IS THE COLOR GLASS CONDENSATE?

The original ideas for the Color Glass Condensate were motivated by the result for the HERA data on the gluon distribution function shown in

CP917, *Particles and Fields*, edited by H. Castilla Valdez, J. C. D'Olivo, and M. A. Perez

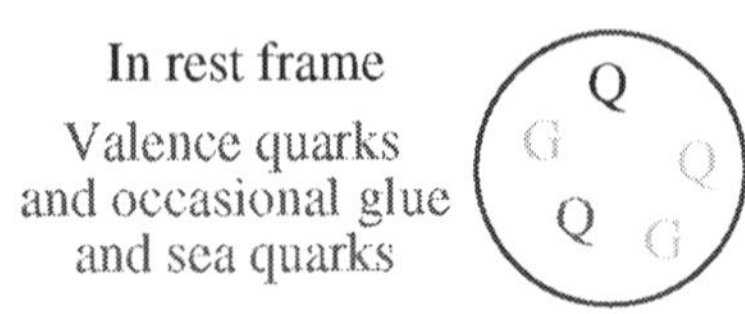


In infinite momentum frame

Wall of lots and lots of glue at near light speed

**FIGURE 1.** The various componenents of the hadron wavefunction.

Fig. 2(a) [1] The gluon density is rising rapidly as a function of decreasing x. This was expected in a variety of theoretical works[2]-[4] and has the implication that the real physical transverse density of gluons must increase.[2]-[3],[5]. This follows because total cross sections rise slowly at high energies but the number of gluons is rising rapidly. The gluons must fit inside the size of the hadron.

This is shown in Fig. 2(b). This led to the conjecture that the density of gluons should become limited, that is, there is gluon saturation. [2]-[3], [5] Actually, I argue that as one goes to higher energy, a hadron becomes a tightly packed system of gluons larger than some size scale. For smaller gluons there are holes. As one increases the energy, one still adds in more gluons, but these gluons are small enough that they fit into the holes. Because in quantum mechanics, we interpret size as wavelength as inversely proportional to momentum, at high energies, the gluons are tightly packed for gluons below some momentum, and are filling in above that momentum. There is therefore a critical momentum, the saturation momentum, which characterizes the filling. This saturation momentum increases as the energy increases, so the total number of gluons can increase without bound.

The low x gluons therefore are closely packed together, and become more closely packed as the energy increases. The strong interaction strength must become weak, $\alpha_S \ll 1$. Weakly coupled systems should be possible to understand from first principles in *QCD*.[5]-[6]

This weakly coupled system is called a Color Glass Condensate (CGC) for reasons we now enumerate:[6]

- **Color** The gluons which make up this matter are colored.

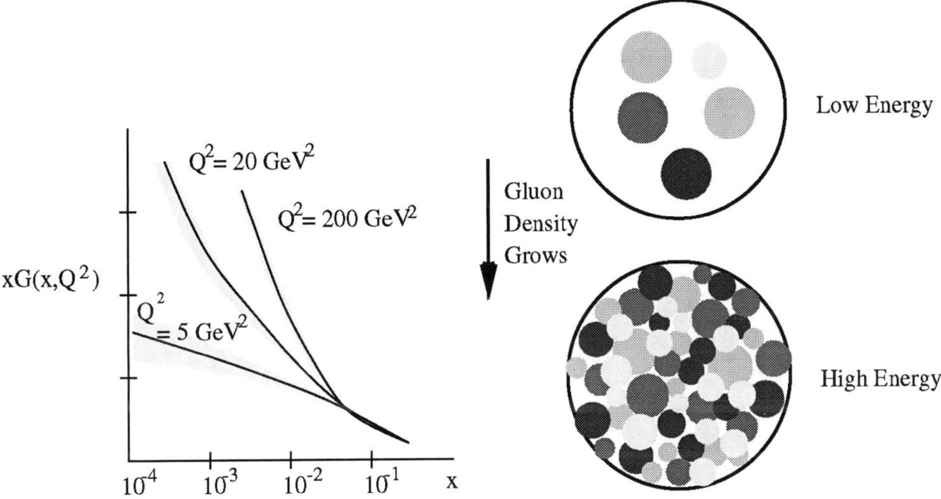


**FIGURE 2.** (a)The HERA data for the gluon distribution function as a function of x for various values of $Q^2$. (b) A physical picture of the low x gluon density inside a hadron as a function of energy.

- **Glass** The gluons at small x are generated from gluons at larger values of x. In the infinite momentum frame, these larger momentum gluons travel very fast and their natural time scales are Lorentz time dilated. This time dilated scale is transferred to the low x degrees of freedom which therefore evolve very slowly compared to natural time scales. This is the property of a glass.
- **Condensate** The phase space density

$$\rho = \frac{1}{\pi R^2} \frac{dN}{dy d^2 p_T} \tag{2}$$

is generated by a trade off between a negative mass-squared term linear in the density which generates the instability, $-\rho$ and an interaction term $\alpha_S \rho^2$ which stabilizes the system at a phase space density $\rho \sim 1/\alpha_S$. Because $\alpha_S << 1$, this means that the quantum mechanical states of the system associated with the condensate are multiply occupied. They are highly coherent, and share some properties of Bose condensates. The gluon occupation factor is very high, of order $1/\alpha_S$, but it is only slowly (logarithmically) increasing when further increasing the energy, or decreasing the transverse momentum. This provides saturation and cures the infrared problem of the traditional BFKL approach.[7]

One can understand the high phase space occupancy $1/\alpha_S$ from simple arguments. The momentum scale in the phase space distribution is the De Broglie wavelength of the gluons which we can interpret as the size of the gluons. The gluons of fixed size will densely occupy the system until there are $1/\alpha_S$ gluons of fixed size closely packing the system. The gluons interact with strength $\alpha_S$, so that when $1/\alpha_S$ sit on top of one another, they act coherently like a hard sphere with interaction strength of order 1.

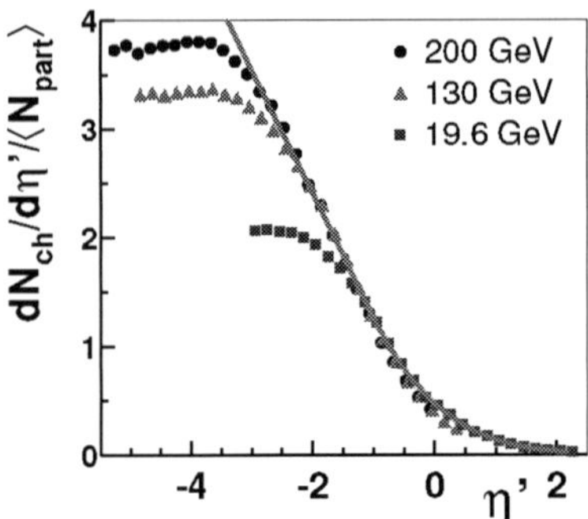


**FIGURE 3.** Limiting fragmentation as measured in the Phobos experiment at RHIC.

Implicit in this definition is a concept of fast gluons which act as sources for the colored fields at small x. These degrees of freedom are treated differently than the fast gluons which are taken to be sources. The slow ones are fields. There is an arbitrary $X_0$ which separates these degrees of freedom. This arbitrariness is cured by a renormalization group equation which requires that physics be independent of $X_0$. In fact this equation determines much of the structure of the resulting theory as its solution flows to a universal fixed point.[6]-[9]

There is evidence which supports this picture. One piece is the observation of limiting fragmentation. This phenomena is that if particles collide at some fixed center of mass energy and the distribution of particles are measured as a function of their longitudinal momentum from the longitudinal momentum of one of the colliding particles, then these distributions do not change as one goes to higher energy, except for the new degrees of freedom that appear. This is true near zero longitudinal momentum in the center of mass frame because new degrees of freedom appear as the center of mass energy is increased. In the analogy with the CGC, the degrees of freedom, save the new ones added in at low longitudinal momentum, are the sources. The fields correspond to the new degrees of freedom. The sources are fixed in accord with limiting fragmentation. One generates an effective theory for the low longitudinal momentum degrees of freedom as fixed sources above some cutoff, and the fields generated by these sources below the cutoff. A recent measurement of limiting fragmentation comes from the Phobos experiment at RHIC shown in Fig. 3 [10]

Of course the perfect scaling of the limiting fragmentation curves is only an approximation. As shown by Jalilian-Marian, the limiting fragmentation curves are given by the total quark, antiquark and gluon distribution functions of the fast particle measured at a momentum scale $Q^2_{sat}$ appropriate for the particle that it collides with.[11] The saturation momentum $Q_{sat}$ will play a crucial role in our later discussion. It is a momentum scale which is determined by the density of gluons in the CGC

$$\frac{1}{\pi R^2}\frac{dN}{dy} \sim \frac{1}{\alpha_S} Q^2_{sat} \qquad (3)$$

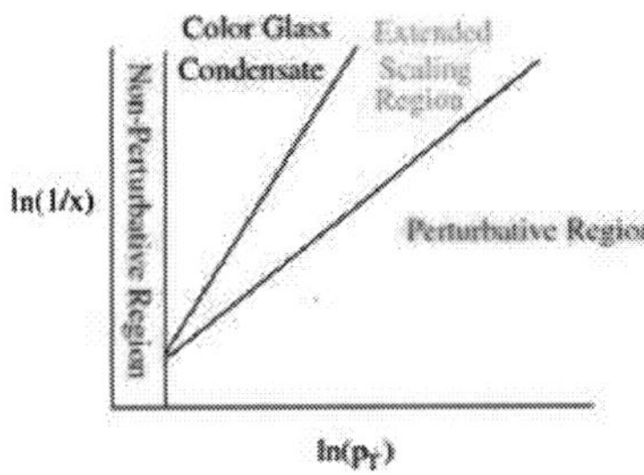


**FIGURE 4.** The kinematic regions of the Color Glass Condensate.

The saturation momentum turns out to depend on the total beam energy because the longitudinal momentum scale of the target particle at fixed $x$ of the projectile will depend upon the beam energy. It is nevertheless remarkable how small these violations appear to be.

The CGC may be defined mathematically by a path integral:

$$Z = \int_{X_0} [dA][dj] exp\,(iS[A,j] - \chi[j]) \tag{4}$$

What this means is that there is an effective theory defined below some cutoff in $x$ at $X_0$, and that this effective theory is a gluon field in the presence of an external source $j$. This source arises from the quarks and gluons with $x \geq X_0$, and is a variable of integration. The fluctuations in $j$ are controlled by the weight function $\chi[j]$. It is $\chi[j]$ which satisfies renormalization group equations which make the theory independent of $X_0$.[6],[8]-[9],[12]-[14]. The equation for $\chi$ is called the JIMWLK equation. This equation reduces in appropriate limits to the BFKL and DGLAP evolution equations.[4], [15] The theory above is mathematically very similar to that of spin glasses.

There are a variety of kinematic regions where one can find solutions of the renormalization group equations which have different properties. There is a region where the gluon density is very high, and the physics is controlled by the CGC. This is when typical momenta are less than a saturation momenta which depends on $x$,

$$Q^2 \leq Q^2_{sat}(x) \tag{5}$$

The dependence of $x$ has been evaluated by several authors, [2],[16]-[18], and in the energy range appropriate for current experiments has been determined by Triantafyllopoulos to be

$$Q^2_{sat} \sim (x_0/x)^{\lambda}\ GeV^2 \tag{6}$$

where $\lambda \sim 0.3$. The value of $x_0$ is not determined from the renormalization group equations and must be found from experiment.

The kinematic region corresponding to the CGC is shown in Fig. 4.

There is also a region of very high $Q^2$ at fixed x, where the density of gluons is small and perturbative QCD is reliable. It turns out there is a third region intermediate between

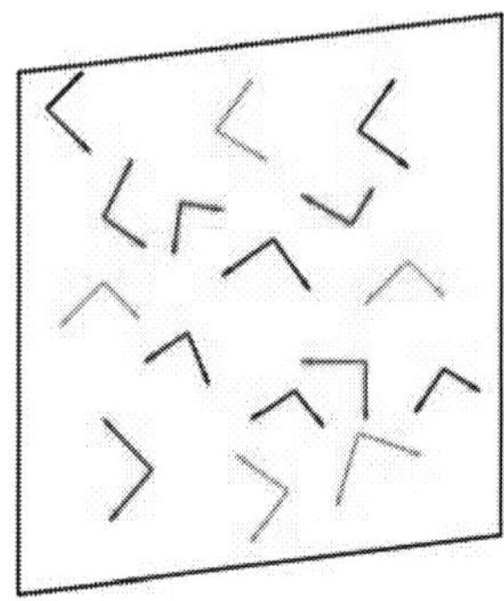

**FIGURE 5.** The color glass field.

high density and low where there are universal solutions to the renormalization group equations and scaling in terms of $Q_{sat}^2$.[17] In this region and in the region of the CGC, distribution functions are universal functions of only $Q^2/Q_{sat}^2(x)$. The extended scaling region is when

$$Q_{sat}^2 \leq Q^2 \leq Q_{sat}^4/\Lambda_{QCD}^2 \tag{7}$$

## WHAT IS THE FORM OF THE COLOR GLASS FIELDS?

One can simply compute the form of the Color Glass fields. If we work in a frame where the hadron has a large momentum, the $z-t \sim 0$. The only big component of $F^{\mu\nu}$ is $F^{i+}$ where

$$x^{\pm} = (z \pm t)/\sqrt{2} \tag{8}$$

If we set $F_{i-} = 0$, then simple algebra tells us that the big field strengths are $E$ and $B$, and that

$$\vec{E} \perp \vec{B} \perp \vec{z} \tag{9}$$

The fields are plane polarized perpendicular to the beam direction. These are the Lienard-Wiechart potentials which correspond to a Lorentz boosted Coulmbg field. They exist within the Lorentz contracted sheet and have a longitudinal extent corresponding to the fast moving sources. (The vector potential corresponding to these field is extended, and the wee gluons corresponding to these fields are extended over a larger longitudinal size scale). The fields have random polarizations and colors. This is shown in Fig. 5.

## WHAT IS THE CGC GOOD FOR?

The CGC provides a unified description of deep inelastic structure functions, of deep inelastic diffraction and of hadron-hadron collisions at high energies. It is the high

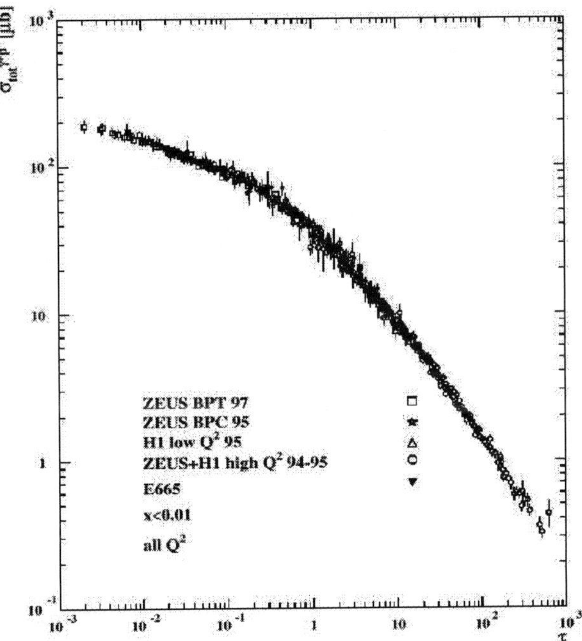


**FIGURE 6.** Geometric Scaling as seen at Hera..

energy limit of QCD. As such, it has many tests to pass before being accepted as a correct description. Over the last several years, there have been many qualitative and semi-quantitative successes of this description. It also provides an intuitively plausible and mathematically consistent description of such phenomena.

One can compute the $x$ dependence of the saturation momentum.[17]-[18] This solutions results from renormalization group equations. The results agree with Hera phenomenology.[19] In particular, within the same description one can compute both deep inelastic structure functions and diffractive structure functions.

In addition, the CGC predicts the existence of geometric scaling.[20] Geometric scaling means that the cross section for deep inelastic scattering of a virtual photon from a hadron depends only upon the scale invariant ratio $Q^2/Q^2_{sat}$, and not independently $Q^2$ and $x$. Such scaling is shown in Fig. 6 for x values of $x < 10^-2$.

The total cross section for hadron-hadron scattering is a slowly varying function of energy as shown in Fig. 7. The Color Glass Condensate provides a heuristic explanation of this.[21]-[23] We assume the distribution of gluons in the transverse plane of a hadron as a function of energy factorizes,

$$\frac{dN}{d^2r_T dx} \sim (x_0/x)^{\lambda} e^{-2m_\pi r_T} \tag{10}$$

The total hadronic cross section:

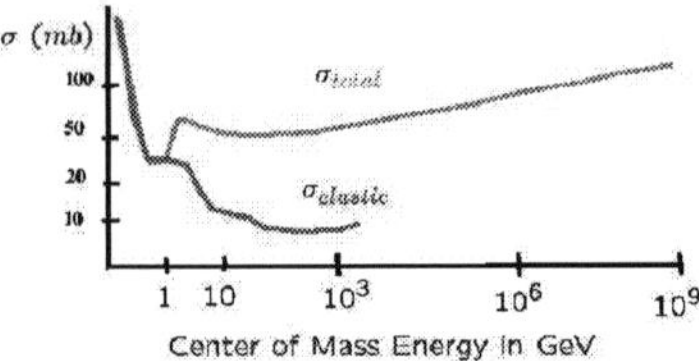


**FIGURE 7.** The total hadronic cross section as a function of energy..

where the exponential fall off should be of the form shown at large $r_T$, since the lowest mass particle exchange with isospin zero is two pions. The cross section for a particle of some size to penetrate the hadron, and have a large probability to scatter occurs when this density is some fixed number. Therefore

$$\sigma \sim R^2 \sim ln^2(1/x) \sim ln^2(E/E_0) \tag{11}$$

This behaviour is the Froissart bound for cross sections, and describes high energy scattering reasonably well. It origin is the trade off in rapidly falling impact parameter profile against rapidly rising density of partons.

One of the remarkable predictions of the Color Glass Condensate was the mutliplicty of particles produced in heavy ion collisions at RHIC.[24] The multiplicty of saturated gluons inside a nucleus should scale as

$$\frac{dN}{dy} \sim \frac{1}{\alpha_S} \pi R^2 Q_{sat}^2 \tag{12}$$

The energy dependence of $Q_{sat}$ is known, and the dependence on centrality should be proportional to $N_{coll}$, the number of nucleons colliding, at not too high an energy. Assuming the hadron multiplicity is proportional to the number of produced gluons gives the plot shown in Fig. 8. [25]

Experiments using deuterons on nuclei in the fragmentation region of the deuteron also test ideas about how the gluon distribution is modified by the Color Glass Condensate. The experimental measurements are in accord with CGC predictions.[25]

In the future years, there will be increasingly stringent tests arising at RHIC, LHC and potentially eRHIC. Theoretically, we are just beginning to understand the properties of this matter. New ideas concerning the structure of the underlying theory and the breadth of phenomena it describes are changing the way we think about high energy density matter.

## WHAT IS THE GLASMA?

When two sheets of colored glass collide, the properties of the matter are changed in the time it take light to propagate across the sheets of colored glass. [26]-[30] In Fig. 9 a, the sheets of colored glass approach one another. The colored fields in the two sheets

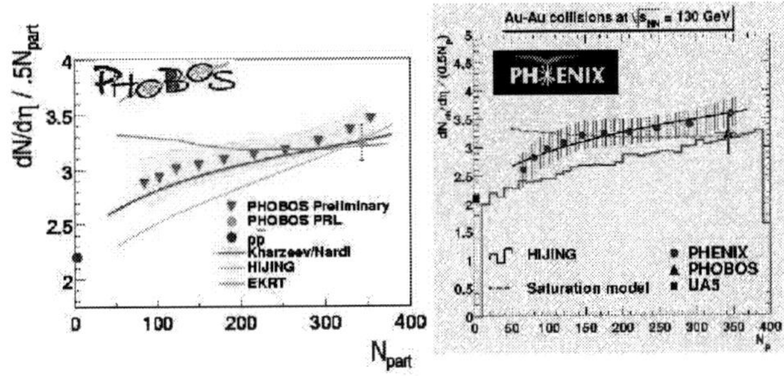


**FIGURE 8.** The multiplicity as a function of energy observed at the RHIC

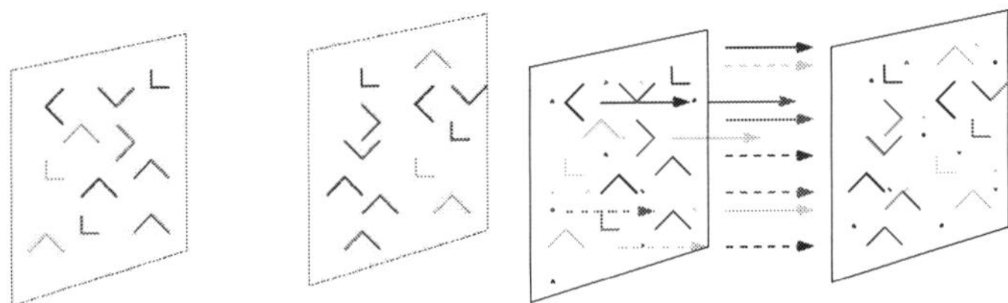

**FIGURE 9.** (a)Two sheets of colored glass approach one another. (b) After the collisions, Glasma is formed in the region between the sheets.

form a condensate of Weizsacker-Willams fields disordered in polarization and color. In the time it takes the sheets to pass through one another, the fast degrees of freedom gain a density of color electric and magnetic charge. The density of charge on each sheet is equal and opposite. This is a consequence of the classical fields which are generated by the source sheets of color glass. Attached to the region away from the collision region is a pure two dimensional transverse vector potential. This potential has no color electric and magnetic field until the nuclei pass through one another, since then the vector potential of one sheet multiplies the field of the other. Then sources are set up according to the Yang-Mill equations

$$\rho_E^a = f^{abc} A^b \cdot E^c \qquad (13)$$

$$\rho_B^a = f^{abc} A^b \cdot B^c \qquad (14)$$

These colored electric and magnetic charges generate longitudinal color electric and magnetic fields as shown in Fig. 9b. The reason why both electric and magnetic fields are made is because of the duality of the Yang-Mill equations under $E \leftrightarrow B$, and because the initial fields of the colored glass have this symmetry.

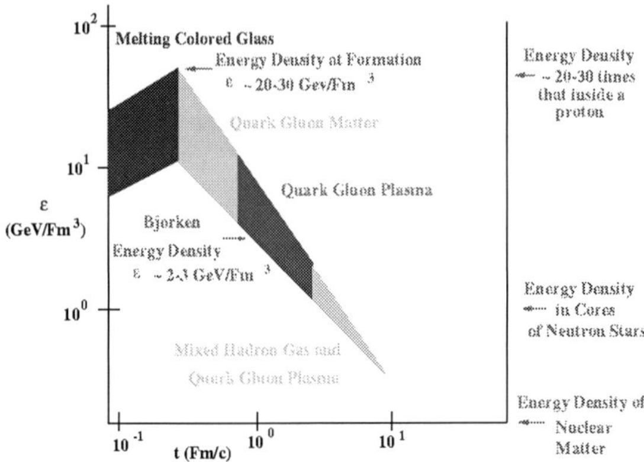


**FIGURE 10.** The Emerging Picture of RHIC collisions

When there is a nonzero $E \cdot B$ means that there is a topological charge induced. This Chern Simons charge is

$$\partial \cdot K = \alpha_S \kappa E \cdot B \tag{15}$$

where $\kappa$ is a constant. This topological charge generates helicity non-conservation. To understand how this works, consider a parallel electric and magnetic field in electrodynamics An electron is accelerated and rotates around a magnetic field in the opposite sense to a positron. Therefore both the electron and positron acquire the same vorticity. The sign of the vorticity depends upon the sign of $E \cdot B$. For an extended charge distribution, corresponding to a hadron, we expect that there will be a similar biasing of the helicity distributions of hadrons.

The classical equations after the collision evolve in time and become dilute as a consequence of the non-linearities of the Yang-Mills equation. There is a simple solution to this problem which has an invariance under Lorentz boosts along the collision access. It has however been recently show that this solution is unstable with respect to small non boost invariant solutions. These solutions grow in magnitude as time evolves, amplifying small initial fluctuations into full scale chaotically turbulent solutions. This turbulence and its rapid onset may be responsible for the early thermalization seen at RHIC.

## THE EMERGING PICTURE OF RHIC COLLISIONS

The emerging picture we have of RHIC collisions is shown in Fig. 10 In the initial state, there are two sheets of colored glass. They collide and produce a glasma, which melts into gluons. During the melting, or perhaps afterwards, the quarks and gluons thermalize. This eventually makes a Quark Gluon Plasma. Data from RHIC indicates this happens very rapidly, on a time scale of the order of 1 $Fm/c$. In Fig. 10, I have presented the typical energy scales and times involved. The scale of energy comes from the measurements of multiplicities and HBT radii at RHIC. The range in estimates comes from making the radical assumptions of complete thermalization and no thermalization of the gluons. The bottom line is that the scales are large and the times probed are very early.

## SUMMARY

In addition to the Quark Gluon Plasma, other interesting new forms of matter are being probed at RHIC. These are the Color Glass Condensate and the Glasma. These forms of matter allow us to test ideas about QCD when the non-linearities of QCD are present, yet to use weak coupling methods. At the LHC, the potential for studying such new form of matter is great due to the larger range in $x$ and typical momentum scales.

## ACKNOWLEDGEMENTS

I gratefully acknowledge conversations with Francois Gelis, Edmond Iancu, Dima Kharzeev, and Tuomas Lappi Genya Levin, and Raju Venugopalan on the subject of this talk

This manuscript has been authorized under Contract No. DE-AC02-98CH10886 with the U. S. Department of Energy.

## REFERENCES

1. J. Breitweg et. al. *Eur. Phys. J.* **67**, 609 (1999).
2. L. V. Gribov, E. M. Levin and M. G. Ryskin, *Phys. Rept.* **100**, 1 (1983).
3. A. H. Mueller and Jian-wei Qiu, *Nucl. Phys.* **B268**, 427 (1986); J.-P. Blaizot and A. H. Mueller, *Nucl. Phys.* **B289**, 847 (1987).
4. L.N. Lipatov, *Sov. J. Nucl. Phys.* **23** (1976), 338;
   E.A. Kuraev, L.N. Lipatov and V.S. Fadin, *Sov. Phys. JETP* **45** (1977), 199;
   Ya.Ya. Balitsky and L.N. Lipatov, *Sov. J. Nucl. Phys.* **28** (1978), 822.
5. L. D. McLerran and R. Venugopalan, *Phys. Rev.* **D49**, 2233(1994); 3352 (1994); **D50**, 2225 (1994).
6. E. Iancu, A. Leonidov and L. D. McLerran, *Nucl. Phys.* **A692**, 583 (2001); E. Ferreiro E. Iancu, A. Leonidov and L. D. McLerran, *Nucl. Phys.* **A710**,373 (2002).
7. E. Iancu and L. McLerran, *Phys. Lett.* **B510**, 145 (2001).
8. J. Jalilian-Marian, A. Kovner, L. McLerran and H. Weigert, *Phys. Rev.* **D55** (1997), 5414.
9. J. Jalilian-Marian, A. Kovner, A. Leonidov and H. Weigert, *Nucl. Phys.* **B504** (1997), 415; *Phys. Rev.* **D59** (1999), 014014.
10. B. Back et. al. *Phys. Rev. Lett.* **91**, 052303 (2003).
11. J. Jalilian-Marian, *Phys. Rev,.* **C70** 027902 (2004)
12. I. Balitsky, *Nucl. Phys.* **B463** (1996), 99.
13. Yu. V. Kovchegov, *Phys. Rev.* **D60** (1999), 034008; *ibid.* **D61** (2000), 074018.
14. A. H. Mueller, *Phys.Lett.* **B523**, 243 (2001)
15. V.N. Gribov and L.N. Lipatov, *Sov. Journ. Nucl. Phys.* **15** (1972), 438; G. Altarelli and G. Parisi, *Nucl. Phys.* **B126** (1977), 298; Yu. L. Dokshitzer, *Sov. Phys. JETP* **46** (1977), 641.
16. E. Levin and K. Tuchin, *Nucl. Phys.* **A691**, 779 (2001)
17. E. Iancu, K. Itakura and L. McLerran, *Nucl. Phys.* **A708**, 327 (2002).
18. A. H. Mueller and V. N. Triantafyllopoulos, *Nucl.Phys.* **B640**, 331 (2002). D. N. Triantafyllopoulos, *Nucl. Phys.* **B648**, 293 (2003).
19. K. Golec-Biernat and M. Wustoff. *Phys. Rev.* **D60** 114023 (1999).
20. K. Golec Biernat, Anna Statso, and J. Kwiecinski, *Phys. Rev. Lett.* **86** 596 (2001).

21. A. Kovner and U. Wiedemann, **Phys. Lett. B551** 311 (2003);
22. E. Ferreiro, E. Iancu, K. Itakura and L. McLerran, *Nucl. Phys.* **A710** 373 (2002)
23. L. McLerran and T. Ikeda, *Nucl. Phys.* **A756** 385 (2005).
24. D. Kharzeev and M. Nardi, *Phys. Lett.* **B507** 121 (2001).
25. Reports of the Star, Phenix, Phobos and Brahms experimental collaborations in *Nucl. Phys* **A757** (2005) Brahms Collaboration p 1; Phobos Collaboration p 28; Star Collaboration p 102; Phenix Collaboration 184.
26. A. Kovner, L. McLerran and H. Weigert, *Phys. Rev.* **D52** 3809 (1995); 6231 (1995).
27. A. Krasnitz and R. Venugopalan, *Nucl. Phys.* **B557** 237 (1999); *Phys. Rev. Lett.* **84** 4309 (2000); Phys. Rev. Lett. **86** 1717 (2001).
28. A. Krasnitz, Y. Nara and R. Venugopalan, *Phys. Rev. Lett.* **87** 192302 (2001); *Nucl. Phys.* **A717** 268 (2003)
29. T. Lappi, *Phys. Rev.* **C67** 054903 (2003).
30. T. Lappi and L. Mclerran, *Nucl. Phys.* **A772** 200 (2006).
31. S. Mroczynski, *Phys. Lett.* **B214** 587 (1988); **B314** 118 (1993); **B363**, 26 (1997).
32. P. Arnold, J. Lenaghan, and G. Moore, *JHEP* **0308** 002 (2003); P. Arnold, J. Lenaghan, G. Moore and L. Yaffe, *Phys. Rev. Lett.* **94** 072302 (2005)
33. P. Romatschke and M. Strikland, *Phys. Rev.* **D68** 036004 (2003); **D70** 116006 (2004).
34. P. Romatshke and R. Venugopalan, *Phys. Rev. Lett.* **96** 062302 (2006) .
35. A. Dumitru and Y. Nara, *JHEP* **0509** 041 (2005); A. Dumitru, Y. Nara and M. Strikland, *Phys. Lett.* **B621** 89 (2005).

# THE ALICE EXPERIMENT AT THE LHC AND THE MEXICAN CONTRIBUTION*

G. HERRERA CORRAL†

*ALICE-Mexico(**)*

*Physics Department, CINVESTAV, P.O. Box 14 740*
*Mexico, D.F. 07300 , Mexico*

The final installation of the detectors that form ALICE has started on year 2005. The first device of ALICE that was completed and set up to work was the Cosmic Ray Detector. The V0A detector will be installed and commissioned on the summer of 2007. These two detectors were designed and built in Mexico. Here we give a very general description of these two devices.

## 1. Introduction

The ALICE collaboration is formed by 1000 people from 80 institutes in 30 countries. The goal of the collaboration is to build a facility to study heavy ion collisions at the highest energy ever achieved [1, 2]. Mexico signed a Memorandum of Understanding in 2005 with the main body of the collaboration but has been working on the design and construction of two of the 16 detection systems [3] of ALICE for several years. Here we will give a general view of the Mexican contribution to the ALICE project.
The mexican teams are grouped around the two main projects: The Cosmic Ray Detector with the participation of the National University, The University of Puebla, University of Sinaloa and Cinvestav; and the V0A detector with the participation of the National University and Cinvestav.
The Cosmic Ray Detector covers the upper part of ALICE magnet (see Fig.1) while the V0A is located inside the magnet on the left side of the interaction point.

* This work is supported by CONACYT

† e-mail: gherrera@fis.cinvestav.mx

** The ALICE-Mexico group is listed at the end of the contribution

CP917, *Particles and Fields,* edited by H. Castilla Valdez, J. C. D'Olivo, and M. A. Perez

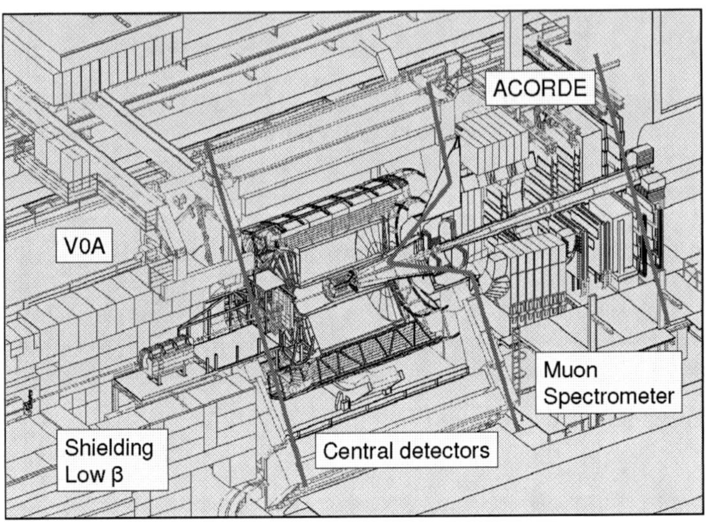


Figure 1. ALICE detector and its components. ACORDE (A Cosmic Ray Detector) on the top of the magnet and the V0A on the left side of the interaction point, just in front of the panel that represents the Photon Multiplicity Detector.

## 2. The V0 Detector

The V0 system consists of two detectors named V0A and V0C [4]. They will be located in the central part of ALICE. The V0A will be installed at a distance of 340 cm from the interaction point as shown in Fig. 2, mounted in two rigid half boxes around the beam pipe. The V0C will be on the right side at a distance of 0.9 mts from the interaction point. Each detector is an array of 32 cells of scintillator plastic, distributed in 4 rings forming a disc with 8 sectors as shown in Fig. 2.
The light produced in the scintillator plastic is collected by wavelength shifting fibers and transported to photomultipliers. Each cell is read out independently. This setup is based on experimental tests and simulations [5].

The V0 detector system has several functions:

- On line vertex determination using the timing information.
- Track distribution consistent with beam-beam interaction to allow suppression of undesirable beam-gas events.
- On-line centrality measurement which provide an online multiplicity measurement.
- Production of a wake up signal for the electronics of the Transition Radiation Detector.
- Luminosity measurement by counting triggered events.

The segments of the V0A detector were built with a megatile construction (see ref. [4]). This technique consists of machining the scintillator plastic and filling the grooves with $TiO_2$ loaded epoxy in order to separate one sector from the other. This

is done on both faces all the way through the plastic. The surface of the slices are then wrapped with teflon tape.
The wavelength shifting fibers that collect the light are then embedded in the plastic in 3 mm deep grooves on both sides of the cell. These fibers bring the light to the PMTs.
The construction is made with scintillator plastic from Bicron (BC404). The wavelength shifting fibers WLS (BC9929AMC) are 1 mm diameter. At the end of the WLS inside the plastic, aluminum coating is used as a reflector.
The photomultiplier tubes (PMT) will be installed inside the magnet not far from the detector. In order to tolerate the magnetic field, fine mesh tubes have been chosen.

A more detailed description of the V0 system can be found in ref. [4]. Further studies of the role and techniques that make use of the information provided by the V0 system can be found in references [ 6, 7, 8 ].

## 3. The Cosmic Ray Detector for ALICE

The cosmic array on top of the ALICE magnet will serve two purposes:

1) As a cosmic ray trigger for the TRD and TPC detectors in the calibration phases.
2) As a device to study rare cosmic ray events in conjunction with the ALICE tracking detectors.

Together with other ALICE devices it will provide interesting information about cosmic rays originated by primaries with an energy of $10^{15-17}$ eV. In particular multi-muon events can be studied with the use of the Time Projection Chamber.

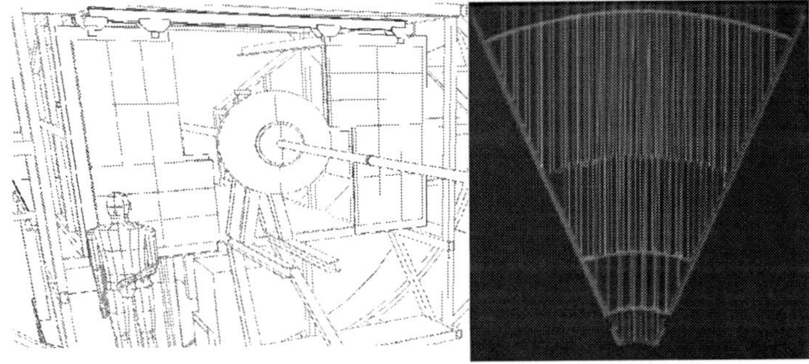

Figure 2. V0A detector in front of the Photon Multiplicity Detector. An octant of V0A before optical isolation.

Figure 3. The Cosmic Ray Detector ( ACORDE ) on the top of the ALICE magnet. The detector has been commissioned in 2005.

The Cosmic Ray Detector consists of an array of 60 scintillator counters located in the upper part of the ALICE magnet. Fig. 3 shows the distribution of the panels in its final position.

Each module has a sensitive area of $1.9 \times .195 m^2$ and is built with two superimposed plastics.

The electronics of the detector was designed in Mexico. It consists of a Front End amplification and then processing of the signals to

- generate a single muon trigger to calibrate the Time Projection Chamber and other components of ALICE,
- generate a multi-muon trigger to study cosmic rays,
- provide a wake-up signal for the Transition Radiation Detector.

The electronics should generate the signal in 100 ns and will provide spatial information of the scintillator fired. It will also store the information on this spatial location.

**Acknowledgements**

I would like to thank the organizers for the invitation to deliver this talk. I also express my gratitude to the members of the Mexican team at ALICE:

G. Contreras (Departamento de Física Aplicada, CINVESTAV).
A. Gago, G. Herrera Corral, L. Montaño, C. Pérez, A. Zepeda (Departamento de Física, CINVESTAV).
A. Ayala, E. Cuautle, I. Domínguez, L. Nellen, G. Paic, P. Podesta, L. Serkin (Instituto de Ciencias Nucleares, UNAM).
R. Alfaro, E. Belmont, V. Grabski, A. Martínez, A. Menchaca, A. Sandoval (Instituto de Física, UNAM).
A. Fernández, R. López, M. I. Martínez, S. Roman, G. Tejeda, M. A. Vargas, S. Vergara, (Universidad de Puebla).

I. León Monzón (Universidad de Sinaloa).

**References**


1. ALICE Collab., ALICE Physics Performance Report, Vol. I, Journal Phys. G: Nucl. Part. Phys. 30 (2004)1517
2. ALICE Collab., ALICE: Physics Performance Report Vol. II, Journal Phys. G: Nucl. Part. Phys. 32(2006)1295
3. ALICE Technical Proposal, Ahmad et al. CERN/LHCC/ 95 71, 1995
4. ALICE Technical Design Report, Forward Detectors FMD, T0, V0 CERN/LHCC-2004-025, 2004.
5. J.R. Alfaro et al. Simulation of the V0A Detector , ALICE-INT-2006-018
6. J.G. Contreras, Measuring leading forward neutrons in pp collisions with the ZDC in ALICE, ALICE-INT-2006-007
7. J. Conrad, J. G. Contreras, C.E. Jorgensen, Minimum bias triggers in proton-proton collisions with VZERO and pixel detectors, ALICE-INT-2005-025
8. R. Alfaro, E. Cuautle, G. Paic, Rejection of beam gas interactions in pp collisions and timing requirements, ALICE-INT-2004-021

# Heavy Ion Physics with CMS

E.García[1] for the CMS Collaboration

*Department of Physics, University of Illinois at Chicago, Chicago, IL 60607-7059, USA*

**Abstract.** The Large Hadron Collider (LHC) will produce heavy ion collisions at the nucleon-nucleon center of mass energy of 5.5 TeV, the highest energy ever available in a controlled environment. This represents an opportunity to study nuclear matter in systems with unprecedented energy densities. Due to the high incident energy, semi-hard and hard processes will be a dominant feature at the LHC. The Compact Muon Solenoid (CMS) heavy-ion program is ideally suited to study the physics of these probes, addressing open questions in the field of Quantum Chromodynamics (QCD). In this paper an overview of the heavy-ion physics capabilities of the CMS detector is presented.

**Keywords:** Relativistic heavy-ion collisions, Compact Muon Solenoid
**PACS:** 25.75.q

## 1. INTRODUCTION

During the last few years, the four experiments at the Relativistic Heavy Ion Collider (RHIC) have collected a large set of data on nuclear collisions over a wide range of incident energy and system size. There is strong evidence of the formation of a high energy density medium in these collisions, which is strongly interactive, seems thermalized and appears almost opaque to jet-like partons. Furthermore, the data is consistent with the creation of a system with energy densities higher than the critical density expected for the transition to a system where quarks and gluons have achieved asymptotic freedom. A summary of the key observations leading to this description is in Fig. 1: Left panel, normalized charge particle distribution as function of the collision energy for several systems [1]. With the point for at GeV, one can estimate that the initial energy density created at $\tau_0$ = 1 cm/$c$ at RHIC is $\sim$ 5 GeV$/fm^3$, about a factor of ten larger than the energy density of nuclear matter [2]. Center panel, elliptic flow signal $v_2$ near mid rapidity as a function of the number of participants (centrality) [3]. The large signal of elliptic flow is evidence of early interactions between the produced particles, also the flow signal is close to the value predicted by relativistic hydrodynamics calculation [4], an indication of fast thermalization. Right panel, suppression of the yield on the opposite side on back to back correlation in high $p_T$ charged hadrons [5].

At the LHC, the energy densities are predicted to be around 20 times higher than at RHIC, implying a factor of 2 higher for the initial temperature [6]. This is well above the phase transition region between hadronic matter and quark gluon plasma (QGP). Furthermore, at the LHC the higher densities of the produced partons will result in more rapid thermalization, so that the time that the matter will be in a QGP phase is increased

[1] email address: ejgarcia@uic.edu

CP917, *Particles and Fields*, edited by H. Castilla Valdez, J. C. D'Olivo, and M. A. Perez

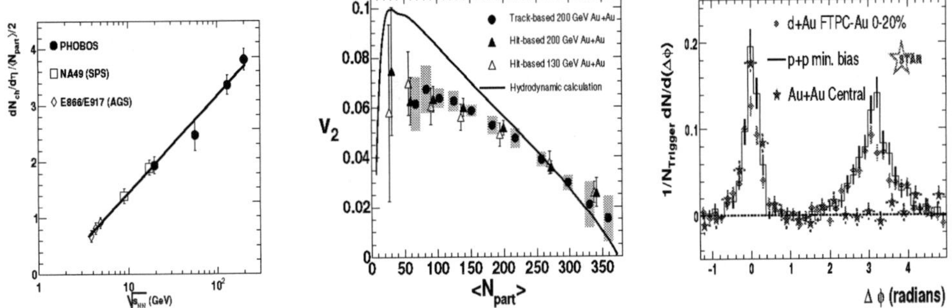


**FIGURE 1.** Left panel evolution of the midrapidity charged particle density $dN_{ch}/d\eta\rfloor_{|\eta|\leq 1}$, per participating nucleon pair, $N_{part}/2$, as a function of center of mass energy [1], the solid line is a logarithmic extrapolation of the data from lower energies to guide the eye. Center panel elliptic flow of charged particles near mid rapidity as function of centrality in Au+Au collisions [3]. The curve shows the prediction from a relativistic hydrodynamics calculation [4]. Right panel shows azimuthal correlations for p+p, central d+Au and central Au+Au collisions (background subtracted) [5].

by almost a factor of three when compared to RHIC. These conditions may allow the creation of a weakly interacting gas-like QGP state in contrast to the strongly interacting liquid-like state believed to be created at RHIC [7]. In addition to the study of the RHIC base-line measurements, the LHC expands the heavy-ion physics possibilities in the kinematic phase space: larger momentum transfer ($Q^2$) and lower momentum fraction ($x$). For example the production of hard probes like high $p_T$ jets, photons, heavy-quark particles ($J/\psi$) and gauge bosons ($W^{\pm}$, $Z^0$) will be available at LHC, while many of them are not experimentally accessible at RHIC.

## 2. OVERVIEW OF THE CMS DETECTOR

CMS offers a high resolution tracking and calorimetry over a uniquely large range in rapidity and $2\pi$-azimuthal coverage. The acceptance of the different elements of the detector is represented in Fig. 2. In addition CMS is complemented with several detector components to characterize particle production at very forward rapidities: CASTOR calorimeter ($5.3 < |\eta| < 6.6$), TOTEM roman pots ($7 < |\eta| < 10$), and the Zero Degree Calorimeter (ZDC) ($|\eta| > 8.3$ for photons and neutrons).

A detailed description of the detector elements is given in the Technical Design Reports [8, 9, 10, 11, 12, 13]. The dominant element of the CMS detector is a superconducting solenoidal coil magnet. It is 13 m long and 6 m in diameter, providing a 4 T field throughout the inner portion of the detector. This inner region holds the highly segmented Si tracker that provides precision tracking of the particle trajectories close to the interaction point. This information can be used to determine the momentum of the particles going to the more exterior calorimeters and muon detectors. Similar to the tracker the electromagnetic and hadronic calorimeters are arranged in barrels and endcaps located inside the magnet coils. The electromagnetic calorimeter (ECAL) consists

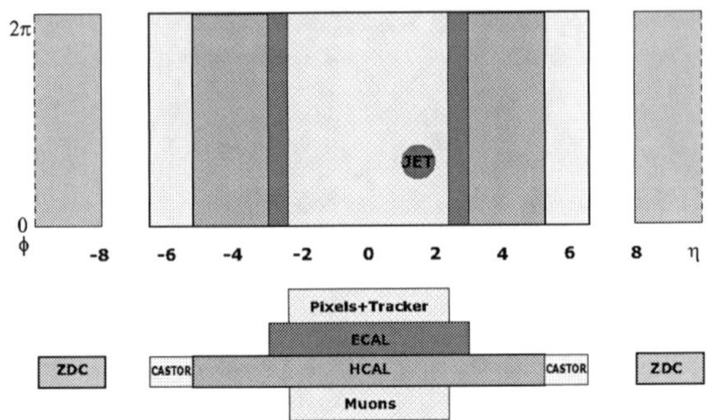


**FIGURE 2.** Acceptance of CMS tracking, calorimetry, and muon identification in pseudorapidity and azimuth. The size of a typical jet cone (R = 0.5) is included as an illustration

of $PbWO_4$ crystals, while the hadronic sampling calorimeters (barrel and endcap) are formed by layers of scintillator and copper. Finally outside the magnet the muon detection system is made of drift tubes with anode wires.

## 3. HEAVY ION PHYSICS CAPABILITIES OF CMS

The capabilities of the CMS detector provide a unique opportunity to study the new physics potential provided by the heavy ion collisions at the LHC. CMS is designed to deal with p+p collisions at luminosities of $10^{34}$ cm$^{-2}$ s$^{-1}$. At full luminosity there will be, on average, 25 p+p collisions per bunch crossing. In this environment the design resolution and granularity of all detector components have been maximized in order to resolve the high momentum ($p_T > 5$ GeV/c) observables. This makes the detector ideal for the high multiplicity conditions in central heavy ion collisions: the technologies chosen for tracking, calorimetry, and muon identification will allow a read out of CMS with a minimum bias trigger at the full expected $10^{27}$ cm$^{-2}$ s$^{-1}$, Pb+Pb luminosity. The CMS and TOTEM detectors will form the largest acceptance system at the LHC. The ZDC and CASTOR will enhance further the acceptance in the forward region, allowing low-x measurements. As for particle identification, the muon system in combination with the silicon tracker will allow studies of the interaction of identified quark jets with the medium. In addition the physics of meson vs. baryon production at large $p_T$ can be studied via the the reconstructed $\pi^0 s$.

### 3.1. Global Observables

One of the first measurements of CMS from heavy ion collision will be the charged particle multiplicity. The performance of the Si tracker for this measurement was fully simulated for central Pb+Pb collisions for a range of maximum rapidity charged particle yields. The reconstructed charge particle pseudorapidity distribution from simulation is shown in the left panel of Fig. 3, this was calculated using the pixel detector and for

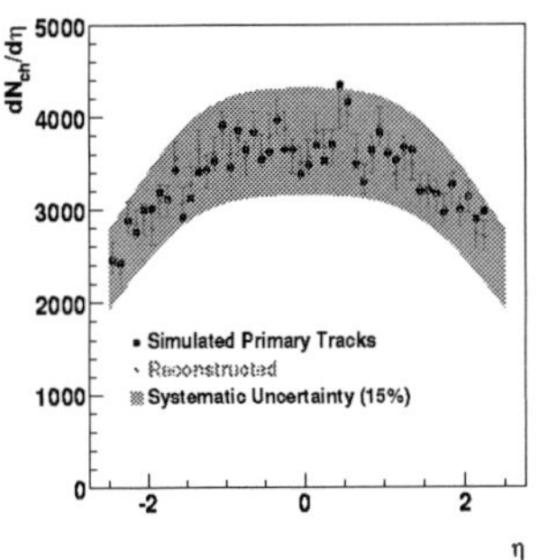

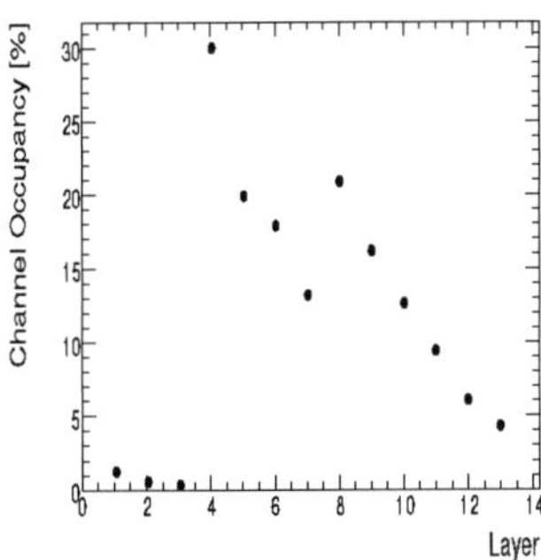


**FIGURE 3.** Left panel is the reconstructed charged particle density $dN_{ch}/d\eta\rfloor_{|\eta|\leq 1}$ for Pb+Pb collisions at 5.5 TeV. Right panel shows the channel occupancy in the tracker as a function of the detector layer, 1-3 correspond to the pixel detector, 4-7 the inner Si-strip and 8-13 the outer strip.

$dN_{ch}/d\eta \sim 4000$. The right panel of Fig. 3 gives the occupancy in the tracker: 2% for the pixel section, 15 - 30% for the inner Si-strip and 5 - 20% for the outer strip.

For the study of different physics topics in heavy ions, as heavy quarkonium or hard jet production, it is important to perform measurements at different centralities. Since the forward rapidity region is almost free of final state re-interactions, the transverse energy deposition in the hadron forward calorimeter (HF) or in the ZDC is determined mainly by the initial nuclear geometry of the collision, rather than by final state dynamical effects. These are then the signal that can be better correlated to the impact parameter of the collision. Figure 4 shows the correlation between the transverse energy deposition in HF and the impact parameter for 1000 minimum bias Ar+Ar and 1000 Pb+Pb collisions at $\sqrt{s_{NN}}$ = 5.5 TeV (left panel), and the resolution of this measurement ($\sigma_{E_T}/E_T$) (right panel) [14]. The resolution is of about 20% up to $\langle E_T \rangle \sim 500$ GeV, for lower values of transverse energy the resolution is degraded due to screening of particles produced in the beam pipe. It is expected however that the ZDC response will help to characterize the centrality for more peripheral collisions.

The rescattering and energy loss of hard partons in an azimuthally asymmetric volume can result in an observable azimuthal anisotropy of high-$p_T$ particles and jets. At RHIC studies of the elliptic flow of the azimuthal distributions of produced hadrons relative to their reaction plane have been used to study the question of thermal equilibration at the early stages of the heavy ion collisions. The CMS calorimeters are well suited to measure energy flow and jet azimuthal anisotropy. The ability to reconstruct the event plane and the possibility to observe jet azimuthal anisotropy is shown in Fig. 5 for Pb+Pb collisions at impact parameter $b = 6$ fm. The left panel is the energy deposition in the barrel and end cap regions, and the right panel the difference between the generated and reconstructed azimuthal reaction plane angle. The jet spectra were generated with PYTHIA5.7 and the hydrodynamics input with CMSIM125+ORCA6.2.0 as described in [15].

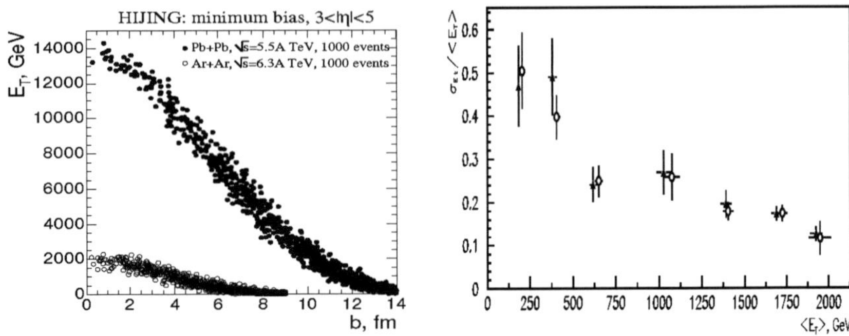


**FIGURE 4.** Left panel Correlation between transverse energy deposition in HF and impact parameter for minimum bias Ar+Ar and Pb+Pb collisions. Right panel resolution of the transverse energy distribution at fixed impact parameter. HIJING predictions (triangles) and reconstructed events (circles).

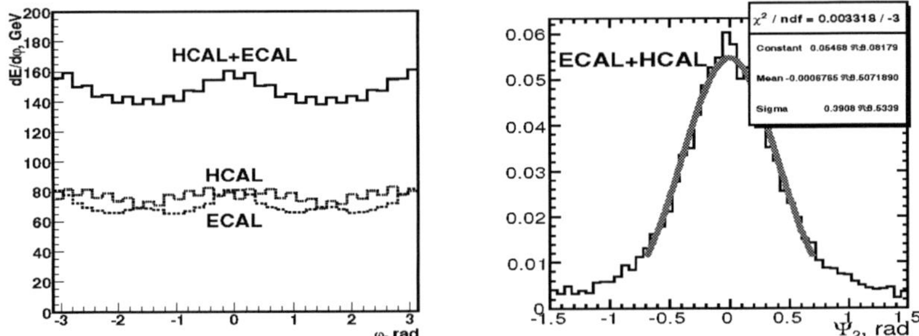


**FIGURE 5.** Left panel energy deposition in barrel and endcap regions. ECAL, dashed histogram, HCAL dotted histograms and total deposition (HCAL+ECAL) solid histogram. Right panel is the difference between the generated and reconstructed azimuthal angle $\Psi_2$ of the reaction plane.

## 3.2. Jets and hadron yields

The study at RHIC of the hadron yield and back-to-back hadron correlations above $p_T \geq 3$ GeV/c indicate a pronounced energy loss of fast partons from their interaction with the dense medium created during the collisions. The kinematic region available at the LHC will allow the study of fully reconstructed jets and will extend significantly the hadron suppression studies in $p_T$. For example at the LHC, $10^7$ jets with $E_T{}^{jet} > 100$ GeV are produced at $|\eta| \leq 2.6$ for Pb+Pb collisions in one month of running at nominal luminosity. Event jets with $E_T > 50$ GeV can be reconstructed with good efficiency and purity using the calorimeters at CMS. The capabilities for jet reconstruction studies will be enhanced with the Si tracker that allows the momentum reconstruction of charged hadrons of $p_T > 600$ to 700 MeV/c depending on the multiplicity, and with good efficiency, low level of contamination and good momentum resolution [16]. Figure 6 shows the tracker reconstruction efficiency (left panel), the jet-finding efficiency and purity

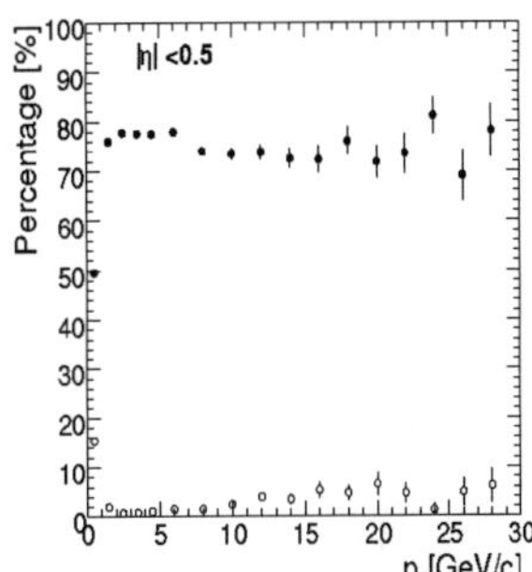


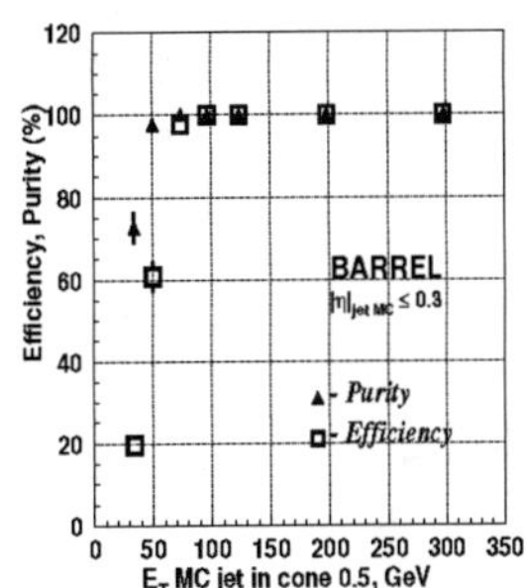


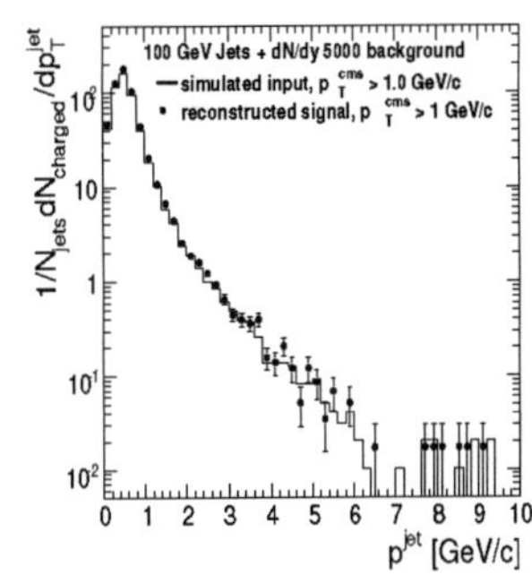


**FIGURE 6.** Left panel tracker reconstruction efficiency (full symbols) and fake track rate (open symbols) as a function of the transverse momentum near mid rapidity for central Pb+Pb collisions with $dN_{ch}/d\eta|_{\eta=0} = 3000$, the quality cuts for this figure are optimized for low fake track rate. Center panel is the jet reconstruction efficiency and purity using barrel calorimeters for PYTHIA generated jets in Pb+Pb events with $dN_{ch}/d\eta|_{\eta=0} = 5000$. Right, Input and reconstructed fragmentation function for individual particle momentum transverse to the jet for 100 GeV jets also within Pb+Pb events, $dN_{ch}/d\eta|_{\eta=0} = 5000$.

for jet reconstruction studies [17] (center), and the reconstructed transverse momentum fragmentation distribution of charged particles for 100 GeV jets (right).

## 3.3. Quarkonia

One of the proposed signatures of the QCD phase transition is the suppression of the quarkonium production, in particular of the charmonium family ($J/\psi,\psi'$).The charmonium suppression has been observed at CERN SPS and at RHIC but competing mechanisms to color deconfinement (hadronic co-movers interactions and charm quark recombination respectively) have been proposed to explain the observed cross sections. The study of the much heavier bottomonia spectroscopy accessible at LHC is free from the distorting hadronic and coalescence contributions, and can help to clarify this issue. $\Upsilon$ resonances can be detected in the $\mu^+\mu^-$ decay mode with good efficiency (33% for $\Upsilon$ vs. 6% for $J/\psi$ in the CMS detector). Furthermore, studies of the relative yield production of the $\Upsilon$ and $\Upsilon$' as a function of $p_T$ have been predicted to be highly sensitive to the temperature and size of the medium created in the collisions addressing information about the dynamics of the high density QCD matter. The feasibility of quarkonia detection shows good resolution efficiency over $|\eta| \leq 2.4$. The expected di-muon invariant reconstructed mass spectra for $J/\psi$ and $\Upsilon$ family after background subtractionare shown in Fig. 7. This reconstruction was studied with a detailed simulation and detector acceptance within central $dN_{ch}/d\eta|_{\eta=0} = 5000$ Pb+Pb collisions [18].

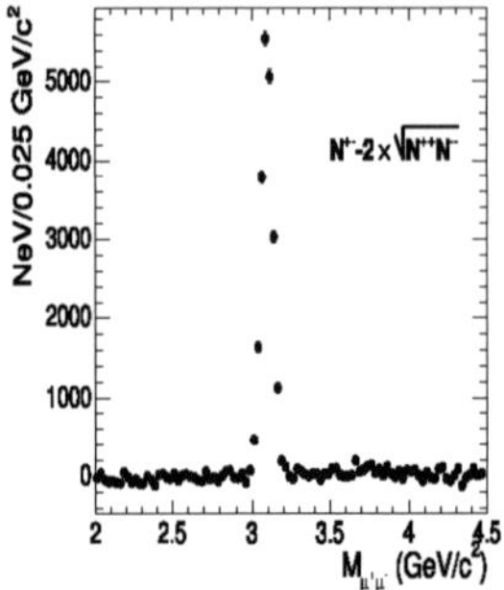


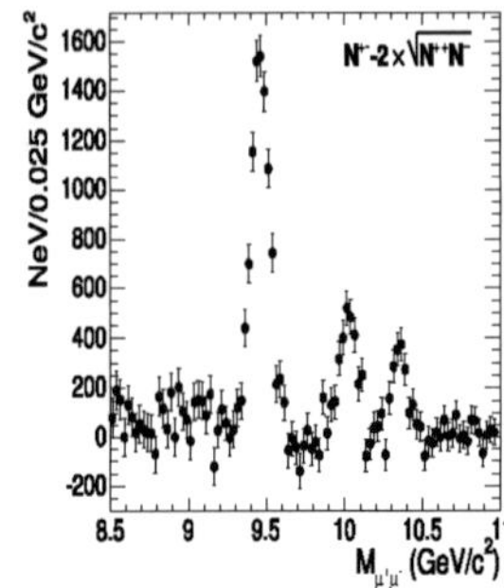


**FIGURE 7.** The signal invariant mass distribution, after background subtraction in $J/\psi$ (left) and $\Upsilon$ (right) mass regions.In each case, both muons have $|\eta| < 0.8$.

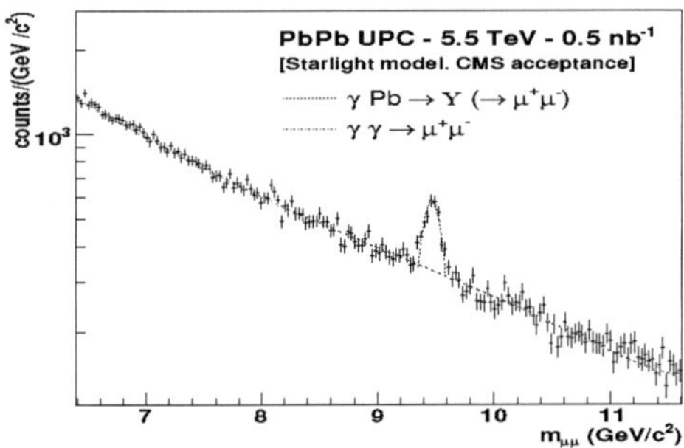


**FIGURE 8.** Expected $\mu^+\mu^-$ invariant mass from $\gamma Pb \to \gamma Pb^* \to \mu^+\mu^- + Pb^*$ and $\gamma\gamma \to \mu^+\mu^-$ predicted by STARLIGHT for UPC Pb+Pb collisions at $\sqrt{s_{NN}} = 5.5$ TeV in the CMS acceptance [20].

## 3.4. Forward Physics

TOTEM and the forward calorimeters CASTOR and the ZDC will enhance the CMS physics program. Parton distribution functions will be mapped out at CMS at low momentum fractions of $x \sim 10^{-6}$ and scales of few $GeV^2$ of momentum transfer. This coverage will allow the study of the predicted gluon saturation in the framework of the Color Glass Condensate model [19]. For example heavy quark production from ultra-peripheral collisions (UPCs) would be reduced in the presence of CGC relative to the predictions using parton distributions of a free proton. The CMS experiment can measure $\Upsilon \to e^+e^-$ and $\Upsilon \to \mu^+\mu^-$ produced in electromagnetic Pb+Pb UPC using the tracker, the ECAL, and the muon chambers tagged with forward neutron detection in the ZDC. Figure 8 shows the expected dimuon invariant mass distributions around the $\Upsilon$ mass predicted by STARLIGHT within the CMS acceptance for integrated Pb+Pb luminosity of 0.5 $nb^{-1}$. The details of the simulation can be found in [20].

## ACKNOWLEDGMENTS

This work was partially supported by the U.S. DOE grant DE-FG02-93ER40802.

## REFERENCES

1. B. B. *et al.*, *Phys. Rev. Lett.* **88**, 22302 (2002).
2. J. D. Bjorken, *Phys. Rev.* **D 27**, 140 (1983).
3. B. B. *et al.*, *Phys. Rev.* **C 72**, 031901 (2004).
4. P. F. K. *et al.*, *Phys. Lett.* **B 500**, 232 (2001).
5. C. A. *et al.*, *Phys. Rev. Lett.* **90**, 082302 (2003).
6. I.Vitev, and M. Gyulassy, *Phys. Rev. Lett* **89**, 252301 (2002).
7. T. D. Lee, and M. Gyulassy, *nucl-th/0403032* (2004).
8. CMS HCAL Design Report CERN/LHCC 97-31 (1997).
9. CMS MUON Design Report CERN/LHCC 97-32 (1997).
10. CMS ECAL Design Report CERN/LHCC 97-33 (1997).
11. CMS Tracker Design Report CERN/LHCC 98-6 (1998).
12. A. Angelis, and A. D. Panagiotou, *Journ. Phys* **G 23**, 18 (1997).
13. CMS TOTEM Technical Proposal CERN/LHC 99-7 (1999).
14. CMS Note 2001/055 (2001).
15. CMS Note 2003/019 (2003).
16. CMS Note 2006/100 (2006).
17. CMS Note 2006/050 (2006).
18. CMS Note 2006/089 (2006).
19. L. McLerran, and R. Venugopalan, *Phys. Rev.* **D 49**, 2233 (1994).
20. D. d'Enterria, *nucl-ex/0610061* (2006).

# SM Precision Constraints at the LHC/ILC

Jens Erler

*Departamento de Física Teórica, Instituto de Física, Universidad Nacional Autónoma de México, México D.F. 04510, México*

**Abstract.** The prospects for electroweak precision physics at the LHC and the ILC are reviewed. This includes projections for measurements of the effective Z pole weak mixing angle, $\sin^2\theta_W^{\rm eff.}$, as well as top quark, $W$ boson, and Higgs scalar properties. The upcoming years may also see very precise determinations of $\sin^2\theta_W^{\rm eff.}$ from lower energies.

**Keywords:** Electroweak interaction; standard model; higgs boson.
**PACS:** 12.15.-y, 13.66.Jn, 14.80.Bn

## INTRODUCTION

Fig. 1 is a summary of current electroweak precision physics. It shows the constraints from different types of observables on the Higgs boson and top quark masses, $M_H$ and $m_t$. The most important input are the $Z$-pole asymmetries from LEP and SLC [1, 2], shown as the dotted (brown) lines. The long-dashed (blue) lines correspond to the $W$-boson mass, $M_W = 80.394 \pm 0.029$ GeV, from LEP 2 [2] ($e^+e^-$), UA2 [3] and the Tevatron [4, 5] ($p\bar{p}$). It is interesting that other (non-asymmetry) $Z$-pole measurements by themselves result in a finite region in the $M_H$–$m_t$ plane, shown as the closed (green) contour. These three types of constraints overlap in a common region and at $m_t$ values consistent with the Tevatron average of kinematic mass measurements [6], $m_t = 171.4 \pm 2.1$ GeV. The only conflicting data set is from low energies (dashed contour), driven by the NuTeV result on deep inelastic neutrino scattering off approximately isoscalar nuclear targets [7] which shows a 2.7 $\sigma$ deviation in the effective four-Fermi coupling for neutrino interactions with left-handed quarks. The combination of all precision data yields the filled (red) ellipse. There is mounting evidence for a relatively light Higgs boson with much of the 90% C.L. ellipse already excluded by direct searches at LEP 2 [8]. Combining these search results with the precision constraints yields the histogram in Fig. 2. The strong peak is due to the significant excess of Higgs-like events observed by the ALEPH Collaboration [10]. Most of the probability is for $M_H$ values below 130 GeV, in perfect agreement with expectations from supersymmetric extensions of the Standard Model (SM). The 95% C.L. upper limit, $M_H \leq 178$ GeV, can also be read off from Fig. 2. In summary, the global fit to all data yields,

$$\begin{aligned} M_H &= 84^{+32}_{-25}\ \mathrm{GeV}, \\ m_t &= 171.4 \pm 2.1\ \mathrm{GeV}, \\ \alpha_s(M_Z) &= 0.1216 \pm 0.0017, \end{aligned} \qquad (1)$$

where the result for $M_H$ is only barely consistent (within 1 $\sigma$) with the lower limit from LEP 2 [8], $M_H > 114.4$ GeV. Some observables show interesting deviations (at the 2–

CP917, *Particles and Fields,* edited by H. Castilla Valdez, J. C. D'Olivo, and M. A. Perez

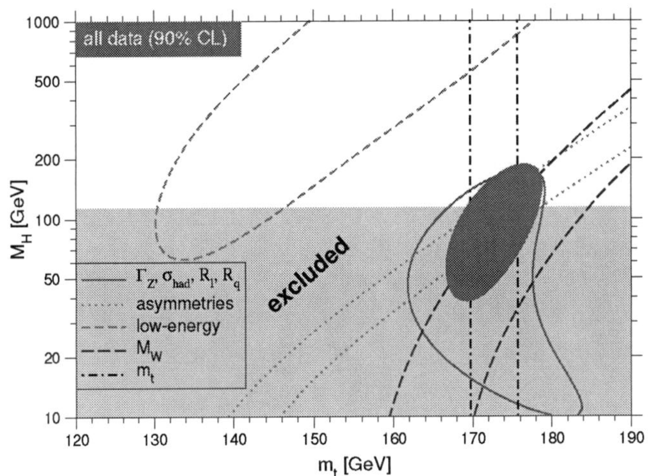


**FIGURE 1.** One-standard-deviation (39.35%) uncertainties in $M_H$ as a function of $m_t$ for various inputs, and the 90% C.L. region allowed by all data. The 95% C.L. direct lower limit from LEP 2 is also shown.

3 $\sigma$ level) from the SM, but the overall goodness of the global fit is reasonable, with a $\chi^2$ of 47.3 for 42 degrees of freedom and a probability for a larger $\chi^2$ of 27%.

When discussing future improvements for the key observables, $m_t$, $\sin^2\theta_W^{\text{eff.}}$, and $M_W$, it is useful to keep some benchmark values in mind. An increase of $M_H$ from 100 to 150 GeV (distinguishing between these values provides a rough discriminator between minimal supersymmetry and the SM) is equivalent to a change in $M_W$ by $\Delta M_W = -25$ MeV. But this 25 MeV decrease can be mimicked by $\Delta m_t = -4$ GeV, and also by an increase of the fine structure constant at the $Z$ scale, $\Delta\alpha(M_Z) = +0.0014$. We know $\alpha(M_Z)$ an order of magnitude better than this — despite hadronic uncertainties in its relation to the fine structure constant in the Thomson limit. On the other hand, improving $m_t$ will be important. The same shift in $M_H$ is also equivalent to $\Delta\sin^2\theta_W^{\text{eff.}} = +0.00021$, which in turn can be mimicked by $\Delta m_t = -6.6$ GeV or by $\Delta\alpha(M_Z) = +0.0006$. Thus, $\sin^2\theta_W^{\text{eff.}}$ is more (less) sensitive to $\alpha(M_Z)$ ($m_t$) compared to $M_W$, demonstrating complementarity and underlining the general advantage of having a diverse portfolio of measurements at ones disposal. Once the Higgs boson has been discovered and its mass determined kinematically, these observables are then free to constrain heavy new particles which cannot be produced or detected directly. An example is the mass of the heavier top squark in the minimal supersymmetric SM [11].

## LHC

The Large Hadron Collider (LHC) is well on its way to produce first collisions in 2007 [12]. Initial physics runs are scheduled for 2008 with several $\text{fb}^{-1}$ of data and the precision program can be expected to take off in 2009. The low luminosity phase with about 10 $\text{fb}^{-1}$ of data (corresponding to 150 million $W$ bosons, 15 million $Z$

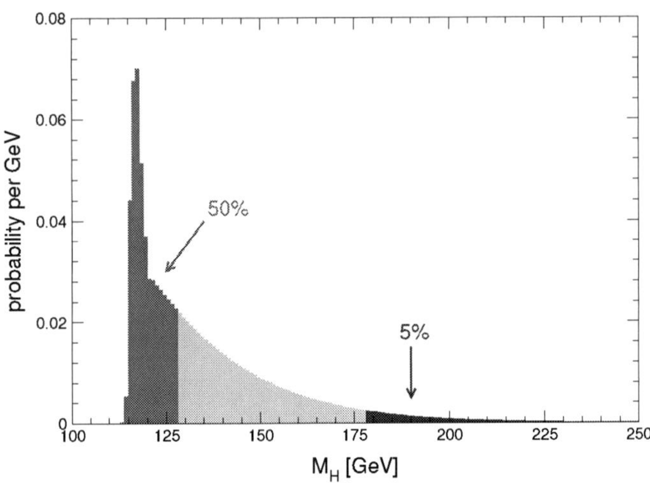


**FIGURE 2.** Probability distribution of $M_H$ from the combination of direct (search) and indirect (precision) data (updated from Ref. [9]).

bosons, and 11 million top quarks) per year and experiment [13] will already allow most precision studies to be performed. Some specific measurements, most notably competitive results on $\sin^2\theta_W^{\rm eff.}$, will probably have to wait for the high luminosity phase with $\mathcal{O}(100\ \mathrm{fb}^{-1})$ per year and experiment. The determination of the Higgs self-coupling would even call for a luminosity upgrade by another order of magnitude. Good knowledge of the lepton and jet energy scales will be crucial. Initially these will be known to 1% and 10%, respectively, but with sufficient data one can use the $Z$ boson mass for calibration, allowing 0.02% and 1% determinations. Furthermore, a 2% measurement of the luminosity and 60% $b$-tagging efficiency can be assumed [13].

LEP and SLC [1] dominate the current average $\sin^2\theta_W^{\rm eff.} = 0.23152 \pm 0.00016$. Via leptonic forward-backward (FB) asymmetries, the Tevatron Run II is expected to add another combined $\pm 0.0003$ determination [14], competitive with the most precise measurements from LEP (the FB asymmetry for $b\bar{b}$ final states) and SLD (the initial state polarization asymmetry). Having $p\bar{p}$ collisions are a crucial advantage here. At the LHC, by contrast, one has to focus on events with a kinematics suggesting that a valence quark was involved in the collision and which proton provided it ($Z$ rapidity tag). This will be possible for a small fraction of events only, requiring high luminosity running. Furthermore, sufficient rapidity coverage of $|\eta| < 2.5$ will be necessary for even a modest $\pm 0.00066$ determination [15]. A breakthrough measurement at the LHC with a statistical error as small as $\pm 0.00014$ [14, 15] is ambitious and will require a much more challenging rapidity coverage of $|\eta| < 4.9$ for jets and missing transverse energy. Thus, it is presently unclear what the impact of the LHC on $\sin^2\theta_W^{\rm eff.}$ will be.

At hadron colliders $M_W$ is determined by kinematic reconstruction. All channels and experiments combined, the Tevatron Run II will likely add a $\pm 30$ MeV constraint to the world average. The huge number of $W$ bosons will enable the LHC to provide further

$\pm 30$ MeV measurements per experiment and lepton channel ($e$ and $\mu$) for a combined $\pm 15$ MeV uncertainty (it is assumed here that the additional precision that can be gained by cut optimization is compensated approximately by common systematics). This kind of measurement is limited by the lepton energy and momentum scales, but these can be controlled using leptonic $Z$ decays. With the even larger data samples of the high luminosity phase, one may alternatively consider the $W/Z$ transverse mass ratio, opening the avenue to a largely independent measurement with an error as low as $\pm 10$ MeV [14], for a combined uncertainty about three times smaller than our benchmark of $\pm 25$ MeV.

The sensitivity of the total $W$ decay width, $\Gamma_W$, to new physics and its complementarity to and correlation with other quantities depends on how it is obtained. It can be extracted indirectly through measurements of cross section ratios,

$$\Gamma_W(\text{indirect}) = \left[\frac{\sigma(pp \to Z \to \ell^+\ell^- X)}{\sigma(pp \to W \to \ell\nu X)}\right]_{\text{exp.}} \times \left[\frac{\sigma(pp \to W)}{\sigma(pp \to Z)}\right]_{\text{th.}} \times \frac{\Gamma_{\text{SM}}(W \to \ell\nu)}{B_{\text{LEP}}(Z \to \ell^+\ell^-)},$$

(CDF currently quotes $\Gamma_W = 2.079 \pm 0.041$ GeV [16]) but the leptonic $W$ decay width, $\Gamma_{\text{SM}}(W \to \ell\nu)$, has to be input from the SM. More interesting is therefore the direct method using the tail of the transverse mass distribution. An average of final Tevatron Run I and preliminary DØ II [17] and LEP 2 [2] results gives, $\Gamma_W = 2.103 \pm 0.062$ GeV. The final Tevatron Run II is expected to contribute $\pm 50$ MeV measurements for each channel and experiment. Detailed studies for the LHC are not yet available, but historically the absolute error in $\Gamma_W$ at hadron colliders has traced roughly the one in $M_W$. If this trend carries over to the LHC, a $\pm 0.5\%$ error in $\Gamma_W$ may be in store.

Some of the Tevatron Run II results are already included in the current $\pm 2.1$ GeV [6] uncertainty in $m_t$, and with the expected total of about 8 $\text{fb}^{-1}$ the error may decrease by another factor of two. The LHC is anticipated to contribute a $\pm 1$ GeV determination from the lepton + jets channels alone [18]. The cleaner but lower statistics dilepton channels may provide another $\pm 1.7$ GeV determination, compared with $\pm 3$ GeV from the systematics limited all hadronic channel [18]. The combination of these channels (all dominated by the $b$ jet energy scale) would yield an error close to the additional irreducible theoretical uncertainty of $\pm 0.6$ GeV from the conversion from the pole mass (which is approximately what is being measured [19]) to a short-distance mass (such as $\overline{\text{MS}}$) which actually enters the loops. Folding this in, the grand total may give an error of about $\pm 1$ GeV, so that the parametric uncertainty from $m_t$ in the SM prediction for $M_W$ would be somewhat smaller than the anticipated experimental error in $M_W$.

With 30 $\text{fb}^{-1}$ the LHC will also be able to determine the CKM parameter, $V_{tb}$, in single top quark production to $\pm 5\%$ [20] (one expects $\pm 9\%$ from the Tevatron[1]). Anomalous flavor changing neutral current decays, $t \to Vq$ (where $V$ is a gluon, $\gamma$, or $Z$, and $q \neq b$), can be searched for down to the $10^{-4} - 10^{-5}$ level [18]. This sensitivity gain by three orders of magnitude over current HERA bounds [22], will be relevant, *e.g.*, for extra $W'$ bosons. Measuring $t\bar{t}$ spin correlations at the 10% level [18] will allow to establish the top as a spin 1/2 particle, to study non-standard production mechanisms (*e.g.* through resonances), and to discriminate between $W^+b$ and charged Higgs ($H^+b$) decays.

---

[1] After the conclusion of this conference, there was an announcement by the DØ Collaboration [21] of the first evidence of single top quark production. This translates into the bound $|V_{tb}| > 0.68$ (95% C.L.).

**TABLE 1.** Results and future expectations for $M_W$. The last column extrapolates the Tevatron Run I precision under the assumption that sensitivities scale as in background dominated types of experiments. The exception is MegaW which refers to a dedicated threshold scan at the ILC with $\mathscr{O}(10^6)$ $W$ pairs and is based on a $\sqrt{L}$-scaling from a similar scan at LEP 2. As can be seen, such scalings provide simple estimates of future precision goals.

| | $\text{fb}^{-1}$ **per experiment** | **value [GeV]** | **error/*goal*** | $\sqrt[4]{L}$**-scaling** |
|---|---|---|---|---|
| Tevatron Run I | 0.11 | 80.452 | **59** | — |
| LEP 2 | 0.70 | 80.376 | **33** | 37 |
| **currently** | **0.81** | **80.394** | **29** | **36** |
| Tevatron Run IIA | 2 | | *31* | 29 |
| Tevatron Run IIB | 8 | | *25* | 20 |
| LHC low luminosity | 10 | | *23* | 19 |
| LHC high luminosity | 400 | | *9* | 8 |
| ILC | 300 | | *10* | 8 |
| MegaW | 70 | | *7* | 4* |

If the Higgs boson exists, its production at the LHC will proceed primarily through gluon fusion, $gg \to H$, and/or vector boson fusion, $qq' \to Hqq'$. Higgs couplings can generally be determined to $10-30\%$ [14]. The top Yukawa coupling is best studied in associated production, $pp \to t\bar{t}H$, to $20-30\%$ precision [18]. Most difficult proves the Higgs self-coupling, $\lambda$, whose measurement would need a luminosity upgrade [23]. With 3 $\text{ab}^{-1}$, $\lambda$ can be measured to $\pm 20\%$, for 150 GeV $< M_H <$ 200 GeV, while only $\pm 70\%$ precision would be possible for a lighter (and weaker coupled) Higgs boson [14].

## ILC

While the hadron colliders are primarily discovery machines with remarkable capabilities for precision studies, the $e^+e^-$ International Linear Collider (ILC) — if built — would be a precision machine par excellence. In its first phase of operation the ILC would operate at center of mass energies from about 200 GeV (the reach of LEP 2) to 500 GeV, which would allow to scan the top and $ZH$ threshold regions. An integrated luminosity of 500 $\text{fb}^{-1}$ is expected in the first 4 years of running. The baseline design includes an at least 80% polarized electron beam. The second phase foresees an energy upgrade to around 1 TeV and the collection of 1 $\text{ab}^{-1}$ of data in 3–4 years. A relative determination of the jet energy scale is expected to within $\pm 0.3/\sqrt{E}$, where $E$ is the center of mass energy in GeV. Heavy quark tagging will also be important and is anticipated with an efficiency at the 50–60% (30–40%) level for $b$ ($c$) quarks.

The ILC comes with a variety of add-on options. For example, one may want to run it in other collision modes such as $\gamma\gamma$, $e^-\gamma$, or $e^-e^-$. Most relevant for precision physics would be the GigaZ mode [11], which would allow to repeat the LEP 1 program with 20 million $Z$ bosons daily. The physics motivation for this option is very high. In particular, the world's best measurements of $\sin^2\theta_W^{\text{eff.}}$ [1] have been provided by SLD ($\pm 0.00029$) from the left-right cross section asymmetry, $A_{LR} = A_e$, and the LEP groups ($\pm 0.00028$) from the forward-backward asymmetry for $b$-quark final states, $A_{FB} = 3/4 A_e A_b$, where $\sin^2\theta_W^{\text{eff.}}$ is extracted from $A_e$, and where $A_b$ is well-known

**TABLE 2.** Results and future expectations for $\sin^2\theta_W^{\text{eff.}}$. Based on $\sqrt{L}$-scaling as appropriate for statistics dominated measurements, the last column extrapolates the Tevatron Run I precision to future hadron colliders. GigaZ refers to two years of data taking with $\mathcal{O}(10^9)$ $Z$ bosons and is scaled from the LEP 1 precision. JLab refers to Qweak [28] (see next Section) and the 12 GeV Møller experiment [27]. $\sqrt{P}$-scaling from E-158 [26] (with $P$ the beam power) is used for Møller scattering at JLab and ILC.

| | $\mathbf{fb^{-1}}$ **per experiment** | **experimental value** | **error/*goal*** | $\sqrt{L}$**-scaling** |
|---|---|---|---|---|
| Tevatron Run I | 0.072 | 0.2238 | **0.0050** | — |
| SLC | 0.05 | 0.23098 | **0.00026** | — |
| LEP 1 | 0.20 | 0.23187 | **0.00021** | — |
| **currently** | | **0.23152** | **0.00016** | — |
| Tevatron Run IIA | 2 | | *0.0008* | 0.0009 |
| Tevatron Run IIB | 8 | | *0.0003* | 0.0005 |
| JLab | $\vec{e}e$, $\vec{e}p$ | | *0.0003* | 0.00024 |
| LHC high luminosity | 400 | | *0.00014* | 0.00008 |
| ILC | Møller | | *0.00007* | 0.00007 |
| GigaZ | 140 | | *0.000013* | 0.000016 |

within the SM. These two measurements contribute greatly to our current knowledge of $M_H$. However, they disagree from each other by 3.1 $\sigma$ and it is important to resolve this discrepancy. It is conceivably due to new physics effects in $A_b$. Assuming this, one can turn $A_{FB}$ into a measurement of $A_b$ (taking $A_e$ from other asymmetries) and combine this with a more direct measurement of $A_b$ from SLD [1]. The result, $A_b = 0.899 \pm 0.013$, deviates by 2.8 $\sigma$ from the SM prediction ($A_b = 0.935$). GigaZ with its ultra-high rates combined with polarized electrons (unlike LEP 1) would be able to determine $A_b$ to $\pm 0.001$! Similarly, a high precision $WW$ threshold scan with $\mathcal{O}(10^6)$ $W$-pairs ("MegaW") would allow to study the $W$ boson with unprecedented precision. See Table 1 for a summary of $M_W$ measurements. The option of $e^+$ polarization ($\gtrsim 50\%$) would allow additional cross-checks (and thus reduction in systematic uncertainties) and to measure new kinds of asymmetries. For more details see, Refs. [24, 25].

Even without the GigaZ option, the ILC might provide an ultra-high precision measurement of the weak mixing angle (more precise than the current or foreseeable world average) in polarized fixed target Møller scattering. This may by then be a third generation experiment, building on the E-158 pioneering measurement at SLAC [26] (which used the 50 GeV SLC electron beam and achieved a precision of $\pm 0.0014$ in $\sin^2\theta_W^{\text{eff.}}$), and a potential effort (e2ePV [27]) at JLab (after the 12 GeV upgrade of CEBAF) with an anticipated error reduction by a factor of five. The ILC would then improve on JLab by an additional factor of four. Table 2 summarizes future prospects for $\sin^2\theta_W^{\text{eff.}}$.

Spectacular improvements (see Table 3) would be possible for $m_t$, because the aforementioned short-distance mass can be extracted directly from the top threshold scan.

Higgs production at the ILC would dominantly proceed through Higgs-strahlung, $e^+e^- \to ZH \to e^+e^-H$ ($\mu^+\mu^-H$). Associated production, $e^+e^- \to t\bar{t}H$, will be important for measurements of the top Yukawa coupling. It will be possible to determine the Higgs couplings $Hb\bar{b}$, $Ht\bar{t}$, $H\tau^+\tau^-$, $HW^+W^-$, and $HZZ$ to high precision, while the $Hc\bar{c}$ will be less precise. The self-coupling $\lambda$ can be obtained to $\pm 20\%$ (for $M_H = 120$ GeV) by using, *e.g.*, the process $e^+e^- \to ZHH$. The total Higgs width would be known to 5%.

**TABLE 3.** Results and future expectations for $m_t$. The last column extrapolates the Tevatron Run I precision assuming that sensitivities scale as in background dominated types of experiments. At hadron colliders, a ±0.6 GeV theory uncertainty has to be added.

| | $\text{fb}^{-1}$ per experiment | value [GeV] | error/*goal* | $\sqrt[4]{L}$-scaling |
|---|---|---|---|---|
| Tevatron Run I | 0.11 | 178.0 | **4.3** | — |
| summer 2005 | 0.43 | 172.7 | **2.9** | 3.1 |
| **currently** | **1** | **171.4** | **2.1** | **2.5** |
| Tevatron Run IIA | 2 | | *2.0* | 2.1 |
| Tevatron Run IIB | 8 | | *1.2* | 1.5 |
| LHC low luminosity | 10 | | *0.9* | 1.4 |
| LHC high luminosity | 400 | | *0.7* | 0.6 |
| ILC | 300 | | *0.1* | — |

# PERSPECTIVE: LOW ENERGY PRECISION MEASUREMENTS

Fixed-target Møller scattering was already mentioned above, both at high and relatively low energy, but all at very low momentum transfer, $\sqrt{Q^2} = \mathcal{O}(100\text{ MeV})$. A similar low-$Q^2$ experiment in elastic $e^-p$ scattering at JLab will determine the so-called weak charge of the proton, $Q_W^p$. With an expected polarization of $85 \pm 1\%$ the Qweak Collaboration [28] anticipates to measure the parity violating asymmetry, $A_{PV} \propto (Q^2 Q_W^p + Q^4 B)$. The $Q^4 B$ term is the leading form factor contribution and will be determined experimentally. The anticipated errors in $Q_W^p$ and the corresponding $\sin^2\theta_W$ are $\pm 0.003$ and $\pm 0.0007$, respectively (for the calculation of the SM prediction, see Refs. [29, 30]).

One can also explore the kinematic regimes of quasi-elastic (QE) and deep inelastic scattering (DIS). The discrepancy in $\nu$-DIS (NuTeV [7]) has already been mentioned. The interpretation of the experiment is hampered, however, by a variety of theoretical issues. An $eD$-DIS experiment is approved at JLab [31] to use the current 6 GeV CEBAF beam and to repeat the historical SLAC experiment [32] with greater precision. One hopes to be able to collect additional data points after the 12 GeV CEBAF upgrade [33]. This would improve the SLAC result and the current world average on the combination of effective four-Fermi quark-lepton couplings $2C_{2u} - C_{2d}$ by factors of 54 and 17, respectively. The issues to be addressed are higher twist effects and charge symmetry violating (CSV) parton distribution functions. Since higher twist effects are strongly $Q^2$ dependent and CSV should vary with the kinematic variable, $x$, while contributions from beyond the SM would be kinematics independent, one can separate all these possible effects by measuring a large array of data points. Thus, a great deal can be learned about the strong and weak interactions at the same time.

Determinations of $\sin^2\theta_W$ from $\nu e$-scattering are not competitive at present ($\pm 0.008$), but may be another future direction, in particular if $\beta$-beams or a $\nu_\mu$-factory became available with well-known $\nu$-flavor compositions and energy spectra. For example, a $\nu_e(\bar\nu_e)$ $\beta$-beam could provide a determination of $\sin^2\theta_W$ to $\pm 0.0008$ ($\pm 0.0005$) [34], while a $\nu_\mu(\bar\nu_\mu)$ factory could achieve a precision of $\pm 0.0001$ ($\pm 0.0003$) [34]. Other electroweak measurements may also be possible at these facilities with high precision.

It should be stressed that these and other low energy tests of the SM (including leptonic anomalous magnetic moments, atomic parity violation, and many SM forbidden or highly suppressed processes) will remain very important even in the LHC era and beyond, because they are not only competitive with but also complementary to high energy experiments. The complementarity refers to experimental uncertainties, as well as to theoretical issues, and the way new physics may enter. Low energy experiments may also play a prominent role in deciphering what may be discovered at the energy frontier, even if no significant SM deviations are seen at low energies.

## ACKNOWLEDGMENTS

It is a pleasure to thank the organizers for the invitation to this meeting. I am particularly grateful to Miguel Ángel Pérez for his support and help. This work was supported by contracts CONACyT (México) 42026–F and DGAPA–UNAM PAPIIT IN112902.

## REFERENCES

1. ALEPH, DELPHI, L3, OPAL, and SLD Collaborations, LEP and SLD Electroweak Working Groups (EWWGs), and SLD Heavy Flavour Group: S. Schael *et al.*, *Phys. Rept.* **427**, 257–454 (2006).
2. ALEPH, DELPHI, L3, OPAL, and LEP EWWG: J. Alcaraz *et al.*, hep-ex/0612034.
3. UA2 Collaboration: J. Alitti *et al.*, *Phys. Lett.* **B276**, 354–364 (1992).
4. CDF Collaboration: T. Affolder *et al.*, *Phys. Rev.* **D64**, 052001 (2001).
5. DØ Collaboration: V. M. Abazov *et al.*, *Phys. Rev.* **D66**, 012001 (2002).
6. Tevatron Electroweak Working Group: E. Brubaker *et al.*, hep-ex/0608032.
7. NuTeV Collaboration: G. P. Zeller *et al.*, *Phys. Rev. Lett.* **88**, 091802 (2002).
8. ALEPH, DELPHI, L3, and OPAL Collaborations, and LEP Working Group for Higgs Boson Searches: R. Barate *et al.*, *Phys. Lett.* **B565**, 61–75 (2003).
9. J. Erler, *Phys. Rev.* **D63**, 071301 (2001).
10. ALEPH Collaboration: A. Heister *et al.*, *Phys. Lett.* **B526**, 191–205 (2002).
11. J. Erler, S. Heinemeyer, W. Hollik, G. Weiglein, and P. M. Zerwas, *Phys. Lett.* **B486**, 125–133 (2000).
12. T. Rodrigo, these proceedings.
13. ATLAS Collaboration: P. Pralavorio *et al.*, *PoS* **HEP2005**, 294 (2006).
14. U. Baur, hep-ph/0511064, to appear in the proceedings of HCP 2005.
15. W. Quayle, in the proceedings of ICHEP 04 (World Scientific, Hackensack, 2005) Vol. 1, 531–534.
16. CDF II Collaboration: D. Acosta *et al.*, *Phys. Rev. Lett.* **94**, 091803 (2005).
17. CDF and DØ Collaborations and Tevatron EWWG: B. Ashmanskas *et al.*, hep-ex/0510077.
18. J. Womersley, *PoS* **TOP2006**, 038 (2006).
19. M. C. Smith and S. S. Willenbrock, *Phys. Rev. Lett.* **79**, 3825–3828 (1997).
20. K. Mazumdar, *AIP Conf. Proc.* **792**, 587–590 (2005).
21. DØ Collaboration: V. M. Abazov *et al.*, hep-ex/0612052.
22. ZEUS and H1 Collaborations: J. Ferrando *et al.*, *Eur. Phys. J.* **C33**, S761–S763 (2004).
23. D. Denegri, these proceedings.
24. H. Yamamoto, talk presented at DPF2006 + JPS2006 ....
25. E. Elsen, these proceedings.
26. SLAC–E158–Collaboration: P. L. Anthony *et al.*, *Phys. Rev. Lett.* **95**, 081601 (2005).
27. D. Mack, talk presented at PAVI06.
28. Qweak Collaboration: D. S. Armstrong *et al.*, *Eur. Phys. J.* **A24S2**, 155–158 (2005).
29. W. J. Marciano and A. Sirlin, *Phys. Rev.* **D27**, 552–556 (1983) and *ibid.* **D29**, 75–88 (1984).
30. J. Erler, A. Kurylov, and M. J. Ramsey-Musolf, *Phys. Rev.* **D68**, 016006 (2003).
31. X. Zheng, talk presented at PAVI06.
32. SLAC–E–122 Collaboration: C. Y. Prescott *et al.*, *Phys. Lett.* **B84**, 524–528 (1979).
33. P. Reimer and P. Souder, talks presented at PAVI06.
34. A. de Gouvea and J. Jenkins, *Phys. Rev.* **D74**, 033004 (2006).

# Grand Unification with and without Supersymmetry

Alejandra Melfo

*CFF, Universidad de Los Andes, Mérida, Venezuela*
*Institute J. Stefan, Ljubljana, Slovenia*

**Abstract.**
Grand Unified Theories based on the group SO(10) generically provide interesting and testable relations between the charged fermions and neutrino sector masses and mixings. In the light of the recent neutrino data, we reexamine these relations both in supersymmetric and non-supersymmetric models, and give a brief review of their present status,

## INTRODUCTION

Of the two main candidates for a unifying gauge group, $SU(5)$ and $SO(10)$, the latter is often cited as the one most suitable in accounting for neutrino masses. This is not at all clear, in particular after the recent suggestion of a non-supersymmetric $SU(5)$ model with a Type III see-saw mechanism, that provides not only a consistent unified theory with neutrino masses, but also predictions directly testable at the LHC [1]. $SU(5)$ models are in addition generally simpler, making proton decay rate calculations feasible. But the remarkable feature of $SO(10)$ is that the 16-dimensional spinor representation can accommodate a complete family of fermions, including the right-handed neutrino. This complete unification of quarks and leptons opens up the possibility of obtaining connections between the charged fermions and the neutrino sector. Furthermore, $SO(10)$ has the Left-Right group as a subgroup [3], making the implementation of the see-saw mechanism [2] very natural in these theories.

The smallness of neutrino mass is also calling for an intermediate scale. Namely, the see-saw mechanism will give neutrinos a small mass after the electroweak symmetry breaking, provided the $B-L$ symmetry is broken at some scale $M_R$, of order $M_W^2/M_R$. With the current limit on $\Delta m_{23}^2 \simeq 2.5\times 10^{-3} eV^2$, the mass of the heaviest neutrino is an indication that $M_R \sim 10^{13} GeV$, below $M_{GUT} \sim 10^{16} GeV$, and even smaller in the case neutrinos are degenerate. It is only an indication, of course, since three orders of magnitude can in principle be accounted for by the coupling constants. In practice however it is not so simple, at has been recently seen in the case of the Minimal Supersymmetric Grand Unified Model, which fails to predict a large enough mass for the heaviest neutrino, as we will review below.

In their turn, intermediate scales are an indication of the possibility of a non-supersymmetric unification. The Minimal Supersymmetric Standard Model (MSSM) gives one-step unification at $M_{GUT}$ with such a good precision [4] that it leaves basically no room for intermediate scales, if the scale of supersymmetry breaking is reason-

CP917, *Particles and Fields*, edited by H. Castilla Valdez, J. C. D'Olivo, and M. A. Perez

ably close to $M_W$, as required by the supersymmetric solution to the hierarchy problem. Assuming an intermediate scale of symmetry breaking allows for unification without supersymmetry, and it is possible in many cases to have this intermediate scale at the required value, as we will see below.

In this talk, we review briefly the current status of supersymmetric and non-supersymmetric theories based on $SO(10)$, with emphasis on the Yukawa sector an the possibility of general relations connecting the charged fermion and neutrino sectors.

## YUKAWA SECTOR

In $SO(10)$, the Yukawa sector is particularly simple. Each family of fermions is in a **16**-dimensional spinor representation, and they can therefore couple to only three fields, $\mathbf{10_H}$, $\mathbf{120_H}$ and $\overline{\mathbf{126}}_\mathbf{H}$ representations, since

$$\mathbf{16} \times \mathbf{16} = \mathbf{10} + \mathbf{120} + \mathbf{126}\,. \tag{1}$$

Furthermore, the Yukawa coupling with $120_H$ is antisymmetric in family space. The Higgs fields decompose under the SU(2)$_L\times$ SU(2)$_R\times$ SU(4)$_C$ Pati-Salam [5] group as

$$\begin{aligned} \mathbf{10} &= (2,2,1)+(1,1,6) \\ \overline{\mathbf{126}} &= (1,3,10)+(3,1,\overline{10})+(2,2,15)+(1,1,6) \\ \mathbf{120} &= (1,3,6)+(3,1,6)+(2,2,15)+(2,2,1)+(1,1,20) \end{aligned} \tag{2}$$

The $\mathbf{126_H}$ provides mass terms for right-handed and left-handed neutrinos:

$$M_{\nu_R} = \langle 1,3,10\rangle Y_{126}, \;\; M_{\nu_L} = \langle 3,1,\overline{10}\rangle Y_{126} \tag{3}$$

which means that one has both type I and type II seesaw:

$$M_N = -M_{\nu_D} M_{\nu_R}^{-1} M_{\nu_D} + M_{\nu_L} \tag{4}$$

In the type I case it is the large vev of $(1,3,10)$ that provides the masses of right-handed neutrino whereas in the type II case, the left-handed triplet provides directly light neutrino masses through a small vev [6, 7]. The disentangling of the two contributions is in general hard.

One can now ask: what is the minimal Yukawa sector one can choose? Clearly only one of the three representations cannot do the job, as then one would have a fixed relation between the down quarks and the charged leptons for all generations.

If one is interested in a renormalizable version of the see-saw mechanism the representation $\overline{\mathbf{126}}_\mathbf{H}$ indispensable, since it breaks the SU(2)$_R$ group and gives a see-saw neutrino mass. By itself it gives no fermionic mixing, so it does not suffice. The realistic fermionic spectrum requires adding either $\mathbf{10_H}$ or $\mathbf{120_H}$. An interesting possibility which we shall not review here is a radiatively-induced see-saw [8].

Let us start by analyzing the non-supersymmetric case, with an extra $\mathbf{10_H}$ field [9]. The most general Yukawa interaction is

$$\mathscr{L}_Y = \mathbf{16_F}\left(\mathbf{10_H} Y_{10} + \overline{\mathbf{126}}_\mathbf{H} Y_{126}\right)\mathbf{16_F} + h.c.\,. \tag{5}$$

where $Y_{10}$ and $Y_{126}$ are symmetric matrices in the generation space. With this one obtains relations for the Dirac fermion masses

$$\begin{aligned} M_D = M_1 + M_0 \quad &, \quad M_U = c_1 M_1 + c_0 M_0 \ , \\ M_E = -3M_1 + M_0 \quad &, \quad M_{\nu_D} = -3c_1 M_1 + c_0 M_0 , \end{aligned} \tag{6}$$

where we have defined

$$M_1 = \langle 2,2,15\rangle^d_{126} Y_{126} \, , \quad M_0 = \langle 2,2,1\rangle^d_{10} Y_{10} \, , \tag{7}$$

and

$$c_0 = \frac{\langle 2,2,1\rangle^u_{10}}{\langle 2,2,1\rangle^d_{10}} , \ c_1 = \frac{\langle 2,2,15\rangle^u_{126}}{\langle 2,2,15\rangle^d_{126}} . \tag{8}$$

In the physically sensible approximation $\theta_q = V_{cb} = 0$, these relations imply

$$c_0 = \frac{m_c(m_\tau - m_b) - m_t(m_\mu - m_s)}{m_s m_\tau - m_\mu m_b} \approx \frac{m_t}{m_b} , \tag{9}$$

Notice that this means that $\mathbf{10_H}$ cannot be real, since in that case one would have $|\langle 2,2,1\rangle^u_{10}| = |\langle 2,2,1\rangle^d_{10}|$, implying $m_t/m_b$ of order one. It is necessary to complexify $\mathbf{10_H}$, just as in a supersymmetric theory. If taking advantage of this fact one decides to impose a Peccei-Quinnsymmetry, thus providing a Dark Matter candidate, the Yukawa sector in non-supersymmetric and supersymmetric models is similar.

In this case, this model has the interesting feature of automatic connection between $b-\tau$ unification and large atmospheric mixing angle in the type II see-saw. From $M_{\nu_L} \propto Y_{126}$ , one has $M_{\nu_L} \propto M_D - M_E$. as shown in [10, 11]. This fact has inspired the careful study of the analogous supersymmetric version where $m_\tau \simeq m_b$ at the GUT scale works rather well [12] . In the non-supersymmetric theory, $b-\tau$ unification fails badly, $m_\tau \sim 2m_b$ [13]. The realistic theory will require a Type I seesaw, or an admixture of both possibilities.

Suppose now that we choose instead $\mathbf{120_H}$ [9]. Since $Y_{120}$ is antisymmetric, this means only 3 new complex couplings on top of $Y_{126}$. On gets in this case

$$\begin{aligned} M_D = M_1 + M_2 \quad &, \quad M_U = c_1 M_1 + c_2 M_2 \ , \\ M_E = -3M_1 + c_3 M_2 \quad &, \quad M_{\nu_D} = -3c_1 M_1 + c_4 M_2 \end{aligned} \tag{10}$$

where $M_1$ and $c_1$ are defined in (7),(8), and:

$$\begin{aligned} M_2 = Y_{120}\left(\langle 2,2,1\rangle^d_{120} + \langle 2,2,15\rangle^d_{120}\right) \quad &, \quad c_2 = \frac{\langle 2,2,1\rangle^u_{120} + \langle 2,2,15\rangle^u_{120}}{\langle 2,2,1\rangle^d_{120} + \langle 2,2,15\rangle^d_{120}} , \\ c_3 = \frac{\langle 2,2,1\rangle^d_{120} - 3\langle 2,2,15\rangle^d_{120}}{\langle 2,2,1\rangle^d_{120} + \langle 2,2,15\rangle^d_{120}} \quad &, \quad c_4 = \frac{\langle 2,2,1\rangle^u_{120} - 3\langle 2,2,15\rangle^u_{120}}{\langle 2,2,1\rangle^d_{120} + \langle 2,2,15\rangle^d_{120}} . \end{aligned} \tag{11}$$

It is easy to see that again there is a need to complexify the Higgs fields, by arguments similar to the case of $\mathbf{10_H}$.

In order to obtain algebraic expressions, from which a clearer physical meaning can be extracted, one can restrict the analysis to the second and third generations. Later, numerical studies could include the effects of the first generation as a perturbation. In the basis where $M_1$ is diagonal, real and non-negative, for the two-generation case one gets:

$$M_1 \propto \begin{pmatrix} \sin^2\theta & 0 \\ 0 & \cos^2\theta \end{pmatrix} \tag{12}$$

and the most general charged fermion matrix can be written as:

$$M_f = \mu_f \begin{pmatrix} \sin^2\theta & i(\sin\theta\cos\theta + \varepsilon_f) \\ -i(\sin\theta\cos\theta + \varepsilon_f) & \cos^2\theta \end{pmatrix}, \tag{13}$$

where $f = D, U, E$ stands for charged fermions and $\varepsilon_f$ vanishes for negligible second generation masses. In other words $|\varepsilon_f| \propto m_2^f/m_3^f$. Furthermore the real parameter $\mu_f$ sets the third generation mass scale. By calculating up to leading order in $|\varepsilon_f|$, we have to the following interesting predictions [9]:

1. type I and type II seesaw lead to the same structure

$$M_N^I \propto M_N^{II} \propto M_1 \tag{14}$$

so that in the selected basis the neutrino mass matrix is diagonal. We see that the angle $\theta$ has to be identified with the leptonic (atmospheric) mixing angle $\theta_A$ up to terms of the order of $|\varepsilon_E| \approx m_\mu/m_\tau$. For the neutrino masses we obtain from (12)

$$\frac{m_3^2 - m_2^2}{m_3^2 + m_2^2} = \frac{\cos 2\theta_A}{1 - \sin^2 2\theta_A/2} + \mathcal{O}(|\varepsilon|) \tag{15}$$

This equation points to an intriguing correlation: the degeneracy of neutrino masses is measured by the maximality of the atmospheric mixing angle.

2. the ratio of tau and bottom mass at the GUT scale is given by:

$$\frac{m_\tau}{m_b} = 3 + 3\sin 2\theta_A \,\mathrm{Re}[\varepsilon_E - \varepsilon_D] + \mathcal{O}(|\varepsilon^2|) \tag{16}$$

This is not correct in principle, the extrapolation in standard model gives $m_\tau \approx 2m_b$. However, several effects modify this conclusion, such as for example the inclusion of the first generation or the running of Yukawa couplings. We would in any case expect that $m_b$ comes out as small as possible.

3. the quark mixing is found to be:

$$|V_{cb}| = |\,\mathrm{Re}\xi - i\cos 2\theta_A \,\mathrm{Im}\xi\,| + \mathcal{O}(|\varepsilon^2|) \tag{17}$$

where $\xi = \cos 2\theta_A\ (\varepsilon_D - \varepsilon_U)$. This equation demonstrates the successful coexistence of small and large mixing angles. In order for it to work quantitatively, $|\cos 2\theta_A|$ should be as large as possible, i.e. $\theta_A$ should be as far as possible from the maximal value 45°. To make a definite numerical statement, again, the effects from the first generation and the loops have to be included.

## UNIFICATION CONSTRAINTS

In the non-supersymmetric case, an intermediate scale is necessary for unification to succeed, and in the over-constrained models discussed here, the Dirac neutrino Yukawa couplings are not arbitrary. Thus one must make sure that the pattern of intermediate mass scale is consistent with a see-saw mechanism for neutrino masses. The lower limit on $M_R$ stems from the heaviest neutrino mass $m_\nu \geq m_t^2/M_R$, which gives $M_R \geq 10^{13}$ GeV or so. One can now turn to the useful table of Ref. [14], where the most general patterns of SO(10) symmetry breaking with two intermediate scales consistent with proton decay limits are presented. Most models are ruled out by the constrain above; the most promising candidates are those with an intermediate $\mathrm{SU}(2)_L \times \mathrm{SU}(2)_R \times \mathrm{SU}(4)_C \times \mathrm{P}$ symmetry breaking scale (that is, the Pati-Salam group with unbroken parity).

The supersymmetric models have been extensively studied in the last few years, in particular in the minimal version the so-called Minimal Supersymmetric GUT (MSGUT). Besides the $\mathbf{10_H}$ and $\overline{\mathbf{126}}_\mathbf{H}$ providing the fermion masses, in this model just one additional representation suffices to break the symmetry all the way to the Left-Right group (recall that $\overline{\mathbf{126}}_\mathbf{H}$ is in charge of breaking it further to the Standard Model group) [15, 16]. It has been studied in detail, and the mass spectrum calculated simultaneously by various groups [17, 18, 19]. It has a number of very interesting features, like the conservation of R-parity at all energies, providing the LSP as candidate for dark matter [21], and the connection between the $b-\tau$ unification and a large atmospheric mixing angle [10].

The MSGUT has so few parameters, 26 in all, that is becomes even too predictive for its own good. Namely, the composition of the light Higgs doublets is no longer arbitrary, but given in terms of very few parameters. In fact, the dependence is such that only one complex parameter is truly relevant for the fitting of the known fermion masses and mixing, in particular for neutrino masses.

The superpotential of the model is (see e.g., [16] for a more explicit notation):

$$\begin{aligned} W &= \mathbf{16_m}(Y_{10}\,\mathbf{10_H} + Y_{\overline{126}}\,\overline{\mathbf{126}}_\mathbf{H})\mathbf{16_m} + m\mathbf{210_H}^2 + \lambda\mathbf{210_H}^3 + M\mathbf{126_H}\,\overline{\mathbf{126}}_\mathbf{H} \\ &\quad + \eta\mathbf{126_H}\,\mathbf{210_H}\,\overline{\mathbf{126}}_\mathbf{H} + m_H\mathbf{10_H}^2 + \mathbf{210_H}\,\mathbf{10_H}\,(\alpha\mathbf{126_H} + \overline{\alpha}\overline{\mathbf{126}}_\mathbf{H}) \end{aligned} \tag{18}$$

where $Y_{10}$ and $Y_{\overline{126}}$ are the two Yukawa couplings of the theory. The rest of the superpotential describes the Higgs sector at high (GUT) scale.

Notice that in fact, this model has only one mass scale. The mass $m_H$ is fixed by the fine-tuning [16, 19] condition:

$$m_H = m \times \frac{\alpha\overline{\alpha}}{2\eta\lambda}\, f(x) \tag{19}$$

with $f(x)$ a known function, that is needed to ensure a pair of light Higgs doublets $H_u$ and $H_d$ in the spectrum of the low energy supersymmetric standard model (MSSM). The parameter $x$ measures one of the vacuum expectation values (VEV) in unities of $m/\lambda$ [16]. The mass $M$ can be calculated in terms of $m$ and the other parameters due to the consistency condition for symmetry breaking:

$$M = m \times \frac{\eta}{\lambda}\frac{3-14x+15x^2-8x^3}{(x-1)^2} \tag{20}$$

(that, if one likes so, can be thought as a replacement of $M$ in favor of $x$). Thus, any new scale, as the right-handed neutrino mass scale, has to arise by the mass $m$ times some adimensional quantity. One can then find the symmetry-breaking VEVs as functions of $x$ [16]. The SM singlets that acquire ves are five: $\langle 1,1,1\rangle_{210}, \langle 1,1,15\rangle_{210}, \langle 1,3,1\rangle_{210}, \langle 1,3,10\rangle_{\overline{126}}$ and $\langle 1,3,10\rangle_{126}$ . The Higgs sector has just 8 real parameters after fine-tuning, which can be chosen as e.g.:

$$m,\ \alpha,\ \overline{\alpha},\ |\lambda|,\ |\eta|,\ \phi = \arg(\lambda) = -\arg(\eta),\ x = \mathrm{Re}(x) + i\mathrm{Im}(x). \tag{21}$$

The parameter $x$ is known to be convenient to describe the VEVs and the masses of the particles; the dependences of observed fermion masses on the other parameters is rather simple, whereas the behaviour in $x$ is usually non trivial [19]. Using this, it is found that because of the absence of an intermediate scale $M_R$, the type II see-saw is readily seen to fail the test [22, 23], the neutrino masses come out just too small. Surprisingly enough, the same happens in the Type I case [23, 24]. Namely, if $\mathbf{126_H}$ is to play any role in the charged fermion masses, it Yukawa couplings cannot be arbitrarily small. The fitting requires them to be quite large, and since the Type I see-saw mass is proportional to the inverse Yukawa matrix, again it turns out too small. The current status is that this model is not viable in its minimal form.

## THE NON-MINIMAL MODELS

One obvious alternative for the supersymmetric case is to enlarge the Yukawa sector by adding a $\mathbf{120_H}$, on top of $\mathbf{10_H}$ and $\mathbf{\overline{126}_H}$. This means three complex additional parameters form the (asymmetric) Yukawa matrix, along with five new complex parameters in the superpotential, one of which could be removed by a redefinition of the new field. Since the 120-dimensional representation contains no Standard Model singlets, the symmetry breaking pattern of the minimal model is maintained, which somehow simplifies the analysis. However, the parameter space is now so large that additional assumptions (such as the possibility of spontaneous CP violation) have to be made in order to attempt the fitting of fermion data.

Once the Yukawa sector becomes maximal, there is very little difference between the minimal model and the most simple alternative, the model with 54 and 45-dimensional representations instead of 210, thoroughly studied in [25]. $\mathbf{54_H}$ and $\mathbf{45_H}$ fail to provide a coupling between $\mathbf{10_H}$ and $\mathbf{\overline{126}_H}$, thus in order to have a rich enough Yukawa structure the simplest possibility is to add a $\mathbf{120_H}$, in which case all the bidoublets mix. This version has only 5 parameters more than the minimal one. A detailed study of the supersymmetry breaking and particle spectrum has been carried out [26]. The superpotential has 5 more parameters than the extended MSGUT, and the symmetry breaking conditions are given in terms of two mass ratios now.

It is reasonable to suppose that the maximal Yukawa sector can provide a good fitting of charged fermions and neutrinos. But this model has also the potential of providing a Type II dominance simply by fine-tuning a small mass for the left-handed triplet (a couple of orders of magnitude below $M_{GUT}$ would suffice). Unification constrains can still be satisfied, provided the triplet's contributions to the RGE is cancelled by other fields in the spectrum, with masses fine-tuned to approximately the same value. At one loop, the contribution to the RGE equations of the $SU(2)_L$ triplets $(1,3,\pm1)$, can be cancelled out by two color sextets $(6,1,\pm1/3)$ present in $126_H$, $\overline{126}_H$ and $120_H$, and a couple of doublets $(1,2,\pm1/2)$, with the same quantum numbers as the light MSSM Higgs. Thus, for example, the mass parameter of the $\mathbf{10_H}$ can be fixed as usual, in order to fine-tune doublets at the electroweak scale. Then the mass of $\mathbf{126_H}, \overline{\mathbf{126}}_\mathbf{H}$, that of $\mathbf{54_H}$ and that of $\mathbf{120_H}$ can be fixed so as to have a light triplet, thus a type-II dominance in the neutrino mass. This means that the right-handed scale entering the neutrino mass becomes arbitrary again, depriving the GUT of one of its most important roles.

## CONCLUDING REMARKS

We have argued that the grand unified model based on the group $SO(10)$ provides, through quark-lepton unification, interesting connection between the charged and non charged fermion sectors. The see-saw mechanism, in its renormalizable version, calls for a $\overline{\mathbf{126}}$ representation, but it alone cannot accomodate the compelte spectrum. Including a 10 dimension or 120 dimensional representations cures this problem, but a complex field has to be used, or one is lead to wrong relations between quark masses at the GUT scale. Compelxifying the Yukawa sector, in turn, leads to the interesting possibility of invoking a Peccei-Quinn symemtry, thus allowing for the axion as a candidate dark matter. If the additional fied is the $\mathbf{10_H}$, one is lead to the connection between the large athmospheric angle and $b-\tau$ unification. If on the other hand the antysymmetric $\mathbf{120_H}$ is added, the two generation analysis suggests that a large athmospheric angle is related to small mixing between second and third generations of quarks, and that it prefers degenrate neutrinos.

Non-supersymmetric models, with an intermediate scale of at least $10^{13}$ GeV, satisfy the unification constrains in the case where the symmetry breaking proceeds through the Pati-Salam subgroup. Supersymmetric models do not allow for an intermediate scale, and that leads to too small neutrino masses in the case of Type II neutrino, but even Type I neutrino masses come too small in the MSGUT, where the symmetry breaking requires only one additional $\mathbf{210_H}$, and the Yukawa sector contains the minimal combination, $\mathbf{10_H}$ and $\overline{\mathbf{126}}_\mathbf{H}$. The restrictions coming form symmetry breaking and unification constraints prove fatal to the fitting of fermion masses. One is forced away from the minimal models into an enlarged Yukawa sector, and then the predictivity of the models becomes questionable. It is no longer clear if the extended MSGUT is to be preferred over the alternative model (with $\mathbf{54_H}$ and $\mathbf{45}$ ).

## ACKNOWLEDGMENTS

I wish to thank my collaborators Borut Bajc, Goran Senjanović, Francesco Vissani and Alba Ramírez. Thanks also to the organizers of the VI Silafae for a truly enjoyable meeting.

## REFERENCES


1. B. Bajc and G. Senjanovic, arXiv:hep-ph/0612029. B. Bajc, M. Nemevsek and G. Senjanovic, arXiv:hep-ph/0703080.
2. P. Minkowski, Phys. Lett. B **67** (1977) 421; T. Yanagida, proceedings of the *Workshop on Unified Theories and Baryon Number in the Universe*, Tsukuba, 1979, eds. A. Sawada, A. Sugamoto; S. Glashow, in *Cargese 1979, Proceedings, Quarks and Leptons* (1979); M. Gell-Mann, P. Ramond, R. Slansky, proceedings of the *Supergravity Stony Brook Workshop*, New York, 1979, eds. P. Van Niewenhuizen, D. Freeman; R. Mohapatra, G. Senjanović, Phys.Rev.Lett. **44** (1980) 912.
3. R. N. Mohapatra and J. C. Pati, Phys. Rev. D **11** (1975) 2558; G. Senjanović and R. N. Mohapatra, Phys. Rev. D **12** (1975) 1502; G. Senjanović, Nucl. Phys. B **153** (1979) 334.
4. S. Dimopoulos, S. Raby, F. Wilczek, Phys. Rev. D **24** (1981) 1681. L.E. Ibáñez, G.G. Ross, Phys. Lett. B **105** (1981) 439. M.B. Einhorn, D.R. Jones, Nucl. Phys. B **196** (1982) 475. W. Marciano, G. Senjanović, Phys.Rev.D **25** (1982) 3092.
5. J. C. Pati and A. Salam, Phys. Rev. D **10** (1974) 275.
6. G. Lazarides, Q. Shafi and C. Wetterich, Nucl. Phys. B **181** (1981) 287.
7. R. N. Mohapatra and G. Senjanović, Phys. Rev. D **23** (1981) 165.
8. B. Bajc and G. Senjanovic, Phys. Lett. B **610** (2005) 80 [arXiv:hep-ph/0411193]. B. Bajc and G. Senjanovic, Phys. Rev. Lett. **95** (2005) 261804 [arXiv:hep-ph/0507169].
9. B. Bajc, A. Melfo, G. Senjanovic and F. Vissani, Phys. Rev. D **73** (2006) 055001 [arXiv:hep-ph/0510139].
10. B. Bajc, G. Senjanović, F. Vissani, Phys. Rev. Lett. **90** (2003) 051802 [hep-ph/0210207], and hep-ph/0110310.
11. B. Bajc, G. Senjanović and F. Vissani, Phys. Rev. D **70** (2004) 093002 [arXiv:hep-ph/0402140].
12. K. S. Babu and C. Macesanu, Phys. Rev. D **72** (2005) 115003 [arXiv:hep-ph/0505200]. S. Bertolini, M. Frigerio and M. Malinsky, Phys. Rev. D **70**, 095002 (2004) [arXiv:hep-ph/0406117].
13. H. Arason, D. J. Castano, E. J. Piard and P. Ramond, Phys. Rev. D **47** (1993) 232 [arXiv:hep-ph/9204225].
14. N. G. Deshpande, E. Keith and P. B. Pal, Phys. Rev. D **47**, 2892 (1993) [arXiv:hep-ph/9211232].
15. T. E. Clark, T. K. Kuo and N. Nakagawa, Phys. Lett. B **115**, 26 (1982); C. S. Aulakh and R. N. Mohapatra, Phys. Rev. D **28**, 217 (1983).
16. C. S. Aulakh, B. Bajc, A. Melfo, G. Senjanovic and F. Vissani, Phys. Lett. B **588** (2004) 196 [arXiv:hep-ph/0306242].
17. C. S. Aulakh and A. Girdhar, Int. J. Mod. Phys. A **20** (2005) 865 [arXiv:hep-ph/0204097].
18. T. Fukuyama, A. Ilakovac, T. Kikuchi, S. Meljanac and N. Okada, J. Math. Phys. **46**, 033505 (2005).
19. B. Bajc, A. Melfo, G. Senjanovic and F. Vissani, Phys. Rev. D **70** (2004) 035007 [arXiv:hep-ph/0402122].
20. C. S. Aulakh, hep-ph/0501025; C. S. Aulakh and A. Girdhar, Nucl. Phys. B **711** (2005) 275. C. S. Aulakh, arXiv:hep-ph/0506291.
21. R. N. Mohapatra, *Phys. Rev.* **D 34**, 3457 (1986). A. Font, L. E. Ibáñez and F. Quevedo, *Phys. Lett.* **B228**, 79 (1989). S. P. Martin, *Phys. Rev.* **D46**, 2769 (1992). C.S. Aulakh, K. Benakli, G. Senjanović, Phys. Rev. Lett. **79** (1997) 2188. C. S. Aulakh, A. Melfo and G. Senjanović, Phys. Rev. D **57**, 4174 (1998); [arXiv:hep-ph/9707256]. C. S. Aulakh, A. Melfo, A. Rašin and G. Senjanović, Phys. Rev. D **58**, 115007 (1998) [arXiv:hep-ph/9712551]. C. S. Aulakh, A. Melfo, A. Rašin and G. Senjanović, Phys. Lett. B **459** (1999) 557. [arXiv:hep-ph/9902409].
22. B. Bajc, A. Melfo, G. Senjanovic and F. Vissani, Phys. Lett. B **634** (2006) 272 [arXiv:hep-ph/0511352].
23. C. S. Aulakh and S. K. Garg, Nucl. Phys. B **757**, 47 (2006) [arXiv:hep-ph/0512224].
24. S. Bertolini, T. Schwetz and M. Malinsky, Phys. Rev. D **73**, 115012 (2006) [arXiv:hep-ph/0605006]. W. Grimus and H. Kuhbock, Phys. Lett. B **643** (2006) 182 [arXiv:hep-ph/0607197].
25. C. S. Aulakh, B. Bajc, A. Melfo, A. Rasin and G. Senjanovic, Nucl. Phys. B **597** (2001) 89 [arXiv:hep-ph/0004031].
26. A. Ramírez and A. Melfo, in preparation.

# Looking For Signals Of New Hadrons In Low Energy Physics

G. Sánchez-Colón* and Augusto García†

*Departamento de Física Aplicada.
Centro de Investigación y de Estudios Avanzados del IPN.
Unidad Merida.
A.P. 73, Cordemex.
Mérida, Yucatán, 97310. MEXICO.
†Departamento de Física.
Centro de Investigación y de Estudios Avanzados del IPN.
A.P. 14-740.
México, D.F., 07000. MEXICO.

**Abstract.** We review the possibility of new hadrons which mix with ordinary ones and which may contribute to non-leptonic decays. We consider the most favorable conditions to indirectly observe this mixing which allows us to set a lower bound on the masses of the new hadrons. The lower bound turns out to be quite large, of the order of $10^3 - 10^4$ TeV.

**Keywords:** mirror matter, mixing, symmetry breaking
**PACS:** 13.30.Eg, 11.30.Er, 11.30.Hv, 12.60.-i

## 1. INTRODUCTION

Before looking for new physics it is very important to have reliable guesses of where it can be found. This was the case of the searches of the W and Z bosons and of the top quark. Their discoveries would have been extremely difficult had it not been that reliable ideas about the values of their masses were available. Analogously, the search for new hadrons (or quarks) requires one be able to establish the values of their masses or some ranges or at least some lower bounds for them. There is a very special opportunity provided by the enormous gap that exists between strong or electromagnetic interactions and weak interactions. Forms of new physics may give signals in low energy physics through their mixings with ordinary physics. If such signals may fall in that gap, then severe constraints may be put on the couplings or masses that produce those mixings. In this paper we shall review the possibility of new hadrons which mix with ordinary ones and which may contribute to non-leptonic decays. The left-right symmetry of ordinary and new quarks may conduct to the existence of mirror hadrons. Their mixing with ordinary ones may produce non-perturbative signals of parity and flavor violations in non-leptonic decays. In Sec. 2 we discuss a general framework that covers both perturbative and non-perturbative parity and flavor violation. Before constructing a phenomenological approach it is necessary to take guidance from extensions of the Standard Model (SM). This we do in Sec. 3. Our approach is presented in Sec. 4. The main results of the comparison with experimental data on several groups of non-leptonic decays

CP917, *Particles and Fields*, edited by H. Castilla Valdez, J. C. D'Olivo, and M. A. Perez

are discussed in Sec. 5. In Sec. 6 these results are translated into a lower bound on the masses of mirror hadrons. The last section is reserved for conclusions.

## 2. GENERAL FRAMEWORK

We shall formulate a framework based on first principles that allows to cover both perturbative and non-perturbative violations of parity and strong flavors (strangeness, charm, etc.). It is a fact of nature that such quantum numbers are not conserved. This means that the total Hamiltonian $H = H_{\rm st} + H_{\rm em} + H_{\rm wk}$ does not commute with the corresponding operators. That is, $[H,P] \neq 0$, $[H,S] \neq 0$, etc. The physical states eigenstates of $H$ (mass eigenstates) cannot be simultaneously eigenstates of $P$, $S$, $C$, etc. This does not mean that we cannot use the eigenvalues of these operators. It only means that we cannot restrict mass eigenstates to single eigenvalues of such operators. This is not new, linear and angular momenta operators do not commute with one other, $[\vec{P},\vec{L}] \neq 0$. Eigenstates of $\vec{P}$ may use the eigenvalues of $L^2$ and $L_3$, but they must use all of them. This leads to the well known partial wave expansion of plane waves.

Analogously, physical states (mass eigenstates) will be linear superpositions of the form

$$|A_{\rm ph}\rangle = \sum_P \sum_F C(A,P,F)|A,P,F\rangle, \tag{1}$$

where the sums cover all eigenvalues of $P$ and of the several strong flavors, although we indicated only one ($F$).

The transitions amplitudes for decays will be matrix elements of $H$ between physical states $A_{\rm ph}$ and $B_{\rm ph}$ (which will be a multiparticle state), namely,

$$\begin{aligned}\langle B_{\rm ph}|H|A_{\rm ph}\rangle &= \sum_{P,F}\sum_{P',F'} C(A,P,F)C(B,P',F')\langle B,P',F'|H|A,P,F\rangle \\ &= \sum_{P,F} C(A,P,F)C(B,P,F)\langle B,P,F|H_{\rm st}+H_{\rm em}|A,P,F\rangle \\ &+ \sum_{P,F}\sum_{P',F'} C(A,P,F)C(B,P',F')\langle B,P',F'|H_{\rm wk}|A,P,F\rangle\end{aligned} \tag{2}$$

the double sums are reduced to single ones with flavor and parity conserving ($\Delta P = 0$, $\Delta F = 0$) amplitudes for $H_{\rm st}$ and $H_{\rm em}$ because these operators conserve those quantum numbers, while such sums remain in matrix elements of $H_{\rm wk}$.

All possibilities are covered in Eqs. (1) and (2). The possibilities which are particularly interesting for our purposes are in the first part of (2) when one coefficient is large (order 1 in some scale) and the other coefficient is very small, say

$$C(A,P,F) \approx 1, \quad C(B,P,F) << 1. \tag{3}$$

These cases, which are essentially non-perturbative, may compete and even exceed the perturbative first order contributions of the matrix elements of $H_{\rm wk}$ of the second part of (2).

It should be clear from the above arguments that one has two paths open to introduce small parity and flavor non-conservation, a non-perturbative one via $H_{\rm st}$ and $H_{\rm em}$ with (3) and a perturbative one via $H_{\rm wk}$.

## 3. GUIDANCE FROM EXTENSIONS OF THE STANDARD MODEL

The SM with its specific assignment of parity and flavors to the ordinary quarks excludes the sums of (2) and, accordingly, it excludes the second path mentioned at the end of Sec. 2. Only the path via perturbations with $H_{\rm wk}$ around $H_{\rm st}+H_{\rm em}$ are allowed by it. It is therefore necessary to consider new hadrons (quarks), i.e., extensions of the SM to be able to implement the second path.

Our purpose is to develop a phenomenological approach to implement this latter path. To do this it is convenient to get guidance from models which extend the SM. One must keep in mind that flavor and parity assignments are a matter of convention. It is therefore essential that whatever extension one wishes to consider respects the conventions of the SM. Otherwise, there would arise severe conflicts with the well established success of the SM.

An extension of the SM which is useful for our purposes is one that doubles the electroweak sector.[1] It uses the gauge group $SU(3)_{\rm c}\otimes[SU(2)\otimes U(Y)\otimes \hat{SU}(2)\otimes\hat{U}(Y)]$ and doubles the quark content. Denoting the quantum numbers in parenthesis as $(T,Y,\hat{T},\hat{Y})$, the ordinary quarks will be $q_{\rm L}(1/2,Y_{\rm L},0,0)$ and new quarks will be $\hat{q}_{\rm R}(0,0,1/2,\hat{Y}_{\rm R})$ and $\hat{q}_{\rm L}(0,0,0,\hat{Y}_{\rm L})$. The up and down character of quarks will be introduced replacing $q$, $\hat{q}$ by $u$, $\hat{u}$, $d$, $\hat{d}$, $c$, $\hat{c}$, etc. Notice that new quarks have the L, R labels reversed with respect to ordinary ones. Masses will be generated with two higgs doublets $\varphi(1/2,Y_\varphi,0,0)$ and $\hat{\varphi}(0,0,1/2,\hat{Y}_{\hat{\varphi}})$, one bidoublet $\Phi(1/2,Y_\Phi,1/2,\hat{Y}_{\hat{\Phi}})$ and several bisinglets of the type $\psi_u(0,Y_u,0,-Y_u)$, $\psi_d(0,Y_d,0,-Y_d)$.

Since we have increased the number of quarks the Cabibbo-Kobayashi-Maskawa (CKM) matrix must be extended. If one considers only the spontaneous breaking (SB) of $\varphi$ and $\hat{\varphi}$ one will get a mass matrix with two blocks in the diagonal and the extended CKM rotations will diagonalize such blocks. When the SB of $\Phi$ and $\psi_u$, $\psi_d$, etc. is allowed then non-diagonal blocks will appear in the mass matrix which connect ordinary and new quarks, a last diagonalization rotation will be required and the physical quarks (mass eigenstates) will become admixtures of ordinary and new quarks. For example, physical $s$ and $d$ quarks will become linear combinations of the type $d_{\rm ph}=d_{0\rm s}+\sigma s_{0\rm s}+\delta s_{0\rm p}+\cdots$, $s_{\rm ph}=s_{0\rm s}-\sigma d_{0\rm s}+\delta' d_{0\rm p}+\cdots$, $\bar{d}_{\rm ph}=\bar{d}_{0\rm p}+\sigma\bar{s}_{0\rm p}-\delta\bar{s}_{0\rm s}+\cdots$, $\bar{s}_{\rm ph}=\bar{s}_{0\rm p}-\sigma\bar{d}_{0\rm p}-\delta'\bar{d}_{0\rm s}+\cdots$, and similar expressions for $\hat{d}_{\rm ph}$, etc. We shall skip the details of the $(4\times4)$ matrices involved because it runs in parallel with the discussions of Sec. 6. In these admixtures we have assumed that the connection between the two sets of quarks is small enough to make the angles of the rotations so small that only the linear contributions in mixings are relevant and are parametrized by $\sigma$, $\delta$, and $\delta'$. The dots stand for further mixings which are not relevant in strange hadron decays.

An important step is the parity assignment to quarks. Ordinary ones carry positive parity (s) and we assumed that new quarks carry negative parity (p). This is a convention that respects the convention of the SM and that extends the concept of left-right symmetry of the model we are discussing. We may refer to this as manifest mirror symmetry, if the strong and electromagnetic couplings of mirror hadrons are of equal magnitude as those of ordinary hadrons.

To get the parity and flavor admixtures at the hadron level one may use the above physical quarks in the non-interacting quark model wave functions of hadrons. This leads to a substitution rule which produces the physical hadrons mass eigenstates. The guidance provided by the previous discussions allows one to produce a phenomenological approach to non-perturbative violations of parity and flavor in low energy physics.

## 4. A PHENOMENOLOGICAL APPROACH TO MANIFEST MIRROR SYMMETRY

Due to our current inability to use Quantum Chromodynamics (QCD) to make the passage from the quark to the hadron level, it is necessary that we introduce a non-perturbative approach to make this passage. In this respect, the use of strong-interaction symmetries is quite useful.

Since in Sec. 3 QCD is common to ordinary and mirror quarks it is possible to assign the strong flavors of ordinary quarks also to the mirror ones. If one does this, then the substitution rule of the last section becomes a $U$-spin ladder operation of strong-flavor $SU(3)$. That is, at the hadron level the last rotation of Sec. 3 will produce physical hadrons states such as

$$p_{\rm ph} = p_{\rm s} - \sigma \Sigma^+_{\rm s} - \delta \Sigma^+_{\rm p} + \cdots, \qquad \Sigma^+_{\rm ph} = \Sigma^+_{\rm s} + \sigma p_{\rm s} - \delta' p_{\rm p} + \cdots,$$

$$\pi^0_{\rm ph} = \pi^0_{\rm p} - \sigma \frac{1}{\sqrt{2}}(K^0_{\rm p} + \bar{K}^0_{\rm p}) + \delta \frac{1}{\sqrt{2}}(K^0_{\rm s} - \bar{K}^0_{\rm s}) + \cdots, \tag{4}$$

where the small angles are given in terms of reduced matrix elements $F$, $\hat{F}_0$, $F_0$, etc. as $\sigma = a(-F/\sqrt{6})$, $\delta = c'(-\hat{F}_0/\sqrt{6})$, $\delta = c(-F_0/\sqrt{6})$. For details see Ref. [2].

For decays, the new contributions to transitions amplitudes are calculated using (2) in the case of the inequalities (3). For example, one has [3, 4, 5, 6] for $\Lambda_{\rm ph} \to p_{\rm ph} \pi^-_{\rm ph}$, with transition amplitude $\bar{u}_p(A_1 - B_1 \gamma_5) u_\Lambda$, that

$$A_1 = \delta' \sqrt{\frac{3}{2}} g^{\rm p,sp}_{n,p\pi^-} + \delta(g^{\rm s,ss}_{\Lambda,pK^-} - g^{\rm s,pp}_{\Lambda,\Sigma^+\pi^-}), \quad B_1 = \sigma(-\sqrt{\frac{3}{2}} g^{\rm s,sp}_{n,p\pi^-} + g^{\rm s,sp}_{\Lambda,pK^-} - g^{\rm s,sp}_{\Lambda,\Sigma^+\pi^-}). \tag{5}$$

An example of a weak radiative decay is $\Sigma^+_{\rm ph} \to p_{\rm ph}\gamma$. Its transition amplitude, $\bar{u}_p(C_1 + D_1\gamma_5) i\sigma^{\mu\nu} q_\nu u_{\Sigma^+}$, has

$$C_1 = \sigma(f^p_{2\rm ss} - f^{\Sigma^+}_{2\rm ss}), \qquad D_1 = -\delta' f^p_{2\rm sp} - \delta f^{\Sigma^+}_{2\rm ps}. \tag{6}$$

Another example is $\Omega^-_{\rm ph} \to \Xi^-_{\rm ph}\pi^0_{\rm ph}$. Its transition amplitude, $\bar{u}_{\Xi^-}(\mathcal{B}_1 + \gamma^5 \mathcal{C}_1) q^\mu u^{\Omega^-}_\mu$, has

$$\mathcal{B}_1 = -\sigma(\sqrt{3} g^{\rm s,sp}_{\Xi^{*-},\Xi^-\pi^0} + \frac{1}{\sqrt{2}} g^{\rm s,sp}_{\Omega^-,\Xi^-\bar{K}^0}), \quad \mathcal{C}_1 = \delta' \sqrt{3} g^{\rm p,sp}_{\Xi^{*-},\Xi^-\pi^0} - \delta \frac{1}{\sqrt{2}} g^{\rm s,ss}_{\Omega^-,\Xi^-\bar{K}^0}. \tag{7}$$

In Eqs. (5) and (7) ordinary strong interaction Yukawa couplings appear such as $g^{\rm s,sp}_{B,B'M}$ and there are new ones such as $g^{\rm p,sp}_{B,B'M}$, the s and p labels indicate when mirror hadrons are coupled to ordinary ones. The same occurs with ordinary electromagnetic transitions magnetic moments $f^B_{2\rm ss}$ and with new ones such as $f^B_{2\rm sp}$ in Eqs. (6).

At this point we introduce the assumption of manifest mirror symmetry. We shall assume that mirror hadrons couple to ordinary ones with the same strong and electromagnetic intensities of ordinary matter coupling. With respect to strong interaction this is reasonable if indeed $SU(3)_{\rm c}$ is common to both sectors and with respect to electromagnetic interactions this is reasonable if the doubled SM of the last section is spontaneously broken to a single $U(1)_{\rm em}$ which is also common to both sectors.

According to these assumptions, the magnitudes of the new Yukawa couplings will be the same as their corresponding ordinary ones, e.g., $|g^{\rm s,sp}_{B,B'M}| = |g^{\rm p,sp}_{B,B'M}|$. However their relative phases may be different. The same applies to electromagnetic transition magnetic moments. However, the strong isospin $SU(2)$ and strong $SU(3)$ approximate symmetries do require the same relative phases of couplings involving hadrons within the same irreducible representation.

The above construction provides a phenomenological framework to produce contributions to weak decays of hadrons that implements the second path discussed at the end of Sec. 2.

## 5. COMPARISON WITH EXPERIMENT

We are now in position to obtain the contributions of non-perturbative parity and flavor violations in low energy physics. As we have seen, new hadrons will produce such violations through their mixings with ordinary ones. For strange hadron decays the relevant parameters are the new CKM-like angles $\sigma$, $\delta$, and $\delta'$. All we can advance about their values is that they must be very small compared to the ordinary CKM angles, otherwise they must be treated as free parameters.

These new contributions will compete with the SM perturbative contributions of the $W^\pm$ boson and they may even exceed them. They may then be relevant in the enhancement phenomenon observed in non-leptonic decays of strange hadrons. This provides a most favorable possibility to detect new physics in low energy physics.

It is therefore imperative to see if indeed these non-perturbative effects contribute to such phenomenon, otherwise we should conclude that SM contributions screen the new physics effects. We will return to this in the discussion section.

We have studied systematically several groups of decays, namely, non-leptonic hyperon decays ($\Lambda \to p\pi^-$, $\Lambda \to n\pi^0$, $\Sigma^- \to n\pi^-$, $\Sigma^+ \to n\pi^+$, $\Sigma^+ \to p\pi^0$, $\Xi^- \to \Lambda\pi^-$, and $\Xi^0 \to \Lambda\pi^0$) [3], weak radiative decays ($\Sigma^+ \to p\gamma$, $\Xi^- \to \Sigma^-\gamma$, $\Lambda \to n\gamma$, $\Xi^0 \to \Lambda\gamma$) [4], $\Omega^-$ two body decays ($\Omega^- \to \Xi^-\pi^0$, $\Omega^- \to \Xi^0\pi^-$, $\Omega^- \to \Lambda K^-$) [5, 6], meson two body decays ($K^+ \to \pi^+\pi^0$, $K_S \to \pi^+\pi^-$, $K_S \to \pi^0\pi^0$) [7], and the rare mode decays $K_L \to \gamma\gamma$ [8] and $K_S \to \gamma\gamma$ [9]. We have used all available data, decay rates, asymmetry coefficients, measured Yukawa couplings, and hyperon magnetic moments. The detailed studies involve many numerical tables. There is no need to reproduce them here, the reader will find them in the references. For our purposes here a short discussion is sufficient.

The main features are (i) the predictions of the $\Delta I = 1/2$ rule are well reproduced (although, in our case this rule corresponds to the strong-isospin SU(2) symmetry limit, i.e., a $\Delta I = 0$ rule), (ii) the observed violations of this rule are obtained with small SU(2) breaking (of less than 1%), and (iii), as must be required, the new angles have the same numerical values in the different groups of decays. Here, it is these numerical values that constitute our main result. They are

$$\sigma = (4.9 \pm 2.0) \times 10^{-6}, \quad \delta = (0.22 \pm 0.09) \times 10^{-6}, \quad \delta' = (0.26 \pm 0.09) \times 10^{-6}. \tag{8}$$

Clearly, the stability of these values is a necessary condition for the non-perturbative contributions to be relevant. If this were not the case we would have a contradiction with the $\Delta I = 1/2$ rule. This would imply that the angles $\sigma$, $\delta$, and $\delta'$ are further suppressed.

## 6. WHERE TO LOOK FOR NEW HADRONS

The numerical values of Eqs. (8) contain very interesting information. They will tell us where to look for new hadrons, specifically, mirror hadrons. The values of the angles may be translated into values of masses of mirror hadrons. The lightest one will be the mirror proton $\hat{m}_p$.

To obtain the value of $\hat{m}_p$, let us work out in detail the CKM-like rotations in the case of $p_{\rm ph}$ and $\Sigma^+_{\rm ph}$. The mass matrix of this system is a $(4 \times 4)$ matrix and the extended CKM procedure leads to a rotation $R$ (at this hadron level) that can be split into the product of two $(4 \times 4)$ matrices, $R = R_1 R_0$.

The first rotation can be written in terms of $(2 \times 2)$ matrices as

$$R_0 = \begin{pmatrix} R_0^a & 0 \\ 0 & R_0^b \end{pmatrix} \tag{9}$$

such that $R_0^a$ diagonalizes the mass matrix of the ordinary flavor eigenstates $p_{\rm s}$ and $\Sigma^+_{\rm s}$ into

$$\begin{pmatrix} \bar{p}_{\rm s} & \bar{\Sigma}^+_{\rm s} \end{pmatrix} \begin{pmatrix} m^0_p & 0 \\ 0 & m^0_{\Sigma^+} \end{pmatrix} \begin{pmatrix} p_{\rm s} \\ \Sigma^+_{\rm s} \end{pmatrix} \tag{10}$$

and $R^b_0$ works analogously for the mirror parity and flavor eigenstates $p_{\rm p}$ and $\Sigma^+_{\rm p}$ giving them the masses $\hat{m}^0_p$ and $\hat{m}^0_{\Sigma^+}$, respectively. At the $(4 \times 4)$ level one still has $(2 \times 2)$ non-diagonal blocks and the action of $R$ is now centered in the action of $R_1$,

$$\begin{pmatrix} \bar{p}_{\rm s} & \bar{\Sigma}^+_{\rm s} & \bar{p}_{\rm p} & \bar{\Sigma}^+_{\rm p} \end{pmatrix} \begin{pmatrix} m^0_p & 0 & \Delta_{11} & \Delta_{12} \\ 0 & m^0_{\Sigma^+} & \Delta_{21} & \Delta_{22} \\ \Delta_{11} & \Delta_{21} & \hat{m}^0_p & 0 \\ \Delta_{12} & \Delta_{22} & 0 & \hat{m}^0_{\Sigma^+} \end{pmatrix} \begin{pmatrix} p_{\rm s} \\ \Sigma^+_{\rm s} \\ p_{\rm p} \\ \Sigma^+_{\rm p} \end{pmatrix} \tag{11}$$

To implement this last rotation one must first recognize that the masses of these mirror hadrons must obey the inequalities $\hat{m}^0_p, \hat{m}^0_{\Sigma^+} \gg \Delta_{ij}, m^0_p, m^0_{\Sigma^+}$, otherwise someone would have already found them. These conditions allow us to work with $R_1$ to first order in its angles $\epsilon_{ij}$, namely, $(R_1)_{ij} \simeq \delta_{ij} + \epsilon_{ij}$. The final result is

$$\begin{pmatrix} \bar{p}_{\rm ph} & \bar{\Sigma}^+_{\rm ph} & \bar{\hat{p}}_{\rm ph} & \bar{\hat{\Sigma}}^+_{\rm ph} \end{pmatrix} \begin{pmatrix} m_p & 0 & 0 & 0 \\ 0 & m_{\Sigma^+} & 0 & 0 \\ 0 & 0 & \hat{m}_p & 0 \\ 0 & 0 & 0 & \hat{m}_{\Sigma^+} \end{pmatrix} \begin{pmatrix} p_{\rm ph} \\ \Sigma^+_{\rm ph} \\ \hat{p}_{\rm ph} \\ \hat{\Sigma}^+_{\rm ph} \end{pmatrix} \tag{12}$$

with $p_{\rm ph} = p_{\rm s} + \epsilon_{12}\Sigma^+_{\rm s} + \epsilon_{13}p_{\rm p} + \epsilon_{14}\Sigma^+_{\rm p}$ and $\Sigma^+_{\rm ph} = \Sigma^+_{\rm s} - \epsilon_{12}p_{\rm s} + \epsilon_{23}p_{\rm p} + \epsilon_{24}\Sigma^+_{\rm p}$. The masses of (11) and (12) and the angles $\epsilon_{ij}$ are connected among themselves. Since we are working to first order, one may assume the equalities $m_p \simeq m^0_p$, $m_{\Sigma^+} \simeq m^0_{\Sigma^+}$, $\hat{m}_p \simeq \hat{m}^0_p$, $\hat{m}_{\Sigma^+} \simeq \hat{m}^0_{\Sigma^+}$. Under these conditions there are only eight connecting equations for the absolute values of $\epsilon_{ij}$. We shall only reproduce three of them, the others are found in [10]. One has

$$|\epsilon_{12}| = \left| \frac{\Delta_{11}\epsilon_{23} + \Delta_{12}\epsilon_{24}}{m_{\Sigma^+} - m_p} \right| = \left| \frac{\Delta_{22}\epsilon_{14} + \Delta_{21}\epsilon_{13}}{m_{\Sigma^+} - m_p} \right|, \quad |\epsilon_{14}| = \left| \frac{\Delta_{12}}{\hat{m}_{\Sigma^+}} \right|, \quad |\epsilon_{23}| = \left| \frac{\Delta_{21}}{\hat{m}_p} \right|. \tag{13}$$

At this point we may return to Sec. 4 and identify $\epsilon_{12} = \sigma$, $\epsilon_{14} = \delta$, and $\epsilon_{23} = \delta'$. However, $\epsilon_{13}$ and $\epsilon_{24}$ are not yet measured and probably will not be for a long time and we ignore $\Delta_{ab}$ with $(ab) = (12) = (21)$, so we cannot determine $\hat{m}_p$. Nevertheless, a careful analysis of all the connecting equations shows that a lower bound can be established. Skipping details [10], one has for a lower bound that $|\epsilon_{13}| \simeq |\epsilon_{12}|$, $|\epsilon_{24}| \simeq |\epsilon_{12}|$. Then one gets

$$\frac{|\Delta_{ab}|}{m_{\Sigma^+} - m_p} \simeq 1 \tag{14}$$

which determines $|\Delta_{12}| \simeq 0.25\text{GeV}$, using $m_{\Sigma^+} - m_p \simeq 0.25\text{GeV}$. From Eqs. (13), one then gets

$$\hat{m}_p \geq \mathcal{O}(10^6\,\text{GeV}). \tag{15}$$

This is the result we were looking for.

## 7. DISCUSSION AND CONCLUSIONS

Our main result is the lower bound (15). It is quite a large lower bound, in TeV's one has $\hat{m}_p \geq 10^3\,\text{TeV}$. Important observations need to be made at this point. In our search for signals of new physics or mirror hadrons, we have assumed that such signal saturate the data. That is, we have assumed that SM model contributions to the enhancement phenomenon are small and we have ignored them.

This can only be approximately true, since we know that SM contributions will be there. It is well known that the description of that phenomenon is still an important unsolved problem within the SM and it is very difficult to assess how much of this phenomenon is covered by those contributions. Many calculations put the SM saturation at the level of $30-50\%$ and it cannot be excluded that a better mastering of QCD in the future may allow a 100% saturation.

Taking these considerations into account we may qualify our lower bound: it may be even more stringent. The SM may screen the new contributions to the extent that the only room left for the latter fall into the experimental errors bars. These are of the order of a few percent ($\sim 2\%$). Since the observables are quadratic in $\sigma$, $\delta$, and $\delta'$, this would bring our lower bound to $10^4\,\text{TeV}$ and thus (15) becomes

$$\hat{m}_p \geq \mathcal{O}(10^3 - 10^4\,\text{TeV}). \tag{16}$$

This is then our final conclusion, manifest mirror hadrons are very far from our ordinary matter world. This is not surprising if one looks back at the smallness of the CKM-like angles of Eqs. (8). It is clearly a reflection of how restricting the enormous gap between strong or e.m. interactions and weak interactions is.

## ACKNOWLEDGMENTS

We would like to thank CONACyT (México) for partial support.

## REFERENCES

1. S. M. Barr, D. Chang, and G. Senjanović, Phys. Rev. Lett. **67** (1991) 2765.
2. A. García, R. Huerta, and G. Sánchez-Colón, J. Phys. G: Nucl. Part. Phys. **24** (1998) 1207.
3. A. García, R. Huerta, and G. Sánchez-Colón, J. Phys. G: Nucl. Part. Phys. **25** (1999) 45.
4. A. García, R. Huerta, and G. Sánchez-Colón, J. Phys. G: Nucl. Part. Phys. **25** (1999) L1.
5. G. Sánchez-Colón, R. Huerta, and A. García, J. Phys. G: Nucl. Part. Phys. **29** (2003) L1.
6. G. Sánchez-Colón, R. Huerta, and A. García, Int. J. Mod. Phys. Lett. A **19** (2004) 775.
7. G. Sánchez-Colón and A. García, Mod. Phys. Lett. A **15** (2000) 1749.
8. G. Sánchez-Colón and A. García, Mod. Phys. Lett. A **40** (2004) 3019.
9. G. Sánchez-Colón and A. García, Int. J. Mod. Phys. Lett. A **21** (2006) 4197.
10. A. García, R. Huerta, and G. Sánchez-Colón, Phys. Lett. B **498** (2001) 251.

# Finite Unified Theories and Low Energy Phenomenology

Myriam Mondragón* and George Zoupanos†

*Instituto de Física,
Universidad Nacional Autónoma de México (IF-UNAM),
Apdo. Postal 20-364, 01000 México D.F., México.
†Physics Department, National Technical University,
GR-157 80 Zografou, Athens, Greece.

**Abstract.** Finite Unified Theories (FUTs) are N=1 supersymmetric Grand Unified Theories, which can be made all-loop finite, both in the gauge and Yukawa couplings and in the soft supersymmetry breaking sector. This remarkable property, based on the reduction of couplings at the quantum level, provides a drastic reduction in the number of free parameters, which in turn leads to an accurate prediction of the top quark mass, and predictions for the Higgs boson mass and the supersymmetric spectrum. Here we examine the predictions of three FUTs taking into account a number of theoretical and experimental constraints.

**Keywords:** Grand Unified Theories, supersymmetry, soft supersymmetry breaking terms, finiteness
**PACS:** 11.30.Pb,12.10.Ktl,11.10.Hi,14.80.Ly,14.80.Cp

## INTRODUCTION

Finite Unified Theories are $N = 1$ supersymmetric Grand Unified Theories (GUTs) which can be made finite even to all-loop orders, including the soft supersymmetry breaking sector. The method to construct GUTs with reduced independent parameters [1, 2] consists of searching for renormalization group invariant (RGI) relations holding below the Planck scale, which in turn are preserved down to the GUT scale. Of particular interest is the possibility to find RGI relations among couplings that guarantee finitenes to all-orders in perturbation theory [3, 4]. In order to achieve the latter it is enough to study the uniqueness of the solutions to the one-loop finiteness conditions [3, 4]. The constructed *finite unified* $N = 1$ supersymmetric SU(5) GUTs using the above tools, predicted correctly from the dimensionless sector (Gauge-Yukawa unification), among others, the top quark mass [5]. The search for RGI relations and finiteness has been extended to the soft supersymmetry breaking sector (SSB) of these theories [6, 7], which involves parameters of dimension one and two. Eventually, the full theories can be made all-loop finite and their predictive power is extended to the Higgs sector and the s-spectrum.

CP917, *Particles and Fields*, edited by H. Castilla Valdez, J. C. D'Olivo, and M. A. Perez

# REDUCTION OF COUPLINGS AND FINITENESS IN $N=1$ SUSY GAUGE THEORIES

We will review here some of the main points and ideas concerning the *reduction of couplings* and *finiteness* in $N=1$ supersymmetric theories. A RGI relation among couplings $g_i$, $\Phi(g_1,\cdots,g_N) = 0$, has to satisfy the partial differential equation $\mu\ d\Phi/d\mu = \sum_{i=1}^{N}\beta_i\partial\Phi/\partial g_i = 0$, where $\beta_i$ is the $\beta$-function of $g_i$. There exist $(N-1)$ independent $\Phi$'s, and finding the complete set of these solutions is equivalent to solve the so-called reduction equations (REs) [2], $\beta_g\,(dg_i/dg)=\beta_i$ , $i=1,\cdots,N$, where $g$ and $\beta_g$ are the primary coupling and its $\beta$-function. Using all the $(N-1)\,\Phi$'s to impose RGI relations, one can in principle express all the couplings in terms of a single coupling $g$. The complete reduction, which formally preserves perturbative renormalizability, can be achieved by demanding a power series solution, whose uniqueness can be investigated at the one-loop level.

Finiteness can be understood by considering a chiral, anomaly free, $N=1$ globally supersymmetric gauge theory based on a group G with gauge coupling constant $g$. The superpotential of the theory is given by

$$W=\frac{1}{2}m^{ij}\Phi_i\Phi_j+\frac{1}{6}C^{ijk}\Phi_i\Phi_j\Phi_k\ , \tag{1}$$

where $m^{ij}$ (the mass terms) and $C^{ijk}$ (the Yukawa couplings) are gauge invariant tensors and the matter field $\Phi_i$ transforms according to the irreducible representation $R_i$ of the gauge group $G$.

The one-loop $\beta$-function of the gauge coupling $g$ is given by

$$\beta_g^{(1)} = \frac{dg}{dt}=\frac{g^3}{16\pi^2}[\sum_i l(R_i)-3C_2(G)]\ , \tag{2}$$

where $l(R_i)$ is the Dynkin index of $R_i$ and $C_2(G)$ is the quadratic Casimir of the adjoint representation of the gauge group $G$. The $\beta$-functions of $C^{ijk}$, by virtue of the non-renormalization theorem, are related to the anomalous dimension matrix $\gamma_i^j$ of the matter fields $\Phi_i$ as:

$$\beta_C^{ijk}=\frac{d}{dt}C^{ijk}=C^{ijp}\sum_{n=1}\frac{1}{(16\pi^2)^n}\gamma_p^{k(n)}+(k\leftrightarrow i)+(k\leftrightarrow j)\ . \tag{3}$$

At one-loop level $\gamma_i^j$ is given by

$$\gamma_i^{j(1)}=\frac{1}{2}C_{ipq}C^{jpq}-2g^2C_2(R_i)\delta_i^j\ , \tag{4}$$

where $C_2(R_i)$ is the quadratic Casimir of the representation $R_i$, and $C^{ijk}=C^*_{ijk}$.

All the one-loop $\beta$-functions of the theory vanish if the $\beta$-function of the gauge coupling $\beta_g^{(1)}$, and the anomalous dimensions $\gamma_i^{j(1)}$, vanish, i.e.

$$\sum_i \ell(R_i)=3C_2(G)\,,\ \frac{1}{2}C_{ipq}C^{jpq}=2\delta_i^j g^2C_2(R_i)\ , \tag{5}$$

where $l(R_i)$ is the Dynkin index of $R_i$, and $C_2(G)$ is the quadratic Casimir invariant of the adjoint representation of $G$.

A very interesting result is that the conditions (5) are necessary and sufficient for finiteness at the two-loop level [8, 9].

The one- and two-loop finiteness conditions (5) restrict considerably the possible choices of the irreducible representations $R_i$ for a given group $G$ as well as the Yukawa couplings in the superpotential (1). Note in particular that the finiteness conditions cannot be applied to the supersymmetric standard model (SSM), since the presence of a $U(1)$ gauge group is incompatible with the condition (5), due to $C_2[U(1)] = 0$. This leads to the expectation that finiteness should be attained at the grand unified level only, the SSM being just the corresponding low-energy, effective theory.

The finiteness conditions impose relations between gauge and Yukawa couplings. Therefore, we have to guarantee that such relations leading to a reduction of the couplings hold at any renormalization point. The necessary, but also sufficient, condition for this to happen is to require that such relations are solutions to the reduction equations (REs) to all orders. The all-loop order finiteness theorem of ref.[3] is based on: (a) the structure of the supercurrent in $N = 1$ SYM and on (b) the non-renormalization properties of $N = 1$ chiral anomalies [3]. Alternatively, similar results can be obtained [4, 10] using an analysis of the all-loop NSVZ gauge beta-function [11].

The above described method of reducing the dimensionless couplings has been extended [6, 7] to the soft supersymmetry breaking (SSB) dimensionful parameters of $N = 1$ supersymmetric theories, based on the work of refs.[12, 13, 14, 15]. In addition, it was found that RGI SSB scalar masses in Gauge-Yukawa unified models satisfy a universal sum rule at one-loop [16]. This result was generalized to two-loops for finite theories [17], and then to all-loops for general Gauge-Yukawa and Finite Unified Theories [18].

In order to obtain a feeling of some of the above results, consider the superpotential given by (1) along with the Lagrangian for SSB terms

$$-\mathscr{L}_{\rm SB} = \frac{1}{6} h^{ijk}\,\phi_i\phi_j\phi_k + \frac{1}{2} b^{ij}\,\phi_i\phi_j + \frac{1}{2}(m^2)^j_i\,\phi^{*i}\phi_j + \frac{1}{2} M\lambda\lambda + \text{H.c.}, \tag{6}$$

where the $\phi_i$ are the scalar parts of the chiral superfields $\Phi_i$ , $\lambda$ are the gauginos and $M$ their unified mass. Since only finite theories are considered here, it is assumed that the gauge group is a simple group and the one-loop $\beta$-function of the gauge coupling $g$ vanishes. It is also assumed that the reduction equations admit power series solutions of the form $C^{ijk} = g \sum_{n=0} \rho^{ijk}_{(n)} g^{2n}$ . According to the finiteness theorem [3], the theory is then finite to all-orders in perturbation theory, if, among others, the one-loop anomalous dimensions $\gamma_i^{j(1)}$ vanish. The one- and two-loop finiteness for $h^{ijk}$ can be achieved by [9]

$$h^{ijk} = -MC^{ijk} + \cdots = -M\rho^{ijk}_{(0)}\, g + O(g^5)\,. \tag{7}$$

An additional constraint in the SSB sector up to two-loops [17], concerns the soft scalar masses as follows

$$\frac{(\,m_i^2 + m_j^2 + m_k^2\,)}{MM^{\dagger}} = 1 + \frac{g^2}{16\pi^2}\Delta^{(2)} + O(g^4) \tag{8}$$

for i, j, k with $\rho_{(0)}^{ijk} \neq 0$, where $\Delta^{(2)}$ is the two-loop correction

$$\Delta^{(2)} = -2\sum_l [(m_l^2/MM^\dagger) - (1/3)]\, T(R_l), \tag{9}$$

which vanishes for the universal choice [9], i.e. when all the soft scalar masses are the same at the unification point.

If we know higher-loop $\beta$-functions explicitly, we can follow the same procedure and find higher-loop RGI relations among SSB terms. However, the $\beta$-functions of the soft scalar masses are explicitly known only up to two loops. In order to obtain higher-loop results, we need something else instead of knowledge of explicit $\beta$-functions, e.g. some relations among $\beta$-functions. Due to space limitations we refer the interested reader to ref. [18].

## FINITE UNIFIED THEORIES

In this section we examine two concrete $SU(5)$ finite models, where the reduction of couplings in the dimensionless and dimensionful sector has been achieved. For other interesting Finite Unified Theories based on cross group structure see ref.[19]. A predictive Gauge-Yukawa unified $SU(5)$ model which is finite to all orders, in addition to the requirements mentioned already, should also have the following properties:

1. One-loop anomalous dimensions are diagonal, i.e., $\gamma_i^{(1)\,j} \propto \delta_i^j$.
2. Three fermion generations, in the irreducible representations $\overline{\mathbf{5}}_i, \mathbf{10}_i$ $(i = 1,2,3)$, which obviously should not couple to the adjoint **24**.
3. The two Higgs doublets of the MSSM should mostly be made out of a pair of Higgs quintet and anti-quintet, which couple to the third generation.

In the following we discuss two versions of the all-order finite model. The model of ref. [5], which will be labeled **A**, and a slight variation of this model (labeled **B**), which can also be obtained from the class of the models suggested by Kazakov *et al.* [12] with a modification to suppress non-diagonal anomalous dimensions[1].

The superpotential which describes the two models takes the form [5, 17]

$$\begin{aligned} W = \sum_{i=1}^{3} \Big[ \frac{1}{2} g_i^u \, \mathbf{10}_i \mathbf{10}_i H_i + g_i^d \, \mathbf{10}_i \overline{\mathbf{5}}_i \overline{H}_i \Big] + g_{23}^u \, \mathbf{10}_2 \mathbf{10}_3 H_4 + g_{23}^d \, \mathbf{10}_2 \overline{\mathbf{5}}_3 \\ \overline{H}_4 + g_{32}^d \, \mathbf{10}_3 \overline{\mathbf{5}}_2 \overline{H}_4 + \sum_{a=1}^{4} g_a^f H_a \mathbf{24} \overline{H}_a + \frac{g^\lambda}{3} (\mathbf{24})^3 , \end{aligned} \tag{10}$$

where $H_a$ and $\overline{H}_a$ $(a = 1,\ldots,4)$ stand for the Higgs quintets and anti-quintets.

[1] An extension to three families, and the generation of quark mixing angles and masses in Finite Unified Theories has been addressed in [20], where several realistic examples are given. These extensions are not considered here.

The non-degenerate and isolated solutions to $\gamma_i^{(1)} = 0$ for the models $\{\mathbf{A}\ ,\ \mathbf{B}\}$ are:

$$\begin{aligned}
(g_1^u)^2 &= \{\frac{8}{5},\frac{8}{5}\}g^2\ ,\ (g_1^d)^2 = \{\frac{6}{5},\frac{6}{5}\}g^2\ ,\\
(g_2^u)^2 &= (g_3^u)^2 = \{\frac{8}{5},\frac{4}{5}\}g^2\ ,\\
(g_2^d)^2 &= (g_3^d)^2 = \{\frac{6}{5},\frac{3}{5}\}g^2\ ,\\
(g_{23}^u)^2 &= \{0,\frac{4}{5}\}g^2\ ,\ (g_{23}^d)^2 = (g_{32}^d)^2 = \{0,\frac{3}{5}\}g^2\ ,\\
(g^\lambda)^2 &= \frac{15}{7}g^2\ ,\ (g_2^f)^2 = (g_3^f)^2 = \{0,\frac{1}{2}\}g^2\ ,\\
(g_1^f)^2 &= 0\ ,\ (g_4^f)^2 = \{1,0\}g^2\ .
\end{aligned} \tag{11}$$

According to the theorem of ref. [3] these models are finite to all orders. After the reduction of couplings the symmetry of $W$ is enhanced [5, 17].

The main difference of the models **A** and **B** is that three pairs of Higgs quintets and anti-quintets couple to the **24** for **B** so that it is not necessary to mix them with $H_4$ and $\overline{H}_4$ in order to achieve the triplet-doublet splitting after the symmetry breaking of $SU(5)$.

In the dimensionful sector, the sum rule gives us the following boundary conditions at the GUT scale [17]:

$$m_{H_u}^2 + 2m_{\mathbf{10}}^2 = m_{H_d}^2 + m_{\overline{\mathbf{5}}}^2 + m_{\mathbf{10}}^2 = M^2 \ \text{ for } \mathbf{A}\ ; \tag{12}$$

$$\begin{aligned}
m_{H_u}^2 + 2m_{\mathbf{10}}^2 &= M^2\ ,\ m_{H_d}^2 - 2m_{\mathbf{10}}^2 = -\frac{M^2}{3}\ ,\\
m_{\overline{\mathbf{5}}}^2 + 3m_{\mathbf{10}}^2 &= \frac{4M^2}{3} \ \text{ for } \mathbf{B},
\end{aligned} \tag{13}$$

where we use as free parameters $m_{\overline{\mathbf{5}}} \equiv m_{\overline{\mathbf{5}}_3}$ and $m_{\mathbf{10}} \equiv m_{\mathbf{10}_3}$ for the model **A**, and $m_{\mathbf{10}} \equiv m_{\mathbf{10}_3}$ for **B**, in addition to $M$.

## PREDICTIONS OF LOW ENERGY PARAMETERS

Since the gauge symmetry is spontaneously broken below $M_{\mathrm{GUT}}$, the finiteness conditions do not restrict the renormalization properties at low energies, and all it remains are boundary conditions on the gauge and Yukawa couplings (11), the $h = -MC$ relation, and the soft scalar-mass sum rule (8) at $M_{\mathrm{GUT}}$, as applied in the two models. Thus we examine the evolution of these parameters according to their RGEs up to two-loops for dimensionless parameters and at one-loop for dimensionful ones with the relevant boundary conditions. Below $M_{\mathrm{GUT}}$ their evolution is assumed to be governed by the MSSM. We further assume a unique supersymmetry breaking scale $M_s$ (which we define as the geometric mean of the stop masses) and therefore below that scale the effective theory is just the SM.

The predictions for the top quark mass $M_t$ are $\sim 183$ and $\sim 173$ GeV in models **A** and **B** respectively. Comparing these predictions with the most recent experimental value $M_t^{exp} = (172.7 \pm 2.9)$ GeV [21], and recalling that the theoretical values for $M_t$ may suffer from a correction of $\sim 4\%$ [22], we see that clearly model **B** is preferred. In addition the value of $\tan\beta$ is found to be $\tan\beta \sim 54$ and $\sim 48$ for models **A** and **B** respectively.

In the SSB sector, besides the constraints imposed by finiteness there are further restrictions imposed by phenomenology. In the case where all the soft scalar masses are universal at the unfication scale, there is no region of $M$ below $O(few\ TeV)$ in which $m_{\tilde{\tau}} > m_{\chi^0}$ is satisfied (where $m_{\tilde{\tau}}$ is the lightest $\tilde{\tau}$ mass, and $m_{\chi^0}$ the lightest neutralino mass, which is the lightest supersymmetric particle). But once the universality condition is relaxed this problem can be solved naturally (thanks to the sum rule). More specifically, using the sum rule (8) and imposing the conditions a) successful radiative electroweak symmetry breaking, b) $m_{\tilde{\tau}}^2 > 0$ and c) $m_{\tilde{\tau}} > m_{\chi^0}$, a comfortable parameter space for both models (although model **B** requires large $M \sim 1$ TeV) is found.

As additional constraints, we consider the following observables: the anomalous magnetic moment of the muon, $(g-2)_\mu$, rare $b$ decays BR($b \to s\gamma$) and BR($B_s \to \mu^+\mu^-$), as well as the density of cold dark matter in the Universe, assuming it consists mainly of neutralinos.

For the branching ratio $BR(b \to s\gamma)$ [23], we take the present experimental value estimated by the Heavy Flavour Averaging Group (HFAG) is [24]

$$\mathrm{BR}(b \to s\gamma) = (3.54^{+0.30}_{-0.28}) \times 10^{-4}, \tag{14}$$

where the error includes an uncertainty due to the decay spectrum, as well as the statistical error. In the case of the anomalous magnetic moment of the muon $a_\mu \equiv (g-2)_\mu$, we compare our different models with

$$a_\mu^{\rm exp} - a_\mu^{\rm theo} = (25.2 \pm 9.2) \times 10^{-10}. \tag{15}$$

For the branching ratio BR($B_s \to \mu^+\mu^-$), the SM prediction is $(3.4 \pm 0.5) \times 10^{-9}$ [25], and the present experimental upper limit from the Fermilab Tevatron collider is $3.4 \times 10^{-7}$ at the 95% C.L. [26], providing the possibility for the MSSM to dominate the SM contribution.

The lightest supersymmetric particle (LSP) is an excellent candidate for cold dark matter (CDM) [27], with a density that falls naturally within the range

$$0.094 < \Omega_{\rm CDM} h^2 < 0.129 \tag{16}$$

favoured by a joint analysis of WMAP and other astrophysical and cosmological data [28].

In fig.1 we show the **FUTA** and **FUTB** results concerning $M_h$, for different values of $M$, for the cases where $\mu < 0$ and $\mu > 0$ and the LSP is a neutralino $\chi^0$.

The results for $\mu > 0$ and $\mu < 0$ are different for **FUTA**: with $\mu < 0$ the spectrum starts with an LSP around 750 GeV, whereas for $\mu > 0$ the spectrum starts around 500 GeV. The main difference, though, is in the value of the bottom mass $m_{bot}$, where we have included the corrections coming from bottom squark-gluino loops and top squark-chargino loops [29]. We give the predictions for the running bottom quark mass evaluated at $M_Z$, $m_{bot}(M_Z) = 2.825 \pm 0.1$ [30], to avoid the large QCD uncertainties

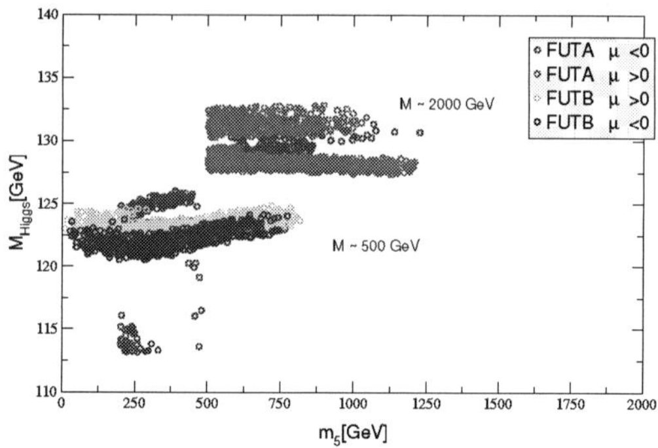


**FIGURE 1.** The lightest Higgs mass, $m_H$, as function of $m_5$ for different values of $M$ for both models.

at the pole mass. The value of $m_{bot}$ depends strongly on the sign of $\mu$, due to the above mentioned susy radiative corrections. Both for models **A** and **B** the values for $\mu > 0$ are above the central experimental value, with $m_{bot}(M_Z) \sim 4.0 - 5.0$ GeV . For $\mu < 0$ on the other hand, model **B** has overlap with the experimental allowed values, $m_{bot}(M_Z) \sim 2.5 - 2.8$ whereas for model **A**, $m_{bot}(M_Z) \sim 2.0 - 2.6$, there is only a small region of allowed parameter space at two sigma level, and only for large values of $M$.

The Higgs mass prediction of the two models is, although the details of each of the models differ, in the following range

$$m_h = \quad \sim 112 - 132 \; GeV, \tag{17}$$

where the uncertainty comes from variations of the gaugino mass $M$ and the soft scalar masses, and from finite (i.e. not logarithmically divergent) corrections in changing renormalization scheme. The one-loop radiative corrections have been included [31] for $m_h$, but not for the rest of the spectrum. In making the analysis, the value of $M$ was varied from $200 - 2000$ GeV. We have also included a small variation, due to threshold corrections at the GUT scale, of up to 5% of the FUT boundary conditions. This variation gives small effects in the results at low energies. The requirement $m_h > 114.4$ GeV [32] (neglecting the theoretical uncertainties) excludes the possibility of $M = 200$ GeV for FUTA, as seen also from the graph.

A more detailed numerical analysis, where the results of our program and of the known programs FeynHiggs [33] and Suspect [34] are combined, and the theoretical uncertainties for the models are estimated, is currently under progress [35].

## FUTS BASED ON PRODUCT GAUGE GROUPS

Let us now examine the possibility of constructing realistic FUTs based on product gauge groups [19]. Consider the gauge group $\mathrm{SU}(N)_1 \times \mathrm{SU}(N)_2 \times \cdots \times \mathrm{SU}(N)_k$ with $n_f$ copies of the supersymmetric multiplet $(N, N^*, 1, \ldots, 1) + (1, N, N^*, \ldots, 1) + \cdots + (N^*, 1, 1, \ldots, N)$. The one-loop $\beta$-function coefficient in the renormalization-group equation of each SU$(N)$ gauge coupling is simply given by

$$b = \left(-\frac{11}{3}+\frac{2}{3}\right)N + n_f\left(\frac{2}{3}+\frac{1}{3}\right)\left(\frac{1}{2}\right)2N = -3N + n_f N. \quad (18)$$

This means that $n_f = 3$ is a solution of the equation $b = 0$, independently of the values of $N$ and $k$. Since $b = 0$ is a necessary condition for a finite field theory, the existence of three families of quarks and leptons is natural in such models. This is true of course only if the matter content is exactly as given above.

The model of this type with best phenomenology is the $SU(3)^3$ model discussed in ref.[19], where the details of the model are given. Let us just point out that the finiteness conditions impose in this model the following boundary conditions for the adimensional couplings

$$f^2 = f_{111}^2 = f_{222}^2 = f_{333}^2 = \frac{16}{9}g^2. \quad (19)$$

where the $f$'s are the Yukawa couplings associated to each generation. For the soft susy breaking sector we have the following conditions

$$m_{H_u}^2 + m_{\tilde{t}^c}^2 + m_{\tilde{q}}^2 = M^2 \quad (20)$$

$$m_{H_d}^2 + m_{\tilde{b}^c}^2 + m_{\tilde{q}}^2 = M^2. \quad (21)$$

A numerical analysis putting as low energy constraints successful radiative electroweak symmetry breaking, $m^2_{\tilde{\tau},\tilde{b},\tilde{t}} > 0$, and correct branching ratio BR$(b \to s\gamma)$ shows that it is possible to find regions of parameter space which comply with all the above requirements. In this case with $\mu < 0$, we obtain $m_t \sim 183$ GeV, $\tan\beta \sim 47 - 54$, and the Higgs mass is $\sim 130 - 132$ GeV. The sparticle spectrum starts with an LSP at $m_{\chi^0} \sim 450$GeV, for a boundary condition of $M \sim 1800$GeV. In the case that $\mu > 0$ we do not find solutions which satisfy all the above requirements.

This work is supported by the EPEAEK programme "Pythagoras" of the Greek Ministry of Education and co-founded by the European Union (75%) and the Hellenic State (25%). We acknowledge also support by CONACyT México, 42026-F, and PAPIIT-UNAM IN115207 grants.

## REFERENCES

1. Kubo, J., Mondragón, M. and Zoupanos, G. (1994) *Nucl. Phys.* **B424** 291.
2. Zimmermann, W., (1985) *Com. Math. Phys.* **97** 211; Oehme, R. and Zimmermann, W. (1985) *Com. Math. Phys.* **97** 569; Ma,E. (1978) *Phys.Rev* **D 17** 623;ibid (1985) **D31** 1143.
3. Lucchesi, C., Piguet, O. and Sibold, K. (1988) *Helv. Phys. Acta* **61** 321; Piguet, O. and Sibold, K. (1986) *Intr. J. Mod. Phys.* **A1** 913; (1986) *Phys. Lett.* **B177** 373; see also Lucchesi, C. and Zoupanos, G. (1997) *Fortsch. Phys.* **45** 129.
4. Ermushev, A.Z., Kazakov, D.I. and Tarasov, O.V. (1987) *Nucl. Phys.* **281** 72; Kazakov, D.I. (1987) *Mod. Phys. Lett.* **A9** 663.
5. Kapetanakis D., Mondragón, M. and Zoupanos, G., (1993) *Zeit. f. Phys.* **C60** 181; Mondragón, M. and Zoupanos, G. (1995) *Nucl. Phys. B* (Proc. Suppl.) **37C**) 98.
6. Kubo, J., Mondragón, M. and Zoupanos, G. (1996) *Phys. Lett.* **B389** 523.

7. Jack, I. and Jones, D.R.T. (1995) *Phys. Lett.* **B349** 294.
8. D.R.T. Jones, L. Mezincescu and Y.-P. Yao, *Phys. Lett.* **B148** 317 (1984).
9. Jack, I. and Jones, D.R.T. (1994) *Phys. Lett.* **B333** 372.
10. Leigh, R.G., and Strassler, M.J. (1995) *Nucl. Phys.* **B447** 95.
11. Novikov, V., Shifman, M., Vainstein, A., and Zakharov, V. (1983) *Nucl. Phys.***B229** 381; (1986)*Phys. Lett.* **B166** 329; Shifman, M. (1996) *Int.J. Mod. Phys.***A11** 5761 and references therein.
12. Avdeev, L.V., Kazakov, D.I. and Kondrashuk, I.N. (1998) *Nucl. Phys.* **B510** 289; Kazakov, D.I. (1999) *Phys.Lett.* **B449** 201.
13. Yamada, Y. (1994) *Phys.Rev.* **D50** 3537.
14. Delbourgo, R. (1975) *Nuovo Cim* **25A** 646; Salam, A. and Strathdee, J. (1975) *Nucl. Phys.* **B86** 142; Fujikawa, K. and Lang, W. (1975) *Nucl. Phys.* **B88** 61; Grisaru, M.T., Rocek, M. and Siegel, W. (1979) *Nucl. Phys.* **B59** 429.
15. Girardello, L. and Grisaru, M.T. (1982) *Nucl. Phys.* **B194** 65; Helayel-Neto, J.A. (1984) *Phys. Lett.* **B135** 78; Feruglio, F., Helayel-Neto, J.A. and Legovini, F. (1985) *Nucl. Phys.* **B249** 533; Scholl, M. (1985) *Zeit. f. Phys.* **C28** 545.
16. Kawamura, T., Kobayashi, T. and Kubo, J. (1997) *Phys. Lett.* **B405** 64.
17. Kobayashi, T., Kubo, J., Mondragón, M. and Zoupanos, G. (1998) *Nucl. Phys.* **B511** 45; in proc. of ICHEP 1998, vol. 2, p. 1597 (Vancouver 1998); Acta Phys. Polon. **B30** (1999) 2013; in proc. of HEP99, p. 804 (Tampere 1999).
18. T. Kobayashi, J. Kubo and G. Zoupanos, (1998) Phys. Lett. **B427** 291; T. Kobayashi et al, in proc. of "Supersymmetry, Supergravity and Superstrings", pp. 242-268, Seoul, 1999; T.Kobayashi et. al (2001) *Surveys High Energy Phys.* **16** 87-129.
19. E. Ma, M. Mondragón and G. Zoupanos, JHEP **0412** (2004) 026 [arXiv:hep-ph/0407236].
20. K. S. Babu, T. Enkhbat and I. Gogoladze (2003) Phys. Lett. B **555** 238 [arXiv:hep-ph/0204246].
21. K. Sato, EPS05, Lisbon, Portugal, to be published in the proceedings.
22. For an extended discussion and a complete list of references see: Kubo, J., Mondragón, M. and Zoupanos, G. (1997) *Acta Phys. Polon.* **B27** 3911.
23. Bertolini, S., Borzumati, F., Masiero, A. and Ridolfi, G. (1991) *Nucl. Phys.* **B353** 591.
24. R. Barate et al. [ALEPH Collaboration], *Phys. Lett.* **B 429** (1998) 169; S. Chen et al. [CLEO Collaboration], *Phys. Rev. Lett.* **87** (2001) 251807; P. Koppenburg et al. [Belle Collaboration], *Phys. Rev. Lett.* **93** (2004) 061803; B. Aubert et al. [BABAR Collaboration], hep-ex/0207074.
25. G. Buchalla and A. Buras, *Nucl. Phys.* **B 400** (1993) 225; M. Misiak and J. Urban, *Phys. Lett.* **B 451** (1999) 161; A. Buras, *Phys. Lett.* **B 566** (2003) 115.
26. M. Herndon [CDF and D0 Collaborations], FERMILAB-CONF-04-391-E, to appear in the proceedings of 32nd International Conference on High-Energy Physics (ICHEP 04), Beijing, China, 16-22 Aug 2004
27. H. Goldberg, *Phys. Rev. Lett.* **50** (1983) 1419; J. Ellis, J. Hagelin, D. Nanopoulos, K. Olive and M. Srednicki, *Nucl. Phys.* **B 238** (1984) 453.
28. C. Bennett et al., *Astrophys. J. Suppl.* **148** (2003) 1; D. Spergel et al. [WMAP Collaboration], Astrophys. J. Suppl. **148** (2003) 175.
29. L. J. Hall, R. Rattazzi and U. Sarid, Phys. Rev. D **50**, (1994) 7048 ; R. Hempfling, Z. Phys. C **63**, (1994) 309; M. Carena, M. Olechowski, S. Pokorski and C. E. M. Wagner, Nucl. Phys. B **426** (1994) 269.
30. J. Erler, private communication.
31. Gladyshev, A.V., Kazakov, D.I., de Boer, W., Burkart, G. and Ehret, R. (1997) *Nucl. Phys.* **B498** 3; Carena, M. et. al., (1995) *Phys. Lett.* **B355** 209.
32. LEP Higgs working group, *Phys. Lett.* **B 565** (2003) 61.
33. S. Heinemeyer, W. Hollik and G. Weiglein, Comput. Phys. Commun. **124** (2000) 76; Higgs bosons in the MSSM: Accurate Eur. Phys. J. C **9** (1999) 343.
34. A. Djouadi, J. L. Kneur and G. Moultaka, arXiv:hep-ph/0211331.
35. Djouadi, A., Heinemeyer, S., Mondragón, M. and Zoupanos, G., work in progress.

# The Supersymmetric Flavor Problem

J.L. Díaz-Cruz

*Facultad de Ciencias Físico-Matemáticas, BUAP.*
*Apdo. Postal 1364, C.P. 72000 Puebla, Pue., México*

**Abstract.** We discuss several aspects of the flavor problem in the Supersymmetry. First, in order to quantify the SUSY flavor problem, we generate randomly the entries of the sfermion mass matrices and determine which percentage of the points are consistent with current bounds on the flavor violating transitions, for which we take as an illustration the lepton flavor violating (LFV) decays $l_i \to l_j \gamma$. In the first instance we apply the mass-insertion method, and study how this percentage changes as one varies the parameters of the model. It is found that for $10^5$ points about 10% of points pass current LFV bounds on $\mu \to e\gamma$ provided the sleptons masses are $\sim 10$ TeV. While bounds on $\tau \to \mu\gamma, e\gamma$ are satisfied for almost 100% of points even for sleptons masses as low as 360 GeV. Then, we consider an ansatz for sfermion masses that can be diagonalized exactly, and compare the results obtained previously for $\tau \to \mu\gamma$. Now, we get that 100% of points satisfy the experimental bounds but with sleptons masses larger than 460 GeV.

**Keywords:** flavor problem, supersymmetry
**PACS:** 11.30.Hv, 11.30.Pb, 13.35.-r

## INTRODUCTION.

Weak-scale supersymmetry (SUSY) [1], has become one of the leading candidates for physics beyond the standard model, notably by supporting the mechanism of electroweak symmetry breaking (EWSB). Being a new fundamental space-time symmetry, SUSY necessarily extends the SM particle content by including superpartners for all fermions. Because the mass spectrum of the superpartners needs to be lifted, SUSY must be softly broken, this is needed so in order to maintain its ultraviolet properties. SUSY breaking is parameterized in the Minimal Supersymmetric SM (MSSM) by the soft-breaking lagrangian [2]; as an outcome, the combined effects of the large top quark Yukawa coupling and the soft-breaking masses, make possible to induce radiatively the breaking of the electroweak symmetry. The Higgs sector of the MSSM includes two higgs doublets, with the light Higgs boson ($m_h \leq 125$ GeV), being perhaps the strongest prediction of the model.

However, the soft breaking sector of the MSSM is often problematic with low-energy flavor changing neutral currents (FCNC) without making specific assumptions about its free parameters. Minimal choices to satisfy those constraints, such as assuming universality of squark masses, have been widely studied in the literature [2]. However, non-minimal flavor structures could be generated in a variety of contexts. For instance, within the context of realistic unification models through the evolution of the soft-terms, from a high-energy GUT scale, to the weak scale. Similarly, models that attempt to address the flavor problem, could induce sfermion soft-terms that reflect the underlying flavor symmetry of the fermion sector [3, 4].

CP917, *Particles and Fields,* edited by H. Castilla Valdez, J. C. D'Olivo, and M. A. Perez

It is not a trivial task to find models of SUSY breaking that can actually generate minimal and safe patterns. This is the so called SUSY flavor problem. The known solutions include:

- *Degeneracy*[5] In this case, sfermions of different families have the same mass,
- *Proportionality* [5] This assumes that trilinear terms are proportional to the Yukawa terms),
- *Decoupling* [6] This option takes the superpartners heavy enough not to affect low energy physics,
- *Alignment* [7] Here, the same physics that explains the pattern of fermion masses and mixing angles forces the sfermion mass matrices to be aligned with the fermion ones, in such a way that the fermion-sfermion-gaugino vertices remain close to diagonal.

Sometimes the SUSY flavor problem is stated by saying that if the sfermion mass matrix entries were generated randomly, most of such points would lead to the exclusion of the MSSM. In this paper, we would like to quantify the above statement, namely we are interested in estimating the *size* of the SUSY flavor problem, and to determine its dependence on the parameters of the MSSM. One would like then to determine what would be left of the SUSY flavor problem after Tevatron and LHC will deliver bounds on the masses of the superpartners, or luckily a signal of their presence! We concentrate mainly in the lepton sector, and in particular we use the LFV decays $l_i \to l_j + \gamma$, to make our point, namely to derive bounds on the parameters of the MSSM and to determine the viability and interplay of the above solutions.

First, we evaluate the above LFV decays using the mass-insertion approximation, both for muon and tau decays. Our procedure will consist of writing first the off-diagonal elements of the slepton mass matrices as the product of $O(1)$ coefficients times an average sfermion mass parameter, then we generate randomly $10^5$ points for the $O(1)$ coefficients, and determine which fraction of such points satisfies current bounds on the LFV transitions. We repeat this procedure for different values of other relevant parameters of the MSSM, such as $\tan\beta$, gaugino masses, $\mu$-parameter and the sfermion mass scale.

Next, to estimate how far can we trust the mass-insertion method, we compare those results with the ones coming from particular models that enable to obtain exact diagonalization for the sfermion mass matrices. Namely, we take into account that the constraints on sfermion mixing coming from low-energy data, mainly suppress the mixing between the first two family sleptons, but still allows large flavor-mixings between the second- and third-family sleptons, *i.e.* the smuon and stau, which can be as large as $O(1)$ [8]. Thus, we consider models where mixing involving the selectrons could be neglected, as it involves small off-diagonal entries in the slepton mass matrices. But the mixing among smuon and stau, will involve large off-diagonal entries in the sfermion mass matrices, which requires at least a partial diagonalization in order to be treated in a consistent manner. Namely, in our models the general $6\times 6$ slepton-mass-matrix will include a $4\times 4$ sub-matrix involving only the 2-3 sector, which can be diagonalized exactly, similarly to the squark case first discussed in Ref.[9]. Since we follow a bottom-up approach, we simply take an ansatz for the $A$-terms valid at the TeV-scale; such large off-diagonal entries can be motivated by considering the large mixing detected with atmospheric neutrinos [10], especially in the framework of GUT models with flavor symmetries. Then, we repeat the above method of random generation, now for the parameters of the sfermion matrices, which will be then diagonalized. Armed with the exact expressions for the mass and mixing matrices, and the interaction lagrangian written in terms of mass eigenstates, we evaluate the fraction of points that satisfy all the LFV constraints coming from the decays $\tau \to \mu + \gamma$. The results with exact diagonalization for LFV tau decays will be compared then with those obtained using the mass-insertion approximation. Recently a similar analysis was presented by P. Paradisi in Ref.[11].

The organization of this paper goes as follows: In Section II we discuss the SUSY flavor problem in the lepton sector, using the mass-insertion approximation. This section includes the evaluation of the radiative LFV loop transitions ($l_i \to l_j + \gamma$), with a random generation of the slepton $A$-terms. Then, in sect. III we present an ansatz for soft breaking trilinear terms, the diagonalization of the resulting sfermion mass matrices, and we repeat the calculation of the previous section.. Finally, our conclusions are presented in sect. IV.

## THE SUSY FLAVOR PROBLEM IN THE SUPER CKM BASIS.

### The slepton mass matrices in the MSSS.

We discuss first the slepton mass matrices, and the form that the gaugino-lepton-slepton interactions. The MSSM soft-breaking slepton sector contains the quadratic mass-terms and trilinear $A$-terms, that contribute to the slepton mass matrices.

For the charged slepton sector, this gives a generic $6 \times 6$ mass matrix, which can be written in terms of 3x3 blocks, namely,

$$M_{\tilde{L}L}^2 = M_{\tilde{L}}^2 + M_l^2 + \frac{1}{2}\cos 2\beta\,(2m_W^2 - m_Z^2)\,, \tag{1}$$

$$M_{\tilde{R}R}^2 = M_{\tilde{E}}^2 + M_l^2 - \cos 2\beta \sin^2\theta_W\, m_Z^2\,, \tag{2}$$

$$M_{\tilde{L}R}^2 = A_l v \cos\beta/\sqrt{2} - M_l\,\mu \tan\beta\,, \tag{3}$$

with $m_{W,Z}$ denotes the masses of the $W^{\pm}$ and $Z^0$ bosons, and $M_l$ being the lepton mass matrix. (For convenience, we will choose a basis where $M_l (= M_l^{diag})$ is diagonal).

In our *minimal* scheme, we consider all large LFV that *solely* come from the non-diagonal entries of the $A_l$-terms in the slepton-sector, in a manner that respects the low-energy constrains and CCB-VS bounds [14]. In the Super CKM basis, the gaugino-slepton-lepton interactions are diagonal in flavor space, while flavor-violation associated with the off-diagonal entries of the slepton mass matrices are treated as perturbations, *i.e* mass-insertions. We shall write the off-diagonal soft-terms as

$$(M_{\tilde{M}N}^2)_{off-diag} = z_{MN}^l \cdot m_0^2, \tag{4}$$

where $M, N : L, R$, and $m_0$ denotes an average slepton mass scale and the coefficients $z_{MN}^l$ will be taken as $O(1)$ random coefficients.

## Bounds on the soft-breaking parameters from the LFV decay $l_i \to l_j \gamma$

Here, we are interested in obtaining bounds on the parameters $z^l_{MN}$ and $m_0$, applying the mass-insertion method to evaluate the LFV transition $\mu \to e + \gamma$ and $\tau \to \mu(e) + \gamma$. Within this method, the expression for the branching ratio $BR(l_i \to l_j + \gamma)$, including the bino contributions, has the form [5]:

$$BR(l_i \to l_j \gamma) = \frac{\alpha^3}{G_F^2} \frac{12\pi}{m_{\tilde{l}}^4} \left\{ \left| M_3(x)(\delta^l_{ij})_{LL} + \frac{m_{\tilde{\gamma}}}{m_{l_i}} \right\} \right| \tag{5}$$

Assuming that solely the term $(\delta^l_{ij})_{LR} = \widetilde{M}^2_{LR}/m_0^2$ contributes to the branching ratio, one obtains

$$BR(l_i \to l_j \gamma) \approx \frac{\alpha^3}{G_F^2} \frac{12\pi}{m_{\tilde{l}}^4} \left(\frac{m_{\tilde{\gamma}}}{m_{l_i}}\right)^2 \left\{ \left| M_1(x)(\delta^l_{ij})_{LR} \right|^2 \right\} BR(l_i \to l_j \nu_i \bar{\nu}_j) \tag{6}$$

In order to get an expression for $(\delta^l_{ij})_{LR}$, we write

$$\left(\widetilde{M}^2_{LR}\right)_{ij} = \frac{v}{\sqrt{2}} \cos\beta \cdot (z^l_{LR})_{ij} \cdot m_0 \tag{7}$$

Therefore

$$(\delta^l_{ij})_{LR} = \frac{\left(\widetilde{M}^2_{LR}\right)_{ij}}{m_0^2} = \frac{\cos\beta}{\sqrt{2}} \frac{v}{m_0} \cdot (z^l_{LR})_{ij} \tag{8}$$

Finally, we obtain

$$BR(l_i \to l_j \gamma) \approx \frac{\alpha^3}{G_F^2} \frac{6\pi}{m_{\tilde{l}}^4} \left(\frac{m_{\tilde{\gamma}}}{m_{l_i}}\right)^2 |M_1(x)|^2 \cos^2\beta \left(\frac{v}{m_0}\right)^2 \cdot (z^l_{LR})^2_{ij} \cdot BR(l_i \to l_j \nu_i \bar{\nu}_j) \tag{9}$$

where $m_{\tilde{l}_i} \approx m_0$ and

$$M_1(x) = \frac{1 + 4x - 5x^2 + 4x\ln(x) + 2x^2\ln(x)}{2(1-x)^4} \tag{10}$$

In order to discuss the processes: $\mu \to e\gamma$; $\tau \to \mu\gamma$; $\tau \to e\gamma$, we shall make use of the following experimental results: $BR(\mu \to e\,\nu_\mu\,\bar{\nu}_e) \approx 100\%$; $BR(\tau \to \mu\,\nu_\tau\,\bar{\nu}_\mu) \approx 17.36\%$; $BR(\tau \to e\,\nu_\tau\,\bar{\nu}_e) \approx 17.84\%$ [15]

Now, our statistical analysis is based on a random generation of the parameters $(z^l_{LR})_{ij}$, ($10^5$ points are generated) and then studying their effects on the LFV transitions. Our results for $\mu \to e\gamma$, assuming $\tan\beta = 15$, and $x = 0.3, x = 1.5$ and $x = 5$, are shown in our paper [17], which illustrates the severity of the SUSY flavor problem for low sfermion masses. One finds that even for $m_0 = 1$ *TeV* almost 100% of the randomly generated points are excluded by experiment, while one needs to have $m_0 \approx 10$ *TeV* in order to get about 10% of the generated points that satisfy current bound on $\mu \to e\gamma$. Having larger gaugino masses helps to ameliorate the problem, but not much. For instance, assuming $x = 5$ and $\tan\beta = 15$, implies that even $m0 = 10\,TeV$, the percentage of acceptable points raises only up to about 18%.

Current bounds on tau decays do not pose such a severe problem. In this case most of the randomly generated points satisfy the bounds on $\tau \to \mu\gamma$ and $\tau \to e\gamma$. For instance, in the case of $\tau \to \mu\gamma$, with $x = 0.3$ and $\tan\beta = 15$, one gets that for $m_{\tilde{l}} = 200$ GeV only around 10% of the points are accepted by experimental data. However, this percentage increases with the slepton mass, and for $m_{\tilde{l}} \geq 400$ GeV about 100% of the points are accepted by experimental data. We also find that a similar behavior is obtained for $\tau \to e\gamma$. Again, for $x = 0.5$ and $\tan\beta = 15$, $\tau \to \mu\gamma$ ($\tau \to e\gamma$) requires lower slepton masses, about $m_{\tilde{l}} \geq 220$ GeV ($m_{\tilde{l}} \geq 180$ GeV) in order to get 100% of the points as acceptable by experimental data.

## THE SUSY FLAVOR PROBLEM BEYOND THE MASS-INSERTION APPROXIMATION.

We shall consider now, SUSY FCNC schemes where the general $6 \times 6$ slepton-mass-matrix reduces down to a $4 \times 4$ matrix involving only the $\mu - \tau$ slepton sector, similarly to the quark sector discussed in Ref. [9]. In this case, the flavor-mixings between the smuon and stau, can be as large as $O(1)$. Although such large mixing could be related to the large $\nu_\mu - \nu_\tau$ mixing observed in atmospheric neutrinos [10], we shall follow a bottom-up approach, where we simply take as an ansatz the following form of the A-terms, taken also to be real and valid at the TeV-scale.

### Diagonalization of fermion mass matrices

The reduction of the slepton mass matrix proceeds, for instance, by considering at the weak scale the following $A$-term:

$$\begin{matrix} 0 & 0 & 0 \\ 0 & 0 & z \\ 0 & y & 1 \end{matrix}$$

where $y$ and $z$ can be of $O(1)$, representing a naturally large flavor-mixing in the $\mu - \tau$ sector. Actually, the zero entries could be of $O(\varepsilon)$, with $\varepsilon << 1$, and their effect could be treated using the mass-insertion approximation. Moreover, identifying the non-diagonal $A_l$ as the only source of the observable LFV phenomena implies the slepton-mass-matrices to be nearly diagonal. For simplicity, we define

$$M_{LL}^2 \simeq M_{RR}^2 \simeq m_0^2 \mathbf{I}_{3\times3}\,, \tag{11}$$

with $m_0$ being a common scale for the scalar-masses.

Within this minimal scheme, we observe that the first family sleptons $\tilde{e}_{L,R}$ decouples from the rest, so that, in the slepton basis, the $6\times 6$ mass-matrix is reduced to a $4\times 4$ matrix. The reduced slepton mass matrix has 6 zero-entries in total and is simple enough to allow an exact diagonalization. Therefore, when evaluating loop amplitudes one can use the exact slepton mass-diagonalization, and compare results with those obtained from the popular but crude mass-insertion approximation.

We have worked out the general diagonalization of the 4x4 slepton mass matrix, for any $(y,z)$. The mass-eigenvalues of the eigenstates $(\tilde{\mu}_1, \tilde{\mu}_2, \tilde{\tau}_1, \tilde{\tau}_2)$ are,

$$M^2_{\tilde{\mu}_{1,2}} = m0^2 \mp \tfrac{1}{2}|\sqrt{\omega_+} - \sqrt{\omega_-}|, \tag{12}$$

$$M^2_{\tilde{\tau}_{1,2}} = m0^2 \mp \tfrac{1}{2}|\sqrt{\omega_+} + \sqrt{\omega_-}|, \tag{13}$$

where $\omega_\pm = X_\tau^2 + (A_y \pm A_z)^2$.

## Bounds on the LFV parameters from $\tau \to \mu + \gamma$

Here, we are interested in determining which fraction of points in Parameter Space satisfy current bounds on LFV tau decays, when the exact slepton mass-diagonalization is applied; again we generate $10^5$ random values of $O(1)$ for the parameter $z$ appearing in the soft-terms, and fix the values of $\widetilde{m}_0$, $\widetilde{M}$ and $\tan\beta$. One can write the general expressions for the SUSY contributions to the decays $\tau \to \mu + \gamma$ given in Ref. [16], using the interaction lagrangian (21). The expression for the decay width $\Gamma(\tau \to \mu + \gamma)$, including the smuon and stau contributions, has the form:

$$\Gamma(\tau \to \mu + \gamma) = \frac{\alpha m_\tau^5}{4\pi}[\sum_\alpha |A_{L\alpha}|^2 + |A_{R\alpha}|^2] \tag{14}$$

where

$$A_{R\alpha} = \frac{1}{32\pi^2 m^2_{\tilde{l}_\alpha}}[\eta^R_{\tilde{l}_\alpha \tau}\eta^R_{\tilde{l}_\alpha \mu} f_1(x_\alpha) + \eta^R_{sl p_\alpha \tau}\eta^L_{\tilde{l}_\alpha \mu}\frac{m_{\tilde{B}}}{m_\tau} f_2(x_\alpha)] \tag{15}$$

with $x_\alpha = m^2_{\tilde{B}}/m^2_{\tilde{l}_\alpha}$, and the functions $f_{1,2}(x_\alpha)$ are given in Ref. [16]. $A_{L\alpha}$ is obtained by making the substitutions $L \to R, R \to L$ in Eq. (23). The expression for the decays $\Gamma(\mu \to e + \gamma)$ and $\Gamma(\tau \to e + \gamma)$ is still given by the mass-insertion approximation.

The decay width depends on the SUSY parameters, and again we shall generate randomly the points and use the current bound $BR(\tau \to \mu + \gamma) < 1.1 \times 10^{-6}$ to determine which percentage is excluded/accepted. We can read from our paper, that starting with values of the scalar mass parameter $m_0 \geq 460$ GeV, about 100% of the generated points are acceptable for $x \geq 0.3$.

## CONCLUSIONS

We have discussed the SUSY flavor problem in the lepton sector, using the mass-insertion approximation. We have evaluated the radiative LFV loop transitions ($l_i \to$

$l_j + \gamma$), with a random generation of the slepton $A$-terms. Our results illustrate the severity of the SUSY flavor problem for low sfermion masses. One can see that even for $m_0 = 1$ TeV almost 100% of the randomly generated points are excluded, while one needs to have $m_0 \approx 10$ TeV in order to get about 10% of the generated points that satisfy the current bound on $\mu \to e\gamma$. Having larger gaugino helps to ameliorate the problem, but not much. On the other hand, we have shown that current bounds on tau decays do not pose such a severe problem. In this case most of the randomly generated points satisfy the experimental bounds on $\tau \to \mu\gamma$ and $\tau \to e\gamma$.

We presented an ansatz for soft breaking trilinear terms, the diagonalization of the resulting sfermion mass matrices, and repeat the previous calculation. The decay width depends on the SUSY parameters, again we generate randomly the points and use the current bound $BR(\tau \to \mu + \gamma) < 1.1 \times 10^{-6}$ to determine which percentage is excluded/accepted. We showed that for values of the scalar mass parameter $m_0 \geq 460$ GeV, 100% of the points are acceptable for $x \geq 0.3$, (to be compared with $m_0 \geq 360$ GeV obtained using the mass-insertion approximation).

**Acknowledgments**
I would like to thank A. Rosado, M. Gomez.Brok and R. Noriega-papaqui for valuable discussions and collaboration. This work was supported by CONACYT and SNI (Mexico).

# REFERENCES

1. See, for instance, recent reviews in "Perspectives on Supersymmetry", ed. G. L. Kane, World Scientific Publishing Co., 1998; H. E. Haber, Nucl. Phys. Proc. Suppl. **101**, 217 (2001), [hep-ph/0103095].
2. D. J. H. Chung, L. L. Everett, G. L. Kane, S. F. King, J. D. Lykken and L. T. Wang, Phys. Rept. **407**, 1 (2005) [arXiv:hep-ph/0312378].
3. E.g., S. Khalil, J. Phys. G**27**, 1183 (2001), [hep-ph/0011330]; D. F. Carvalho, M. E. Gomez, S. Khalil, hep-ph/0104292; and references therein.
4. A. Masiero and H. Murayama, Phys. Rev. Lett. **83**, 907 (1999), [hep-ph/9903363].
5. F. Gabbiani, E. Gabrielli, A. Masiero and L. Silvestrini, Nucl. Phys. B**477**, 321 (1996).
6. N. Arkani-Hamed and H. Murayama, Phys. Rev. D **56**, 6733 (1997) [arXiv:hep-ph/9703259].
7. E.g., Y. Nir and N. Seiberg, Phys. Lett. B**309**, 337 (1993), [hep-ph/9304307].
8. For review, M. Misiak, S. Pokorski, J. Rosiek, "Supersymmetry and FCNC Effects", hep-ph/9703442, in *Heavy Flavor II,* pp. 795, eds., A. J. Buras and M. Lindner, Advanced Series on Directions in High Energey Physics, World Scientific Publishing Co., 1998, and references therein.
9. J. L. Diaz-Cruz, H. J. He and C. P. Yuan, "Soft SUSY breaking, stop-scharm mixing and Higgs signatures," Phys. Lett. B **530**, 179 (2002) [arXiv:hep-ph/0103178].
10. Super-Kamiokande Collaboration (Y. Fukuda et al.), Phys. Rev. Lett. 81 (1998) 1562 [ArXiv:hep-ex/0009001].
11. P. Paradisi, JHEP **0510**, 006 (2005) [arXiv:hep-ph/0505046].
12. J. L. Diaz-Cruz, JHEP **0305**, 036 (2003) [arXiv:hep-ph/0207030].
13. J. L. Diaz-Cruz and J. J. Toscano, Phys. Rev. D **62**, 116005 (2000) [arXiv:hep-ph/9910233].
14. J. A. Casas and S. Dimopolous, Phys. Lett. B**387**, 107 (1996), [hep-ph/9606237].
15. S. Eidelman *et al.*, (*Review of Particle Physics*), Phys. Lett. B **592**, 1 (2004).
16. J. Hisano, T. Moroi, K. Tobe and M. Yamaguchi, "Lepton-Flavor Violation via Right-Handed Neutrino Yukawa Couplings in Supersymmetric Standard Model," Phys. Rev. D **53**, 2442 (1996) [arXiv:hep-ph/9510309].
17. J. L. Diaz-Cruz, M. Gomez-Bock, R. Noriega-Papaqui and A. Rosado, arXiv:hep-ph/0512168.

# Naturalness of Electroweak Symmetry Breaking while Waiting for the LHC

J.R. Espinosa

*IFT-UAM/CSIC Fac. de Ciencias UAM, Cantoblanco, 28049 Madrid (SPAIN)*

**Abstract.** After revisiting the hierarchy problem of the Standard Model and its implications for the scale of New Physics, I consider the finetuning problem of electroweak symmetry breaking in several scenarios beyond the Standard Model: SUSY, Little Higgs and "improved naturalness" models. The main conclusions are that: New Physics should appear on the reach of the LHC; some SUSY models can solve the hierarchy problem with acceptable residual tuning; Little Higgs models generically suffer from large tunings, many times hidden; and, finally, that "improved naturalness" models do not generically improve the naturalness of the SM.

**Keywords:** Hierarchy problem, Naturalness, Supersymmetry, Little Higgs

## HIERARCHY PROBLEM OF THE SM

It is well known that although the Standard Model (SM) works extremely well (at the per mille level) it is probably not fundamental but rather an approximate description of particle physics valid up to some high energy scale $\Lambda$, where a more fundamental theory takes over. We do have some clues about the value of $\Lambda$ coming from the SM electroweak symmetry breaking (EWSB) sector. This breaking is described in the SM by a fundamental doublet scalar with a mexican-hat potential

$$V = \frac{1}{2}m^2h^2 + \frac{1}{4}\lambda h^4 . \quad (1)$$

The Higgs mass parameter $m^2$ is assumed to be negative so that $h$ develops a non-zero vev that breaks the EW gauge symmetry spontaneously. As $m^2$ is the only mass scale that appears in the SM Lagrangian it sets the scale of EWSB, with $v^2 = -m^2/\lambda$ fixed to $\simeq (246\ \mathrm{GeV})^2$ to get right the masses of $Z^0$ and $W^\pm$. The spectrum contains a Higgs boson with mass squared $m_h^2 = 2\lambda v^2$ that is unknown but expected to be light from loop effects in fits to precision EW data.

There is a theoretical problem with this picture of EWSB: $m^2$ is sensitive to high energy scales through quantum corrections. Loops of $W^\pm$ and $Z^0$ gauge bosons, the top quark and the Higgs itself give the one-loop quadratically divergent correction to $m^2$ [1]:

$$\delta_q m^2 = \frac{3\Lambda^2}{64\pi^2}(3g^2 + g'^2 + 8\lambda - 8\lambda_t^2) + ... \quad (2)$$

If one believes that the SM is valid all the way up to the Planck mass, $\Lambda \sim M_{Pl}$, $\delta m^2$ is huge and has to be balanced with extreme precision against the tree-level value of $m^2$. This is the Big Hierarchy Problem [2]. If, following indications from the absence of

CP917, *Particles and Fields,* edited by H. Castilla Valdez, J. C. D'Olivo, and M. A. Perez

indirect effects of some non-renormalizable operators, one believes the SM is valid up to $\Lambda \sim 10$ TeV one still has a problem (the Little Hierarchy Problem [3]) albeit softer.

Turning the argument around, naturalness of EWSB requires $\delta m^2 \sim \lambda v^2 \sim m^2$ and an upper bound on the scale of New Physics follows. For instance,

$$\frac{\delta m^2}{m^2} < 10 \Longrightarrow \Lambda \lesssim 2 \text{ TeV}\,, \tag{3}$$

for $m_h \sim 130$ GeV. The importance of this figure is obvious: it implies that New Physics beyond the SM should be on the reach of LHC.

This naive estimate can be refined [4, 5] by taking into account higher order effects [for moderate $\Lambda$ the dominant loop corrections are summed up to leading-log order [6] by Eq. (2), but with couplings evaluated at the scale $\Lambda$] and the sensitivity of $m^2$ to other parameters besides $\Lambda$, like $\lambda$ and $\lambda_t$. Before presenting the final result let me remind you how to estimate numerically the finetuning associated to EWSB. Consider your favourite model for EWSB, which should give $v$ as a function of some input parameters $p_\alpha$ (usually these are defined at some UV scale). Following [7], we measure the finetuning associated to $p_\alpha$ by the quantity $\Delta_{p_\alpha}$ defined by $\delta m^2/m^2 = \Delta_{p_\alpha}(\delta p_\alpha/p_\alpha)$ , where $\delta m^2$ is the change induced in $m^2$ by a change $\delta p_\alpha$ in $p_\alpha$. For a given model, we obtain a global finetuning figure by adding the different $\Delta_{p_\alpha}$ in quadrature, $\Delta \equiv (\sum_\alpha \Delta^2_{p_\alpha})^{1/2}$ [8], with the tuning being of order 1 in $\Delta$ (see [9] for discussions and refinements).

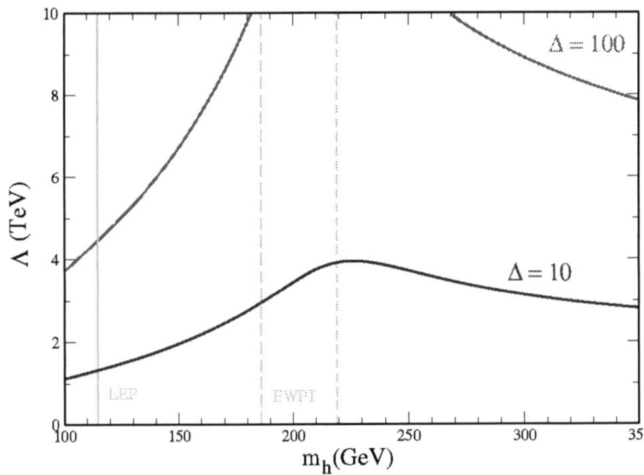


**FIGURE 1.** Upper bound on the scale of New Physics ($\Lambda$) from the absence of 1/10 (1/100) tuning. (Improved calculation, including sensitivity to $\lambda$ and $\lambda_t$, see [10].)

For the SM, fig. 1 shows the finetuning $\Delta = \{\Delta^2_\Lambda + \Delta^2_{\lambda_t} + \Delta^2_\lambda\}^{1/2}$, from which $\Lambda \lesssim 3-4$ TeV (2.5 TeV on average) to avoid more than 10% finetuning. So, one recovers the naive expectation that New Physics beyond the SM should be on the reach of the LHC. (In fact this bound is conservative as it underestimates New Physics effects [5]. For further qualifications on the meaning of this kind of analysis see [11].) For later use let me point out now that the finetuning in the SM for a cutoff $\Lambda = 10$ TeV, *i.e.* the tuning associated to the little hierarchy problem, is at least of 1%. (This estimate includes the effects on $\delta m^2/m^2$ of varying $\Lambda$, $\lambda_t$ and $\lambda$ [8].)

Besides the general bound, $\Lambda \lesssim 3-4$ TeV, more concrete (and solid) implications from naturalness can be obtained in particular scenarios of New Physics. In what follows I consider three scenarios which are well motivated: SUSY, Little Higgses and "improved-naturalness" models. All of them try to reduce the sensitivity of $m^2$ to high scales $\Lambda$ by introducing new particles (so that new loop corrections cancel the SM dangerous contributions) and new symmetries (so that the required cancellation is natural).

## THE SUSY FINETUNING PROBLEM

In SUSY, the new particles introduced are the superpartners of SM particles and Supersymmetry ensures the cancellation of quadratic divergences to all orders! The Higgs mass is protected because SUSY relates the Higgs to chiral fermions, whose mass is under control. However, SUSY must be broken, with superpartner masses $\sim \tilde{m} \lesssim 1$ TeV. As a result, superpartners do not cancel completely the SM quadratic divergences which are replaced by corrections proportional to the soft SUSY breaking mass scale $\tilde{m}$. For instance, the Higgs one-loop correction from top and stop loops goes like

$$\delta m^2 \sim -\frac{\lambda_t^2 \tilde{m}^2}{16\pi^2} \log \frac{M_{mes}^2}{\tilde{m}^2}\,, \tag{4}$$

where $M_{mes}$ is the scale at which SUSY is transmitted to the observable sector (in gravity mediated SUSY breaking $M_{mes} \simeq M_{Pl}$). It is clear that the improvement in naturalness with respect to the SM case (with $\Lambda \sim M_{mes}$) is enormous. In addition, the negative contribution in (4) explains EWSB dinamically (*i.e.* gives a reason for $m^2 < 0$).

Focusing on the Minimal Supersymmetric Standard Model (MSSM), it is known that EWSB suffers from a residual finetuning problem. In order for EWSB to be natural, one needs $\tilde{m} \lesssim 1$ TeV, which implies superpartner masses below $\lesssim$ few hundred GeV and the available experimental data already forces the ordinary MSSM to be somewhat finetuned [7, 9, 12]. Consider *e.g.* the upper bound on the lightest Higgs boson mass

$$m_h^2 \leq M_Z^2 \cos^2 2\beta + \frac{3 m_t^4}{2\pi^2 v^2} \log \frac{M_{\rm SUSY}^2}{m_t^2} + \ldots \tag{5}$$

where $m_t$ is the (running) top mass ($\simeq 166$ GeV for $M_t = 173$ GeV). As the experimental bound, $(m_h)_{\rm exp} \geq 115$ GeV, exceeds the tree-level contribution, radiative corrections must be responsible for the difference, and this translates into a lower bound on $M_{\rm SUSY}$, $M_{\rm SUSY} \gtrsim 3.6\ m_t$ (where this figure corresponds to $m_h = 115$ GeV and large $\tan\beta$, the best choice for finetuning). The last equation implies sizable soft terms, $\tilde{m} \gtrsim 2m_t$ and leads to large tunings. The typical tuning in this model is shown in fig. 2.

There are three main reasons for such a large finetuning: *1)* In the MSSM, $\lambda$ is calculable but quite small: $\lambda_{\rm MSSM} = (1/8)(g^2 + g_Y^2)\cos^2 2\beta \simeq (1/15)\cos^2 2\beta$, amplifying whatever cancellations are taking place inside $m^2$ in $v^2 = -m^2/\lambda$. *2)* Although for a given size of $\tilde{m}$ radiative corrections reduce the tuning, sizable radiative corrections

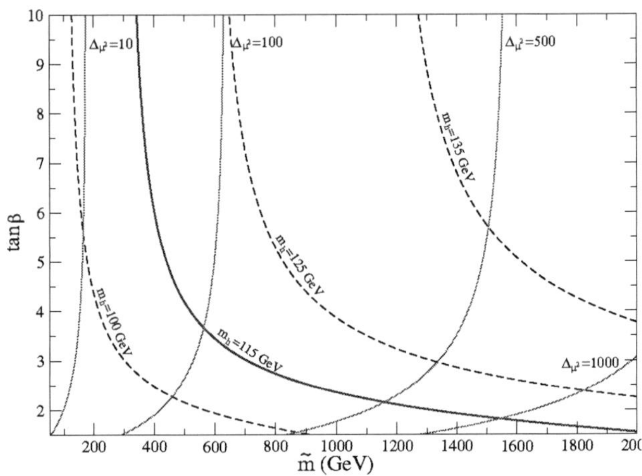


**FIGURE 2.** Fine tuning (measured by $\Delta_{\mu^2}$, solid lines) in the MSSM with universal soft masses, in the $(\tilde{m}, \tan\beta)$ plane. Dashed lines are contour lines of constant Higgs mass (LEP bound in solid).

need large soft terms, which is bad for the finetuning. A given increase in $\tilde{m}^2$ affects $m^2$ linearly and $\lambda$ only logarithmically, so the finetuning usually gets worse. *3)* Typically, the large logarithms in (4) compensate the one-loop suppresion, so that the residual corrections to $m^2$ can be quite sizable.

It is not difficult to come up with alternative SUSY models that are better than the MSSM concerning EWSB finetuning. In fact it is fair to say that SUSY is the most powerful tool for controlling the naturalness of EWSB. One popular model is the Next-to-MSSM (NMSSM) which adds a singlet chiral multiplet to the MSSM. In this model the Higgs quartic coupling gets larger thanks to additional $F$-term contributions from the singlet, improving point 1) of the list above [13]. In other scenarios the breaking of SUSY occurs at a low-scale (not far from the TeV scale) and it is natural [14] to have tree-level $\not{S}\!\!\!\!/\,$USY contributions to $\lambda$ that can make it larger [point 1)]. This helps to evade the LEP Higgs mass bound without large radiative corrections [point 2)] and in such models RG effects are expected to be small since the cut-off scale is much closer to the electroweak scale [point 3)]. All these three improvements can cooperate to make EW breaking much more natural than in the MSSM [14, 10].

## FINETUNING IN LITTLE HIGGS MODELS

Little Higgs (LH) models try to solve the Little Hierarchy problem, that is, to explain the smallness of the Higgs mass compared with 10 TeV. The models in the market typically have the following structure: below $\Lambda \sim 10$ TeV (where some UV completion takes over) new particles appear (heavy gauge bosons, fermions and scalars) sitting together with the SM particles into multiplets of some global symmetry $G$ spontaneously broken at the scale $f \sim \Lambda/(4\pi) \sim 1$ TeV. In this process the new particles gain masses of order $f$ but the SM Higgs is special: it is a (pseudo)-Goldstone boson with $G$ explicitly broken in a "collective" way so that the mass gained by the Higgs is suppressed and under control. Diagrammatically, loops of the new heavy particles cancel the quadratic divergences of SM loops (the equality of couplings necessary for this cancellation is ensured by $G$). As a result, $\delta m^2$ is not of order $\Lambda^2/(16\pi^2)$ but rather $f^2/(16\pi^2)$ which is of electroweak size. Parametrically, this solves the Little Hierarchy problem.

However, a closer look reveals some difficulties. In contrast with SUSY, the LH cancellation occurs only at one-loop. *E.g.*, the top loop divergent correction is cancelled by a heavy Top of mass $M_T \sim f$ giving

$$\delta_T m^2 \sim -\frac{\lambda_t^2 M_T^2}{16\pi^2} \log \frac{\Lambda^2}{M_T^2}, \tag{6}$$

where now $\Lambda \lesssim 10$ TeV [compare with (4)]. These models are also able to explain $m^2 < 0$ as the result of the large negative correction shown above. But this correction causes the first problem: **1)** As in this model $M_T^2 \geq 2\lambda_t^2 f^2$, one gets $\delta_T m^2/m^2 \geq 33$. This number is quite large and points immediately to a finetuning problem.

This large $\delta_T m^2$ has to be compensated by some other correction, for instance from new heavy scalars with a sufficiently large mass $M_\phi$. In practice, **2)** Both $\lambda$ and $M_\phi$ come from the same sector of the model and it is difficult to achieve a large value of $M_\phi$ while keeping $\lambda$ small [8]. Some cancellation is required and this further worsens the total finetuning. **3)** Finally, EWSB requires that some parameters, call them $c$ generically, are numerically much smaller than its natural value (estimated by looking at the radiative corrections they receive). Schematically one has $c = c_0 + c_{\rm rad} \ll c_{\rm rad}$. This is a further finetuning problem. As a result, the final numbers for the finetuning are quite large [8].

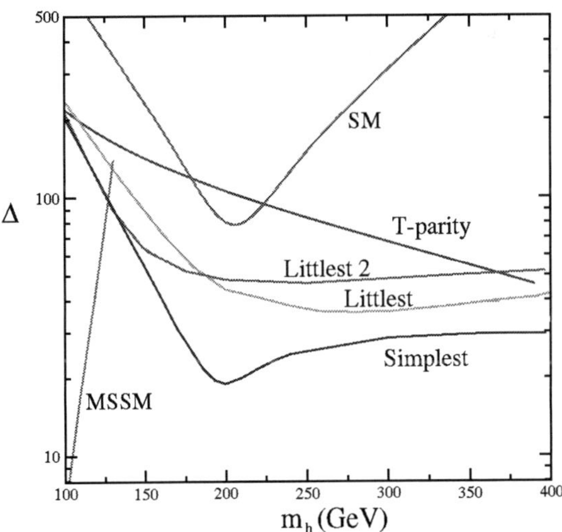


**FIGURE 3.** Finetuning performance of different Little Higgs models, compared with the SM with $\Lambda = 10$ TeV and the MSSM (with $m_{\tilde{t}} = 1$ TeV).

The problems just discussed seem to be generic and the most popular Little Higgs models, analyzed in [8], suffer from similar finetuning problems. This is shown by fig. 3 which compares the finetuning performance of several Little Higgs models as a function of $m_h$. These include the Littlest Higgs [15], a modified version of it [16] (curve labelled "Littlest 2"), a Littlest Higgs model with $T$-parity [17] and the so-called Simplest Little Higgs model [18]. Each curve gives the minimum value of $\Delta$ accessible by varying the parameters of the model. For comparison, the curve labelled "SM" gives the fine-tuning of the Little Hierarchy problem in the SM (*i.e.* with $\Lambda = 10$ TeV) and the "MSSM" line shows the fine-tuning of the MSSM (for large $\tan\beta$ but disregarding stop-mixing).

Usually, only in a marginal area of the parameter space of each model is the fine-tuning close to the lower bound shown, so the LH curves in fig. 3 are very conservative estimates of the fine-tuning in the corresponding LH models. Generically, we see from fig. 3 that the value of $\Delta$ for all these models is $\geq \mathcal{O}(100)$ in most of parameter space, and larger that $20-30$ in all cases. Such finetuning is larger than the MSSM one, at least for the especially interesting range $m_h \lesssim 130$ GeV. Notice here that $m_h \gtrsim 135$ GeV is not accesible in the MSSM if SUSY masses are $\lesssim 1$ TeV. This limitation does not hold for other supersymmetric models, e.g. those with low-scale SUSY breaking, as discussed in ref. [10, 14], which are definitely in better shape than LH models concerning finetuning.

I should also emphasize that in order to compare different LH models we chose $f = 1$ TeV ($f_1 = f_2 = 1$ TeV in the Simplest Model). In some models such value is already too low and causes problems with precision electroweak data, which tend to favour larger values of $f$ (that is, they prefer heavier extra particles) [19]. If one takes into account the constraints from precision electroweak data in the finetuning analysis the results would be much worse: generically the finetuning grows with $f$ as $\Delta \propto f^2$.

In conclusion, although LH models solve parametrically the Little Hierarchy problem generically they have a substantial finetuning built-in, usually much higher than suggested by the rough considerations commonly made. This does not demonstrate that all LH models are necessarily finetuned, but stresses the need of a rigorous analysis in order to claim that a particular model is not finetuned.

## "IMPROVED NATURALNESS" MODELS

A different approach to the little-hierarchy problem consists in modifying the Higgs sector of the SM to improve its UV sensitivity. An interesting question in such "improved naturalness" models is whether it is possible to raise the new physics scale $\Lambda$ beyond the reach of LHC while keeping the (modified) model natural. There are several examples of such attempts in the literature [20, 21], all of them enlarging the Higgs sector by adding a second scalar $SU(2)$ doublet. We will discuss explicitly the proposal of [20]. The scalar potential is that of a two-Higgs doublet model, respecting the parity $H_1 \to -H_1$:

$$\begin{aligned} V &= \mu_1^2|H_1|^2 + \mu_2^2|H_2|^2 + \lambda_1|H_1|^4 + \lambda_2|H_2|^4 + \lambda_3|H_1|^2|H_2|^2 \\ &+ \lambda_4|H_1^\dagger H_2|^2 + \frac{\lambda_5}{2}[(H_1^\dagger H_2)^2 + \text{h.c.}]\,. \end{aligned} \tag{7}$$

Furthermore, only $H_2$ is coupled to the top quark. The parameters in this potential are arranged in such a way that *a)* both Higgses get non-zero VEVs ($v_1$ and $v_2$) and *b)* the two neutral Higgses in the spectrum, $h^0$ and $H^0$, with $m_h \leq m_H$, are nearly aligned with the interaction eigenstates, with $h^0 \simeq \sqrt{2}\mathrm{Re}H_1^0$ and $H^0 \simeq \sqrt{2}\mathrm{Re}H_2^0$. This set-up seems promising because the light Higgs, being nearly decoupled from the top, does not suffer from its large quadratically divergent corrections. They affect the heavy Higgs, but if $m_H$ is large, its relative impact is less important. Explicitly, the quadratically divergent corrections to $\mu_1^2$ and $\mu_2^2$ are

$$\begin{aligned}
\delta_q\mu_1^2 &= \frac{3\Lambda^2}{64\pi^2}\left[(3g^2+g'^2)+8\lambda_1+\frac{4}{3}(2\lambda_3+\lambda_4)\right], \\
\delta_q\mu_2^2 &= \frac{3\Lambda^2}{64\pi^2}\left[(3g^2+g'^2)+8\lambda_2+\frac{4}{3}(2\lambda_3+\lambda_4)-8\lambda_t^2\right]. \qquad (8)
\end{aligned}$$

In this way one hopes to be able to raise the naturalness bound on $\Lambda$ by a factor of a few,

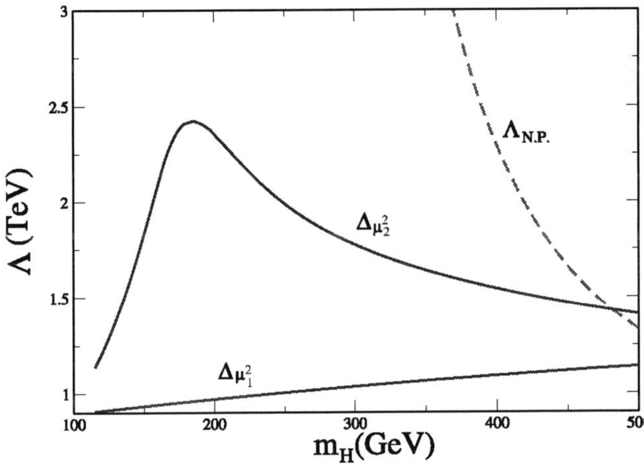


**FIGURE 4.** Upper bound on the scale of New Physics ($\Lambda$) from the absence of 1/10 tuning in $\mu_1^2$ or $\mu_2^2$.

which might be enough to push it beyond the reach of LHC. A careful analysis [11] finds several problems with this idea: *A)* The need to have $v_{1,2} \leq v$ makes naturalness harder by at least a factor of $\sqrt{2}$; *B)* Arranging the eigenstate alignment mentioned above is certainly possible but requires small values of $\lambda_3+\lambda_4+2\lambda_5$ while a sufficiently large mass for the charged Higgs (to avoid problems with $b \to s\gamma$ constraints) requires large enough $\lambda_4+2\lambda_5$. Meeting both demands requires sizeable $\lambda_3$, introducing potentially dangerous $\lambda_3\Lambda^2$ corrections in (8); *C)* Large values of $m_H^2 \simeq 2\lambda_2 v_2^2$ also requires large $\lambda_2$ (which grows even larger at the scale $\Lambda$, relevant for naturalness estimates). In fact $\lambda_2$ would hit a Landau pole near the TeV scale suggesting that any UV completion of this low-energy effective theory will be strongly coupled. This would presumably cause difficulties with precision electroweak tests. Even disregarding such problem, the detailed analysis of finetuning presented in [11] shows that the naturalness bound for this model (especially the one coming from $\mu_1^2$, due to effects from $\lambda_3\Lambda^2$ corrections) is comparable to that of the SM, see fig. 4. In fact, one should multiply the tunings for both $\mu_1^2$ and $\mu_2^2$, in which case the bound would be even tighter than in the SM.

These negative results seem to be quite generic in this type of "improved naturalness" models [11]. The models proposed [20, 21] can have interesting implications but improving the EWSB naturalness of the SM is not a motivation for them.

## ACKNOWLEDGMENTS

I thank the organizers for their invitation and for a stimulating atmosphere and to Alberto Casas and Irene Hidalgo for a very fruitful and enjoyable collaboration. This work is supported in part by a Spanish Ministry for Education and Science project (FPA2004-02015) and by a Comunidad de Madrid project (HEPHACOS; P-ESP-00346).

## REFERENCES

1. R. Decker and J. Pestieau, Lett. Nuovo Cim. **29** (1980) 560; M. J. G. Veltman, Acta Phys. Polon. B **12** (1981) 437.
2. S. Weinberg, Phys. Rev. D **13** (1976) 974; Phys. Rev. D **19** (1979) 1277; L. Susskind, Phys. Rev. D **20** (1979) 2619; G. 't Hooft, PRINT-80-0083 (UTRECHT) *Cargese Summer Inst., Cargese, 1979.*
3. See e.g. R. Barbieri and A. Strumia, [hep-ph/0007265]; R. Barbieri, A. Pomarol, R. Rattazzi and A. Strumia, Nucl. Phys. B **703**, 127 (2004) [hep-ph/0405040].
4. C. F. Kolda and H. Murayama, JHEP **0007** (2000) 035 [hep-ph/0003170].
5. J. A. Casas, J. R. Espinosa and I. Hidalgo, JHEP **0411** (2004) 057 [hep-ph/0410298].
6. M. B. Einhorn and D. R. T. Jones, Phys. Rev. D **46** (1992) 5206.
7. R. Barbieri and G. F. Giudice, Nucl. Phys. B **306** (1988) 63.
8. J. A. Casas, J. R. Espinosa and I. Hidalgo, JHEP **0503** (2005) 038 [hep-ph/0502066].
9. B. de Carlos and J. A. Casas,, Phys. Lett. B **309** (1993) 320 [hep-ph/9303291]. M. Olechowski and S. Pokorski, Nucl. Phys. B **404** (1993) 590. G. W. Anderson and D. J. Castaño, Phys. Lett. B **347** (1995) 300. P. Ciafaloni and A. Strumia, Nucl. Phys. B **494** (1997) 41 [hep-ph/9611204].
10. J. A. Casas, J. R. Espinosa and I. Hidalgo, JHEP **0401** (2004) 008 [hep-ph/0310137].
11. J. A. Casas, J. R. Espinosa and I. Hidalgo, [hep-ph/0607279].
12. P. H. Chankowski, J. R. Ellis and S. Pokorski, Phys. Lett. B **423** (1998) 327 [hep-ph/9712234]; R. Barbieri and A. Strumia, Phys. Lett. B **433** (1998) 63 [hep-ph/9801353]; P. H. Chankowski, J. R. Ellis, M. Olechowski and S. Pokorski, Nucl. Phys. B **544** (1999) 39 [hep-ph/9808275]; G. L. Kane and S. F. King, Phys. Lett. B **451** (1999) 113 [hep-ph/9810374]; M. Bastero-Gil, G. L. Kane and S. F. King, Phys. Lett. B **474** (2000) 103 [hep-ph/9910506].
13. M. Bastero-Gil, C. Hugonie, S. F. King, D. P. Roy and S. Vempati, Phys. Lett. B **489** (2000) 359 [hep-ph/0006198].
14. A. Brignole, J. A. Casas, J. R. Espinosa and I. Navarro, [hep-ph/0301121].
15. N. Arkani-Hamed, A. G. Cohen, E. Katz and A. E. Nelson, JHEP **0207** (2002) 034 [hep-ph/0206021].
16. M. Perelstein, M. E. Peskin and A. Pierce, Phys. Rev. D **69** (2004) 075002 [hep-ph/0310039].
17. H. C. Cheng and I. Low, JHEP **0408** (2004) 061 [hep-ph/0405243].
18. M. Schmaltz, JHEP **0408**, 056 (2004) [hep-ph/0407143].
19. C. Csaki, J. Hubisz, G. D. Kribs, P. Meade and J. Terning, Phys. Rev. D **67**, 115002 (2003) [hep-ph/0211124]; Phys. Rev. D **68**, 035009 (2003) [hep-ph/0303236]; J. L. Hewett, F. J. Petriello and T. G. Rizzo, JHEP **0310** (2003) 062 [hep-ph/0211218]; T. Han, H. E. Logan, B. McElrath and L. T. Wang, Phys. Rev. D **67**, 095004 (2003) [hep-ph/0301040]; M. C. Chen and S. Dawson, Phys. Rev. D **70** (2004) 015003 [hep-ph/0311032].
20. R. Barbieri and L. J. Hall, [hep-ph/0510243].
21. R. Barbieri, T. Gregoire and L. J. Hall, [hep-ph/0509242]; R. Barbieri, L. J. Hall and V. S. Rychkov, [hep-ph/0603188].

# Precise QCD Predictions for Higgs Production at the LHC

Daniel de Florian

*Dpto. de Física, FCEyN-UBA, Buenos Aires, Argentina*

**Abstract.** The status of QCD corrections to Higgs boson production in hadronic colliders is summarized. Results are presented for the transverse momentum distribution at the LHC including the resummation of the dominating logarithmic contributions to all orders in the coupling constant.

## HIGGS PRODUCTION AND QCD

The search for the Higgs boson is among the highest priorities of the LHC physics program [1]. A significant amount of work has been devoted not only to build and design new accelerators and detectors but also to refining the necessary theoretical predictions for the various Higgs production channels and the corresponding backgrounds.

Actually during the last decades, there has been an impressive improvement in perturbative QCD calculations. Several observables have been computed to next-to-leading order (NLO) accuracy and, forced by the precision achieved by the experiments, a great effort is being performed in order to reach the same status at one order higher, NNLO.

For Higgs production, a full NNLO calculation is very complicated since it means counting, among other things, with a two-loop contribution on top of the heavy-quark loop coupling to the Higgs at the lowest order. Nevertheless, a great simplification can be achieved in the large-$M_t$ approximation ($M_t$ being the mass of the top quark), where the effective Lagrangian

$$\mathscr{L}_{ggH} = -\frac{H}{4v} C(\alpha_s) G^a_{\mu\nu} G_a^{\mu\nu} \tag{1}$$

involving only a gluon-gluon-Higgs vertex, as depicted in Figure 1 can be introduced. The LO cross section factors out the dependence on the top mass and the Wilson

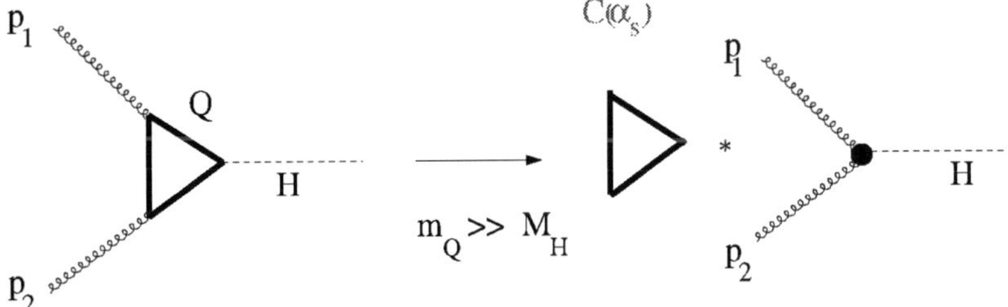


**FIGURE 1.** Hgg vertex in the large-$M_t$ approximation

coefficient $C(\alpha_s)$ is known up to $\alpha_s^5$ [3, 4].

CP917, *Particles and Fields,* edited by H. Castilla Valdez, J. C. D'Olivo, and M. A. Perez

While the approximation is in principle expected to be accurate only for low Higgs masses, it has been found excellent to compute the *relative* contribution of the higher order QCD corrections, within a few percent. Within this approach, NNLO results for Higgs boson production in hadronic colliders have been obtained first for the total rate [5], and more recently for fully exclusive distributions [6]. Higher order contributions are found to be quite large for the total cross-section, with NLO corrections as large as the Born contribution and a non-negligible but more moderate increase at the NNLO level. For less inclusive observables, like the transverse momentum distribution, the size of the corrections can be drastically large, at the point of completely spoiling the perturbative expansion.

This is actually a general feature in perturbative QCD: for some observables, at least under certain extreme kinematical conditions, fixed order expansions in the strong coupling constant $\alpha_s$ are bounded to fail. In order to have a reasonably convergent series, not only the coupling constant should be small, but also the higher order coefficients should be of $\mathscr{O}(1)$. A problem related to the last condition usually occurs in processes involving two or more energy scales at the boundaries of the phase space, where the scales take very different values. Typically a fixed order expansion for a cross section $\sigma = (\mathscr{C}_0 + \mathscr{C}_1\alpha_s + \mathscr{C}_2\alpha_s^2 + \ldots)$ depends on coefficients which contain terms proportional to logarithms of the energies involved $\mathscr{C}_n \simeq \log^m \frac{E_1}{E_2}$ with $m \leq 2n-1$.

When $E_1 \sim E_2 \gg \Lambda$ the coefficient is not affected by the presence of the logs ($\mathscr{C}_n \sim \mathscr{O}(1)$) and, at the same time, the coupling constant (evaluated at any of the scales) is small enough such that the pQCD series can converge.

On the other hand, if one of the scales is much smaller than the other, the large logs in the coefficients can spoil the convergence of the expansion no matter how small $\alpha_s$ is, since the power of the logarithms grows twice as fast as the power of the coupling constant. The origin of these large logarithmic contributions is well known; they arise due to an non-complete cancellation of infrared singularities between real and virtual contributions due to restrictions in the emission of particles in the boundaries of phase space. The appearance of double logs is just related to the fact that the singularities can arise due to both collinear and soft gluon radiation.

Typical examples of processes affected by such a problem are the threshold region for inclusive processes, like Drell-Yan, Higgs and event shape observables in $e^+e^-$ and the production of high mass systems with small transverse momentum in hadronic collisions. As an example of the serious consequences of the appearance of large logarithmic contributions, we show in Figure 2 the LO and NLO transverse momentum distributions of the Higgs boson with $M_H = 125$ GeV at the LHC. While at small $q_T$ the LO cross section increases without limit, diverging to $+\infty$, the NLO becomes negative pointing to $-\infty$, showing that the fixed order expansion makes no-sense at all at small transverse momentum.

In order to allow for a precise QCD description for such a process, the large logarithms should be *resummed* to all orders in $\alpha_S$. The situation is depicted in Table 1. There, a fixed order calculation corresponds to adding all possible terms corresponding to a given *file*, i.e. to a fix power of the coupling constant. When the logarithm involved in the process is large, say when $\alpha_S \log \sim 1$, each term in a file below the one considered can be larger by a factor of $\log \sim 1/\alpha_S$ than the one evaluated at the previous order,

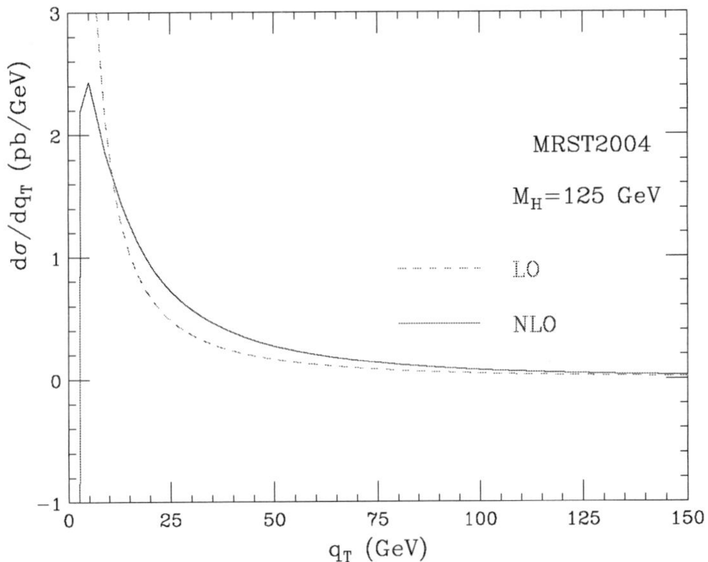


**FIGURE 2.** LO and NLO Higgs $q_T$ distribution

**TABLE 1.** Structure of fixed order and resummed contributions

| $\alpha_S \log$ | . | | | | $\mathcal{O}(\alpha_S)$ LO |
|---|---|---|---|---|---|
| $\alpha_S^2 \log^3$ | $\alpha_S^2 \log^2$ | $\alpha_S^2 \log$ | . | | $\mathcal{O}(\alpha_S^2)$ NLO |
| . | . | . | . | . | . |
| . | . | . | . | . | . |
| $\alpha_S^n \log^{2n-1}$ | $\alpha_S^n \log^{2n-2}$ | . | . | . | $\mathcal{O}(\alpha_S^n) N...LO$ |
| . | . | . | . | . | . |
| . | . | . | . | . | . |
| LL | NLL | NNLL | ... | ... | |

even though they seem to be formally $\alpha_S$ suppressed! On the other hand, if all terms in a given *column* are added, the next column to the right contains terms that are actually $1/\log \sim \alpha_S$ suppressed, allowing to define a convergent expansion (if those terms can be summed to all orders).

The *resummation* can be formally achieved for a certain type of observables by showing that the large logarithms exponentiate in a Sudakov form factor such that the cross section can be written as $\sigma \sim C(\alpha_S)\exp\{\mathcal{G}(\alpha_S, \log)\}\ \sigma_{born}$ plus non-logarithmic contributions.

The Sudakov exponent is then organized according to its logarithmic accuracy; at leading-logarithmic accuracy(LL) only the highest power of the logarithm is included ($\alpha_S^n \log^{n+1}$). The second tower of logarithms in the Sudakov exponent $\alpha_S^n \log^n$ are taken into account at next-to-leading-logarithmic accuracy (NLL) and so on.

Since the resummation is relevant in the region where the logarithms are large and, on the other hand, the fixed order expansion is valid in the opposite kinematical region, it is convenient to match both predictions in order to count with the most general description of a given observable. The matching has to be done in a consistent way in order to avoid double counting. This is achieved by defining the matched cross-section as

$$\sigma^{match} = \sigma^{res} + \sigma^{f.o.(\alpha_s^n)} - \sigma^{res}_{f.o.(\alpha_s^n)} \tag{2}$$

i.e., adding the fixed order and resummed cross-section and subtracting the fixed order expansion of the resummed contribution, which has the same small $q_T$ limit as the fixed order contribution, therefore rendering a finite matched cross-section.

## TRANSVERSE MOMENTUM DISTRIBUTIONS

An example were the resummation of the large logarithms is mandatory corresponds to the production of a Higgs boson with small transverse momentum, where the scales are the transverse momentum $q_T$ and the mass of the Higgs $M_H$.

In this contribution we focus on the dominant SM Higgs production channel, gluon–gluon fusion. When the transverse momentum $q_T$ of the Higgs boson is of the order of its mass $M_H$, the perturbative series is controlled by a small expansion parameter, $\alpha_S(M_H^2)$, and the fixed order prediction is reliable. The leading order (LO) calculation [7] shows that the large-$M_t$ approximation works well as long as both $M_H$ and $q_T$ are smaller than $M_t$. In the framework of this approximation, the NLO QCD corrections have been computed [8, 9, 10, 6].

The small-$q_T$ region ($q_T \ll M_H$) is the most important, because it is here where the bulk of events is expected. In this region the convergence of the fixed-order expansion is spoiled, since the coefficients of the perturbative series in $\alpha_S(M_H^2)$ are enhanced by powers of large logarithmic terms, $\ln^m(M_H^2/q_T^2)$. To obtain reliable perturbative predictions, these terms have to be systematically resummed to all orders in $\alpha_S$ as discussed in the previous section.

According to the QCD factorization theorem the corresponding transverse-momentum differential cross section[1] $d\hat{\sigma}_F/dq_T^2$ can be written as

$$\frac{d\sigma_F}{dq_T^2}(q_T,M,s) = \sum_{a,b}\int_0^1 dx_1 \int_0^1 dx_2\, f_{a/h_1}(x_1,\mu_F^2)\, f_{b/h_2}(x_2,\mu_F^2)\, \frac{d\hat{\sigma}_{F\,ab}}{dq_T^2}(q_T,M,\hat{s};\alpha_S(\mu_R^2))\,, \tag{3}$$

where $f_{a/h}(x,\mu_F^2)$ ($a = q_f, \bar{q}_f, g$) are the parton densities of the colliding hadrons at the factorization scale $\mu_F$, $d\hat{\sigma}_{F\,ab}/dq_T^2$ are the partonic cross sections, $\hat{s} = x_1x_2s$ is the partonic centre-of-mass energy, and $\mu_R$ is the renormalization scale.

To correctly enforce transverse-momentum conservation, the resummation has to be carried out in $b$ space, where the impact parameter $b$ is the variable conjugate to $q_T$. The resummed component of the partonic transverse-momentum cross section in Eq. (3) is

[1] To be precise, when the system $F$ is not a single on-shell particle of mass $M$, what we denote by $d\hat{\sigma}_F/dq_T^2$ is actually the differential cross section $d\hat{\sigma}_F/dM^2dq_T^2$.

then obtained by performing the inverse Fourier (Bessel) transformation with respect to the impact parameter $b$. We write

$$\frac{d\hat{\sigma}_{F\,ab}^{(\text{res.})}}{dq_T^2}(q_T,M,\hat{s};\alpha_S(\mu_R^2)) = \int \frac{d^2\mathbf{b}}{4\pi}\, e^{i\mathbf{b}\cdot q_T}\, \mathscr{W}_{ab}^F(b,M,\hat{s};\alpha_S(\mu_R^2)) \tag{4}$$

$$= \int_0^\infty db\, \frac{b}{2}\, J_0(bq_T)\, \mathscr{W}_{ab}^F(b,M,\hat{s};\alpha_S(\mu_R^2))\,, \tag{5}$$

where $J_0(x)$ is the 0th-order Bessel function.

The perturbative and process-dependent factor $\mathscr{W}_{ab}^F$ embodies the all-order dependence on the large logarithms $\ln M^2b^2$ at large $b$, which correspond to the $q_T$-space terms $\ln M^2/q_T^2$ that are logarithmically enhanced at small $q_T$ (the limit $q_T \ll M$ corresponds to $Mb \gg 1$, since $b$ is the variable conjugate to $q_T$). Resummation of these large logarithms is better expressed by defining the $N$-moments $\mathscr{W}_N$ of $\mathscr{W}$ with respect to $z = M^2/\hat{s}$ at fixed $M$:

$$\mathscr{W}_{ab,N}^F(b,M;\alpha_S(\mu_R^2)) \equiv \int_0^1 dz\, z^{N-1}\, \mathscr{W}_{ab}^F(b,M,\hat{s}=M_H^2/z;\alpha_S(\mu_R^2))\,. \tag{6}$$

The logarithmic terms embodied in $\mathscr{W}_{ab,N}^F$ are due to final-state radiation of partons that are soft and/or collinear to the incoming partons. Their all-order resummation can be organized in close analogy to the cases of soft-gluon resummed calculations for hadronic event shapes in hard-scattering processes and for threshold contributions to hadronic cross sections. We write

$$\mathscr{W}_N^F(b,M;\alpha_S(\mu_R^2),\mu_R^2,\mu_F^2) = \mathscr{H}_N^F\left(M,\alpha_S(\mu_R^2);M^2/\mu_R^2,M^2/\mu_F^2,M^2/Q^2\right) \times \exp\{\mathscr{G}_N(\alpha_S(\mu_R^2),L;M^2/\mu_R^2,M^2/Q^2)\}\,, \tag{7}$$

The function $\mathscr{H}_N^F$ does not depend on the impact parameter $b$ and, therefore, it contains all the perturbative terms that behave as constants in the limit $b \to \infty$. The function $\mathscr{G}$ includes the complete dependence on $b$ and, in particular, it contains all the terms that order-by-order in $\alpha_S$ are logarithmically divergent when $b \to \infty$. This factorization between constant and logarithmic terms involves some degree of arbitrariness, since the argument of the large logarithms can always be rescaled as $\ln M^2b^2 = \ln Q^2b^2 + \ln M^2/Q^2$, provided that $Q$ is independent of $b$ and that $\ln M^2/Q^2 = \mathscr{O}(1)$ when $bM \gg 1$. To parametrize this arbitrariness, on the right-hand side of Eq. (7) we have introduced the scale $Q$, such that $Q \sim M$, and we have defined the large logarithmic expansion parameter, $L$, as

$$L \equiv \ln \frac{Q^2b^2}{b_0^2}\,, \tag{8}$$

where the coefficient $b_0 = 2e^{-\gamma_E}$ ($\gamma_E = 0.5772\ldots$ is the Euler number) has a kinematical origin.

Finally, the Sudakov form factor is given by

$$\mathcal{G}_N(\alpha_S(\mu_R^2), L; M^2/\mu_R^2, M^2/Q^2) = -\int_{b_0^2/b^2}^{Q^2} \frac{dq^2}{q^2} \left[ A(\alpha_S(q^2)) \ln \frac{M^2}{q^2} + \widetilde{B}_N(\alpha_S(q^2)) \right] , \tag{9}$$

where $A(\alpha_S)$ and $\widetilde{B}_N(\alpha_S)$ are perturbative functions

$$A(\alpha_S) = \frac{\alpha_s}{\pi} A^{(1)} + \left(\frac{\alpha_s}{\pi}\right)^2 A^{(2)} + \left(\frac{\alpha_s}{\pi}\right)^3 A^{(3)} + \sum_{n=4}^{\infty} \left(\frac{\alpha_S}{\pi}\right)^n A^{(n)} , \tag{10}$$

$$\widetilde{B}_N(\alpha_S) = \frac{\alpha_s}{\pi} \widetilde{B}_N^{(1)} + \left(\frac{\alpha_s}{\pi}\right)^2 \widetilde{B}_N^{(2)} + \sum_{n=3}^{\infty} \left(\frac{\alpha_S}{\pi}\right)^n \widetilde{B}_N^{(n)} . \tag{11}$$

The coefficients $A^{(n)}$ and $\widetilde{B}_N^{(n)}$, free of large logarithms, are related to the customary coefficients of the Sudakov form factors and of the parton anomalous dimensions.

In the case of the Higgs boson, the calculation of the coefficients needed to perform the resummation has been explicitly worked out at leading logarithmic (LL), next-to-leading logarithmic (NLL) [13], [14] and next-to-next-to-leading logarithmic (NNLL) [15] level.

In the following we present predictions for the Higgs boson $q_T$ distribution at the LHC within the formalism described above. In particular, we include the most accurate description that is available at present: NNLL resummation at small $q_T$ and NLO calculation at large $q_T$. An important feature of our formalism is that a unitarity constraint on the total cross section is automatically enforced, such that the integral of the spectrum reproduces the known inclusive results at NLO [17] and NNLO [5]. More details can be found in Refs. [18] and [19].

For the sake of brevity we concentrate on the quantitative results at NLL+LO (resummed+fixed order) and NNLL+NLO accuracy. At NLL+LO (NNLL+NLO)the NLL (NNLL) resummed result is matched to the LO (NLO) perturbative calculation valid at large $q_T$. As for the evaluation of the fixed-order results, the Monte Carlo program of Ref. [8] has been used. The numerical results are obtained by choosing $M_H = 125$ GeV and using the MRST2002 set of parton distributions [21]. At NLL+LO, LO parton densities and 1-loop $\alpha_S$ have been used, whereas at NNLL+NLO we use NLO parton densities and 2-loop $\alpha_S$.

The NLL+LO results at the LHC are shown in Fig. 3. In the left panel, the full NLL+LO result (solid line) is compared with the LO one (dashed line) at the default scales $\mu_F = \mu_R = M_H$. We see that the LO calculation diverges to $+\infty$ as $q_T \to 0$ while the matched cross-section remains finite in the same limit. The effect of the resummation starts to be relevant below $q_T \sim 100$ GeV. In the right panel we show the NLL+LO band obtained by varying $\mu_F = \mu_R$ between $1/2M_H$ and $2M_H$.

The corresponding NNLL+NLO results are shown in Fig. 4. In the left panel, the full result (solid line) is compared with the NLO one (dashed line) at the default scales $\mu_F = \mu_R = M_H$. The NLO result diverges to $-\infty$ as $q_T \to 0$ and, at small values of $q_T$, it has an unphysical peak that is produced by the numerical compensation of negative leading and positive sub-leading logarithmic contributions. Notice that at $q_T \sim 50$ GeV, the $q_T$ distribution sizably increases when going from LO to NLO and from NLO to NLL+LO.

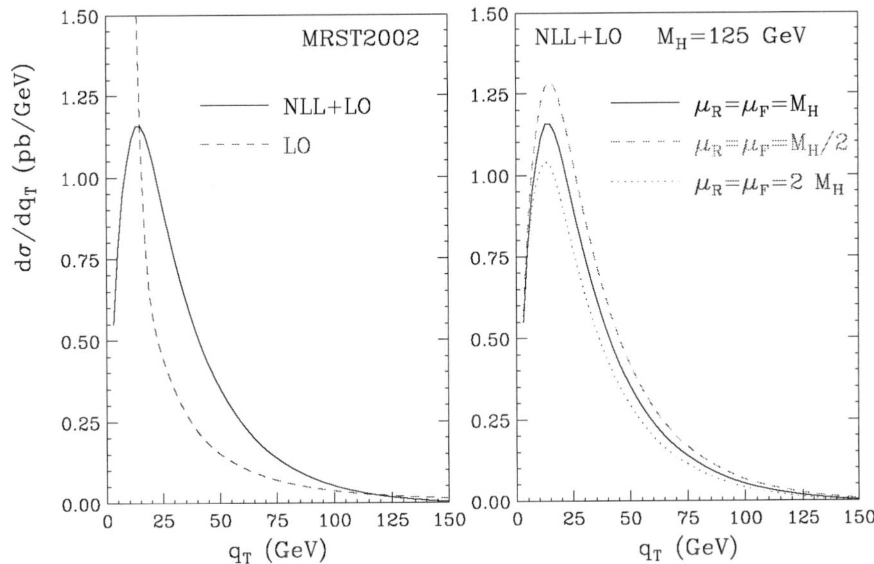


**FIGURE 3.** LHC results at NLL+LO accuracy.

This implies that in the intermediate-$q_T$ region there are important contributions that have to be resummed to all orders rather than simply evaluated at the next perturbative order. The $q_T$ distribution is (moderately) harder at NNLL+NLO than at NLL+LO accuracy while the height of the NNLL peak is a bit lower than the NLL one. This is mainly due to the fact that the total NNLO cross section (computed with NLO parton densities and 2-loop $\alpha_S$), which fixes the value of the $q_T$ integral of our resummed result, is slightly smaller than the NLO one, whereas the high-$q_T$ tail is higher at NNLL order, thus leading to a reduction of the cross section at small $q_T$. The resummation effect starts to be visible below $q_T \sim 100$ GeV, and it increases the NLO result by about 40% at $q_T = 50$ GeV. The right panel of Fig. 4 shows the scale dependence computed as in Fig. 3. Comparing Figs. 3 and 4, we see that the NNLL+NLO band is smaller than the NLL+LO one and overlaps with the latter at $q_T \lesssim 100$ GeV. This suggests a good convergence of the resummed perturbative expansion allowing to reach a theoretical accuracy of the order of 10% for Higgs boson production at the LHC.

## REFERENCES


1. CMS Coll., *Technical Proposal*, report CERN/LHCC/94-38 (1994); ATLAS Coll., *ATLAS Detector and Physics Performance: Technical Design Report*, Vol. 2, report CERN/LHCC/99-15 (1999).
2. *The Higgs Working Group: Summary Report*, to appear in the Proceedings of the Workshop on Physics at TeV Colliders, Les Houches, France, 2003, hep-ph/0406152.
3. Y. Schroder and M. Steinhauser, JHEP **0601**, 051 (2006).
4. K. G. Chetyrkin, J. H. Kuhn and C. Sturm, Nucl. Phys. B **744**, 121 (2006).
5. S. Catani, D. de Florian and M. Grazzini, JHEP **0105** (2001) 025; R. V. Harlander and W. B. Kilgore, Phys. Rev. D **64** (2001) 013015, Phys. Rev. Lett. **88** (2002) 201801; C. Anastasiou and K. Melnikov,

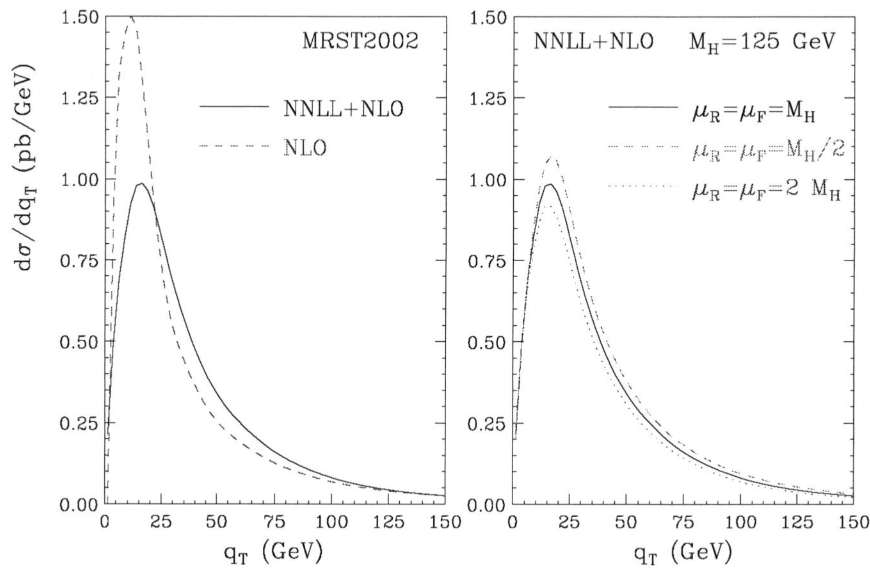


**FIGURE 4.** LHC results at NNLL+NLO accuracy.


Nucl. Phys. B **646** (2002) 220; V. Ravindran, J. Smith, W. L. van Neerven, Nucl. Phys. B **665** (2003) 325.

6. C. Anastasiou, K. Melnikov and F. Petriello, Phys. Rev. Lett. **93**, 262002 (2004); Nucl. Phys. B **724**, 197 (2005); S. Catani and M. Grazzini, arXiv:hep-ph/0703012.
7. R. K. Ellis, I. Hinchliffe, M. Soldate and J. J. van der Bij, Nucl. Phys. B **297** (1988) 221; U. Baur and E. W. Glover, Nucl. Phys. B **339** (1990) 38.
8. D. de Florian, M. Grazzini and Z. Kunszt, Phys. Rev. Lett. **82** (1999) 5209.
9. V. Ravindran, J. Smith and W. L. Van Neerven, Nucl. Phys. B **634** (2002) 247.
10. C. J. Glosser and C. R. Schmidt, JHEP **0212** (2002) 016.
11. G. Parisi and R. Petronzio, Nucl. Phys. B **154** (1979) 427; Y. L. Dokshitzer, D. Diakonov and S. I. Troian, Phys. Rep. **58** (1980) 269; J. C. Collins, D. E. Soper and G. Sterman, Nucl. Phys. B **250** (1985) 199.
12. S. Catani et al., hep-ph/0005025, in the Proceedings of the CERN Workshop on *Standard Model Physics (and more) at the LHC*, eds. G. Altarelli and M.L. Mangano (CERN 2000-04, Geneva, 2000), p. 1.
13. S. Catani, E. D'Emilio and L. Trentadue, Phys. Lett. B **211** (1988) 335.
14. R. P. Kauffman, Phys. Rev. D **45** (1992) 1512.
15. D. de Florian and M. Grazzini, Phys. Rev. Lett. **85** (2000) 4678, Nucl. Phys. B **616** (2001) 247.
16. S. Catani, D. de Florian and M. Grazzini, Nucl. Phys. B **596** (2001) 299.
17. S. Dawson, Nucl. Phys. B **359** (1991) 283; A. Djouadi, M. Spira and P. M. Zerwas, Phys. Lett. B **264** (1991) 440; M. Spira, A. Djouadi, D. Graudenz and P. M. Zerwas, Nucl. Phys. B **453** (1995) 17.
18. G. Bozzi, S. Catani, D. de Florian and M. Grazzini, Phys. Lett. B **564** (2003) 65.
19. G. Bozzi, S. Catani, D. de Florian and M. Grazzini, Nucl. Phys. B **737** (2006) 73.
20. C. Balazs and C. P. Yuan, Phys. Lett. B **478** (2000) 192; E. L. Berger and J. w. Qiu, Phys. Rev. D **67** (2003) 034026; A. Kulesza, G. Sterman, W. Vogelsang, Phys. Rev. D **69** (2004) 014012.
21. A. D. Martin, R. G. Roberts, W. J. Stirling and R. S. Thorne, Eur. Phys. J. C **28** (2003) 455.

# Higgs diffractive production through intrinsic flavors

Boris Kopeliovich and Ivan Schmidt

*Departamento de Física, Universidad Técnica Federico Santa María, Casilla 110-V, Valparaíso, Chile*

**Abstract.** Higgs particles can be produced diffractively in the fragmentation region of one of the colliding protons. This mechanism will provide a clear experimental signal for Higgs production, due to the small background in this kinematic region. The key assumption underlying our analysis is the presence of intrinsic heavy flavor components of the proton bound state, whose existence at high light-cone momentum fraction $x$ has growing experimental and theoretical support.

**Keywords:** Higgs,diffraction
**PACS:** 14.80.Bn, 12.38.Bx

## INTRODUCTION

The main problems for detecting Higgs particles are: completely unknown mass, small production probabilities, and large backgrounds. Since this particle gives masses to quarks, its coupling is proportional to the quark mass. It is therefore very difficult to produce a Higgs particle in a collision of protons, which consist mainly of light quarks. However, the proton contains some amount of intrinsic heavy flavors as well. Although the fraction is tiny, its smallness is compensated by a huge enhancement proportional to the heavy quark mass squared. In order to reduce the background from multiple production of hadrons, we propose to search for a Higgs particle produced diffractively in the fragmentation region of one of the colliding protons. The presence of a large rapidity gap and the possibility to perform a missing mass measurement, which needs detection of only two protons, substantially increases the chances to detect the Higgs particle. In this case the magnitude of the cross section of Higgs production reaches a level which can be measured at LHC with a proper trigger system.

## LARGE RAPIDITY GAP PROCESSES

Perhaps the most novel production process for the Higgs is the double large rapidity gap exclusive diffractive reaction, $pp \to p + H + p$ [1], where the + sign stands for the large rapidity gap (LRG) between the produced particles. If both protons are detected, the mass and momentum distribution of the Higgs can be determined. The TOTEM detector [2] proposed for the LHC will have the capability to detect exclusive diffractive channels. The detection of the Higgs via the exclusive diffractive process $pp \to p + H + p$, has the advantage that it does not depend on a specific decay mechanism for the Higgs. The branching ratios for the decay modes of the Higgs can then be individually

CP917, *Particles and Fields,* edited by H. Castilla Valdez, J. C. D'Olivo, and M. A. Perez

determined by combining the measurement of $\sigma(pp \to p+H+p)$ with the rate for a specific diffractive final state $B_f\ \sigma(pp \to p+H_{\to f}+p)$. This is in contrast to the standard inclusive measurement, where one can only determine the product of the cross section and branching ratios $B_f\ \sigma(pp \to H_{\to f}X)$.

The existing theoretical estimates for diffractive Higgs production are based on the gluon-gluon fusion subprocess, where two hard gluons couple to the Higgs ($gg \to H$) [1]. A third gluon is also exchanged in order that both projectiles remain color singlets. Perturbative QCD then predicts $\sigma(pp \to p+H+p) \simeq 3$ fb for the production of a Higgs boson of mass 120 GeV at LHC energies, with a factor of 2 uncertainty [1]. Since the annihilating gluons each carry a small fraction of the momentum of the proton, the Higgs is primarily produced in the central rapidity region.

Here we will specifically consider the single LRG exclusive diffractive production reaction $pp \to pM+p$, where $M$ stands for $J/\psi, \chi_c, \Upsilon, \chi_b, Z^0$ or $H$. The final state $M$ will be produced in the projectile proton's fragmentation region with a significant fraction $x_F$ of the incident proton's momentum, since the sum of the momenta of two heavy quarks contribute to the momentum of $M$. This has an important advantage of providing a distinctive signal with relatively small background. This production process is analogous to the positron-antiproton coalescence reaction by which anti-hydrogen was first detected [3, 4, 5].

## INTRINSIC HEAVY FLAVORS

It was originally suggested in Refs. [6, 7] that there is a $\sim 1\%$ probability of intrinsic charm Fock states in the nucleon; more recently, the operator product expansion has been used to show that the probability for Fock states in light hadron to have an extra heavy quark pair of mass $M_Q$ decreases only as $\Lambda^2_{QCD}/M^2_Q$ in non-Abelian gauge theory [8].

Therefore we assume the presence in the proton of an intrinsic heavy flavor (IQ) component, a $\bar{Q}Q$ pair, which is predominantly in a color-octet state, and which has either a nonperturbative or perturbative origin. In the former case this heavy component can interact strongly with the $3q$ valence quark component. As in charmonium, the mean $\bar{Q}Q$ separation should be considerably larger than the transverse size $1/m_Q$ of perturbative $\bar{Q}Q$ fluctuations. For instance, if the binding potential has the oscillator form, the mean distance is

$$\langle r^2_{\bar{Q}Q}\rangle = \frac{2}{\omega m_Q}\,, \tag{1}$$

where $\omega \sim 300$ MeV is the oscillation frequency. Alternatively, the IQ component can be considered to derive perturbatively from the minimal gluonic couplings of the heavy quark pair to two valence quarks of the proton; this is likely the dominant mechanism at the largest values of $x_Q$. In this case the transverse separation of the $\bar{Q}Q$ is controlled by the energy denominator, $\langle r^2_{\bar{Q}Q}\rangle = 1/m^2_Q$, and is much smaller than the estimate given by Eq. (1). In this paper we will consider only this last possibility, which turns out to be the dominant one for Higgs production.

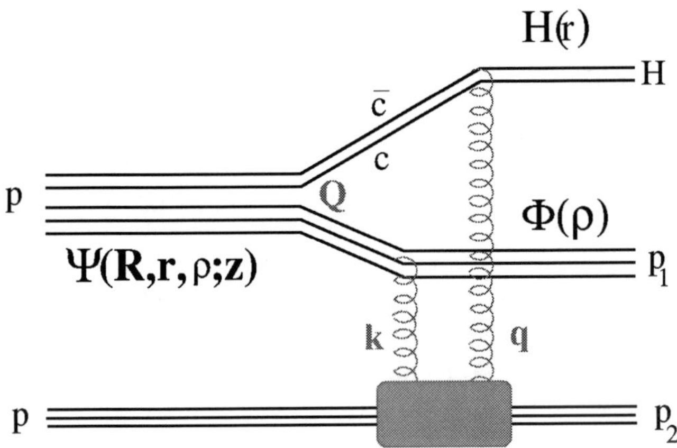


**FIGURE 1.** The two-gluon exchange diagram for Higgs exclusive production.

## CALCULATION OF EXCLUSIVE DIFFRACTION

The cross section of exclusive diffractive production of the Higgs, $pp \to Hp + p$, can be estimated in the light cone (LC) dipole approach [9]. The Born graph for this process is shown in Fig. 1.

As noted above, we shall assume that the projectile (upper) proton has an approximate 1% probability to fluctuate to an IQ Fock component with the color structure $|[uud]_{8_C}[\bar{Q}Q]_{8_C}>$. This virtual state has a long coherence length in a high energy collision $\propto s/\mathcal{M}^2 M_p$, where $\mathcal{M}$ is the total invariant final-state mass. In a $pp$ collision, two soft gluons must be exchanged in order to keep both protons intact and to create a rapidity gap, mimicking pomeron exchange. For example, as shown in Fig. 1, one of the exchanged gluon can be attached to the $d$ valence quark spectator in $|[uud]_{8_C}[\bar{Q}Q]_{8_C}>$, changing its color, and the other one can be attached to the $\bar{Q}$, also changing its color. The net effect of this color rearrangement is the same as single-gluon exchange between the two color-octet clusters. The $\bar{Q}Q$ and the $uud$ can thus emerge as color singlets because of the gluonic exchange. The $[\bar{Q}Q]_{1_C}$ can couple to the $J/\psi$, or to a $Z^0$ or to a $H$. Meanwhile the color-singlet $uud$ gives rise to the scattered proton, thus producing the rapidity gap in the final state. Notice that the $x_F$ distribution of the produced particle is approximately the same as the distribution of the $[\bar{Q}Q]$ inside the proton, and therefore $x_F \simeq x_Q + x_{\bar{Q}}$. The sum of couplings of the gluon to all of the quarks, as dictated by gauge invariance, brings in a form factor which vanishes at zero momentum transfer, thus giving an important suppression factor.

The diffractive cross section has the form,

$$\frac{d\sigma(pp \to ppH)}{dx_2\, d^2p_1\, d^2p_2} = \frac{1}{(1-x_2)16\pi^2} \left|A(x_2, \vec{p}_1, \vec{p}_2)\right|^2 , \tag{2}$$

where the diffractive amplitude in Born approximation reads,

$$
\begin{aligned}
A(x_2,\vec{p}_1,\vec{p}_2) &= \frac{8}{3\sqrt{2}}\int d^2Q\frac{d^2q}{q^2}\frac{d^2k}{k^2}\,\alpha_s(q^2)\alpha_s(k^2)\,\delta(\vec{q}+\vec{p}_2+\vec{k})\,\delta(\vec{k}-\vec{p}_1-\vec{Q}) \\
&\times \int d^2\tau\,|\Phi_p(\tau)|^2\left[e^{i(\vec{k}+\vec{q})\cdot\vec{\tau}/2}-e^{i(\vec{q}-\vec{k})\cdot\vec{\tau}/2}\right]\int d^2R\,d^2r\,d^2\rho\,H^\dagger(\vec{r})\,e^{i\vec{q}\cdot\vec{r}/2} \\
&\times \left(1-e^{-i\vec{q}\cdot\vec{r}}\right)\Phi_p^\dagger(\vec{\rho})e^{i\vec{k}\cdot\vec{\rho}/2}\left(1-e^{-i\vec{k}\cdot\vec{\rho}}\right)\Psi_p(\vec{R},\vec{r},\vec{\rho},z)\,e^{i\vec{Q}\cdot\vec{R}}. \qquad (3)
\end{aligned}
$$

Here $\Psi_p(\vec{R},\vec{r},\vec{\rho},z)$ is the light-cone wave function of the IQ component of the projectile proton with transverse separations $\vec{R}$ between the $\bar{Q}Q$ and $3q$ clusters, $\vec{r}$ between the $c$ and $\bar{c}$, $\vec{Q}$ is the relative transverse momentum of the $3q$ and $\bar{Q}Q$ clusters in the projectile and $\vec{\rho}$ is the transverse separation of the quark and diquark which couple to the final-state proton $p_2$. The density $|\Phi_p(\tau)|^2$ is the wave function of the target proton which we also treat as a color dipole quark-diquark with transverse separation $\tau$. (The extension to three quarks is straightforward [9]). The fraction of the projectile proton light-cone momentum carried by the $\bar{Q}Q$, $z \approx 1-x_1$. This wave function is normalized as,

$$
\int_0^1 dz\int d^2R\,d^2r\,d^2\rho\,\left|\Psi_p(\vec{R},\vec{r},\vec{\rho},z)\right|^2 = P_{IQ}\,, \qquad (4)
$$

where $P_{IQ}$ is the weight of the IQ component of the proton, which is suppressed as $1/m_Q^2$, and is assumed to be $P_{IC}\sim 1\%$ for the case of charm. The amplitudes $H(\vec{r})$ and $\Phi_p(\vec{\rho})$ denote the wave functions of the produced Higgs and the outgoing proton, respectively, in accordance with Fig. 2.

The phase factors in Eq. (3) correspond to different attachments of the exchange gluons to quarks in Fig. 1. Thus, attaching the gluon either to the $Q$, or to the $\bar{Q}$ quarks one gets factor $[\exp(i\vec{q}\cdot\vec{r}/2)-\exp(-i\vec{q}\cdot\vec{r}/2)]$. An analogous factor corresponding to the second gluon coupling to the proton $p_1$ is also included in Eq. (3). The transverse coordinates of the quark and diquark in the target proton are $\tau/2$ and $-\tau/2$ (relative to its center of gravity). The phase factor in the square brackets in Eqn. (3) thus includes two terms corresponding to attachment of the exchanged gluons to the same or different valence quark or diquark in $p_2$.

In order to proceed further, we make the following simplifying steps:

1) assume a factorized form of the proton wave function,

$$
\Psi_p(\vec{R},\vec{r},\vec{\rho},z) = \Psi_{IQ}(\vec{R},z)\,\Psi_{\bar{Q}Q}(\vec{r})\,\Psi_{3q}(\vec{\rho})\,. \qquad (5)
$$

Here $\Psi_{\bar{Q}Q}$ and $\Psi_{3q}$ are the $\bar{Q}Q$ and $3q$ wave functions normalized to unity, whereas $\Psi_{IQ}(\vec{R},z)$ is the wave function describing the relative motion of the $\bar{Q}Q$ and $3q$ clusters, where $z$ is the fraction of the longitudinal momentum carried by the $\bar{Q}Q$. This wave function is normalized as,

$$
\int d^2R\,\left|\Psi_{IQ}(\vec{R},z)\right|^2 = P_{IQ}(z)\,, \qquad (6)
$$

where $P_{IQ}(z)$ is the $z$-distribution of $\bar{Q}Q$, related to the $x_1$ distribution of the produced protons, since with a very high precision $z = 1 - x_1 = M_H^2/s(1-x_2)$ (unless $x_1$ is as small as $x_1 \sim 2m_p/\sqrt{s}$).

2) perform the calculations in Eq. (3) only for forward diffraction, i.e. $p_2 = 0, \vec{q} = -\vec{k}$, and we assume for the Pomeron the typical Gaussian $t$-dependence ($t = -p_2^2$),

$$\frac{d\sigma}{d^2p_1\, d^2p_2} \propto e^{-B(s')p_2^2} \,, \tag{7}$$

so the $t$-integrated cross section then reads,

$$\frac{d\sigma}{d^2p_1\, dx_2} = \frac{\pi}{B(s')} \frac{d\sigma}{d^2p_1\, d^2p_2\, dx_2}\bigg|_{p_2=0} \,. \tag{8}$$

Here the slope $B(s') \sim B_0 + 2\alpha'_{I\!P} \ln(s'/M_0^2)$, where $B_0 = 4\,\mathrm{GeV}^{-2}$, $\alpha'_{I\!P} = 0.25\,\mathrm{GeV}^{-2}$, $s'/M_0^2 = s/M_H^2$ and $M_0 = 1\,\mathrm{GeV}$.

3) replace the two-gluon proton vertex, represented by the integral over $\vec{\tau}$ in Eq. (3), by the unintegrated gluon density, $\mathcal{F}(x,k^2) = \partial G(x,k^2)/\partial(\ln k^2)$, where $G(x,Q^2) = x\,g(x,Q^2)$. This preserves the infra-red stability of the cross section, since $\mathcal{F}$ vanishes at $k^2 \to 0$. The phenomenological gluon density fitted to data includes by default all higher order corrections and supplies the cross section with an energy dependence important for extrapolation to very high energies

Finally, combining all the above modifications and performing the $p_1$-integration in Eq. (3), we arrive at,

$$\begin{aligned} \frac{d\sigma^{IQ}(pp \to ppH)}{dx_2} &= \frac{32\pi P_{IQ}(z)}{9B(s')(1-x_2)} \bigg| \int \frac{d^2k}{k^4}\, \alpha_s(k^2)\, \mathcal{F}(x,k^2) \\ &\times \int d^2r H^\dagger(\vec{r}) e^{-i\vec{k}\cdot\vec{r}/2} \left(1 - e^{i\vec{k}\cdot\vec{r}}\right) \Psi_{\bar{Q}Q}(\vec{r}) \\ &\times \int d^2\rho\, \Phi_p^\dagger(\vec{\rho}) e^{-i\vec{k}\cdot\vec{\rho}/2} \left(1 - e^{i\vec{k}\cdot\vec{\rho}}\right) \Psi_{3q}(\vec{\rho}) \bigg|^2 \,. \end{aligned} \tag{9}$$

Here

$$z = [x_F^H + \sqrt{(x_F^H)^2 + 4M_H^2/s}]/2 \approx x_F^H \approx 1 - x_1 = M_H^2/s(1-x_2) \,. \tag{10}$$

This relation receives sizeable corrections only at very small Higgs Feynman $x_F^H \sim 2M_H/\sqrt{s}$. Notice that the expansion of the exponentials in Eq. (9) contains only odd powers of $\vec{k}\cdot\vec{r}$ and $\vec{k}\cdot\vec{\rho}$. This signals a change of orbital momentum of the quark configurations participating in the one-gluon exchange process. In order to obtain a nonzero result of the integration over $\vec{r}$, either the initial or the final $\bar{Q}Q$ wave function must contain a factor $\vec{\nabla}_r$, i.e. it must be a $P$-wave. Since we assume that the Higgs is a scalar, its $\bar{Q}Q$ component must be in a $P$-wave state, while the primordial $\bar{Q}Q$ in the projectile IQ state should be in an $S$-wave. This is vice versa for the proton $p_1$: the final $|3q\rangle$ system is in an $S$-wave, but $\Psi_{3q}(\vec{\rho})$ must be a $P$-wave.

Notice that both the scalar Higgs and $\chi$ states may be produced from the same IQ component of the proton containing $S$-wave $\bar{Q}Q$. However, the production of $J/\psi$, $\Upsilon$,

$Z^0$ requires an IQ component containing a $P$-wave $\bar{Q}Q$, which is presumably more suppressed.

## HIGGS WAVEFUNCTION

The $P$-wave LC wave function of Higgs in impact parameter representation is given by the Fourier transform of its Breit-Wigner propagator:

$$H(\vec{r}) = i\frac{\sqrt{N_c G_F}}{2\pi} m_Q \bar{\chi}\vec{\sigma}\chi\frac{\vec{r}}{r}\left[\varepsilon Y_1(\varepsilon r) - \frac{ir}{2}\Gamma_H M_H Y_0(\varepsilon r)\right] . \tag{11}$$

Here $G_F$ is the Fermi constant, $\chi$ and $\bar{\chi}$ are the spinors for $c$ and $\bar{c}$ respectively and

$$\varepsilon^2 = \alpha(1-\alpha)M_H^2 - m_Q^2 , \tag{12}$$

where $\alpha$ is the fraction of the LC momentum of the Higgs carried by the $c$-quark. The functions $Y_{0,1}(x)$ in Eq. (11) are the second order Bessel functions and $\Gamma_H$ is the total width of the Higgs. Assuming $\Gamma_G \ll M_H$, we neglect the second term in Eq. (11).

The LC wave function Eq. (12) assumes that the Higgs mass is much larger than the quark masses, which is probably true for charm and bottom. However, it is quite probable that for top-antitop in the Higgs $2m_t > M_H$, then the wave function is different,

$$H_{\bar{t}t}(\vec{r}) = \frac{\sqrt{N_c G_F}}{2\pi} m_t \bar{\chi}\vec{\sigma}\chi\frac{\vec{r}}{r}\varepsilon_t K_1(\varepsilon_t r) , \tag{13}$$

where $K_1(x)$ is the modified Bessel function and

$$\varepsilon_t^2 = m_t^2 - \alpha(1-\alpha)M_H^2 . \tag{14}$$

## IQ WAVEFUNCTION

Since the Higgs is produced from an $S$-wave $\bar{Q}Q$, the perturbative distribution amplitude is ultraviolet stable and can be normalized to one, in order to correspond to $P_Q$ as a probability to have such a heavy quark pair in the proton

$$\Psi^{pt}_{\bar{Q}Q}(\vec{r}) = \frac{m_Q}{\sqrt{\pi}} K_0(m_Q r) . \tag{15}$$

Here the modified Bessel function $K_0(m_Q r)$ is the Fourier transform of the energy denominator associated with the $\bar{Q}Q$ fluctuation. We assume the $c$ and $\bar{c}$ quarks to carry equal fractional momenta. For fixed $\alpha_s$ the energy denominator governs the probability of the fluctuation in momentum space, since perturbatively one treats the heavy quarks as free particles.

## FEYNMAN $x_F^H$ DISTRIBUTION OF HIGGS PARTICLES

The $x_F^H$ distribution of the cross section is related to the LC wave function $\Psi_{IQ}(R,z)$ of the system $3q-\bar{Q}Q$, namely to the function $P_{IQ}(z)$ defined in Eq. (6). The momentum fraction $z$ is related to $x_{1,2}$ and $x_F^H$ by Eq. (10).

The light-cone wave function of a perturbative fluctuation $p \to |3q\bar{Q}Q\rangle$ in momentum representation is controlled by the energy denominator,

$$\Psi_{IQ}(Q,z,\kappa) \propto \frac{z(1-z)}{Q^2+z^2m_p^2+M_{\bar{Q}Q}^2(1-z)} . \tag{16}$$

Momentum $\vec{Q}$ was defined in Fig. 1 and Eq. (3). The effective mass of the $\bar{Q}Q$ depends on the intrinsic transverse momentum of the $\bar{Q}Q$ pair, $M_{\bar{Q}Q}^2 = 4(\kappa^2+m_Q^2)$. It is controlled by the convolution of the IQ $\bar{Q}Q$ wave function with the $P$-wave $\bar{Q}Q$ wave function in the Higgs and the one-gluon exchange amplitude (see Fig. 1), which has the form,

$$\int\limits_0^\infty d\kappa^2\,\Psi_{IQ}(Q,z,\kappa)\left[H_{\bar{Q}Q}(\vec{\kappa}+\vec{k}/2)-H_{\bar{Q}Q}(\vec{\kappa}-\vec{k}/2)\right]$$

$$\propto \quad z(1-z)\,\frac{\ln\left[\frac{|M_H^2-4m_Q^2|(1-z)}{Q^2+4m_Q^2(1-z)+m_p^2z^2}\right]}{M_H^2(1-z)+Q^2+m_p^2z^2} . \tag{17}$$

This expression peaks at $1-z\sim m_p/M_H$, therefore the logarithmic factor hardly varies as function of $Q^2$ which is restricted by the proton form factor. Making use of this we arrive at the following $z$-distribution,

$$\frac{P_{IQ}(z)}{P_{IQ}} = Nz(1-z)\,\frac{\left\{\ln\left[\frac{|M_H^2-4m_Q^2|(1-z)}{4m_Q^2(1-z)+m_p^2z^2}\right]\right\}^2}{M_H^2(1-z)+m_p^2z^2} , \tag{18}$$

where $N$ is a constant normalizing to one the integral over $z$.

## FINAL RESULTS

Putting all the previous results together, and including absorptive corrections we calculated the total cross sections for diffractive Higgs production, $pp \to Hpp$, from the IQ components. More details can be found in Ref. [10]. The results at the energy of LHC, $\sqrt{s} = 14\,\text{TeV}$, are plotted as function of Higgs mass in the Fig. 2. We assume a $1/m_Q^2$ scaling for the intrinsic component weights, and a 1% probability of IC.

## REFERENCES


1. A. De Roeck, V. A. Khoze, A. D. Martin, R. Orava and M. G. Ryskin, Eur. Phys. J. C **25**, 391 (2002) [arXiv:hep-ph/0207042], and references therein.

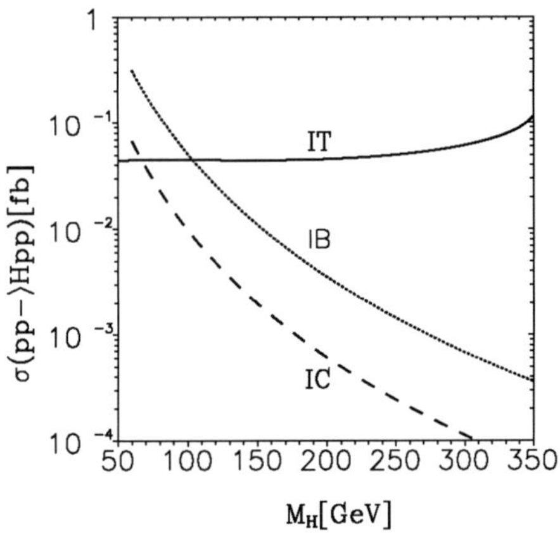


**FIGURE 2.** The cross section of the reaction $pp \to Hp + p$ as function of the Higgs mass. Contributions of IQ (dashed) of IB (dotted) and IT (solid).


2. M. Deile [TOTEM Collaboration], arXiv:hep-ex/0410084.
3. C. T. Munger, S. J. Brodsky and I. Schmidt, Phys. Rev. D **49**, 3228 (1994).
4. G. Baur *et al.*, Phys. Lett. B **368**, 251 (1996).
5. G. Blanford, D. C. Christian, K. Gollwitzer, M. Mandelkern, C. T. Munger, J. Schultz and G. Zioulas [E862 Collaboration], Phys. Rev. Lett. **80**, 3037 (1998).
6. S. J. Brodsky, C. Peterson and N. Sakai, Phys. Rev. D **23**, 2745 (1981).
7. S. J. Brodsky, P. Hoyer, C. Peterson and N. Sakai, Phys. Lett. B **93**, 451 (1980).
8. M. Franz, V. Polyakov and K. Goeke, Phys. Rev. D **62**, 074024 (2000) [arXiv:hep-ph/0002240].
9. B.Z. Kopeliovich, L.I. Lapidus and A.B. Zamolodchikov, Sov. Phys. JETP Lett. **33**, 612 (1981).
10. S. J. Brodsky, B.Z. Kopeliovich, Ivan Schmidt and Jacques Soffer, Phys. Rev. **D73**, 113005 (2006).

# Results of Amplitude Analyses from D-meson Decays

Carla Göbel[1]

*Depto. de Física, Pontifícia Universidade Católica, Rio de Janeiro, Brazil*

**Abstract.**
The decays of charmed mesons can provide valuable information about hadron physics and espectroscopy. D-meson hadronic decays proceed mainly through resonant intermediate states. To study these processes, amplitude analysis techniques have been widely used. Here we present some results on this subject coming from Fermilab experiments E791 and FOCUS. In particular, we present studies of the $\pi\pi$ and $K\pi$ S-wave systems from 3- and 4-body D-meson decays.

**Keywords:** Amplitude analysis, Dalitz plot, D-meson decays
**PACS:** 13.25.Ft,13.20.Fc,14.40.Cs

## INTRODUCTION

It is well known that hadronic decays of D mesons often proceed through the production of intermediate resonances. Usually, by the use of amplitude analysis, one can access information on the contributing resonant channels, extracting, for instance, relative strenghts and phases.

A few years ago the Fermilab E791 Collaboration published a series of 3-body D decay analyses [1, 2, 3] where the Dalitz plots were fitted not only to extract strenghts and phases, but also resonant parameters, e.g., masses and widths for not-well established or controversial states. From $D^+ \to \pi^-\pi^+\pi^+$ [2] and $D^+ \to K^-\pi^+\pi^+$ [3] decays evidence for low mass, broad scalar states, identified as the $\sigma$ and $\kappa$, was presented.

Since then, and with the rising statistics of charm samples, the decays of charm mesons has become a new source of information for the study of light meson spectroscopy, complementary to that from scattering experiments. This is particularly relevant for the understanding of the scalar sector. Various analyses from different experiments have followed [4, 5, 6, 7, 8, 9] where charm-meson decays were used to bring information on light meson espectroscopy.

A nice feature of D-decays it that the 2-body system ($K\pi$ and $\pi\pi$, for example) can be studied from threshold up to about 1.7 GeV. This is a big advantage compared to scattering processes. For instance, LASS [10] data on $K\pi$ scattering starts at 825 GeV, about 200 MeV above threshold. Besides, the scalar component in hadronic D-decays seems to be favored in 3-body final states where a pair of identical particles appears. This is for instance what happens in $D^+, D_s^+ \to \pi^-\pi^+\pi^+$ and $D^+ \to K^-\pi^+\pi^+$.

---

[1] e-mail:gobel@fis.puc-rio.br. Member of FOCUS and E791 Collaborations.

CP917, *Particles and Fields,* edited by H. Castilla Valdez, J. C. D'Olivo, and M. A. Perez

Here, we overview the main results obtained by Fermilab E791 and FOCUS collaborations on various amplitude analyses from D-decays, including 3- and 4-body hadronic and semileptonic channels. From those, interesting informations were obtained to help the understanding of the $\pi\pi$ and $K\pi$ S-wave systems.

## $\pi\pi$ S-WAVE FROM D-DECAYS

The possible existence of a light scalar state $\sigma$ has been under debate for the last 40 years or so. The difficulty in establish its existence has relied on the lack of clear evidence for its production from scattering processes. One well known reason for this is the presence of the Adler zeroes, which suppress the production of low-mass scalar resonance. This effect is absent in production processes, like D decays.

The first evidence for the production of a light $\pi\pi$ scalar state, $\sigma$, in charm decays came from Fermilab E791, in the study of the decay $D^+ \to \pi^-\pi^+\pi^+$ [2]. The Dalitz plot was initially fitted using a coherent sum of all known $\pi\pi$ resonances and a non-resonant term, yielding a bad description of the data. The inclusion of an extra scalar state improves the fit substantially. The mass and the width of the $\sigma$, parametrized as a Breit-Wigner, were measured to be $m_\sigma = 478^{+24}_{-23} \pm 17$ MeV and $\Gamma_\sigma = 324^{+42}_{-40} \pm 21$ MeV. The channel $\sigma\pi^+$ was found to contribute to 50% of the decay rate. Soon after, CLEO [4], BaBaR [5], Belle [6] and BES [7, 8] also found further evidence for the production of this state in charm meson decays.

In all of the above analyses the S-wave component of the decay amplitude included a Breit-Wigner for the $\sigma$. Although the Breit-Wigner is not expected to be a good representation for broad and near threshold states, very good fits to the data are obtained in all cases, so this form can be seen as an effective representation of the amplitude for real values of energy, where experimental data lies. This means that, even though there is strong agreement for the $\sigma$ Breit-Wigner parameters from different D decays analyses, one should not used them to get directly the $\sigma$ pole position.

An attempt to extract the low-mass $\pi\pi$ phase variation in a model independent way was performed by Miranda and Bediaga [11] using the E791 data. The approach they have developed, the Amplitude Difference Method, uses the interference between the $f_2(1270)$ and the low-mass S-wave in the decay $D^+ \to \pi^-\pi^+\pi^+$ to extract the S-wave phase. The result confirmed the resonant behavior at low $\pi\pi$ mass, although the statistics was limited. This method could be used when larger samples become available.

In order to address the limitation of the description of broad scalar states as simple Breit-Wigners, FOCUS has used the K-matrix approach to describe the $\pi\pi$ system in the $D^+ \to \pi^-\pi^+\pi^+$ channel [12]. The K-matrix model works nicely in representing overlapping resonances; the total S-wave propagator should include all possible poles from the scattering matrix, and it imposes 2-body unitarity. The poles are obtained to a fit to $\pi\pi$ scattering data. By using this approach to fit the data, a good representation is obained. The same data can be also fitted with a $\sigma$ Breit-Wigner, yielding mass and width which are very similar to the E791 parameters. The use of K-matrix to fit the $\pi\pi$ S-wave from $D^+ \to \pi^-\pi^+\pi^+$ assumes that the $\pi\pi$ system is well isolated from the third pion. Under this assumption, the S-wave phase motion in $D^+ \to \pi^-\pi^+\pi^+$ should match that from scattering, as a consequence of the Watson theorem [13]. Nevertheless, this assumption is far from trivial, since the $\pi\pi$ is embedded in a 3-body, strongly interacting final state.

## $K\pi$ S-WAVE FROM D-DECAYS

### $D^+ \to K^-\pi^+\pi^+$ Decay from E791

Fermilab experiment E791 reported on the indication of a light scalar $K\pi$ resonance [3], $\kappa$, based on a full Dalitz plot analysis of the decay $D^+ \to K^-\pi^+\pi^+$. As in the case of $D^+ \to \pi^-\pi^+\pi^+$, the first approach to fit the data included the known $K^*$ resonances plus a constant NR term, yielding a very bad description of the data. The inclusion of an extra scalar state, identified as the $\kappa$ resonance, improved the fit substantially. The measured Breit-Wigner mass and width of this state were found to be $797 \pm 19 \pm 43$ MeV/c$^2$ and $410 \pm 43 \pm 87$ MeV/c$^2$, respectively. In a complementary study [14], E791 tried to model the S-wave following LASS [10] in their analysis of $K\pi$ scattering data. The S-wave phase included only the well established scalar $K^{*0}(1430)$ plus a non-resonant component described through an effective-range form. E791 data could not be fitted imposing $K\pi$ unitarity constraint, while by relaxing this condition a good fit was obtained, with parameters for the supposed non-resonant decay playing the role of the missing $\kappa$.

Further evidence for the $\kappa$ appeared from BES [9] by analysing $J/\Psi \to \overline{K}^*(892)^0 K^+\pi^-$ decays. They also fitted this scalar state using a Breit-Wigner parametrization and obtained values consistent to those from E791.

Recently, a model independent partial wave analysis (MIPWA) of the $D^+ \to K^-\pi^+\pi^+$ decay was performed by E791 [15]. In this work, the S-wave amplitude was parameterized by a generic complex function, $\mathscr{A} = f(s)e^{i\phi(s)}$, with $s$ being the $K\pi$ invariant mass squared. The P- and D-waves were parameterized in the framework of the isobar model, as in [3]. $K\pi$ mass spectrum was divided in bins and, for each bin, the magnitude and the phase of $\mathscr{A}(s_k) = a_k e^{i\phi_k}$ were determined from the fit, where the reference channel was the $\overline{K}^*(892)^0$. In Fig. 1 we show the resulting S-wave phase variation. In the same figure we plot the $K\pi$ S-wave phase variation obtained by LASS for isospin 1/2. A direct comparison between the two can be made if one assumes that the $K\pi$ S-wave from $D^+ \to K^-\pi^+\pi^+$ is mainly $I = 1/2$ (see however [16]).

### $D^+ \to K^-\pi^+\mu^+\nu$ Decay from FOCUS

The $D^+ \to K^-\pi^+\mu^+\nu$ decay is a high statistics channel since it is a Cabibbo favored process. It is well know than this decay is highly dominated by the $\overline{K}^*(892)^0$ resonance formation. FOCUS Collaboration has show [17, 18], however, that a small (about 5%) scalar component is needed to describe the observed interference pattern.

Four-body decays of spinless particles are described by five kinematical variables. For $D^+ \to K^-\pi^+\mu^+\nu$ decay, the variables chosen are the $K\pi$ invariant mass ($m_{K\pi}$), the square of the $\mu^+\nu$ mass ($q^2$), and three decay angles: the angle between the $\pi^+$

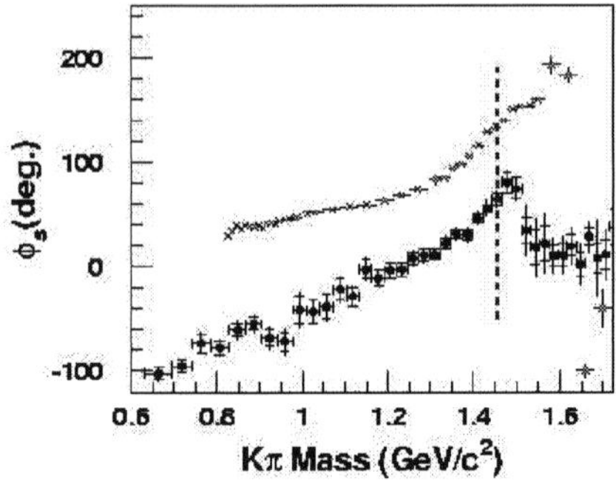


**FIGURE 1.** $K\pi$ S-wave phase variation from $D^+ \to K^-\pi^+\pi^+$, from E791 data.

and the $D^+$ direction in the $K^-\pi^+$ rest frame ($\theta_v$), which defines one decay plane, the angle between the $\nu$ and the $D^+$ direction in the $\mu^+\nu$ rest frame ($\theta_\ell$), which defines a second decay plane, and the acoplanarity angle ($\chi$) between these two planes. The presence of the S-wave was observed by an asymmetry in the $\cos\theta_V$ distribution and this contribution was found to be well described by a constant amplitude $\mathscr{A}_0 = 0.36e^{i\pi/2}$ [17]. A subsequent study fitted the $K\pi$ mass spectrum to get the $K\pi$ lineshape, which is totally dominated by the $K^*(892)^0$ . Breit-Wigner parameters were measured for this resonance, leading to values of mass and width about $1\sigma$ and $2\sigma$ below the PDG average. The lineshape is rather insensitive to the form of the S-wave, so one has to look at the $\cos\theta_V$ distribution in order to distinguish between models. This distribution is well represented if, instead of a constant term $\mathscr{A}_0$ as above, one uses the effective-range formula used by LASS [18], as expected by the Watson theorem. Nevertheless, this phase behavior is different from that of $K\pi$ S-wave from $D^+ \to K^-\pi^+\pi^+$ decays (again assuming $K\pi$ in I=1/2).

## $D^0 \to \pi^-\pi^+\pi^-\pi^+$: AMPLITUDE ANALYSIS IN A 4-BODY DECAY

The $D^0 \to \pi^-\pi^+\pi^-\pi^+$ decay is an interesting process to study. From simple spectator model ($W$-radiation) the virtual $W$ couples to a vector (V) or axial-vector (A) meson, implying that this kind of process have large branching fractions compared to cases in which the $W$ is coupled to a pseudo-scalar (P) meson. In the case of the $4\pi$ final state, one then can expect the $a_1(1260)^+\pi^-$ channel to be the dominant resonant component. Indeed, this decay can be used to study the resonant substructure of the axial-vector meson $a_1(1260)^+$, as well as its line shape, from threshold up to 1.72 GeV/$c^2$. An additional motivation for this study is that any information coming from it is very relevant for analogous studies in the B-system (this is true for other decay channels as well). For example, the CKM angle $\gamma$ can be extracted from the decay $B_d \to \rho\rho$

[20, 21]. The $B_d \to \pi^-\pi^+\pi^-\pi^+$ final state has the same dominant tree-level amplitudes as those of the $D^0 \to \pi^-\pi^+\pi^-\pi^+$ decay. One can thus expect the $a_1(1260)^+\pi^-$ channel to be also the dominant resonant component in the $B_d \to \pi^-\pi^+\pi^-\pi^+$ decay. A correct measurement of the angle $\gamma$ depends on a correct measurement of $B \to \rho\rho$ contribution and, for that, a full amplitude analysis is required. One can see the $D^0 \to \pi^-\pi^+\pi^-\pi^+$ decay as the prototype of the corresponding $B$ decay (in fact, all the systematics related to modeling - form factors, resonance lineshapes, representation of the S-wave component, angular distributions, etc. - are common to both decays).

For the first time, a full coherent analysis of the $D^0 \to \pi^-\pi^+\pi^-\pi^+$ decay was made recently by FOCUS [19]. This is a very complex analysis. The decay dynamics is described in terms of 5 invariants (5D phase space), instead of the two Dalitz invariants in a 3-body decay. There are many possible resonant combinations that can, in principle, contribute to the final state making it difficult to establish which, in fact, contribute. The physics model uses Lorentz invariant amplitudes and/or helicity amplitudes, and there is some freedom in how to define the angular functions.

The data sample consisted of about 6K events, with a background/signal ratio of 8%. It was found contributions of three types. The dominant contribution was found to be $D^0 \to a_1^+(1260)\pi^-$, accounting for 60% of the total decay rate. The decay $D^0 \to \rho^0(770)\rho^0(770)$ accounted for about 25%, while 3-body non-resonant decays $D^0 \to R\pi^+\pi^-$ ($R = \sigma$, $f_0(980)$, $f_2(1270)$) had a combined fraction of 20%.

A good description of the $a_1^+(1260)$ meson was achieved with the following contributions: a dominant $a_1^+ \to \rho\pi^+$ S-wave, a small $a_1^+ \to \rho\pi^+$ D-wave, and the $a_1^+ \to \sigma\pi^+$. This contrasts with the CLEO analysis [22] of the $\tau^- \to \pi^-\pi^0\pi^0\nu_\tau$, which required contributions from seven different subresonant modes. The best description of the data lead to the $a_1^+(1260)$ Breit-Wigner parameters, $\sqrt{s_0^{a_1}}$ and $\Gamma_0^{a_1}$, in the range $1230 < \sqrt{s_0^{a_1}} < 1270\ MeV/c^2$ and $520 < \Gamma_0^{a_1} < 680\ MeV/c^2$, with optimum values at $\sqrt{s_0^{a_1}} = 1240\ MeV/c^2$ and $\Gamma_0^{a_1} = 560\ MeV/c^2$.

The $\rho\rho$ component of the decay was described by the helicity formalism. Using the transversity basis states, FOCUS were able to measure the ratio of the longitudinal polarization to the total $D \to \rho\rho$ rate, $P_L = (70 \pm 4 \pm 2)\%$.

## CONCLUSIONS

Since the last 5 years or so, charm decays have contributed in an unexpected (but welcome) way to the study of light meson espectroscopy, especially on the scalar sector where, over the last decades, there has been continuous interest from both theory and experiment.

Fermilab experiments E791 and FOCUS have been very active on this subject, using different analysis techniques to study $\pi\pi$ and $K\pi$ systems from $D$ modes. In particular, it seems that the S-wave $K\pi$ phase from $K\pi\pi$ (if assumed coming from I=1/2) is different from those from elastic scattering and semileptonic $K\pi\mu\nu$. Possible explanations for that are the presence of 3-body Final State Interactions, or a significant I=3/2 component in $D^+ \to K^-\pi^+\pi^+$ .

Recently, the first full coherent analysis on a 4-body decay, $D^+ \to 4\pi$, was released by FOCUS. Other 4-body studies, as $D^+ \to K_s 3\pi$ and $D^0 \to K3\pi$, will come soon. For the very much discussed mode $D^+ \to K^-\pi^+\pi^+$ , FOCUS expects to release very soon a K-matrix analysis, and a model independent analysis for the $K\pi$ S-wave.

## ACKNOWLEDGMENTS

I wish to acknowledge the kind invitation from the IV SILAFAE organizers, especially Alberto Sánchez and Miguel Angel Pérez. It was really a pleasure to attend the conference. This work was supported in part by the Brazilian Conselho Nacional de Desenvolvimento Científico e Tecnológico (CNPq).

## REFERENCES

1. E.M.Aitala *et al.* (E791 Collaboration), Phys. Rev. Lett. **86**, 765 (2001) .
2. E.M.Aitala *et al.* (E791 Collaboration), Phys. Rev. Lett. **86**, 770 (2001).
3. E.M.Aitala *et al.* (E791 Collaboration), Phys. Rev. Lett. **89**, 121801 (2002).
4. H.Muramatsu *et al.* (CLEO Collaboration), Phys. Rev. Lett. **89**, 251802 (2002).
5. B. Aubert *et al.* (BaBar Collaboration), hep-ex/0408088.
6. K.Abe *et al.* (Belle Collaboration), hep-ex/0308043.
7. M.Ablikim *et al.* (BES Collaboration), Phys. Lett. B **598**, 149 (2004).
8. M.Ablikim *et al.* (BES Collaboration), hep-ex/0610023.
9. M.Ablikim *et al.* (BES Collaboration), Phys. Lett.B **633**, 681 (2006).
10. D. Aston *et al.* (LASS Collaboration), Nucl. Phys. **B296**, 493 (1988).
11. I.Bediaga and J.Miranda, Phys. Lett.B **633**, 167 (2006) .
12. J.M.Link *et al.* (FOCUS Collaboration), Phys. Lett.B **585**, 200 (2004).
13. K. M. Watson, Phys. Rev. **88**, 1163 (1952).
14. C. Göbel (for E791 Collaboration), AIP Conf. Proc. **688**, 266 (2004), hep-ex/0307003.
15. E.M.Aitala *et al.* (E791 Collaboration), Phys. Rev. D **73**, 032004 (2006) .
16. L. Edera and M. Pennington, Phys. Lett. B **623** 55 (2005).
17. J. M. Link *et al.* (FOCUS Collaboration), Phys. Lett. B **535**, 43 (2002).
18. J. M. Link *et al.* (FOCUS Collaboration), Phys. Lett. B **621**, 72 (2005).
19. J. Link *et al.* (FOCUS Collaboration), hep-ex/0701001.
20. K.Abe *et al.* (Belle Collaboration), Phys. Rev. Lett. **91** 221801 (2003).
21. B. Aubert *et al.* (BaBar Collaboration), Phys. Rev. Lett. **93** 231801 (2004).
22. D.M. Asner *et al.*(CLEO Collaboration), Phys. Rev. D **61**, 012002 (1999).

# A Decade with SELEX - New Results from an Old Experiment

James S. Russ

*Carnegie Mellon University*
*Pittsburgh, PA, USA 15213*

**Abstract.** I review the ideas about using baryon beams to produce new baryons that underlie the SELEX experiment at Fermilab. Single charm production shows strong leading particle effects. Unexpectedly, SELEX discovered anomalously large production of double charm baryons in baryon-baryon collisions. I present recent new results on two new decay modes of the $\Xi_{cc}^{+}(3520)$, the first reported double charm baryon. I also show an excited Q=2 double charm baryon and summarize the present state of knowledge about these systems.

**Keywords:** Charmed Baryons, Double Charm Baryons
**PACS:** 13.85.Ni,14.20.Lq

## 1. INTRODUCTION TO SELEX

The SELEX experiment (Fermilab E781) uses the Fermilab Hyperon Beam in a unique study of charm hadroproduction by $\pi^{-}$,$\Sigma^{-}$ and proton beams with energy about 600 GeV in the same apparatus. The collaboration includes about 110 physicists from 19 institutes in 9 countries, including physicists from four Latin American institutes [1]. The Fermilab Hyperon beam produces many high energy $\Sigma^{-}$ hyperons (600 GeV) at zero degrees from an incident 800 GeV proton beam. How does the proton change charge and strangeness in the collision to emerge in the forward direction? This question was the genesis of SELEX, a study of charm hadroproduction emphasizing charm baryons. The beam composition at 600 GeV has roughly equal numbers of $\Sigma^{-}$ and $\pi^{-}$. By changing polarity, we also produced a diffracted proton beam at 570 GeV. The goal was to study the beam particle dependence of the differential cross sections for the various charm hadrons.

### 1.1. Apparatus for Forward Production

The experiment emphasized high momentum, small angle scattering. The apparatus had to have excellent angular resolution, excellent momentum resolution, and excellent particle identification for the high momentum scattered particles in order to identify the different charm hadrons in inclusive scattering [2]. The beam and vertex spectrometers were all silicon detectors. There were two analyzing magnets, each followed by a multiwire proportional chamber system. After the second analysis system there was a RICH detector that separated $\pi$ from K mesons by $2\sigma$ at 165 GeV/c. The particle ID with

CP917, *Particles and Fields,* edited by H. Castilla Valdez, J. C. D'Olivo, and M. A. Perez

the RICH limits combinatoric background to events occuring near the primary vertex, a feature that is very useful in the analysis.

For charm hadroproduction studies the essential features of the apparatus are robust tracking in the vertex area (4 stereo views, high efficiency, 5 $\mu$m track resolution in each view for tracks above 10 GeV/c) and good momentum resolution at all momenta ( 0.5% at 100 GeV/c). Because hadronic collisions produce many tracks in the forward hemisphere, fake tracks are generally a problem. A fake track by definition doesn't point toward its production point, so fake tracks can seed false charm reconstructions. Having four views with high efficiency and redundancy minimizes the fake track problem in SELEX. The high precision of the SELEX system lets us achieve proper time resolution on typical charm particles of 20 fs, substantially better than any other experiment.

### 1.2. Physics of Forward Production

Forward charm production generates particles at small transverse momentum ($p_T < 3$ GeV/c) and large momentum ( > 15% of the beam momentum in SELEX) along the beam direction. The invariant differential cross section factorizes as $\frac{d^2\sigma}{dp_T^2 dx_F}$, where Feynman $x_F = p_{||}^*/p_{max}^*$ is the ratio of the outgoing hadron longitudinal momentum to the maximum allowed in the process. This is usually defined for massless partons, in which case $p_{max}^* = \sqrt{s}/2$. For our case the hadron mass is not negligible compared to $\sqrt{s}/2$, so the correct definition of $x_F$ is not perfectly clear. Since SELEX observes $\Lambda_c$ events in the vicinity of $x_F \sim 1$, the issue of definition is not purely academic. Traditionally the invariant cross section is parametrized as

$$\frac{d^2\sigma}{dp_T^2 dx} = (1-x)^n \, e^{-bp_T^2} \tag{1}$$

These forms are empirical, taken from parton discussions in an infinite momentum frame. Nevertheless they are widely used and represent the data reasonably well for comparison purposes.

The fragmentation picture for particle production is really oriented toward $e^+e^-$ or large-$p_T$ production in which the produced hadrons are well-separated from one another at hadronization time. For low-$p_T$ collisions in fixed target experiments, this picture is not fulfilled, and rescattering effects can be significant. Part of the SELEX program was to evaluate these effects for different beam hadrons, as will be discussed below. Recently the old recombination model has been revisited by Braaten and Mehen for charm hadroproduction [4]. An alternative picture that applies particularly to production at small $p_T$ and large $x_F$, i.e., to the SELEX data, is the intrinsic charm model of Brodsky and collaborators [5].

## 2. $\Lambda_c^+$ PRODUCTION

To illustrate the features of the data I'll concentrate on $\Lambda_c^+$ production, which has also been studied in $\pi$N and $\gamma$N collisions at lower energies. The SELEX $x_F$ data at left

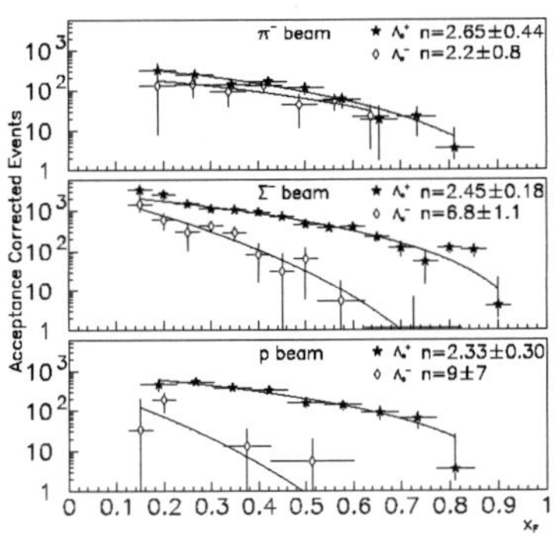


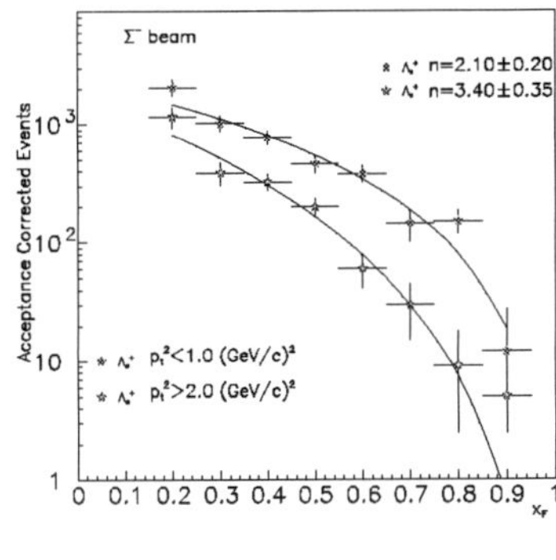


**FIGURE 1.** left: $x_F$ distribution for $\Lambda_c^+$ production by $\Sigma^-$ integrated over all $p_T$; right: the same events separated into low $p_T^2$ ($< 1\text{GeV}/c^2$ ) and large $p_T^2$ ($> 2\text{GeV}/c^2$). The $x_F$ dependence is steeper for the large $p_T^2$ events.

in Fig. 1 show that all three beam hadrons produce $\Lambda_c^+$ baryons with similar n values, suggesting similar mechanisms. In the simple quark picture, the outgoing $\Lambda_c^+$ shares at least one valence quark in common with all three beam hadrons, so similar behavior is expected. In contrast, the $\overline{\Lambda_c^-}$ antibaryon shares a valence quark only with the $\pi^-$. You see the striking difference in antibaryon production between the $\pi^-$ beam and the baryon beams, which show an n $\sim$ 7 dependence suggestive of gg production instead of qg production. These kinds of effects are handled well by the recombination model mentioned earlier, both for the $\Lambda_c^+$ case and for charm meson production.

Only SELEX has the data to study the joint $p_T$ and $x_F$ behavior in the large $x_F$ region, to look for intrinsic charm effects. In the right panel of Fig. 1 we have separated the $\Sigma^-$ data into small $p_T$ and large $p_T$ sets. There is an apparent difference in the $x_F$ dependence, especially for $x_F > 0.5$, for the two sets. The harder spectrum at small $p_T$ is what the intrinsic charm model would expect. The recombination model is not expected to be sensitive to $p_T$ changes of this order.

## 3. CHARM LIFETIMES

The SELEX spectrometer was designed to measure charm lifetimes. Despite the relatively small number of charm events from SELEX because of our choice to work at large $x_F$, the SELEX lifetime measurements in the PDG are competitive with the best results from other experiments.

The theoretical analysis of charm lifetimes is based on the joint expansion in heavy quark mass and powers of $\alpha_s$ [3]. Charm and beauty meson lifetimes follow the expectations well, as do the all the charm baryons except $\Xi_c^+$. The $\Omega_c$ lifetime is the shortest charm baryon lifetime and is a challenge to measure. The best $\Xi_c^+$ lifetimes from CLEO and FOCUS disagree by $2\sigma$ but in both cases are much above the theoretical expectation.

SELEX has recently measured both lifetimes using the reduced proper time method. For each event we know the error $\sigma$ on the decay length L, the distance between the production and decay vertices, from which we compute the proper time t = L/$\beta\gamma$c. However, we accept only events that pass cuts, e.g., decay length significance L/$\sigma$ > N, where the integer N is in the range 6-10 for single charm particles. For each event there is a *minimum* proper time $t_{min}$ that the state has to live in order to pass the cuts. If $t_{min}$ is independent of proper time, then we can just subtract it in the probability distribution function and fit in *reduced proper time* $t_r = t - t_{min}$. This analysis is robust and gives well-understood systematic and statistical errors.

Theory predicts that the $\Xi_c^+$ lifetime should be about 1.5 times longer than that of the $\Lambda_c^+$, i.e., around 300 fs. The SELEX result, based on 301 events in three decay modes, is 419 $\pm$ 24 $\pm$ 9 fs, almost identical in precision to the FOCUS measurement and in good agreement with it. This confirms the disagreement with theory. The new SELEX $\Omega_c$ lifetime comes from measurements in two decay channels, $\Omega\pi$ and $\Omega 3\pi$. The mass and lifetime results from photoproduction or $e^+e^-$ annihilation come only from the $\Omega\pi$ channel - an interesting discrepancy in the kinds of events seen. There is no explanation for that at the moment. Lifetime measurements in both modes agree well for SELEX. The combined lifetime from 76 $\pm$ 10 events is 65 $\pm$ 13 $\pm$ 9 fs. Combined with the similar FOCUS measurement, it gives a precision of about 12% on this shortest known lifetime, which agrees with theory. The failure of the theory to predict the $\Xi_c^+$ lifetime properly is surprising and possibly suggestive of some new features of charm baryon decay.

## 4. DOUBLE CHARM BARYONS

Soon after the discovery of charm potential models were used to compute not only the $\bar{c}$c masses but also the cc binding potential and double charm baryon masses. The ground state mass from various models lay in the range 3.5-3.7 GeV/c. The standard Cabibbo-favored decay would produce a baryonic final state containing a charm quark and a strange quark. Given the very clean SELEX $\Lambda_c^+$ sample, we began a search in the $\Lambda_c^+ K^- \pi^+$ channel using the previously- selected $\Lambda_c^+$ events with no new cuts. The additional $K^-\pi^+$ pair had to form a vertex that lay between the primary vertex and the $\Lambda_c^+$ vertex and be separated from each of them. We found the first double charm baryon candidate, the $\Xi_{cc}^+$(3520), in this search [6]. The state was characterized by a very short lifetime, far shorter than the theoretical expectations.

$\Xi_{cc}^+$ production involved almost 40% of the $\Lambda_c^+$ events in the SELEX sample, corresponding to a much higher double charm cross section than fragmentation calculations predicted. Moreover, it was *not* produced by $\pi^-$ interactions but only by baryon-baryon interactions, in contrast to a fragmentation picture. It appears that the production mechanism is not simple fragmentation. The failure to observe this state in photoproduction (FOCUS) or $e^+e^-$ annihilation (Belle, BaBar) reinforces the non-fragmentation nature of the production. Interestingly intrinsic charm predicts double charm baryon production at about this level from baryon beams and not electromagnetic interactions.

It is important to confirm this finding. SELEX set about looking in other possible decay modes. One completely different channel separates the baryon from the charm

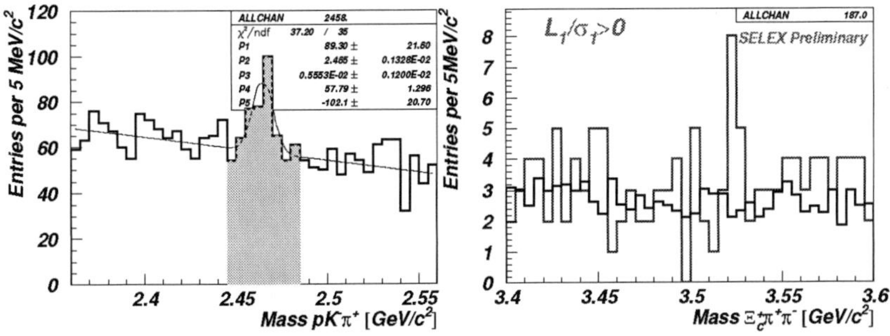


**FIGURE 2.** left: Mass distribution for $pK^-\pi^+$ events showing the Cabibbo-suppressed $\Xi_c^+$ peak; right: Mass distribution for $\Xi_c^+\pi^+\pi^-$ candidates compared to the absolutely normalized combinatorial background from mixed events.

and strange quark content in the final state $p\mathrm{D}^+\mathrm{K}^-$. The signal in this channel was reported by SELEX in Ref. [7]. We have just completed a new analysis for the final state $\Xi_c^+\pi^+\pi^-$, in which the final state charm and strange quarks are both in the outgoing baryon. Both in this new analysis and in the pDK analysis we use a new background estimator based on our observation that the SELEX background is all combinatoric. We make an absolute measurement of the combinatoric background by combining charm candidates from one event with two-prong vertices from a different event that does not contain charm. For all three double charm candidates, this event-mixed background describes both the sideband background shape and normalization well. In this new analysis we use only $\Xi_c^+$ decays to the Cabibbo-suppressed mode pK$\pi$, shown at left in Fig 2. Because this decay mode is Cabibbo-suppressed, it has a higher background, but it gives the best momentum and angle resolution for the $\Xi_c^+$. The event-mixing background procedure handles the higher background under the signal peak nicely.

The double charm mass distribution is shown at right in Fig. 2. The event-mixed combinatoric background is overlaid with the signal events. The excess at 3520 MeV/c$^2$ is clearly seen. Further work on this channel is underway, but this plot is clear evidence for the third decay mode of the $\Xi_{cc}^+(3520)$.

We have extended the double charm search in a blind analysis to the Q=2 system. The most striking feature of the data is shown in Fig. 3. There is a 6.3$\sigma$ peak at 3780 MeV/c$^2$. The absolutely-normalized mixed-event background describes the distribution well outside the signal region. The signal itself is wider than the Monte Carlo resolution. Either the state has a natural width or the signal may contain an HQET pair of states akin to the $\Lambda_c^+(2593)$, $\Lambda_c^+(2632)$ pair. If this state is a typical excited baryon, it should decay via pion emission to the ground state. We have observed such decays. The details are still under study.

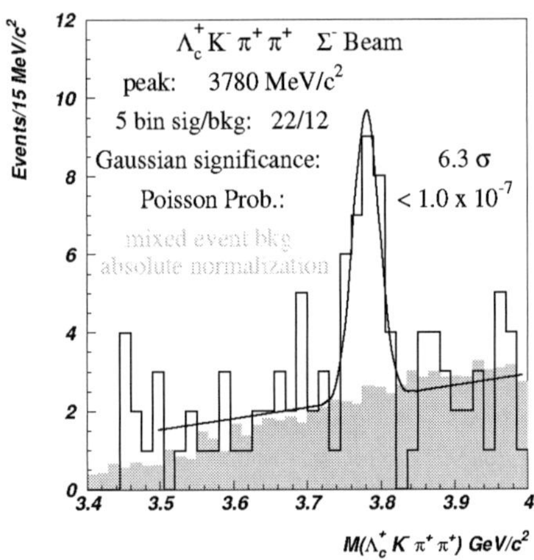


**FIGURE 3.** Mass distribution for $\Lambda_c^+ K^- \pi^+ \pi^+$ candidates

## 5. SUMMARY

The SELEX experiment was designed to probe charm hadroproduction at large $x_F$ and small $p_T$ using a variety of hadron beams, to probe a region of production phase space that had not been thoroughly explored before. The single charm results require significant recombination effects to account for the strong production asymmetries between charm and anticharm hadrons. The unexpected double charm baryon signals introduce a new spectroscopy for studying QCD and present the interesting question about why the SELEX production cross section is as large as we observe. After a decade of SELEX analysis, we have more work to do to uncover all the riches in the SELEX data.

## REFERENCES


1. The SELEX Collaboration -Ball State University, Bogazici University, Carnegie-Mellon University, Centro Brasileiro de Pesquisas Físicas, Fermi National Accelerator Laboratory, Institute for High Energy Physics(Protvino), Institute of High Energy Physics(Beijing), Institute of Theoretical and Experimental Physics, Max-Planck-Institut für Kernphysik, Moscow State University, Petersburg Nuclear Physics Institute, Tel Aviv University, Universidad Autónoma de San Luis Potosí, Universidade Federal da Paraíba, University of Bristol, University of Iowa, University of Michigan-Flint, University of Rome "La Sapienza" and INFN, Rome, University of São Paulo, University of Trieste and INFN, Trieste.
2. J.S. Russ hep-ex/9812031
3. I. Bigi, N. Uraltsev and A. Vainstein, *Phys. Lett.* **B 293**, 430 (1992).
4. E. Braaten, Y. Jia, and T. Mehen, *Phys. Rev. Lett.***89**,122002,2002.
   E. Braaten, M. Kusunkoki, Y. Jia, and T. Mehen, *Phys. Rev.***D70**,054021,2004.
5. R. Vogt, S.J. Brodsky, and P. Hoyer, *Nucl. Phys.* **B 383**, 643 (1992).
6. M. Mattson, *et al., Phys. Rev. Lett* **89**, 112002, 2002.
7. A. Ocherashvili, *et al., Phys. Lett.* **B628**, 18, 2005.

# Recent Results from NA48/2 Experiment at CERN-SPS

Simone Bifani[1]

*University of Torino, Experimental Physics Department*
*via Pietro Giuria 1, I-10125 - Torino (Italy)*

**Abstract.** Preliminary results of the measurement of charged CP violating asymmetry $A_g = \frac{g^+ - g^-}{g^+ + g^-}$ in both $3\pi$ decay modes, based on the whole accumulated statistics in 2003 and 2004, are presented.

A new unexpected anomaly in the $2\pi^0$ mass subsystem spectrum in $K^\pm \to \pi^\pm\pi^0\pi^0$ decay, so called "cusp", due to charge exchange re-scattering process ($\pi^+\pi^- \to \pi^0\pi^0$) is discussed.

A preliminary result indicates the observation of direct emission amplitude and first observation of an interference term in the non-leptonic radiative rare decay $K^\pm \to \pi^\pm\pi^0\gamma$ is presented.

**Keywords:** <charged kaon, direct cp violation, scattering length, radiative decay>
**PACS:** <13.25.Es>

## INTRODUCTION

The NA48/2 experiment at CERN-SPS collected a very high statistics of charged kaon decays with the main purpose of searching for direct CP violation in $K^\pm \to \pi^\pm\pi^+\pi^-$ and $K^\pm \to \pi^\pm\pi^0\pi^0$ final states. Data were taken in 2003 and 2004 for a total amount of $4 \times 10^9$ $K^\pm \to \pi^\pm\pi^+\pi^-$ and $0.1 \times 10^9$ $K^\pm \to \pi^\pm\pi^0\pi^0$. The total accumulated statistics allows to measure rare decays down to $\sim 10^{-9}$ branching ratios.

## NA48/2 EXPERIMENTAL SETUP

NA48/2 is a fixed target experiment located at the CERN-SPS which adds a novel beam line system and a new beam spectrometer to the existing NA48 detector in order to reach a high accuracy in the measurement of the CP violating charge asymmetry parameter $A_g$ and rare kaon decays. The layout of the beams and detectors is shown schematically in Fig.1.

Two opposite in charge beams, $K^+$ and $K^-$, are produced with protons at 400 $GeV/c$ from SPS impinging on a beryllium target and then momentum selected in the band $(60 \pm 3)$ $GeV/c$. Both beams are sent in the *KAonBEamSpectrometer*, a Micromega type detector operating in TPC mode which provides a 1% momentum measurement. The beams travel superimposed in space within $\sim 1mm$ and then sent to the decay tank ($\sim 114m$). The $K^+/K^-$ flux ratio is $\sim 1.8$.

[1] on behalf of the NA48/2 Collaboration: Cambridge, CERN, Chicago, Dubna, Edinburgh, Ferrara, Firenze, Mainz, Northwestern, Perugia, Pisa, Saclay, Siegen, Torino, Vienna

CP917, *Particles and Fields,* edited by H. Castilla Valdez, J. C. D'Olivo, and M. A. Perez

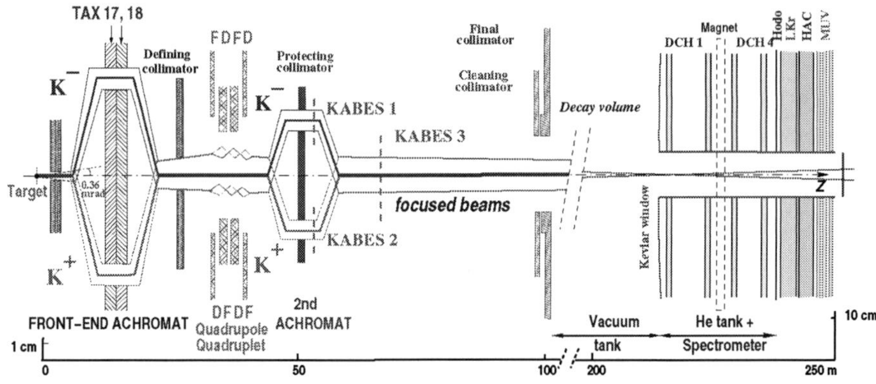


**FIGURE 1.** Schematic lateral view of the NA48/2 beam line, decay volume and detector (the vertical scale changes for the two parts of the figure).

Charged particles are measured using the magnetic spectrometer which is built with four drift chambers (DCH) and one dipole magnet (horizontal momentum kick of 120 $MeV/c$) with a momentum resolution of $\sigma(p)/p = 0.01 \oplus 0.00044 \cdot p$ ($p$ in $GeV/c$). The magnetic spectrometer is followed by a scintillator hodoscope (HODO) consisting of two planes segmented into horizontal and vertical strips and arranged in four quadrants. Neutral particles are reconstructed by a liquid krypton electromagnetic calorimeter (LKr), a quasi-homogeneous ionization chamber ($\sim 27X_0$), which has an energy resolution of $\sigma(E)/E = 0.032/\sqrt{E} \oplus 0.09/E \oplus 0.0042$ and a spatial resolution of $\sigma_x = \sigma_y = 0.42/\sqrt{E} \oplus 0.06cm$ ($E$ in $GeV$). At a depth of $\sim 9.5X_0$ inside the active volume of the calorimeter, a hodoscope consisting of a plane of scintillating fibres is installed. A two level trigger is used to reduce the event rate from $\sim 500kHz$ to $\sim 10kHz$.

A detailed description of the detector can be found elsewhere [3].

## CP VIOLATING CHARGE ASYMMETRY

More than thirty years have passed since the first discovery of indirect CP violation in the neutral kaon system [1] to also establish the existence of direct CP violation [2, 3, 4]. A measurement of that kind of violation in all systems where it is possible is quite relevant to test the Standard Model. In the kaon system the most promising complementary observables are decay rates of GIM suppressed rare kaon decays proceeding through flavour-changing neutral currents ($K^\pm \to \pi^\pm \nu \bar{\nu}$) and the charge asymmetry between $K^+$ and $K^-$ decays in three pions.

The standard phenomenological description of $K^\pm \to 3\pi$ decays [5] is made in terms of the bi-dimensional Dalitz plot parameters $u$ and $v$, related respectively to the energy sharing to the "odd" pion (charge opposite with respect to the other two) and among the two "even" pions (same charge):

$$u = \frac{s_3 - s_0}{m_\pi^2} \qquad v = \frac{s_2 - s_1}{m_\pi^2}$$

where $s_i = (P_K - P_{\pi_i})^2$, $i = 1,2,3$, $s_0 = \frac{s_1+s_2+s_3}{3}$ being $P_K$ and $P_{\pi_i}$ the kaon and pion four-momenta, the indexes $i = 1,2$ correspond to the two identical pions and the index $i = 3$ to the pion of different charge.

The matrix element is usually described in a polynomial expansion of the two Dalitz variables and parametrized in terms of slopes [5]:

$$|M(u,v)|^2 \approx 1 + gu + hu^2 + kv^2$$

where $g(\pi^\pm\pi^+\pi^-) = -0.2154 \pm 0.0035$, $g(\pi^\pm\pi^0\pi^0) = -0.638 \pm 0.020$ and $|h|$, $|k| \ll |g|$. A linear term in $v$ is forbidden due to symmetry considerations.

A difference of slope parameters $g^+$ and $g^-$ describing positive and negative kaon decays respectively, is a manifestation of direct CP violation usually defined by the corresponding slope asymmetry:

$$A_g = \frac{g^+ - g^-}{g^+ + g^-} \approx \frac{\Delta g}{2g}$$

where $\Delta g$ is the slope difference and $g$ is the average slope. SM predictions [6] for that asymmetry vary between few $10^{-6}$ to few $10^{-5}$. Models beyond the SM [7] predict a substantial enhancement of $A_g$ that, while being out of reach for previous experiments [8] with lower statistics, are potentially accessible to NA48/2.

The measurement method is based on comparing the reconstructed $u$ distributions of $K^+$ and $K^-$ decays ($N^+(u)$ and $N^-$(u)): in the $K^\pm \to \pi^\pm\pi^+\pi^-$ case the ratio $R(u) = N^+(u)/N^-(u)$ is in good approximation proportional to $(1 + \Delta g \cdot u)$, so $\Delta g$ can be extracted from a linear fit and $A_g = \Delta g/2g$ can be evaluated.

Charge symmetrization of the experimental conditions is to a large extent achieved by using simultaneous and collinear $K^+$ and $K^-$ beams with similar momentum spectra and reversing all the magnets polarities during the data taking: beam line magnets on weekly basis while spectrometer magnet once per day in 2003 and once in 3 hours in 2004. Data collected over a period with all the four possible magnets setup configurations represent a "SuperSample", which is treated as an independent and self-consistent set of data for asymmetry measurement (nine SuperSamples were collected in two years of data taking). To measure the charge asymmetry the following "quadruple ratio" composed as a product of four $R(u) = N^+(u)/N^-(u)$ ratios with opposite kaon sign, and deliberately chosen setup configurations in numerator and denominator, is considered:

$$R_4(u) = R_{US}(u) \cdot R_{UJ}(u) \cdot R_{DS}(u) \cdot R_{DJ}(u)$$

where the indices $U$ ($D$) denote the beam line polarities corresponding to $K^+$ passing along the upper (lower) path in the achromats, respectively, while the indices S (J) represent spectrometer magnet polarities (opposite for $K^+$ and $K^-$) corresponding to the "even" pions being deflected to negative (positive) horizontal coordinate, i.e. towards the Salève (Jura) mountains, respectively.

The quadruple ratio technique logically completes the procedure of magnet polarity reversal, and allows a three-fold cancelation of systematic biases:

- beam line differences: by comparing $K^+$ and $K^-$ traveling along the same paths;
- detector asymmetries: by comparing $K^+$ and $K^-$ illuminating the detector in the same way;
- global time-dependent effects: by the simultaneous detection of $K^+$ and $K^-$ events.

The result remains sensitive only to time variations of asymmetries in the experimental conditions which have a characteristic time smaller than corresponding field alternation period, and in principle should be free of systematic biases.

The method is independent of the $K^+/K^-$ flux ratio and the relative sizes of the samples collected with different magnet configurations, however the statistical precision is limited by the smallest of the samples involved, so the balance of sample sizes was controlled during the data taking.

Due to the method described above, no Monte-Carlo corrections to the acceptance are expected to be needed, nevertheless a detailed *GEANT*-based Monte-Carlo simulation was developed as a tool for systematic studies, including full detector geometry and material description, simulation of time-variable local DCH inefficiencies, time variations of the beam geometry and DCH alignment.

## "Charged" Mode: $K^{\pm} \to \pi^{\pm}\pi^{+}\pi^{-}$

Tracks are reconstructed from hits in DCHs using the measured magnetic field map of the spectrometer magnet rescaled according to the recorded current. Three-track vertices compatible with a $K^{\pm} \to \pi^{\pm}\pi^{+}\pi^{-}$ decay topology are reconstructed by extrapolation of track segments from the spectrometer upstream to the decay volume, taking into account the small magnetic field due to residual vacuum tank magnetization and the Earth's field, which were measured before the run. Event selection includes requirements on vertex charge, quality, position, limits on the reconstructed $3\pi$ momentum ($54GeV/c < p_K < 66GeV/c$) and invariant mass ($|m_{\pi\pi\pi} - m_K < 9MeV/c^2|$). The reconstructed $3\pi$ mass distribution tails are dominated by events in which a pion undergoes a $\pi \to \mu\nu$ decay but no muon rejection is applied to avoid introducing instrumental asymmetries. The selection leaves a practically background free sample ($3.11 \times 10^9$ $K^{\pm}$ decays selected from the total 2003 and 2004 data sample), as $K^{\pm} \to \pi^{\pm}\pi^{+}\pi^{-}$ is the dominant three-track $K^{\pm}$ decay mode.

Possible sources of systematic effects were studied and evaluated such as the fine alignment of the spectrometer, the geometrical acceptance seen by the two beams, the dependence on the way the $u$ variable is calculated or the fitting limits, effects due to uncertainty on the knowledge of the magnetic fields, pile-up effects, inhomogeneities in the spectrometer alignment and trigger efficiencies (Tab.1).

After applying the above corrections, $\Delta g$ is extracted by fitting the quadruple ratio of the $u$ spectra for each SuperSample: the independent obtained results are compatible with $\chi^2/ndf = 10.0/8$ (Fig.2).

**TABLE 1.** Systematic uncertainties.

| Systematic effect | Correction, uncertainty $\delta(\Delta g) \times 10^4$ |
|---|---|
| Spectrometer alignment | $\pm 0.1$ |
| Acceptance and beam geometry | $\pm 0.2$ |
| Momentum scale | $\pm 0.1$ |
| Pion decay | $\pm 0.4$ |
| Pile-up | $\pm 0.2$ |
| Resolution and fitting | $\pm 0.3$ |
| Total systematic uncertainty | $\pm 0.6$ |
| Level 1 trigger | $\pm 0.3$ |
| Level 2 trigger | $0.1 \pm 0.3$ |

The difference in $K^{\pm} \rightarrow \pi^{\pm}\pi^{+}\pi^{-}$ Dalitz plot slope parameter is found to be:

$$\Delta g = (0.6 \pm 0.7_{stat} \pm 0.4_{trig} \pm 0.6_{syst}) \times 10^{-4} = (0.6 \pm 1.0) \times 10{-}4$$

leading to a CP violating charge asymmetry using $g = -0.2154 \pm 0.0035$ [5]:

$$A_g = (-1.3 \pm 1.5_{stat} \pm 0.9_{trig} \pm 1.4_{syst}) \times 10^{-4} = (-1.3 \pm 2.3) \times 10^{-4}$$

which does not contradict the SM, and due to high precision can be used to constrain SM extensions predicting enhancements of the charge asymmetry. The result has $\sim 17$ times better precision than the best measurement before NA48/2, and the precision is still limited mainly by the available statistics (the uncertainty due to the trigger inefficiency is of statistical nature).

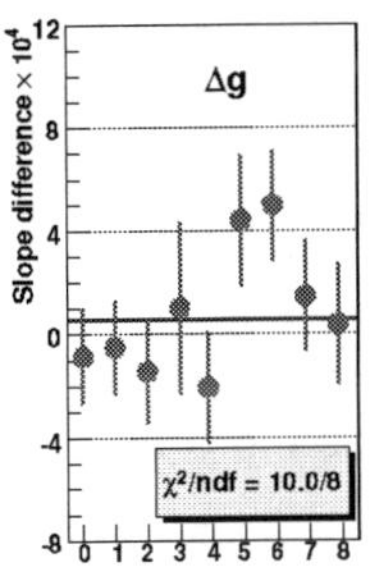


**FIGURE 2.** $\Delta g$ measurement in the nine SuperSamples.

## "Neutral" Mode: $K^{\pm} \rightarrow \pi^{\pm}\pi^{0}\pi^{0}$

An analogous analysis in the "neutral" mode using $91 \times 10^{6}$ $K^{\pm}$ decays, from the combined 2003 and 2004 data sample, gives the result:

$$A_g = (2.1 \pm 1.6_{stat} \pm 1.0_{syst} \pm 0.2_{ext}) \times 10^{-4} = (2.1 \pm 1.9) \times 10^{-4}$$

The statistical precision is similar with respect to the "charged" mode because, even though the $K^{\pm} \rightarrow \pi^{\pm}\pi^{+}\pi^{-}$ decay is statistically favored due to higher branching ratio and acceptance, the population density of the Dalitz plot is less favorable and $|g(\pi^{\pm}\pi^{0}\pi^{0})| \sim 3 \cdot |g(\pi^{\pm}\pi^{+}\pi^{-})|$.

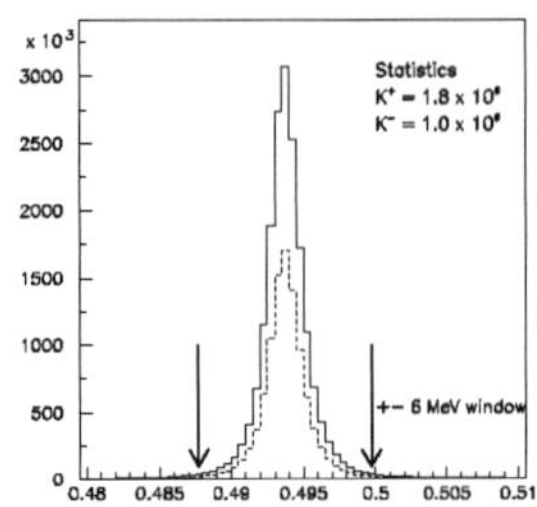

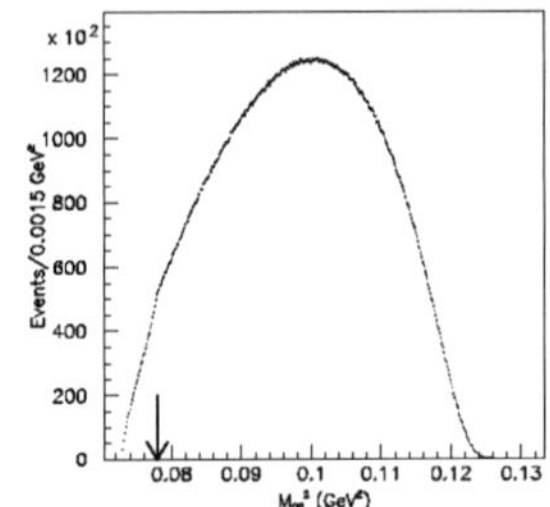

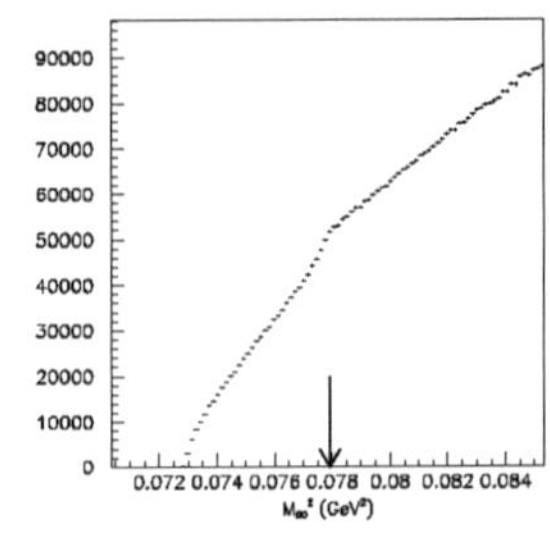


**FIGURE 3.** (Left) Invariant mass distribution of $K^{\pm} \to \pi^{\pm}\pi^{0}\pi^{0}$ candidate events. (Middle) $\pi^{0}\pi^{0}$ invariant mass squared distribution. (Right) Zoom of the previous in the region around the value $M_{00}^{2} = (2m_{+})^{2} = 0.07792(GeV/c^{2})^{2}$ indicated by the arrow.

## "CUSP" EFFECT IN $K^{\pm} \to \pi^{\pm}\pi^{0}\pi^{0}$ DECAY

Using a partial sample of selected $K^{\pm} \to \pi^{\pm}\pi^{0}\pi^{0}$ events $(27.33 \times 10^{6})$ the invariant mass of the $\pi^{0}\pi^{0}$ system has been investigated in order to study the formation of pionium atoms. Thanks to the high statistics, the very good $M_{00}^{2}$ resolution and the proper $M_{00}$ reconstruction strategy, the data revealed an anomalous structure in the region $M_{00} = 2m_{+}$, where $m_{+}$ is the mass of the charged pion (Fig.3). The cusp structure in the distribution has not been observed in previous experiments. Several checks against instrumental effects have been performed. No evidence for either resolution effects or acceptance non-linearities has been found in the cusp region. In addition, variation in shape of photon energy distribution across the cusp agrees with MC prediction without cusp and no difference is observed between $K^{+}$ and $K^{-}$ nor between data taken with opposite directions of the magnetic field. The deficit of events in the data in the region $M_{00} < 2m_{+}$ is due to a real physical effect.

The observed change of slope suggests the presence of a threshold cusp effect from the decay $K^{\pm} \to \pi^{\pm}\pi^{+}\pi^{-}$ contributing to the $K^{\pm} \to \pi^{\pm}\pi^{0}\pi^{0}$ amplitude through the charge exchange process $\pi^{+}\pi^{-} \to \pi^{0}\pi^{0}$.

The phenomenon has been recently discussed by Cabibbo [9] who computed the $K^{\pm} \to \pi^{\pm}\pi^{0}\pi^{0}$ amplitude taking into account the 1-loop diagram (Fig.4). In the Cabibbo theory, the $K^{\pm} \to \pi^{\pm}\pi^{0}\pi^{0}$ decay amplitude is given by the sum of two terms: the

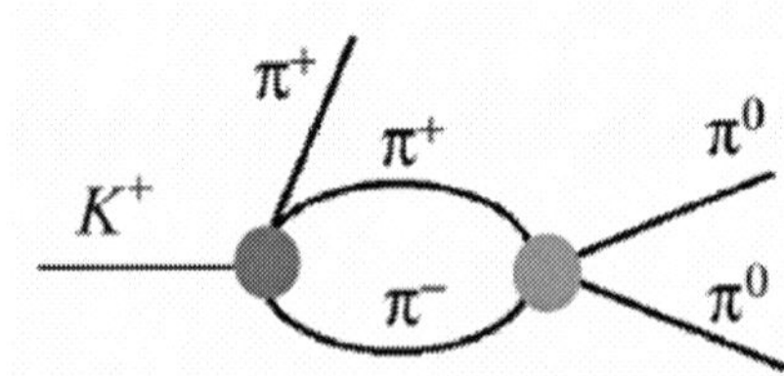


**FIGURE 4.** $\pi\pi$ 1-loop re-scattering diagram.

"unperturbed" amplitude $\mathcal{M}_0$, á la PDG, plus a new term, $\mathcal{M}_1$, proportional to the $I=0$ and $I=2$ S-wave $\pi\pi$ scattering lengths (in the limit of exact isospin symmetry). The new term changes from real to imaginary at $M_{00}=2m_+$. The destructive interference of $\mathcal{M}_0$ and $\mathcal{M}_1$ in the total amplitude causes the cusp and the lack of events below the threshold. More recently Cabibbo and Isidori [10] have computed $\mathcal{O}(a_i^2)$ corrections to $K^\pm \to 3\pi$ amplitudes including other one-loop and two-loop re-scattering processes diagrams. This model has been used to extract $(a_0-a_2)$ from the $\pi^0\pi^0$ invariant mass distribution with high precision.

Taking into account all systematic and external uncertainties, the final result is:

$$\begin{aligned}(a_0-a_2)\cdot m_+ &= 0.268\pm 0.010_{stat}\pm 0.004_{syst}\pm 0.013_{ext}\\ a_2\cdot m_+ &= -0.041\pm 0.022_{stat}\pm 0.014_{syst}\end{aligned}$$

The external error is an additional theoretical error of $\sim 5\%$ on $(a_0-a_2)$ due to neglecting higher order terms in the re-scattering model [10]. This uncertainties have no significant effect on $a_2$. The two statistical errors from the fit are strongly correlated ($\rho=-0.858$). The result obtained performing the fit with constraints imposed on $a_0$ and $a_2$ by analyticity and chiral symmetry [11] is:

$$a_0\cdot m_+ = 0.220\pm 0.006_{stat}\pm 0.004_{syst}$$

which corresponds to

$$(a_0-a_2)\cdot m_+ = 0.264\pm 0.006_{stat}\pm 0.004_{syst}\pm 0.013_{ext}$$

This analysis gives the first direct measurement of $a_2$, though not as precise as that of $(a_0-a_2)$. This result is in very good agreement with Chiral Perturbation Theory ($\chi PT$) calculations [12], which predict $(a_0-a_2)\cdot m_+ = 0.265\pm 0.004$.

## RARE DECAYS: $K^\pm \to \pi^\pm\pi^0\gamma$

Chiral Perturbation Theory, as a low-energy effective field theory of strong interactions, is a very useful tool to study kaon interactions. The $K^\pm \to \pi^\pm\pi^0\gamma$ decay is important to test high order $\chi PT$ calculations: the dominant part of the amplitude is related to the Inner Bremsstrahlung (IB), however a contribution of the Direct Emission (DE) could be present as well, together with the Interference term (INT) between these two amplitudes.

In the $K^\pm \to \pi^\pm\pi^0\gamma$ decay IB, INT, and DE can be distinguished kinematically using the variable W which is defined as follows [13]:

$$W = \frac{(P_K\cdot P_\gamma)(P_\pi\cdot P_\gamma)}{(m_K m_\pi)^2}$$

being $P_K$, $P_\gamma$, $P_\pi$ the four-momenta of the kaon, radiative gamma and charged pion, and $m_K$, $m_\pi$ the kaon and charged pion masses. The decay rate depends only on $T^*_\pi$, the energy of the pion in the kaon rest frame, and $W$. Integrating over $T^*_\pi$ an expression

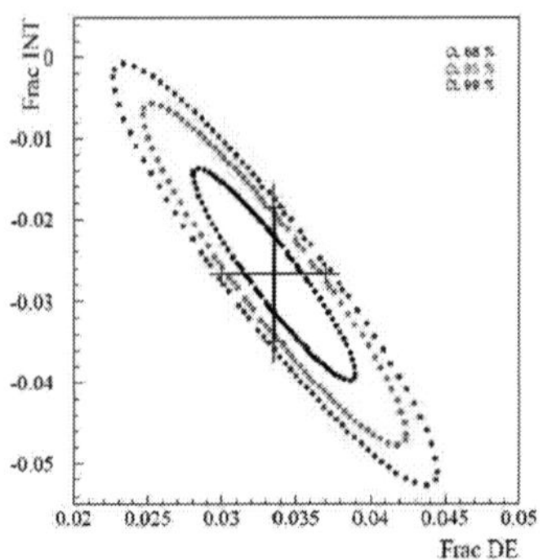


**FIGURE 5.** Contour plot for DE and INT components.

that separates the different contributions into terms with different powers of W can be obtained:

$$\frac{d\Gamma^{\pm}}{dW} \approx \left(\frac{d\Gamma^{\pm}}{dW}\right)_{IB} \cdot \left[1+2\left(\frac{m_\pi}{m_K}\right)^2 W^2|E|\cos((\delta_1)-\delta_0\pm\phi)+\left(\frac{m_\pi}{m_K}\right)^4 W^4(|E|^2+|M|^2)\right]$$

The three terms represent IB, INT and DE contribution respectively.

An analysis was performed using part of the 2003 statistics: $\sim 2.1\times 10^5$ decays were selected in the regions $T^*_\pi \leq 80MeV$ and $0.2 < W < 0.9$. At trigger level these events are selected requiring multiple signals in the electromagnetic calorimeter and applying a kinematical cut against $K^\pm \to \pi^\pm\pi^0$ decays. The event reconstruction requires a single charged track ($\pi^\pm$), two photons coming from the $\pi^0$ decay and an extra photon in time. The global invariant mass has to be compatible with the mass of the charged kaon. Background is reduced to less than 1% of the DE component applying the following selection criteria: $K^\pm \to \pi^\pm\pi^0\pi^0$ decays are suppressed cutting against coalesced gamma events and narrowing the cut around the kaon mass ($|m_{\pi^\pm\pi^0\gamma}-m_K < 10MeV/c^2$); $K^\pm \to \pi^\pm\pi^0$ decays are already suppressed at trigger level but to avoid effects occurring at threshold all events have to fulfil the requirement $T^*_\pi \leq 80MeV$.

To obtain the relative contributions of IB, INT and DE a maximum likelihood technique is applied to fit the measured W distribution to a weighted sum of Monte-Carlo simulated W distributions for IB, INT and DE. The biggest systematic uncertainty is due to the measurement of the trigger efficiencies.

The preliminary results for the measured fractions are (Fig.5):

$$\begin{aligned} Frac(DE) &= (3.35\pm0.35_{stat}\pm0.25_{syst})\% \\ Frac(INT) &= (-2.67\pm0.81_{stat}\pm0.73_{syst})\% \end{aligned}$$

This is the first measurement of a non-vanishing interference term in the $K^\pm \to \pi^\pm\pi^0\gamma$ decay.

## REFERENCES


1. J. H. Christenson *et al.*, Phys. Rev. Lett. **13**, 138 (1964)
2. G. Barr *et al.* (The NA31 Collaboration), Phys. Lett. **B317**, 233 (1993)
3. V. Fanti *et al.* (The NA48 Collaboration), Phys. Lett. **B465**, 335 (1999)
   A. Lai *et al.* (The NA48 Collaboration), Eur. Phys. J. **C22**, 231 (2001)
   J. R. Batley *et al.* (The NA48 Collaboration), Phys. Lett. **B544**, 97 (2002)
4. A. Alavi-Harati *et al.* (The KTeV Collaboration), Phys. Rev. Lett. **83**, 22 (1999)
   A. Alavi-Harati *et al.* (The KTeV Collaboration), Phys. Rev. **D67**, 012005 (2003);
   Erratum: Phys. Rev. **D70**, 079904 (2004)
5. S. Eidelman *et al.* (PDG), Phys. Lett. **B592** (2004)
6. A.A. Belkov, A.V. Lanyov, G. Bohm, hep-ph/0311209
   E. Gamiz, J. Prades, I. Scimemi, JHEP 10, 042 (2003)
   G. D'Ambrosio, G. Isidori, Int. J. Mod. Phys. **A13**, 1 (1998)
   E.P. Shabalin, Phys. Atom. Nucl. 68, 88 (2005)
7. E.P. Shabalin, ITEP preprint 8-98 (1998)
   G. D'Ambrosio, G. Isidori, G. Martinelli, Phys. Lett. **B480**, 164 (2000)
8. W.T. Ford *et al.*, Phys. Rev. Lett. **25**, 1370 (1970)
   K.M. Smith *et al.*, Nucl. Phys. **B91**, 45 (1975)
   G.A. Akopdzhanov *et al.* (TNF-IHEP), Eur. Phys. J. **C40**, 343 (2005)
9. N. Cabibbo, Phys. Rev. Lett. **93**, 121801 (2004)
10. N. Cabibbo and G. Isidori, JHEP **503**, 21 (2005)
11. G. Colangelo, J. Gasser, H. Leutwyler, Phys. Rev. Lett. **86**, 5008 (2001)
12. G. Colangelo, J. Gasser and H. Leutwyler, Phys. Lett. **B488**, 261 (2000)
    G. Colangelo, J. Gasser and H. Leutwyler, Nucl. Phys. **B603**, 125 (2001)
13. J. D. Good, Phys. Rev. **113**, 352 (1959)

# Results from the HyperCP Experiment (FNAL E871)

Christopher White for the HyperCP Collaboration

*Illinois Institute of Technology*
*Division of Physics, Chicago, IL 60616*

**Abstract.** HyperCP is an experiment that collected data during the 1997 and 1999 fixed-target runs at Fermilab. We used an 800 GeV/c proton beam to create a secondary beam of charged hyperons and kaons. The primary goal of the experiment was to conduct a sensitive search for CP violation in hyperon decay; however, we also performed a number of searches for rare and forbidden decays. One such result is evidence for the second-order weak decay $\Sigma^+$ to a proton and a pair of muons.

**Keywords:** Hyperon
**PACS:** 14.20.Jn

## INTRODUCTION

After decades of intense experimental effort and many beautiful experiments we still know little about CP violation: the origin of CP violation remains unknown and there is little hard evidence that it is fully explained by the Standard Model. CP violation is too important, and experimental evidence is too meager, not to examine every possible manifestation of the effect. Hyperons are sensitive to sources of CP violation that are not probed in other systems and these sources are thought to be experimentally accessible. Furthermore, beyond-the-standard-model physics almost always predicts large new sources of CP violation.

While searching for new sources of CP violation, we amassed the world's largest sample of charged hyperon decays, enabling many searches for rare and forbidden decays, including evidence for a second-order weak decay of a baryon. We have definitively observed (for the first time) parity violation in the decay of the $\Omega^-$ baryon. In addition, we searched for but did not find evidence for pentaquarks.

## THEORY OF CP VIOLATION IN HYPERON DECAY

The standard model commonly predicts CP asymmetries in weak interactions, although they are typically small. To date, CP asymmetries have been seen only in the decays of K [1] and B [2] mesons. Although the asymmetries observed are consistent with standard-model predictions, additional sources of CP violation have not been ruled out and it is vital to search for novel, new, or exotic sources of CP violation. The Standard Model does allow for CP violation in hyperon decays. Furthermore, CP violation in hyperons is possible through channels not otherwise excluded by any previous experimental observation. The easiest method to observe CP-violation in

CP917, *Particles and Fields*, edited by H. Castilla Valdez, J. C. D'Olivo, and M. A. Perez

spin-1/2 hyperons is to search for differences in the degree to which parity is violated in their non-leptonic 2-body decays. Parity violation manifests itself in the angular decay distribution of the daughter baryon, $dN / d\Omega = N_0 (1 + \alpha \boldsymbol{P}_p \cdot \hat{p}_d) / 4\pi$, where $\boldsymbol{P}_p$ is the parent polarization vector, $\hat{p}_d$ is the direction vector for the daughter baryon, and $\alpha$ is the parity violating decay parameter ($\alpha = 0$ if parity is conserved). CP invariance requires that $\alpha = -\overline{\alpha}$ [3]. In HyperCP we study the angular decay distribution for the daughter protons from the decay $\Lambda \to p\,\pi^-$. The lambdas were produced in the decay $\Xi^- \to \Lambda\,\pi^-$, where the $\Xi$ is unpolarized. Since the $\Xi$ is unpolarized, the Lambda baryon is in a helicity state with a polarization value equal to $\alpha_\Xi$ [4]. The resulting proton decay distribution is thus a function of both $\alpha_\Xi$ and $\alpha_\Lambda$, allowing for sensitivity to CP violation in either of the decays. The proton decay distribution is thus given by:

$$\frac{dN}{d\cos\theta} = \frac{N_0}{2}(1 + \alpha_\Xi \alpha_\Lambda \cos\theta) \tag{1}$$

Since the alpha parameter changes sign under CP, the proton decay distribution for the decay chain $\Xi^- \to \Lambda\,\pi^- \to p\pi^-\pi^-$ should be identical to the distribution for the decay chain $\overline{\Xi}^+ \to \overline{\Lambda}\,\pi^+ \to \overline{p}\pi^+\pi^+$, if CP is conserved. The defined asymmetry parameter is:

$$A_{\Xi\Lambda} \equiv \frac{\alpha_\Xi \alpha_\Lambda - \overline{\alpha}_\Xi \overline{\alpha}_\Lambda}{\alpha_\Xi \alpha_\Lambda + \overline{\alpha}_\Xi \overline{\alpha}_\Lambda} \approx A_\Xi + A_\Lambda, \tag{2}$$

A recent standard-model calculation for the combined asymmetry, $A_{\Xi\Lambda}$, is $-0.5\times10^{-4} < A_{\Xi\Lambda} < 0.5\times10^{-4}$ [5]. This prediction uses theoretical calculations of the final-state scattering phase-shift differences rather than using recent HyperCP experimental measurements as input [6]. Non-standard-model calculations, such as left-right symmetric models [7] and supersymmetric models [8, 9], allow for much larger asymmetries. The supersymmetric calculation of He *et al.* [8] generates values of $A_{\Xi\Lambda}$ as large as $19\times10^{-4}$. Bounds from $\varepsilon$ and $\varepsilon'/\varepsilon$ in $K^0$ decays limit $A_{\Xi\Lambda}$ to be less than $97\times10^{-4}$ [10].

## HYPERCP DETECTOR

Data were taken at Fermilab using a high-rate spectrometer (Fig. 1) built in the meson center beamline [11]. The hyperons were produced by an 800 GeV/c proton beam incident at $0^\circ$ on a 2mm × 2mm Cu target. Immediately after the target was a 6.1m long curved collimator embedded in a dipole magnet. Charged particles following the central orbit of the collimator exited upward at 19.5 mrad to the incident proton beam direction with a central momentum of 157 GeV/c. Following the collimator was a 13m long evacuated pipe. The momenta and decay vertices of charged particles were measured using nine multiwire proportional chambers (MWPCs), four in front and five behind two dipole magnets. At the rear of the spectrometer were two scintillator hodoscopes used in the trigger. A hadronic calorimeter was used to trigger on the energy of the proton or antiproton. Behind the calorimeter were muon detectors with modest acceptance on either side of the secondary beam.

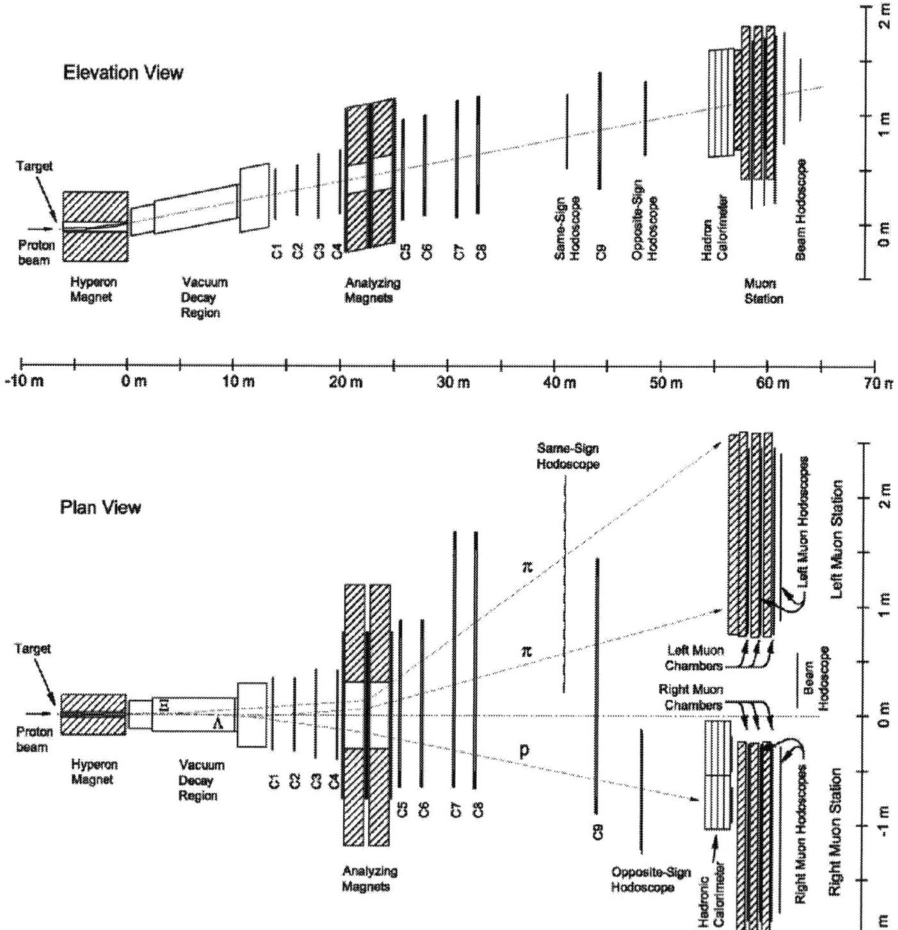


**FIGURE 1.** The HyperCP apparatus.

**TABLE 1. HyperCP data summary.**

| | 1997 Data Set | 1999 Data Set |
|---|---|---|
| Total Triggers ($10^9$) | 58 | 173 |
| Cascade Triggers ($10^9$) | 39 | 90 |
| Data Volume (TB) | 38 | 82 |
| Number of Tapes (8mm) | 8980 | 20,240 |

**TABLE 2. Fully Reconstructed Events (total)**

| Decay Mode | Positive Charge Mode | Negative Charge Mode |
|---|---|---|
| $\Xi \to \Lambda\,\pi$ | $458 \times 10^6$ | $2490 \times 10^6$ |
| $\Omega \to \Lambda\,K$ | $4.86 \times 10^6$ | $19.0 \times 10^6$ |
| $K \to \pi\pi\pi$ | $391 \times 10^6$ | $555 \times 10^6$ |
| $K_s \to \pi^+\pi^-$ | $2025 \times 10^6$ | $2718 \times 10^6$ |

# CP VIOLATION ANALYSIS

This published analysis [12] was based on a subset of the full dataset and included 19% of all the good $\overline{\Xi}^+$ events (41.4 million) and 14% of all the good $\Xi^-$ events (117.3 million) taken during the 1999–2000 running period. The data were divided into 18 analysis sets of roughly equal size, each containing at least three positive and one negative run taken closely spaced in time. Runs with anomalously low (< 95%) hodoscope, wire chamber, or calorimeter efficiencies were not used; these were less than 5% of the total. The criteria used to select the final event samples were:

1. that the $p\pi$ and $p\pi\pi$ invariant masses be, respectively, within ±5.6 MeV/$c^2$ (3.5 σ) and ±3.5 MeV/$c^2$ (3.5 σ) of the mean values of the Ξ and Λ masses;
2. that the z coordinate of the Ξ and Λ decay vertices lie within the vacuum decay region and that the Λ decay vertex precede the Ξ decay vertex by no more than 0.50 m;
3. that the reconstructed Ξ trajectory trace back to within ±2.45 mm (3.3 σ) and ±3.26mm (3.4 σ), respectively, in x and y, of the center of the target;
4. that the Ξ trajectory trace back to within +8.2/-8.4mm and ±6.5mm, respectively in x and y, from the center of the exit of the collimator; and
5. that the $\pi\pi\pi$ invariant mass be greater than 0.5 GeV/$c^2$ (to remove K –> $\pi\pi\pi$ decays).

Cuts on the particle momenta and the numbers of SS and OS hodoscope hits were also made. Events satisfying these criteria had a background to signal ratio of (0.43±0.03)% for the $\Xi^-$ data and (0.41±0.03)% for the $\overline{\Xi}^+$ data. In order to compare the proton and antiproton cosθ distributions directly (without acceptance corrections), the momentum and spatial distributions at the exit of the collimator for the $\Xi^-$ and $\overline{\Xi}^+$ events are made identical. This was accomplished by weighting the events in three momentum dependent variables: the event momentum, the y position at the collimator exit, and the y-slope of the Ξ. There were 100 bins for each variable forming a 100 x 100 x 100 three-dimensional array with 1 millions weights. The ratio of the weighted proton and antiproton cosθ distributions was then made. Any nonzero slope in that ration was evidence of CP violation. The ratio was fit to the following function:

$$R = C\frac{1+\alpha_\Xi\alpha_\Lambda\cos\theta}{1+(\alpha_\Xi\alpha_\Lambda-\delta)\cos\theta} \tag{3}$$

The asymmetry parameter δ is defined as $\alpha_\Xi\alpha_\Lambda - \overline{\alpha}_\Xi\overline{\alpha}_\Lambda$, which is approximately equal to 2 $\alpha_\Xi\alpha_\Lambda A_{\Xi\Lambda}$. The cosθ distribution for one of the data subsets is shown in Figure 2. The average asymmetry over all 18 analysis subsets is consistent with zero. No acceptance or efficiency corrections were required for this method. This is an improvement on the previously published result by about a factor of twenty [13]. Common biases were suppressed by taking the ratio of the p and $\overline{\mathrm{p}}$ cosθ distributions. Also, since the polar axis changes from event to event in the Λ helicity frame, there is only a weak correlation between any particular region of the apparatus and the value of cosθ, minimizing biases due to localized differences in detector efficiencies. The largest systematic error was due to uncertainty in the magnetic field strength of the analysis magnets. The next largest error resulted from differences in calorimeter efficiency between positive and negative data sets. No systematic effect was found due to parent momentum, time, or beam intensity. We have measured $A_{\Xi\Lambda}$ to be $[0.0 \pm 5.1(\text{stat}) \pm 4.4(\text{syst})] \times 10^{-4}$. The analysis of the full data set is ongoing and we hope to improve upon this result in the near future.

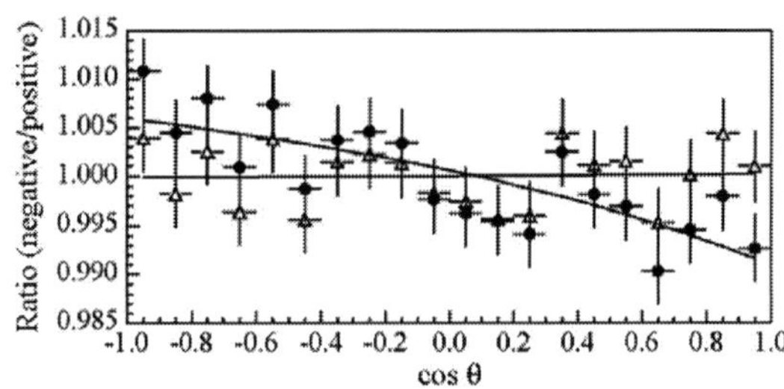


**FIGURE 2.** Ratios of p to $\overline{p}$ cosθ distributions from analysis set 1, both unweighted (filled circles) and weighted (open triangles), with fits to the form given in Eq.(3).

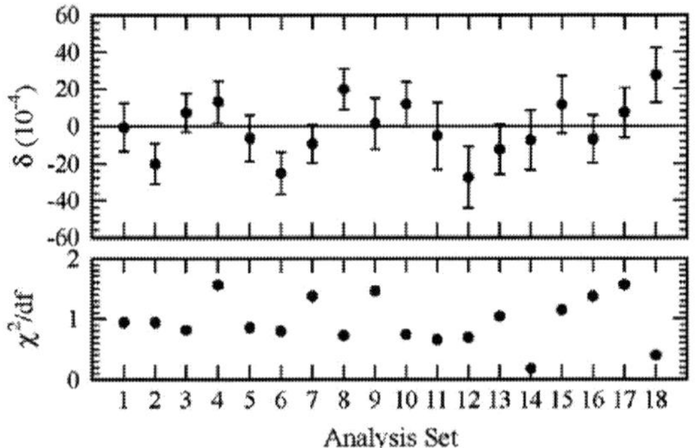


**FIGURE 3.** Asymmetry δ (top) and $\chi^2$/dof (bottom) versus analysis set.

## EVIDENCE FOR THE DECAY $\Sigma^+ \to p\,\mu^+\,\mu^-$

The decay $\Sigma^+ \to p\mu^+\mu^-$ is allowed within the Standard Model. In [14], HyperCP provides evidence for the first observation for this second-order weak decay, the first such observation of second-order weak processes in baryon decay. It is not possible to detail the analysis here. In brief, we select events with two identified muons of opposite charge plus a track consistent with a proton. The tracks must form a high quality 3-prong event vertex well within the vacuum decay region and the parent track must intersect the target. Three data events were thus reconstructed with an invariant mass consistent with a $\Sigma^+$ hyperon. No other events are seen within 20 standard deviations. Two additional cuts eliminate the high-mass background while leaving the three event candidates. The first required that the proton carry at least 68% of the total event momentum. The second removed possible $K^+ \rightarrow \pi\,\mu^+\mu^-$ background events. After all selection, we observe three events with essentially no background (Fig. 4).

When we calculate the di-muon invariant mass, we find that the three events have a common value within 1 MeV/$c^2$ (Fig. 5). This is quite unexpected and would appear to be highly improbable (less than 1% random chance for this to occur based on various reasonable SM phase-space models). Some have suggested that a new particle may mediate this decay [15], (*e.g.* $\Sigma^+ \rightarrow pP^0 \rightarrow p\mu^+\mu^-$). It has even been suggested that this data may be consistent with the observation of a light Higgs [16].

The branching ratio assuming a SM decay is $[8.6{+}6.6{-}5.4(\mathrm{stat}) \pm 5.5(\mathrm{syst})]\times 10^{-8}$, but changes (due to acceptance variation) to $[3.1{+}2.4{-}1.9(\mathrm{stat}) \pm 1.5(\mathrm{syst})]\times 10^{-8}$ if we assume that the decay is mediated by a neutral particle with a mass of 214 MeV/$c^2$. If we assume the three events are all background, then we set an upper limit at the 90% confidence level of $3.4 \times 10^{-7}$.

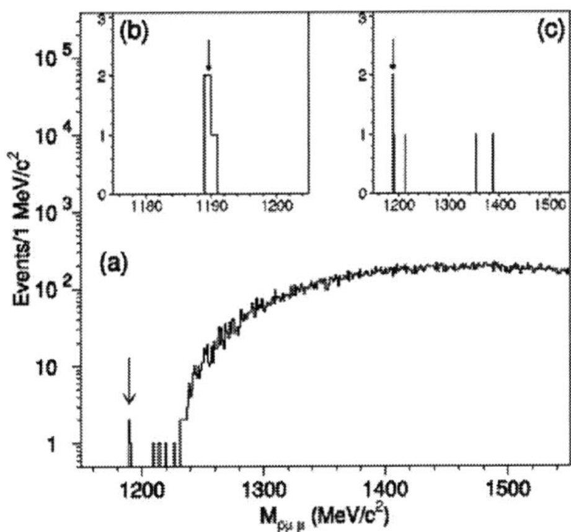


**FIGURE 4.** The invariant mass distribution is shown for pμμ candidates after initial selection (a), close-up around $\Sigma^+$ mass region (b), and after all selection (c).

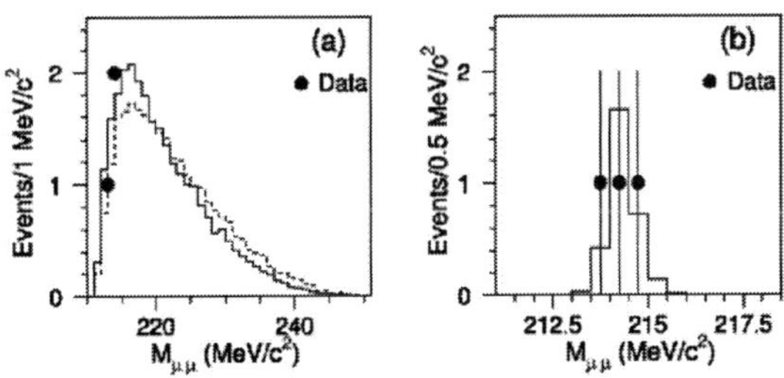


**FIGURE 5.** Di-muon invariant mass distributions are shown. Expected Standard Model distributions are shown at left (with the three data events show as black dots). A close-up of the three data events is shown at right with an MC prediction assuming the muons come from the decay of a neutral, long-lived particle.

## OTHER MEASUREMENTS

We have recently published several other results. A few are summarized here.

Measurement of the Asymmetry in the Decay $\overline{\Omega}^+ \rightarrow \overline{\Lambda} K^+$,
Phys. Rev. Lett. **96**, 242001 (2006).
The asymmetry in the $\overline{p}$ angular distribution in the sequential decay $\overline{\Omega}^+ \rightarrow \overline{\Lambda} K^+$, $\overline{\Lambda} \rightarrow \overline{p}\pi^+$ has been measured to be $\overline{\alpha}_\Omega\overline{\alpha}_\Lambda$ = $[+1.16\pm0.18(stat)\pm0.17(syst)]\times10^{-2}$ using $1.89\times10^6$ unpolarized $\overline{\Omega}^+$ decays. Using the known value of $\alpha_\Lambda$ and assuming that $\alpha_\Lambda = -\overline{\alpha}_\Lambda$, $\overline{\alpha}_\Omega$ = $[-1.81\pm0.28(stat)\pm0.26(syst)]\times10^{-2}$. A comparison between this measurement of $\overline{\alpha}_\Omega\overline{\alpha}_\Lambda$ and recent measurements of $\alpha_\Omega\alpha_\Lambda$ made by HyperCP shows no evidence of a violation of CP symmetry.

Observation of Parity Violation in the Decay $\Omega^- \rightarrow \Lambda K^-$
Phys. Lett. B **617**, 11 (2005).
The $\alpha$ decay parameter in the process $\Omega^- \rightarrow \Lambda K^-$ has been measured from a sample of 4.50 million unpolarized $\Omega^-$ decays and found to be $[1.78+0.19(stat)\pm0.16(syst)]\times10^{-2}$. This is the first unambiguous evidence for a non-zero $\alpha$ decay parameter, and hence parity violation in the $\Omega^- \rightarrow \Lambda K^-$ decay.

Search for the Lepton-Number-Violating Decay $\Xi^- \rightarrow p\, \mu^-\, \mu^-$,
Phys. Rev. Lett. **94**, 181801 (2005).
A sensitive search for the lepton-number-violating decay $\Xi^- \rightarrow p\, \mu^-\mu^-$ has been performed using a sample of $\sim10^9$ $\Xi^-$ hyperons. We obtain $B(\Xi^- \rightarrow p\mu^-\mu^-) < 4.0 \times 10^{-8}$ at 90% confidence, improving on the best previous limit by four orders of magnitude.

Search for $\Delta S = 2$ Nonleptonic Hyperon Decays,
Phys. Rev. Lett. **94** 101804 (2005).
A sensitive search for the rare decays $\Omega^- \rightarrow \Lambda\, \pi^-$ and $\Xi^0 \rightarrow p\, \pi^-$ has been performed using data from the 1997 run of the HyperCP experiment. Limits on other processes do not exclude the possibility of observable rates for $\Delta S = 2$ nonleptonic hyperon decays, provided the decays occur through parity-odd operators. We obtain the branching-fraction limits $B(\Omega^- \rightarrow \Lambda\, \pi^-) < 2.9 \times 10^{-6}$ and $B(\Xi^0 \rightarrow p\, \pi^-) < 8.2 \times 10^{-6}$, both at 90% C.L.

A High Statistics Search for the $\Theta^+(1.54)$ Pentaquark,
Phys. Rev. D **70**, 111101(R) (2004).
We have searched for $\Theta^+(1.54) \rightarrow K^0 p$ decays using data from the 1999 run of the HyperCP experiment at Fermilab. We see no evidence for a narrow peak in the $K^0 p$ mass distribution near 1.54 GeV/$c$ among 106,000 $K^0 p$ candidates, and obtain an upper limit for the fraction of $\Theta^+(1.54)$ to $K^0 p$ candidates of <0.3% at 90% confidence.

New Measurement of $\Xi^-$ -> $\Lambda\ \pi^-$ Decay Parameters,
Phys. Rev. Lett. **93**, 011802 (2004).
Based on a sample of 144 $\times 10^6$ polarized $\Xi^-$ -> $\Lambda\pi^-$, $\Lambda$ -> $p\pi^-$ decays collected by the HyperCP experiment, we report a new measurement of the $\Xi^-$ decay-parameter angle $\phi_\Xi$ = (-2.39±0.64±0.64)° from which we deduce the decay parameters $\beta_\Xi$ = -0.037±0.011±0.010 and $\gamma_\Xi$ = 0.888±0.0004±0.006. Assuming that the CP-violating phase difference between s and p waves is negligible, the strong phase-shift difference, $\delta p - \delta s$, for $\Lambda\pi$ scattering is determined to be (4.6±1.4±1.2)°.

## ACKNOWLEDGMENTS

I would like to thank the organizers for a wonderful conference full of many interesting presentations. Of course the location was beautiful and memorable. I also thank the editors for their patience during the preparation of this document. This report would not have been possible without the hard work of the FNAL staff that provided the necessary experimental facilities to conduct this research and the dedication of the HyperCP collaboration. This work was supported by the U.S. Department of Energy and the National Science Council of Taiwan.

## REFERENCES

1. For example, see J. Christenson *et al.*, Phys. Rev. Lett. **13**, 138 (1964); A. Alavi-Harati *et al.*, Phys. Rev. Lett. **83**, 22 (1999); V. Fanti *et al.*, Phys. Lett. B**465**, 335 (1999).
2. For a review, see T. E. Browder and R. Faccini, Annu. Rev. Nucl. Part. Sci. **53**, 353 (2003).
3. A. Pais, Phys. Rev. Lett. **3**, 242 (1959).
4. T.D. Lee and C.N. Yang, Phys. Rev. **108**, 1645 (1957).
5. J. Tandean and G. Valencia, Phys. Rev. D **67**, 056001 (2003).
6. M. Huang *et al.*, Phys. Rev. Lett. **93**, 011802 (2004); A. Chakravorty *et al.*, Phys. Rev. Lett. **91**, 031601 (2003).
7. J. F. Donoghue and S. Pakvasa, Phys. Rev. Lett. **55**, 162 (1985); J. F. Donoghue, X.G. He, and S. Pakvasa, Phys. Rev. D **34**, 833 (1986); D. Chang, X.G. He, and S. Pakvasa, Phys. Rev. Lett. **74**, 3927 (1995).
8. X.G. He, H. Murayama, S. Pakvasa, and G. Valencia, Phys. Rev. D **61**, 071701 (2000).
9. C.H. Chen, Phys. Lett. B **521**, 315 (2001).
10. J. Tandean, Phys. Rev. D **69**, 076008 (2004).
11. R. Burnstein *et al.*, Nucl. Instrum. Methods A **541**, 516 (2005).
12. T. Holmstrom *et al.*, Phys. Rev. Lett. **93**, 262001 (2004).
13. K.B. Luk *et al.*, Phys. Rev. Lett. **85**, 4860 (2000).
14. H. Park *et al.*, Phys. Rev. Lett. **94**, 021801 (2005).
15. X.G. He, J. Tandean, G. Valencia, Phys. Lett. B**631**, 100-108 (2005).
16. X.G. He, J. Tandean, G. Valencia, Phys. Rev. Lett. **98**:081802, (2007).

# Recent CLEO-c Results

H. Mendez
(for the CLEO Collaboration)

*Department of Physics, University of Puerto Rico, Mayaguez, PR 00681*
*E-mail: mendez@charma.uprm.edu*

**Abstract.** Recent CLEO-c results on open and closed charm physics at center-of-mass energy of 3773 MeV ($\psi(3770)$ resonance), 4170 MeV and 3686 MeV ($\psi(2S)$ peak) are reviewed. Measurements of absolute hadronic branching ratios of $D^0$, $D^+$ and $D_s^+$ mesons as well as charmonium spectroscopy are discussed. An outlook and future prospects for the experiment at CESR is also presented.

**Keywords:** charmed decays; charm branching fractions; mesons; charmonium; CLEO-c
**PACS:** 13.25.Ft, 14.40.Lb, 14.40.-n

## 1. Introduction

Recent advances on precise numerical calculations of non-perturbative QCD, formulated on a space-time lattice (LQCD) [1] as well as the development of heavy quark effective theory [2] have produced a wide variety of non-perturbative results with high accuracies for D meson decay constants and form factors. High precision charm experiments [3] provide crucial data to validate forthcoming LQCD calculations at the few percent level and help to guide the development of QCD calculations techniques for a full understanding of non-perturbative QCD effects. These calculations, useful to improve the accuracy of the Cabibbo-Kobayashi-Maskawa (CKM) unitary matrix [4], make possible alternate methods to improve the precision of the Standard Model predictions of charm behavior. Thus, experiments on charm decays are an excellent laboratory for testing the existing theory on heavy flavor physics.

CLEO-c is a dedicated program of charm physics at the Cornell Electron Storage Ring (CESR) $e^+e^-$ collider located at the Laboratory for Elementary Particles Physics (LEPP) at Cornell University at Ithaca, New York. The experiment [3] is designed to make very high precision measurements of charmed mesons ($D^0$, $D^+$ and $D_s^+$)[1] and to test quantum chromodynamics (QCD), which included a complete set of measurements for hadronic, leptonic and semi-leptonic charm decays, a detailed studies on the lowest and highest mass charmonium states and search for evidence of new physics beyond the Standard Model by searching for rare D and $\tau$ decays, $D\overline{D}$ mixing, and CP violating decays. CLEO-c precision measurements on charmed meson decays to leptonic and semi-leptonic final states are crucial test of the LQCD techniques to compute important heavy quark processes. Hadronic decays play an important role for B physics branching

[1] Charge conjugate particles are implicitly assumed in this paper, unless otherwise noted.

CP917, *Particles and Fields,* edited by H. Castilla Valdez, J. C. D'Olivo, and M. A. Perez

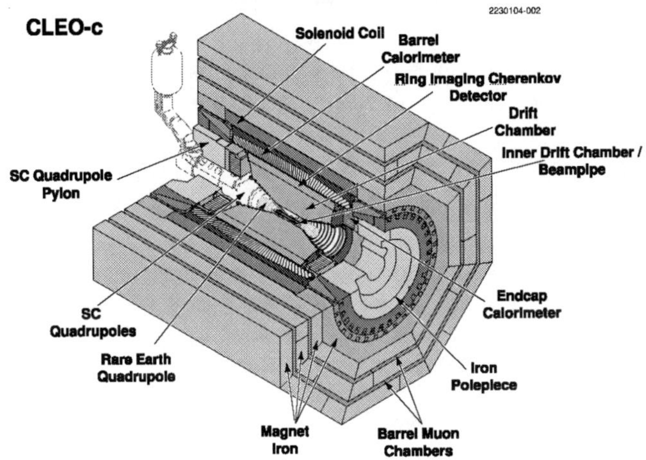


**FIGURE 1.** CLEO-c Detector

fractions normalization as well as in the study of final state which included strong interactions.

The experimental conditions in the charm system at threshold are optimal. Charm events produced at threshold are extremely clean and pure $D\overline{D}$ events, signal/background ratio is high and neutrino reconstruction is clean. In this report, we review some of our recent results on absolute hadronic branching ratio on $D^0$, $D^+$ and $D_s^+$ (section 3 and 4), charmonium (section 5) and a brief description of the detector (section 2).

## 2. Experimental Setup and Data Sample

In order to achieve this broad physics program, both the CESR accelerator and the CLEO III detector were upgraded. The CESR $e^+e^-$ collider was converted to operate from ~10 GeV to a lower center-of-mass energies (~3.6-4.3 GeV) by the addition of 18 meters of superconducting wiggler magnets to enhance the transverse cooling of the beam. In the CLEO III detector, the silicon vertex detector was replaced with a 6-layer vertex drift chamber and the solenoid field was reduced from 1.5 to 1.0 T, which was achieved by a simple reduction of the magnet current.

The CLEO-c spectrometer (described extensively elsewhere [5–10]), shown in Fig. 1 is composed of the following basic sections:

1. a tracking system composed of the inner vertex drift chamber, a 47-layer central drift chamber [10] and a superconducting solenoid running at 1.0 T magnetic field, oriented along the beam axis. The system covers 93% of $4\pi$ and has a momentum resolution of $\sigma_p/P = 0.6\%$ at 1 GeV/c,
2. a cylindrical Ring Imaging Cerenkov (RICH) [8] detector, surrounding the central drift chamber, for charged particle identification. The RICH cover 80% of $4\pi$ and

its efficiency to identified kaon is higher than 90% with 0.2% pion fake at P = 0.9 GeV/c,

3. an Electromagnetic Calorimeter [7] for neutral particle reconstruction covering 93% of $4\pi$. The detector has an energy resolution of $\sigma_E/E = 2.2\%$ at 1 GeV and 5% at 100 MeV,
4. and a muon chambers covering 85% of $4\pi$ for P > 1 GeV/c.

The CLEO-c detector is fully operational and presently taking data at the $\psi(3770)$ center-of-mass energy.

A detailed GEANT [11] based Monte Carlo detector modeling was used to simulate the performance of the detector. Physics events were generated by the event generator EvtGen [12] and final state radiation (FSR) modeled by PHOTOS [13]. Monte Carlo events were treated in the same manner as data.

Charged particle candidates were required to be well measured and to satisfy criteria based on the track fit quality. They must also be consistent with coming from the interaction point, except for the charged pions coming from $K_S^0$ decay. Particle identification to separate charged $\pi$ from K was accomplished by combining the ionization energy loss (dE/dx), measured by the drift chamber, with the RICH detector information.

In addition, we applied mass bound cut and kinematic fit to select $K_S^0$, $\pi^0$, $\eta$ and $\eta'$. We selected $K_S^0$ candidates from oppositely-charged tracks consistent with pions and constrained to the $K_S^0$ vertex ($K_S^0 \to \pi^+\pi^-$). The $\pi^0$ candidates are form by a photon pair ($\pi^0 \to \gamma\gamma$) kinematically fitted to the nominal $\pi^0$ mass [14]. Both of the invariant masses, for the $K_S^0$ ($\pi^+\pi^-$) and for the $\pi^0$ ($\gamma\,\gamma$) candidates are required to be within 3 standard deviations ($\sigma$) of their known mass [14]. For $K_S^0$, we required $\sigma \approx 6.3\ MeV/c^2$ and for $\pi^0$, we required $\sigma \approx 5-7\ MeV/c^2$ depending of the photon location in the calorimeter. We form $\eta$ candidates from photon pairs kinematically fitted to the known $\eta$ mass [14] and $\eta'$ candidates are detected via their decays mode $\eta' \to \pi^+\pi^-\eta$.

The experiment has already collected 281 $\text{pb}^{-1}$ of data at center-of-mass energy $\sqrt{s} = 3.77$ GeV to study $\psi(3770)$ decays, approximately 300 $\text{pb}^{-1}$ at $\sqrt{s} = 4.17$ GeV to study $D_s^{\pm}$ mesons and approximately 50 $\text{pb}^{-1}$ at $\sqrt{s} = 3.68$ GeV to study $\psi(2S)$ decays and charmonium spectrum, including $J/\psi$, $\chi_{cJ}$ ($J = 0,1,2$) and the properties of the CLEO-c observed $h_c(^1P_1)$ state. A small fraction (2.74 $\text{pb}^{-1}$) of the $\psi(2S)$ integrated luminosity was taken with the CLEO III detector while its conversion to CLEO-c took place. In addition, the experiment has recorded a significant amount of continuum data to study its contribution to these resonances.

## 3. D Physics at $\psi(3770)$

The 3773 MeV dataset, taken at the peak of the $\psi(3770)$ resonance, provides a very clean environment for studying with high precision a great variety of D meson decays. Many results based on an initial sample of 56 $\text{pb}^{-1}$ have already been published. These results, among others, included absolute branching fraction measurements for D decays into leptonic [15], semi-leptonic [16, 17], and hadronic [18] final states. In this contribution, we report preliminary hadronic branching fraction measurements

**TABLE 1.** Preliminary absolute branching fractions for $D^0$ and $D^+$ based on 281 $pb^{-1}$ and $D_s^+$ based on the 195 $pb^{-1}$ sample. The first error is statistical and the second is systematics.

| $D^0$ **Mode** | **Branching Fraction (%)** |
|---|---|
| $D^0 \to K^-\pi^+$ | $3.87 \pm 0.04 \pm 0.08$ |
| $D^0 \to K^-\pi^+\pi^0$ | $14.6 \pm 0.1 \pm 0.4$ |
| $D^0 \to K^-\pi^+\pi^+\pi^-$ | $8.3 \pm 0.1 \pm 0.3$ |
| $D^+$ **Mode** | |
| $D^+ \to K^-\pi^+\pi^+$ | $9.2 \pm 0.1 \pm 0.2$ |
| $D^+ \to K^-\pi^+\pi^+\pi^0$ | $6.0 \pm 0.1 \pm 0.2$ |
| $D^+ \to K_S^0\pi^+$ | $1.55 \pm 0.02 \pm 0.05$ |
| $D^+ \to K_S^0\pi^+\pi^0$ | $7.2 \pm 0.1 \pm 0.3$ |
| $D^+ \to K_S^0\pi^+\pi^+\pi^-$ | $3.13 \pm 0.05 \pm 0.14$ |
| $D^+ \to K^+K^-\pi^+$ | $0.93 \pm 0.02 \pm 0.03$ |
| $D_s^+$ **Mode** | |
| $D_s^+ \to K_S^0K^+$ | $1.50 \pm 0.09 \pm 0.05$ |
| $D_s^+ \to K^+K^-\pi^+$ | $5.57 \pm 0.30 \pm 0.19$ |
| $D_s^+ \to K^+K^-\pi^+\pi^0$ | $5.62 \pm 0.33 \pm 0.51$ |
| $D_s^+ \to \pi^+\pi^-\pi^+$ | $1.12 \pm 0.08 \pm 0.05$ |
| $D_s^+ \to \pi^+\eta$ | $1.47 \pm 0.12 \pm 0.14$ |
| $D_s^+ \to \pi^+\eta'$ | $4.02 \pm 0.27 \pm 0.30$ |

based on the total $\psi(3770)$ sample of 281 $pb^{-1}$ available at the present. This dataset include approximately 1 million of $e^+e^- \to \psi(3770) \to D^0\overline{D^0}$ and 0.8 million of $e^+e^- \to \psi(3770) \to D^+D^-$ events.

The $\psi(3770)$ produced in the $e^+e^-$ annihilation decays to pairs of D mesons only, either $D^0\ \overline{D^0}$ or $D^+\ D^-$. At this energy, pure $D\overline{D}$ final states are produced, no additional hadron accompanied the D mesons, therefore, the reconstructed D energy is safely replaced by the beam energy ($E_D \equiv E_{Beam}$) and as a consequence, the reconstructed D invariant mass is a beam constrained quantity ($m_{bc} = \sqrt{E_{Beam}^2 - \vec{P_D}^2}$), which has better resolution than the invariant mass calculated using directly the energy of the D. For equal mass particles of mass M as this case, the kinematic variables $m_{bc}$ peak at M and the energy difference $\Delta E \equiv E_D - E_{Beam}$ peak at zero. Both variables were used in the analysis of the $\psi(3770)$ data. $\Delta E$ around zero was required for the D candidates.

CLEO-c uses a double and single tagging technique, pioneered by the MARK III collaboration [19, 20], to measure absolute branching fractions. It relies on fully reconstructed $D\overline{D}$ decays, in which both D are reconstructed (double tags) in the event. Single tag mode reconstructs at least one D per event. A full description of the tagging method used by the experiment as well as the hadronic analysis based on 56 $pb^{-1}$ can be found in [18]. The tagging technique obviated the need for knowledge of the luminosity or the $e^+e^- \to \psi(3770) \to D\overline{D}$ production cross section. The absolute branching fractions obtained using double tag [21] are listed in Table 1.

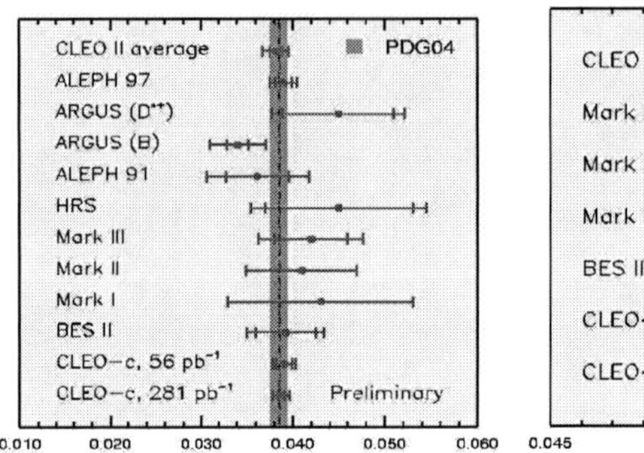

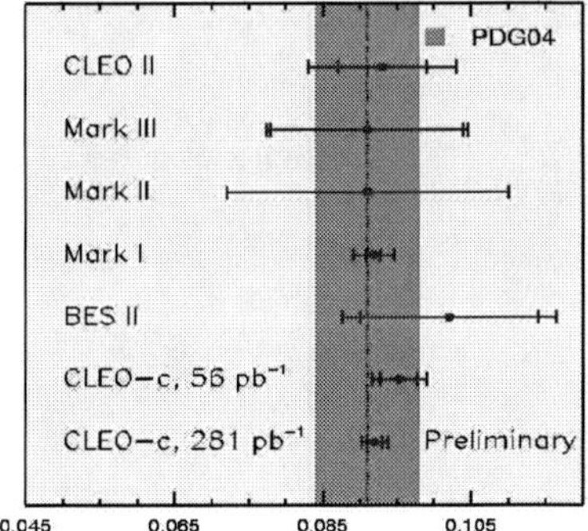


**FIGURE 2.** CLEO-c absolute $D^0 \to K^-\pi^+$ (left) and $D^+ \to K^-\pi^+\pi^+$ (right) branching fractions compared with other measurements.

These results are dominated by the systematic errors and included the FSR correction, which has been included in the Monte Carlo. Without this effect, the branching fractions decreased in average by 2%.

Two of these branching fractions, Br($D^0 \to K^-\pi^+$) and Br($D^+ \to K^-\pi^+\pi^+$) shown in Table 1, are particularly important because they have been used to normalized practically all other $D^0$ and $D^+$ modes. Figure 2 shows a comparison of our results (based on 56 pb$^{-1}$ and 281 pb$^{-1}$), for these two modes with other experimental measurements. These results already represent significant improvements with respect to the world averages. To date, they are the most precise measurements of hadronic branching fractions for D mesons. Other results based on the latest $\psi(3770)$ data sample are also available, such as the semi-leptonic inclusive decays [22]. These high precision results are possible in part by our detector performance.

## 4. The $D_S$ Energy Scan

In order to select the optimal energy to study $D_s$, a scan on the energy region from 3970-4260 MeV was performed [23]. The scan consisted on 12 points (Fig. 3) with a total luminosity of 60 pb$^{-1}$. The last energy point at $E_{cm} = 4260$ MeV, was taken to study the Y(4260) state (See section 5). At each energy point the first task was to make quick determination of the cross section for each of the two charmed meson final states that were accessible at that energy. Several methods were used to measure the cross section. It was measured by counting inclusive hadronic, inclusive D, and exclusive DD* final states events. Fig. 4 shows these cross section using exclusive final states. A peak cross section of about 1.0 nb for $D_s\,\overline{D_s^*}$ is observed around 4170 MeV, which was then selected as the optimal energy for CLEO-c to carry out the $D_s$ physics program. At this energy, we produce $D_s^+\,D_s^-$ pairs, where one of the $D_s$, accompanied by a $\gamma$ or $\pi^0$, is in general the daughter of a $D_s^*$ decay. In order to avoid a large efficiency loss and contributions to systematic uncertainties that would arise from the soft photon, we make no attempt to find either the $\gamma$ or the $\pi^0$. Thus, the main variable used in this analysis is

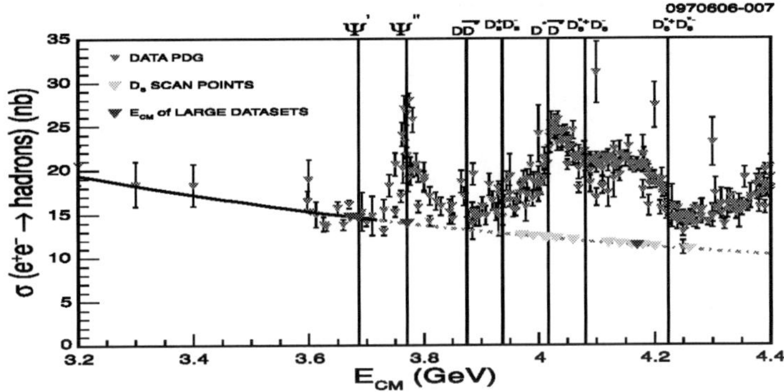


**FIGURE 3.** Data on hadron production cross section as a function of the $e^+e^-$ center-of-mass energy. Inverted green triangles indicates the CLEO-c scan run points.

the reconstructed invariant mass of the $D_s$ instead of the $m_{bc}$.

An initial sample of 195 $pb^{-1}$ of data at the center-of-mass energy $\sqrt{s} = 4170$ MeV were recorded. High precision $D_s$ inclusive decays analysis have been already published [24] and preliminary measurements on exclusive absolute hadronic branching fractions have already reported for several $D_s^+$ decay modes (see Table 1) by using single $D_s$ tags [25]. These measurements are shown in Fig. 5 and while these results are preliminary, they are more precise than the one published on PDG 2006 [14] as is shown in Fig. 5.

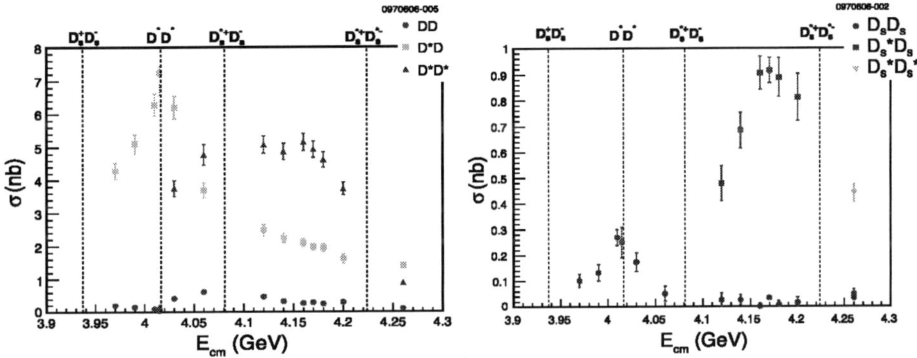


**FIGURE 4.** Preliminary measurement production cross section for $D\overline{D}$, $D\overline{D^*}$, $D^*\overline{D^*}$ on the left and $D_s\overline{D_s}$, $D_s\overline{D_s^*}$, $D_s^*\overline{D_s^*}$ on right as a function of the $e^+e^-$ center-of-mass energy.

Our Br($D_s^+ \to K^+K^-\pi^+$) include the Br($D_s^+ \to \phi\pi^+$) which is one of the largest $D_s$ branching fractions [14]. It has often used as a reference for other $D_s$ decays. This measurement has been essentially derived from a narrow mass region cut around the $\phi$ in the $K^+K^-$ invariant mass ($\phi \to K^+K^-$). However, there is a strong evidence [26] for a broad contribution from $f_0(980)$ or $a_0(980)$ under the $\phi\pi^+$ region. This scalar contribution accounts for approximately 5%, which is comparable to our experimental errors for the Br($D_s^+ \to K^+K^-\pi^+$). CLEO-c will soon exceed this level of precision and the Br($D_s^+ \to \phi\pi^+$) will be measured as soon as the new data is gathered.

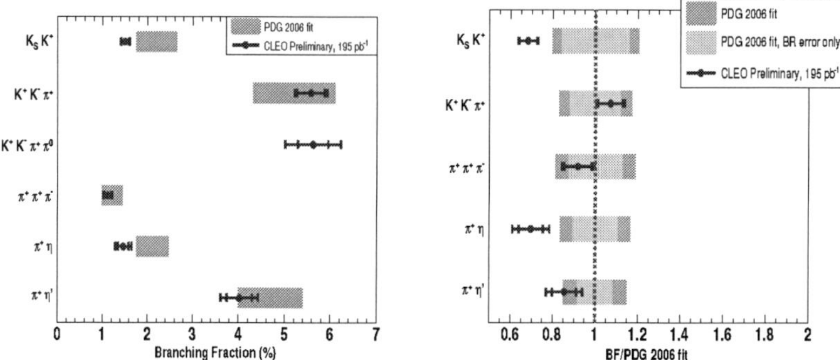


**FIGURE 5.** Absolute $D_s^+$ branching ratio on the left and our results normalized with respect to the PDG 2006 values on the right.

## 5. Charmonium and Charmonium-Like States

The spectrum below the $D\overline{D}$ is well known, however, in the last few years there has been a renewed interest in heavy quark spectroscopy. More than a dozen of unexpected charmonium-like states have been reported by high luminosity experiments and several long elusive states have been observed, including the $h_c(^1P_1)$ [27, 28] and the $\eta_c'(2^1S_0)$ [29]. Many new theoretical models have been proposed to explain all these resonances.

The last energy point of our $D_s$ energy scan, $E_{cm}$ = 4260 MeV, with a total integrated luminosity of 13.2 pb$^{-1}$ was used primarily to investigate the Y(4260) state discovered by the BaBar Collaboration [30]. This observation was based on a sample of 233 fb$^{-1}$ collected at the $\Upsilon(4S)$ in an initial state radiation events as $e^+e^- \to \gamma\,(J/\psi\ \pi^+\pi^-)$. CLEO-c made the first confirmation [31], at $11\sigma$ significance, of this new charmonium decay mode ($J/\psi\ \pi^+\pi^-$) of the Y(4260) state, made the first observation of Y(4260) $\to J/\psi\ \pi^0\pi^0$ at $5.1\sigma$ and find the first evidence for Y(4260) $\to J/\psi\,K^+K^-$ at $3.7\sigma$ significance. The measured CLEO-c cross section as well as the integrated luminosity are shown on Fig. 6. This signal was also confirmed by using the CLEO III dataset of 13.3 fb$^{-1}$ collected at $\Upsilon(1S)-\Upsilon(4S)$ resonances and the extracted signal parameters are consistent with BaBar's results.

The 3686 MeV dataset, taken at the $\psi(2S)$ peak included a sample of approximately 26 million $\psi(2S)$ decays acquired during 2006 to explore charmonium spectroscopy and related states. In the meantime this sample is being prepared for a detailed analysis, a small $\psi(2S)$ sample of ~5.6 pb$^{-1}$, corresponding approximately to 3 million $\psi(2S)$ decays, split equally between CLEO III and CLEO-c, has produced a great well known variety of results which included $J/\psi$ leptonic decays [32], $h_c(^1P_1)$ [27] discovery, etc.

The singlet P-state of charmonium $h_c(^1P_1)$ has been observed in the isospin forbidden reaction $\psi(2S) \to \pi^0\ h_c(^1P_1)$, where $h_c(^1P_1) \to \gamma\ \eta_c$. The $\eta_c$ was identified by two methods: first, it was fully reconstructed in 7 exclusive modes, and second, it was re-

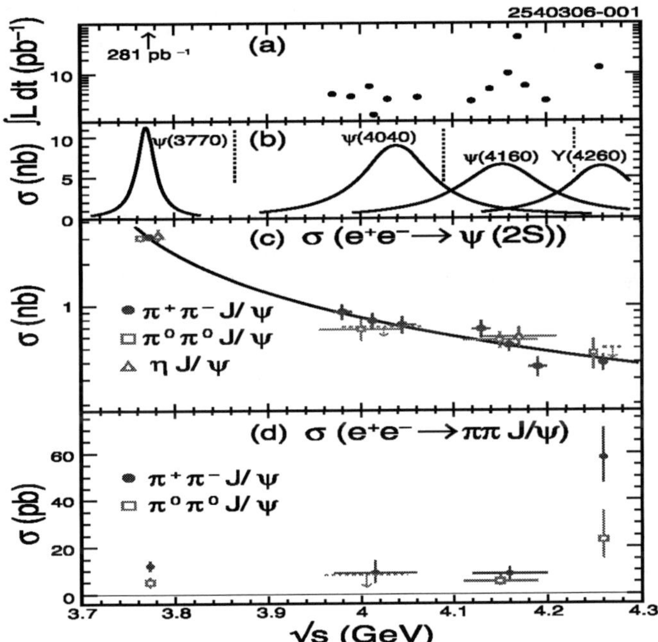


**FIGURE 6.** (a) Integrated luminosity vs. $\sqrt{s}$, (b) Born-level Breit-Wigner cross section for established charmonium and Y(4260) states. (c) $e^+e^- \to \gamma\, \psi(2S)$ cross section vs. energy for three different $\psi(2S)$ decay mode. (d) $e^+e^- \to J/\psi\, \pi\pi$ vs. $\sqrt{s}$. Some points at (c) and (d) are shifted by 10 MeV for clarity.

constructed inclusively. The exclusive mode has the advantage of signal purity while the inclusive mode has the advantage of larger signal yield. Both methods yields consistent results and the average measured mass is $3524.4 \pm 0.6 \pm 0.4$ MeV. This mass leads to a hyperfine splitting of $1 \pm 0.6 \pm 0.4$ MeV, which is consistent with zero as expected. With this discovery, the charmonium family below $\psi(3770)$ is complete, and the mass values can be used in potential models to predict higher states. A detailed analysis on the $h_c(^1P_1)$ observation made by the experiment can be found at [27].

## 6. Summary

Running at charm threshold with D tagging provides a powerful and "background free" environment that allow very precise measurements. Our experimental measurements on leptonic, semi-leptonic and hadronic D decays are producing a major impact on many charm decays with broad implications for all flavor physics. These high precision measurements are more precise than previous world average [33].

The experiment is scheduled to run until March 31 of 2008. In addition to the charmonium sample, we plan to further increase our samples at $\psi(3770)$ and at 4170 MeV to continue the open and closed charm measurements. These high statistics samples from the decays of charmed mesons and charmonium data represent a rich source of important physics on the study of charm quarks. It permits detailed studies of weak and strong interaction physics.

## ACKNOWLEDGMENTS

I would like to thank the organizer for the invitation to give this talk and for their hospitality during the conference. I thank my colleagues from the CLEO-c Collaboration for the many useful comments to this paper. This work was supported in part by the U.S. Department of Energy.

## REFERENCES

1. C. Aubin, et al., *Phys. Rev. Lett.* **94**, 011601 (2005).
2. T. Mannel, Review of heavy quark effective theory (1996), `hep-ph/9611411`.
3. R. A. Briere, et al., *(CLEO Collaboration), CLEO-c and CESR-c. A New Frontier of Weak an Strong Interactions*, CLNS 01/1742. Internal Document (Yellow Book), Nov 2001.
4. M. Kobayashi, and T. Maskawa, *Prog. Theor. Phys.* **49**, 652–657 (1973).
5. G. Viehhauser, et al., *(CLEO Collaboration), Nucl. Instrum. Meth.* **A462**, 146–151 (2001).
6. T. S. Hill, et al., *(CLEO Collaboration), Nucl. Instrum. Meth.* **A418**, 32–39 (1998).
7. Y. Kubota, et al., *(CLEO Collaboration), Nucl. Instrum. Meth.* **A320**, 66–113 (1992).
8. M. Artuso, et al., *(CLEO Collaboration), Nucl. Instrum. Meth.* **A554**, 147–194 (2005).
9. M. Artuso, et al., *(CLEO Collaboration), Nucl. Instrum. Meth.* **A502**, 91–100 (2003).
10. D. Peterson, et al., *(CLEO Collaboration), Nucl. Instrum. Meth.* **A478**, 142–146 (2002).
11. R. Brun, et al., *Computer Code GEANT 3.21* (CERN Report No. W5013, 1993).
12. D. Lange, *Nucl. Instrum. Methods Phys.* **Sect. A 462**, 152 (2001).
13. E. Barberio, and Z. Was, *Comput. Phys. Commun.* **79**, 291 (1994).
14. W. M. Yao, et al., *(PDG 2006), J. Phys.* **G33**, 1–1232 (2006).
15. G. Bonvicini, et al., *(CLEO Collaboration), Phys. Rev.* **D70**, 112004 (2004).
16. G. S. Huang, et al., *(CLEO Collaboration), Phys. Rev. Lett.* **95**, 181801 (2005).
17. T. E. Coan, et al., *(CLEO Collaboration), Phys. Rev. Lett.* **95**, 181802 (2005).
18. Q. He, et al., *(CLEO Collaboration), Phys. Rev. Lett.* **95**, 121801 (2005).
19. R. M. Baltrusaitis, et al., *(MARK-III Collaboration), Phys. Rev. Lett.* **56**, 2140 (1986).
20. J. Adler, et al., *(MARK-III Collaboration), Phys. Rev. Lett.* **60**, 89 (1988).
21. D. Cassel, "D and Ds Decays and Dalitz Analyses," in *Heavy Quarks and Leptons 2006*, October 18,2006, Munchen, Germany (CLEO Web Page Public Results).
22. N. E. Adam, et al., *(CLEO Collaboration), Phys. Rev. Lett.* **97**, 251801 (2006).
23. R. Poling, "CLEO-c hot topics," in *Flavor Physics and CP Violation*, April 9-12,2006, Vancouver, B.C., Canada, `hep-ex/0606016`.
24. G. S. Huang, et al., *(CLEO Collaboration), Phys. Rev.* **D74**, 112005 (2006).
25. N. E. Adam, et al., *Measurement of absolute hadronic branching fractions of Ds mesons*, presented at the $33^{rd}$ International Conference on High Energy Physics, Jul 26-Aug 2, 2006, Moscow, Russia, `hep-ex/0607079`.
26. P. L. Frabetti, et al., *(E687 Collaboration), Phys. Lett.* **B351**, 591–600 (1995).
27. J. L. Rosner, et al., *(CLEO Collaboration), Phys. Rev. Lett.* **95**, 102003 (2005).
28. M. Andreotti, et al., *(E835 Collaboration), Phys. Rev.* **D72**, 032001 (2005).
29. D. M. Asner, et al., *(CLEO Collaboration), Phys. Rev. Lett.* **92**, 142001 (2004).
30. B. Aubert, et al., *(BaBar Collaboration), Phys. Rev. Lett.* **95**, 142001 (2005).
31. Q. He, et al., *(CLEO Collaboration), Phys. Rev. D.* **74**, 091104 (2006).
32. G. S. Adams, et al., *(CLEO Collaboration), Phys. Rev.* **D73**, 051103 (2006).
33. S. Eidelman, et al., *(PDG 2004), Phys. Lett.* **B592**, 1 (2004).

# Physics of Top

C.-P. Yuan

*Department of Physics and Astronomy, Michigan State University, East Lansing, MI 48824-1116, U.S.A.*

**Abstract.** I will briefly review the physics of top quark at high energy colliders. A new discovery of single-top event at the Fermilab Tevatron is expected. At the CERN Large Hadron Collider, detailed top quark properties can be measured and new physics ideas in which top quark plays a special role can be tested. I will also discuss a few phenomenological methods for analyzing experimental data to study top quark interactions.

**Keywords:** Top Quark, Electroweak interaction, QCD corrections, New physics models
**PACS:** 14.65.Ha 12.15.-y 12.38.-t 11.30.Qc

## INTRODUCTION

Prior to the discovery of top quark in 1995, a wide range of its mass was predicted [1], which signals our ignorance about the origin of mass. Hence, the breaking mechanisms of electroweak symmetry (for generating $W$ and $Z$ boson masses) and flavor symmetry (for generating a wide spectrum of fermion masses) remain to be two of the major mysteries in the elementary particle physics. Nevertheless, through the process of "guessing" what the top quark mass is, we learned that only experimental data has the final say about the mother Nature; the interaction between experimentalists and theorists is essential for the advance of science; and theorists should not give up any probable ideas. In this talk, I will discuss a few theory ideas in which top quark plays a special role such that studying the interaction of top quark might help revealing the mechanism of electroweak symmetry breaking (EWSB) and flavor symmetry breaking mechanisms. I will also discuss a few phenomenological methods for analyzing experimental data to study top quark interactions.

## IMPACT OF $m_t$ MEASUREMENT

First, let us briefly review what we know so far about the top quark. Its mass ($m_t$) has been measured with better accuracy around 174 GeV from studying the top quark pair events produced via quark and gluon fusion processes at the Fermilab Tevatron [2]. It is quite challenging to measure $m_t$ from the $bjj$ invariant mass to be better than a couple of GeV at the Tevatron and the CERN Large Hadron Collider (LHC), because of the uncertainty in jet energy resolution (related to under-lying hadronic activities in the event) and the limited accuracy in current theory calculations when the accuracy in the $m_t$ measurement ($\delta m_t$) is required to be less than the width of top quark. Other means for measuring $m_t$ are: studying the fraction of longitudinal polarization of the $W$ boson (or the transverse momentum ($p_T$) distribution of the lepton) and the invariant

CP917, *Particles and Fields,* edited by H. Castilla Valdez, J. C. D'Olivo, and M. A. Perez

mass distribution of $b$ and $\ell$ in the decay of top quark $t \to bW(\to \ell\nu)$ [3]. How well do we need to know about $m_t$ in order to test the Standard Model (SM) beyond the tree level? At the Tevatron Run-II, $\delta m_t \sim 2-3$ GeV, and it is no longer a dominant error in precision test until $\delta M_W$ is reduced to about 20 MeV. At the LHC, $\delta m_t \sim 1.5$ GeV, which is about the same as the top quark width ($\Gamma_t$), and the counterpart in the measurement of $M_W$ needs to be about 10 MeV. Again, we see that $\delta m_t$ is not the dominant theoretical error in testing the SM from comparing its predictions to precision data.

At the future International Linear Collider (ILC), $\delta m_t$ can be reduced to a couple of hundred MeV (of the order of $\Lambda_{QCD}$) from studying the top quark pair production at threshold (from the measurement of total cross section, peak of transverse momentum distribution, and the forward-backward asymmetry), and around 500 MeV from directly reconstructing the mass of top quark produced at continuum [4]. With the increasing precision in the determination of the top quark mass, it becomes important to understand the exact relation between the measured experimental observable (say, the peak position of the invariant mass distribution) and the theoretical parameter $m_t$ defined in the SM theory Lagrangian. Obviously, down to the accuracy of a couple of hundred MeV, this relation also depends on the order of perturbative calculation performed in the theory prediction when comparing to the experimental data.

## TOP QUARK DECAY

Because its mass is at the weak scale, top quark decays like a bare quark (without first forming hadrons) so that we could study its spin property in addition to measure its decay branching ratios. In the SM, top decays into $bW$ mode almost 100% of time. With new physics coupling to top quark, many new decay channels become possible, such as $t \to bH^+, \tilde{t}\chi_0, c\gamma, cZ, cg, ch^0$, predicted in the Minimal Supersymmetric Standard Model (MSSM). Thus, it is important to measure the decay branching ratio (BR) of $t \to bW$. Unfortunately, in the SM, the Cabibbo-Kobayashi-Maskawa (CKM) matrix element $V_{tb}$ is so much larger than $V_{ts}$ and $V_{td}$ that the ratio of BR($t \to bW$) to BR($t \to bq$) cannot effectively measure the magnitude of the $W$-$t$-$b$ coupling. Moreover, the total decay width of the top quark cannot be accurately measured from the $bjj$ invariant mass distribution in top decays, for the experimental uncertainty due to jet energy resolution is much larger than $\Gamma_t$ [5].

It is however possible that new physics effect might not change the value of BR($t \to bW$), for not having additional new light fields with mass less than $m_t$, but could still modify $\Gamma_t$ when the interaction of $t$-$b$-$W$ is strongly modified. Hence, we need to directly measure the interaction strength of $t$-$b$-$W$ coupling. This can be done by studying single-top production rates initiated from weak charged current processes. In particular, the SM tree level t-channel single-top inclusive production rate ($\sigma_t$) is proportional to the decay width of top quark. Hence, $\sigma_t$ can be used to determine the CKM matrix element $V_{tb}$ in the SM, and to determine the partial decay width $\Gamma(t \to bW)$ [6]. When combining the measurement of BR($t \to bW$) from $t\bar{t}$ events and $\Gamma(t \to bW)$ from t-channel singe-top event, one can determine the life-time of top quark from their ratio.

## SINGLE-TOP PRODUCTION

At the Tevatron and the LHC, single-top events can be produced from t-channel, s-channel and $Wt$ associate production processes [7]. Their production rates can be largely modified by new physics interactions, either with new heavy resonances, such as $W',Z',H^{\pm},\pi'$, or with flavor changing neutral currents (FCNC), such as $tcZ,tuZ,tcg,tc\gamma$, or with new flavor changing charged currents (FCC), such as $tsW,tdW,tbH^{+}$. It turns out that the s-channel mode is sensitive to charged resonances. The t-channel mode is more sensitive to FCNC and new interactions. The $tW$ mode is a more direct measure of top quark coupling to $W$ and a down-type (down, strange, bottom) quark, such as FCC couplings. From a theoretical point of view, they are sensitive to different new Physics. From an experimental point of view, they have different signatures and different systematics [7]. Hence, they should be separately measured experimentally. Furthermore, as a proton-antiproton ($p\bar{p}$) collider, Tevatron offers a special chance to measure the amount of (direct) CP-violation in top quark system by measuring the asymmetry in the inclusive single-top versus single-antitop production rates [8]:

$$A_t^{CP} = \frac{\sigma(p\bar{p}\to tX) - \sigma(p\bar{p}\to \bar{t}X)}{\sigma(p\bar{p}\to tX) + \sigma(p\bar{p}\to \bar{t}X)}. \tag{1}$$

This is because under the CP symmetry, the initial hadron state ($p\bar{p}$) is invariant. This is the unique opportunity at Tevatron that the LHC cannot offer. To probe CP-violation in top quark interaction at the LHC, one has to measure the CP-violating (more generally, time-reversal violating) observables that make use of the spin information of the produced (anti)top quarks, such as measuring the expectation values of $< \vec{s}_t \cdot \vec{p}_b \times \vec{p}_{\ell^+} >$ and $< \vec{s}_{\bar{t}} \cdot \vec{p}_{\bar{b}} \times \vec{p}_{\ell^-} >$ from the decay of top and anti-top, respectively, in the single-top and single-antitop events [8]. Needless to say that CP-violation can also be tested at the LHC in the $t\bar{t}$ events by comparing the production rates of $t_L\bar{t}_L$ and $t_R\bar{t}_R$ events, for under the CP operation, a left-handed top ($t_L$) becomes a right-handed anti-top ($\bar{t}_R$). One way to measure this asymmetry is to detect the asymmetry in the energy distributions of $\ell^+$ versus $\ell^-$ in the inclusive $t\bar{t}$ events [9].

## SINGLE-TOP PRODUCTION AND DECAY AT NLO QCD

In order to reliably compare the theory prediction with experimental data on the production rate of the single-top events and the distributions of its final state particles, a next-to-leading order (NLO) QCD calculation has been performed for the s- and t-channel single-top processes [10]. In Ref. [10], we separated the single-top processes into a few smaller gauge invariant sets to organize our calculations, which include corrections to the initial state, final state and top decay in the s-channel process, and corrections to the light quark line, heavy quark line and top decay in the t-channel process. Keeping track on each individual contributions is useful for comparing event generators with exact NLO predictions. One can also study the effect from, for example, having correct implementation of top quark spin correlation between its production and decay at the full NLO in QCD.

**TABLE 1.** Efficiencies of identifying correct $b$-jet ($\varepsilon_b$) and picking up correct $p_z^\nu$ ($\varepsilon_\nu$) in both best-jet algorithm and leading-jet algorithm.

| | best-jet algorithm | | | leading $b$-tagged jet algorithm | | |
|---|---|---|---|---|---|---|
| | $s$-channel | $t$-channel | | $s$-channel | $t$-channel | |
| | inc. 2-jet | incl. 2-jet | excl. 3-jet | incl. 2-jet | incl. 2-jet | excl. 3-jet |
| $\varepsilon_b$ | 80% | 80% | 72% | 55% | 95% | 90% |
| $\varepsilon_\nu$ | | 70% | | | 84% | |

Given the small single-top production rate at the Tevatron, as predicted by the SM, it is important to study the acceptance of the signal event after imposing the necessary kinematic cuts for detecting them experimentally. We found that the signal acceptances are sensitive to the kinematics cuts. A large $R$-separation cut reduces acceptances significantly because $\Delta R(\ell j)$ is typically less than 1. With tight kinematic cuts, leading order (LO) and NLO acceptances are almost the same, but, with lose cuts, they are quite different. Hence, in order to maximize the signal acceptance, we must impose lose cuts and consequently the NLO acceptance of the signal events cannot be correctly modelled by a scaled-up (multiplied by the $K$-factor extracted from inclusive rate calculations) tree level LO calculation.

To identify the single-top signal events and to test the polarization of top quark, by studying spin correlations among the final state particles, we need to reconstruct the top quark in the single-top events. To do so, we shall first identify the $b$-jet and reconstruct the $W$ boson from top decay. In Table 1, we show the efficiency of finding the correct $b$-jet ($\varepsilon_b$) in two different algorithms: best-jet algorithm and leading $b$-tagged jet algorithm. The "best-jet" is defined to be the $b$-tagged jet which gives an invariant mass closest to the true top mass when it is combined with the reconstructed $W$ boson after determining the longitudinal momentum $p_z^\nu$ of the neutrino from $W$ decay. The leading $b$-tagged jet algorithm picks the leading $b$-tagged jet as the correct $b$-jet to reconstruct the top quark after combining with the reconstructed $W$ boson. As shown in Refs. [11, 12], we find that the best-jet algorithm shows a higher efficiency (about 80%) in picking up the correct $b$-jet than the leading-jet algorithm (about 55%) for the $s$-channel single-top events. On the other hand, for the $t$-channel single-top events, the leading $b$-tagged jet algorithm picks up the correct $b$-jet with a higher efficiency, about 95% for inclusive 2-jet events and 90% for exclusive 3-jet events., cf. Table 1. The reason that the leading $b$-tagged jet algorithm works well in the exclusive 3-jet $t$-channel single-top events is due to the distinct kinematic difference between $b$ and $\bar{b}$-jets. To reconstruct the top quark in the signal events, we also need to reconstruct the $W$ boson. The $W$ boson is reconstructed with the help of using its mass constraint: $M_W^2 = (p_l + p_\nu)^2$. Which one of the two-fold solutions in $p_z^\nu$ to be taken depends on the $b$-jet algorithm we used. In the case of best-jet algorithm, we find the one with the smaller magnitude from solving the $W$ mass constraint give the best efficiency in $W$ boson reconstruction. In the case of leading $b$-tagged jet algorithm, we use the top quark mass constraint $M_t^2 = (p_b + p_l + p_\nu)^2$ to pick up the best $p_z^\nu$ value. The efficiency for picking up the correct $p_z^\nu$ value ($\varepsilon_\nu$), at LO and NLO, respectively, is presented in Table 1.

After reconstructing the top quark, one can study the effect of NLO QCD corrections

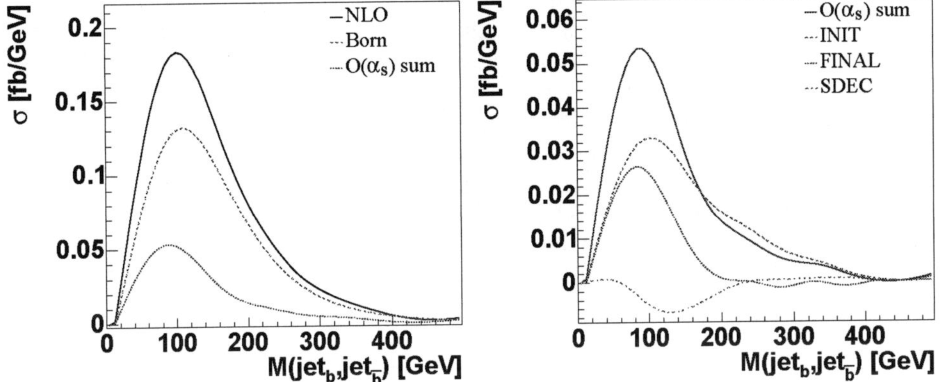


**FIGURE 1.** Invariant mass of the ($b$-jet, $\bar{b}$-jet) system after selection cuts, comparing Born-level to $O(\alpha_s)$ corrections. In the legend, INIT, FINAL and SDEC denotes the contributions from initial state, final state and top quark decay corrections, respectively.

to the measurement of top quark polarization. We found that higher order QCD corrections blur the spin correlation effect. Furthermore, the apparent advantage of some polarization basis at the parton level is washed away after modelling the reconstruction of the top quark as described above.

## s-channel single-top events and Higgs search

The $s$-channel single top quark process also contributes as one of the major backgrounds to the SM Higgs searching channel $q\bar{q} \to WH$ with $H \to b\bar{b}$. In this case it is particularly important to understand how the $O(\alpha_s)$ corrections change distributions around the Higgs mass region. Because of the scalar property of the Higgs boson, its decay products $b$ and $\bar{b}$ have symmetric distributions. Fig. 1 shows the invariant mass distribution of the ($b$-jet, $\bar{b}$-jet) system. For a Higgs signal, this invariant mass of the two reconstructed $b$-tagged jets would correspond to a plot of the reconstructed Higgs mass. Thus, understanding this invariant mass distribution will be important to reach the highest sensitivity for Higgs boson searches at the Tevatron. The figure shows that at $O(\alpha_s)$, the invariant mass distribution not only peaks at lower values than at Born level, it also drops off faster. This change in shape is particularly relevant in the region focused on by SM Higgs boson searches of $80\,\text{GeV} \leq m_{b\bar{b}} \leq 140\,\text{GeV}$ which is also at the *fb* level. In particular, the NLO contribution from the decay of top quark, while small in its overall rate, has a sizable effect in this region of the invariant mass and will thus have to be considered in order to make reliable background predictions for the Higgs boson searches.

Other kinematic distributions are also changing in shape when going from Born-level to $O(\alpha_s)$. In Ref. [11], we also showed the distribution of $\cos\theta$ for the two $b$-tagged jets, where $\theta$ is the angle between the direction of a $b$-tagged jet and the direction of the ($b$-jet, $\bar{b}$-jet) system, in the rest frame of the ($b$-jet, $\bar{b}$-jet) system. Experiments cannot distinguish between the $b$- and the $\bar{b}$-jets, we therefore include both the $b$-jet and the $\bar{b}$-jet in the graph. This distribution is generally flat at Born-level, with a drop-off at high $\cos\theta$ due to jet clustering effects, and a drop-off at negative $\cos\theta$ due to kinematic selection cuts. The $O(\alpha_s)$ corrections change this distribution significantly and result in a more forward peak of the distribution, similar to what is expected in Higgs boson production. In other words, a flatter $\cos\theta$ distribution in the s-channel single-top events make it more difficult to separate the $WH$ events from the s-channel single-top events in experimental analysis.

## t-channel single-top events and Higgs search

The unique signature of the $t$-channel single top process is the spectator jet in the forward direction, which can be utilized to suppress the copious SM backgrounds, such as $Wbb$ and $t\bar{t}$ events [13]. Studying the kinematics of this spectator jet is important to have a better prediction of the acceptance of $t$-channel single top quark events. The impact of the NLO QCD corrections on the kinematic properties of the spectator jet has been reported in Ref. [12] for the Tevatron Run-II phenomenology. We found that the NLO QCD corrections to the light-quark (LIGHT) and heavy-quark (HEAVY) lines show almost opposite behavior in the rapidity distribution of the spectator jet. LIGHT shifts the spectator jet to the forward direction while HEAVY shifts it to the central region, and the NLO corrections originated from the decay of top quark does not modify the rapidity distribution of the spectator jet.

One of the most important tasks at the LHC is to find the Higgs boson, denoted as $H$. It has been shown extensively in the literature that the production mechanism of Higgs boson via weak gauge boson fusion is an important channel for the search of Higgs boson. Furthermore, to test whether it is a SM Higgs boson after its discovery, one needs to determine the coupling of $H$-$V$-$V$, where $V$ denotes either $W^{\pm}$ or $Z$, by measuring the production rate of $q\bar{q}(VV) \to Hq'\bar{q}'$ via weak boson fusion processes. In order to suppress its large background rates, one usual trick is to tag on the two forward-jets resulted from emitting vector boson $V$ to produce Higgs boson via $VV \to H$. Prior to the discovery of Higgs boson, one can learn about the detection efficiency of the forward jet from studying the $t$-channel single-top process. This is because in the $t$-channel single-top process, the forward jet also results from emitting a $W$ boson which interacts with the $b$ quark from the other hadron beam to produce the heavy top quark. As pointed out in Ref. [13], in the effective-$W$ approximation, a high-energy $t$-channel single top quark event is dominated by a longitudinal $W$ boson and the $b$ quark fusion diagram. It is the same effective longitudinal $W$ boson that dominates the production of a heavy Higgs boson at high energy colliders via the $W$-boson fusion process. For a heavy SM Higgs boson, the longitudinal $W$ boson fusion process dominates the Higgs boson production rate. Hence, it is important to study the kinematics of the spectator jet in $t$-channel single top quark events in order to have a better prediction for the kinematics of Higgs boson events via the $WW$ fusion process at the LHC.

## GENERAL ANALYSIS OF $t$-$b$-$W$ COUPLINGS

As discussed above, the $t$-$b$-$W$ couplings can affect the spin correlations among the decay particles of (anti)top quarks produced in the $t\bar{t}$ events and the production cross section of the single-top events. Therefore, it is desirable to find a systematic way to analyze both $t\bar{t}$ and single-top experimental data to extract out the information on the $t$-$b$-$W$ couplings. This was performed in Ref. [6], in which the most general formulation of the $t$-$b$-$W$ couplings in any weak-scale effective theory was formulated. It also summarized what we have learned from indirect measurements whose conclusions unavoidably depend on the specific assumption made about the underlying new physics models. The most interesting question to ask is "How to directly measure $t$-$b$-$W$ couplings at the Tevatron and the LHC?" as well as "How to distinguish models of EWSB from the results of these direct measurements?".

In the language of electroweak chiral Lagrangian, there are in general 8 different operators at the weak scale to describe the $t$-$b$-$W$ couplings which do not require either $t$, $b$ or $W$ to be on their mass-shells. For on-shell $t$ and $b$ quarks, the above 8 operators reduce to 6 independent ones. Since we will consider only the case that the $W$ boson in the $t$-$b$-$W$ couplings couples to massless fermions (quarks or leptons), the number of independent operators is reduced to 4. In the Unitary gauge, it reads as [3]

$$\mathscr{L}_{tbW} = \frac{g}{\sqrt{2}} W_\mu^- \bar{b}\gamma^\mu \left(f_1^L P_L + f_1^R P_R\right) t - \frac{g}{\sqrt{2}M_W} \partial_\nu W_\mu^- \bar{b}\sigma^{\mu\nu} \left(f_2^L P_L + f_2^R P_R\right) t \;\; + h.c. \quad (2)$$

At tree level, the SM predicts $f_1^L = V_{tb} \simeq 1$, and $f_1^R = f_2^L = f_2^R = 0$. Since these form factors can take different values depending on the underlying new physics models that lead to the above effective Lagrangian at the weak scale, we proposed a general analysis to determine these four independent $t$-$b$-$W$ couplings. The idea is to use four experimental observables to determine the four independent form factors $f_{1,2}^{L,R}$. They are the experimental measurements on the degrees of longitudinal ($f_0$) and left-handed ($f_-$) polarizations of the $W$ bosons from top decays measured in the $t\bar{t}$ events and the s- ($\sigma_s$) and t-channel ($\sigma_t$) single-top production rates.

In summary, the above four observables can be expressed in terms of the effective $t$-$b$-$W$ couplings as:[1]

$$f_0 = \frac{a_t^2(1+x_0)}{a_t^2(1+x_0)+2(1+x_m+x_p)}, \qquad f_- = \frac{2(1+x_m)}{a_t^2(1+x_0)+2(1+x_m+x_p)}, \quad (3)$$

$$\Delta\sigma_t = a_0x_0 + a_mx_m + a_px_p + a_5x_5\,, \qquad \Delta\sigma_s = b_0x_0 + b_mx_m + b_px_p + b_5x_5\,, \quad (4)$$

where $\Delta\sigma$ stands for the variation from the SM NLO prediction, and the degree of right-handed polarization of the $W$ boson from top decay is obtained from $f_+ = 1 - f_- - f_0$. The numerical values of the $a_i$ and $b_i$ coefficients were given in Ref. [6] for the Tevatron and the LHC, respectively. They were obtained by integrating over the parton luminosities which are evaluated using CTEQ6L1 parton distribution functions.

---

[1] There are typos in Eqs. (5) and (6) of Ref. [6], in which $x_t$ should be $a_t$.

Furthermore, in the above equations,

$$x_0 = (f_1^L + f_2^R/a_t)^2 + (f_1^R + f_2^L/a_t)^2 - 1, \qquad x_m = (f_1^L + a_t f_2^R)^2 - 1,$$
$$x_p = (f_1^R + a_t f_2^L)^2, \qquad x_5 = a_t^2 (f_2^{L2} + f_2^{R2}), \qquad \text{with} \quad a_t = m_t/M_W.$$

In case that a new light resonance, either a scalar or vector boson, is found, the s-channel process could be significantly enhanced and its production rate may not be dominated by a virtual $W$-boson (s-channel) diagram. The above formulas may also apply to models with extra heavy fermion ($T$), such as the top quark partner in the Little Higgs (LH) Models that couples to the SM $b$ quark and $W$ boson. The $TbW$ coupling in general has the same form of the general $tbW$ coupling given above, and the expressions for single-$T$ production cross sections are exactly the same as single-top except for the heavy mass $m_T$. The size of the coefficients in the production cross sections decrease drastically with a greater mass $m_T$. For instance, at $m_T = 500$ GeV the $a_0$ coefficient decreases one order of magnitude with respect to the value for $m_T = 178$ GeV at the LHC. Furthermore, in the t-channel single-$T$ process, the $a_0$ coefficient, corresponding to longitudinal $W$ boson contribution, dominates its production cross section. This is in analogy with the SM t-channel single-top production. in which longitudinal $W$ boson contribution dominates the inclusive cross section.

Finally, we note that with enough precision in determining the four individual $tbW$ couplings at the LHC, we could start distinguishing models of EWSB [6], such as the MSSM and the Technicolor assisted Topcolor (TC2) mode.

## TOP QUARK AND EWSB

### MSSM and TC2

With its mass around the weak scale, top quark could very well be a special quark that might play an important role in the EWSB. The bottom-up approach is to construct the most general electroweak chiral Lagrangian to describe the interaction of top quark and then compare the effective theory predictions with experimental data to extract out information on the coefficients of the relevant effective operators. Since these coefficients depend on the underlying new physics models, it is possible to discriminate models of EWSB from their values. Usually, two classes of EWSB models are considered in the literature: weakly interacting models and strongly interacting models. The former consists of elementary Higgs boson(s) that originate from spontaneously symmetry breaking (such as MSSM and LH models), and the latter may predict composite Higgs boson(s) due to some strongly interacting dynamical symmetry breaking mechanism (such as TC2 models). In the MSSM, the EWSB is generated radiatively due to heavy top quark contribution in loops. In the TC2 model, top quark condensate relates the heavy top quark mass with the weak scale ($v \sim 246$ GeV) at which the $W$ and $Z$ bosons become massive gauge bosons. It is interesting to note that numerically $m_t \sim v/\sqrt{2} \sim M_W + M_Z$.

Because bottom quark is the isospin partner of top quark, its interactions can also be sensitive to new physics models of EWSB. For example, in the MSSM, two Higgs doublets are required by supersymmetry. When $\tan\beta$ (the ratio of vacuum expectation

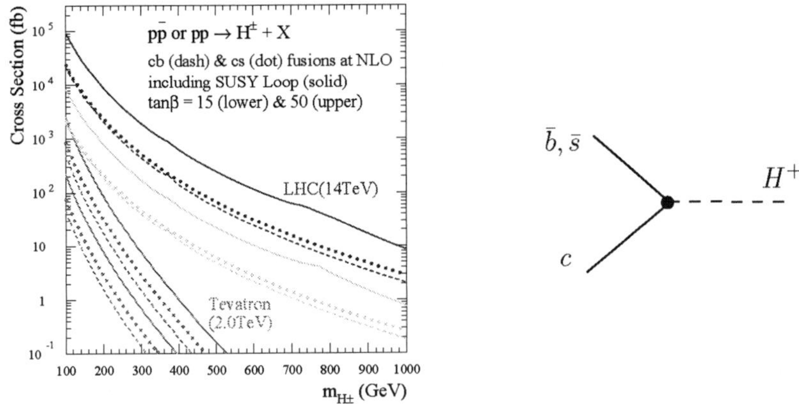


**FIGURE 2.** The MSSM $H^{\pm}$ production via $cb$ (and $cs$) fusions at the Tevatron and the LHC.

values of the two Higgs doublets) is large, the bottom quark Yukawa coupling can become large enough that the production rates of $bbH$ events can become measurable at the Tevatron and the LHC [14]. Assuming a horizontal U(1) flavor symmetry in the soft-breaking sector of MSSM for generating the scalar quark masses, the (32) and (23) entries of the trilinear coupling matrix $A_u$ for up-squarks can have the same size as the (33) entry of $A_u$ and lead to large flavor mixing between stop ($\tilde{t}$) and scharm ($\tilde{c}$) [15]. As shown in Ref. [15], large $\tilde{t}$-$\tilde{c}$ mixing can enhance the s-channel charged Higgs boson production via $cb \to H^+$, cf. Fig. 2, as well as the FCNC decay process $h^0 \to t\bar{c}$. We note that the s-channel $H^+$ production can be tested in the single-top events when $H^+$ decays into a pair of $b$ and $W$ at the Tevatron and the LHC. The decay branching ratio BR($h^0 \to t\bar{c}$) can range from $10^{-5}$ to $10^{-3}$ and is sensitive to the mass of the lightest stop and the mixing of stop and scharm. Furthermore, the chirality of $H^{+}$-$b$-$c$ couplings can be determined at future photon-photon linear collider via $\gamma\gamma \to (b\bar{b})(c\bar{c}) \to H^+b\bar{c}$ by polarizing the polarization of the incoming photon beams [16].

Similarly, in the TC2 model, top-pion ($\pi'$) can be produced at resonance via the s-channel process $cb \to \pi' \to tb$ due to the presence of a large $t_R - c_R$ mixing that is consistent with all the existing FCNC data [17]. We note that the CKM matrix is determined by the product of the left-handed up- and down-type quark rotation matrices (to diagonalize the quark mass matrices) and the right-handed rotation matrices are not fixed by the CKM matrix. Hence, a large $t_R$-$c_R$ mixing does not contradict with the existing experimental data. In the Topflavor model [18], the production of an s-channel $W'$ heavy boson can also produce single-top signature through $q\bar{q}' \to W' \to tb$ at hadron colliders.

In summary, carefully studying the interactions of top quark in experimental data can help distinguishing models of EWSB. A few examples are given in Table 2 to illustrate the idea.

**TABLE 2.** Discriminate models of EWSB by testing the interaction of top and bottom quarks to Higgs boson.

| model | top Yukawa coupling | bottom Yukawa coupling |
|---|---|---|
| SM | $\sim 1$ | $\sim 1/40$ |
| MSSM ($\tan\beta = 40$) | $\sim 1/40$ | $\sim 1$ |
| TC2 | $\sim 1$ | $\sim 1$ |

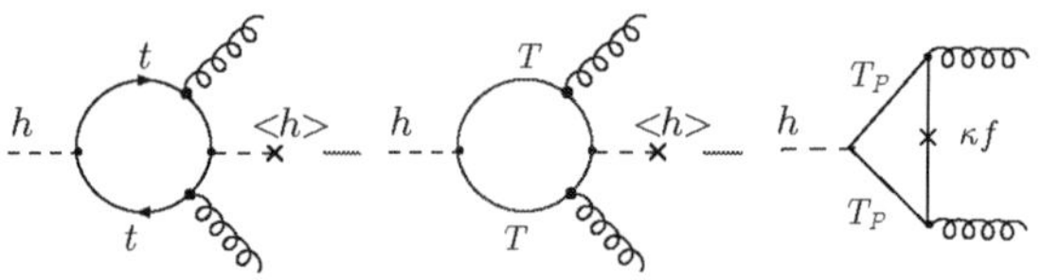

**FIGURE 3.** Representative Feynman diagrams for $gg \to h$ production.

## Little Higgs model

In the Little Higgs model, the Higgs boson mass is naturally at the weak scale, because large quadratic correction to the Higgs boson mass term induced by the top quark loop is cancelled by the fermionic partner ($T$) of top quark ($t$) due to the approximate global symmetry which relates T with t (i.e., Little Higgs mechanism) [19]. Furthermore, to ensure $\rho$-parameter to be one at tree level, a discrete symmetry called T-parity was introduced in the Little Higgs model with T-parity (LHT). Consequently, the effective cutoff scale of the model $\Lambda = 4\pi f$ can be as low as 10 TeV and the masses of new heavy resonances can be of sub-TeV [20]. The LHT model is particularly interesting because it also provides a dark matter candidate which is the lightest T-odd particle $A_H$, the heavy bosonic T-partner of photon. Another important feature of this model is that new Higgs couplings are induced in the part of effective theory that generates the masses of the extra heavy T-partner (either T-odd or T-even) fermions needed for protecting the Higgs boson mass at the weak scale. In Ref. [21] we showed that these new Higgs couplings can lead to non-decoupling effect and alter our conclusions on the collider phenomenology of Higgs boson.

For example, the tree level couplings of Higgs boson to weak gauge bosons and fermions are all suppressed relative to their SM values by a factor $1 - c(v_{SM}^2/f^2)$ where $v_{SM} \sim 246$ GeV and the coefficient $c$ depends on the specific coupling and model scenario. In Fig. 3, we show some representative Feynman diagrams contributing to the production process $gg \to h$. We found that the production rate of Higgs boson via gluon-gluon fusion is also suppressed relative to the SM rate. This can be understood as follows. In the Littlest Higgs model [22], the contribution from the T-partner ($T$) of top quark partially cancel the top quark loop contribution, similar to the effect of cancelling the quadratic divergencies in Higgs boson mass correction. The additional contribution induced by the T-odd heavy fermions further suppress the production rate of $gg \to h$ due to the non-decoupling effect originated from the mass generation mechanism for those heavy T-odd fermions [21]. With $f = 700$ GeV, which is consistent with low

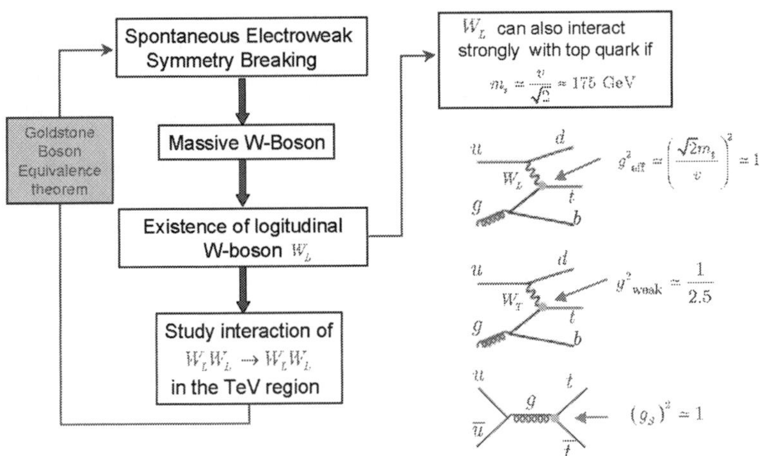


**FIGURE 4.** Schematical diagrams to show the motivation of study done in Ref. [13] for t-channel single-top production.

energy precision data, the cross section $\sigma(gg \to h)$ can be reduced by about 35% for $m_h$ around 115 GeV. It is important to note that the total decay width of Higgs boson in the LHT is always smaller than that predicted by the SM, for the cancellation of quadratic divergencies in Higgs boson mass corrections are among particles with the same spin statistics. While the partial decay width of $\Gamma(h \to \gamma\gamma)$ does not change very much from the SM prediction, the decay branching ratio BR($h \to \gamma\gamma$) can increase by as much as 30% for the total decay width of Higgs boson is largely reduced by the smaller bottom quark Yukawa coupling. Consequently, the discovery potential of the LHC for a light Higgs boson with mass around 100 GeV can be dramatically altered from its usual conclusion. For instance, the rate of $gg \to h \to \gamma\gamma$ can be reduced by about 14%, while the $W$-boson fusion rate increases by about 22% as compared to the SM prediction, cf. Table 1 of Ref. [21]. Hence, in the LHT model, the $W$-boson fusion process becomes the main discovery channel of a 100 GeV Higgs boson at the LHC [21].

## Higgsless model

If Higgs boson exists, discovering the Higgs boson and studying its interaction is essential to probe the electroweak symmetry breaking and the flavor symmetry breaking. Otherwise, we need to carefully study the interaction among longitudinal $W$ (representing $W^\pm$ or $Z$) bosons ($W_L$) in the TeV region [23] as well as the interaction of longitudinal $W$ boson to heavy fermions (top and bottom) [24]. What motivated my 1990 single-top paper was to find a new way to study the interaction of longitudinal $W$ boson to heavy top quark (with its mass around 180 GeV) [13]. Schematically, the relation between the $W_L W_L \to W_L W_L$ scattering and the t-channel single-top process is shown in Fig. 4.

In the effective-$W$ approximation, the t-channel single-top production is dominated by the interaction of longitudinal $W$ boson ($W_L$) to heavy top. While the interaction of transverse $W$ boson ($W_T$) to top quark has the typical weak coupling strength, the interaction of $W_L$ and $t$ is effectively of the same strength as $O(1)$ Yukawa coupling of top quark based on the Goldstone Equivalence Theorem [25].

In some models of extra-dimension, say, Higgsless model, there is neither elementary nor composite Higgs boson to regulate the bad high energy behavior of the $W_LW_L \to W_LW_L$ scattering amplitudes in the TeV region. The breakdown of the unitarity of scattering amplitudes is delayed by the extra heavy Kaluza-Klein gauge bosons predicted in the four-dimensional effective theory [26]. In such kind of models, the interaction of top quark (and its partners) with longitudinal $W$ bosons can become more important. Hence, we need to study the production of top quark (and its partners) from $W$-boson fusion process in the TeV region. It is also possible that the extra bosons needed for delaying the unitarity breakdown in the $W_LW_L \to W_LW_L$ scattering amplitudes can lead to interesting top quark phenomenology. For example, $Z'$ can modify the $t\bar{t}$ event distributions and $W'$ can induce extra production rate for single-top events at the LHC.

## SUMMARY

Because of its heavy mass, top quark may very well be a special quark that could provide hints on electroweak (and flavor) symmetry breaking mechanism. Though theorists have explored many probable ideas in which top quark plays a crucial role in the construction of the theory model, it is entirely up to the experimental data to tell us which theory model is the closest to true Nature. Needless to say that we need experimental data to advance our acknowledge. Currently, we are waiting for the exciting new discovery of single-top events at the Tevatron Run-II. With the huge production rate of top quarks at the LHC, we are able to study many details about its property. The physics of top quark is indeed very rich, and it is also a sure thing that we will learn much more with the realization of future ILC.

**Note Added:** In December 2006, after the presentation of this talk, the D0 Collaboration at Fermilab announced the evidence for production of single top quarks [27].

## ACKNOWLEDGMENTS

I would like to thank the local organizers for their hospitality and for making the Workshop to be a productive one. I also thank my collaborators in the past years who helped me to discover the beauty in physics of top quark and to invent new ways to study its phenomenology. They are G.L. Kane, G.A. Ladinsky, D.O. Carlson, E. Malkawi, T.M.P. Tait, F. Larios, H.-J. He, L. Diaz-Cruz, S. Mrenna, Q.-H. Cao, K. Tobe, and C.-R. Chen. I apologize that due to the limited space in this write-up, I am not being able to cite all the references in the literature that are relevant to top quark physics, but they can be found collectively in the upcoming TeV4LHC report. This work is supported in part by the U. S. National Science Foundation under award PHY-0555545.

# REFERENCES


1. C. Quigg, *Top-ology, Phys. Today* **50N5** (1997) 20.
2. CDF and D0 Collaborations, *Precision Measurement of Top QUark Mass in Dilepton Channel,* preprint no. `Fermilab-conf-06-030-E`, Jan 2006.
3. G.L. Kane, G.A. Ladinsky, and C.–P. Yuan, *Using the Top Quark for Testing Standard Model Polarization and CP Predictions, Phys. Rev.* **D45** (1992) 124.
4. A. Juste, *Top Quark Physics at the ILC,* talk given at *Top Quark Symposium*, Ann Arbor, April 7-8, 2005.
5. S. Mrenna and C.–P. Yuan, *QCD Radiative Decay of the Top Quark Produced in Hadron Collisions, Phys. Rev.* **D46** (1992) 1007.
6. C.-R. Chen, F. Larios and C.–P. Yuan, *General Analysis of Single Top Production and W Helicity in Top Decay, Phys. Lett.* **B631** (2005) 126.
7. T.M.P. Tait and C.–P. Yuan, *Single Top Quark Production as a Window to Physics Beyond the Standard Model, Phys. Rev.* **D63** (2001) 014018.
8. C.–P. Yuan, *Strategies for Probing CP Properties in the Top Quark System, Mod. Phys. Lett.* **A10**, No. 8 (1995) 627.
9. C.R. Schmidt, M.E. Peskin, *A Probe Of Cp Violation In Top Quark Pair Production At Hadron Supercolliders, Phys. Rev. Lett.* **69** (1992) 410.
10. Q.-H. Cao and C.–P. Yuan, *Single top quark production and decay at next-to-leading order in hadron collision, Phys. Rev.* **D71** (2005) 054022.
11. Q.-H. Cao, R. Schwienhorst and C.–P. Yuan, *Next-To-Leading Order Corrections to Single Top Quark Production and Decay at Tevatron. 1. S-Channel Process, Phys. Rev.* **D71** (2005) 054023.
12. Q.-H. Cao, R. Schwienhorst, R. Brock, J. Benitez and C.–P. Yuan, *Next-To-Leading Order Corrections to Single Top Quark Production and Decay at Tevatron. 2. t-Channel Process, Phys. Rev.* **D72** (2005) 094027.
13. C.–P. Yuan, *A New Method to Detect a Heavy Top Quark at the Fermilab Tevatron, Phys. Rev.* **D41** (1990) 42.
14. J.L. Diaz-Cruz, H.-J. He, T. Tait, and C.–P. Yuan, *Higgs Bosons with Large Bottom Yukawa Coupling at Tevatron and LHC, Phys. Rev. Lett.* **80** (1998) 4641.
15. J.L. Diaz-Cruz, H.-J. He and C.–P. Yuan, *Soft SUSY Breaking, Stop-Scharm mixing and Higgs Signatures, Phys. Lett.* **B530** (2002) 179.
16. H.-J. He, S. Kanemura and C.–P. Yuan, *Determining the Chirality of Yukawa Couplings via Single-Charged Higgs Boson Production in Polarized Photon Collision, Phys. Rev. Lett.* **89** (2002) 101803.
17. H.-J. He and C.–P. Yuan, *New Method for Detecting Charged Pseudoscalars at Colliders, Phys. Rev. Lett.* **83** (1999) 28.
18. E. Malkawi, T. Tait, and C.–P. Yuan, *A Model of Strong Flavor Dynamics for the Top Quark, Phys. Lett.* **B385** (1996) 304.
19. N. Arkani-Hamed, A.G. Cohen and H. Georgi, *Electroweak symmetry breaking from dimensional deconstruction, Phys. Lett.* **B513** (2001) 232.
20. H. C. Cheng and I. Low, *TeV symmetry and the Little hierarchy problem, JHEP* **0309** (2003) 051.
21. C.-R. Chen, K. Tobe and C.–P. Yuan, *Higgs Boson Production And Decay In Little Higgs Models With T-Parity,* e-Print Archive: `[hep-ph/0602211]`.
22. N. Arkani-Hamed, A.G. Cohen, E. Katz and A.E. Nelson, *The Littlest Higgs, JHEP* **0207** (2002) 034.
23. Jon Bagger, *et. al., LHC Analysis of the Strongly Interacting WW System: Gold-Plated Modes, Phys. Rev.* **D52** (1995) 3878.
24. F. Larios, T. Tait and C.–P. Yuan, *Anomalous $W^+W^-t\bar{t}$ Couplings at the $e^-e^+$ Linear Collider, Phys. Rev.* **D57** (1998) 3106.
25. H.-J. He, Y.-P. Kuang, and C.–P. yuan, *Equivalence Theorem and Probing the Electroweak Symmetry Breaking Sector, Phys. Rev.* **D51** (1995) 6463.
26. C. Csaki, C. Grojean, L. Pilo, and J. Terning, *Towards A Realistic Model Of Higgsless Electroweak Symmetry Breaking, Phys. Rev. Lett* **92** (2004) 101802.
27. D0 Collaboration (V.M. Abazov et al.), *Evidence for production of single top quarks and first direct measurement of $|V_{tb}|$*, FERMILAB-PUB-06-475-E, Dec 2006, e-Print Archive: hep-ex/0612052.

# Top Quark Physics at DØ

G. Gutierrez, for the DØ collaboration.

*Fermilab, PO Box 500, Batavia, IL 60510*

**Abstract.**
This note describes some of the most recent top quark measurements produced by the D0 collaboration. At the same time it tries to give the reader an understanding of why after more that 10 years since the top quark discovery only the top quark pair production cross section and the top quark mass have been measured with relevant accuracy. The rewards that an increase in statistics will bring in the quest to completely characterize the top quark are also discussed.

**Keywords:** Top Physics
**PACS:** 14.65.Ha

## 1. INTRODUCTION

The top quark with its very large mass and Yukawa couplings close to unity is expected to play a central role in the Standard Model (SM). The top quark mass together with the $W$ boson mass provide today's best constraints on the Higgs boson mass. Any deviations from the top quark's SM predictions will be a signal for new physics. But even after more than a decade since its discovery little is known today about the top quark. The decay width, spin and parity have not yet been measured, the top quark production via the weak force has only recently been observed. Measurements of the lifetime, branching fractions and charge are ongoing but are still far from testing the SM expectations [1, 2]. Only the top quark mass and the pair production cross section have been measured with relevant accuracy. Measurements of the top quark decay vertex $V-A$ structure will only achieve relevant accuracy as the Tevatron approaches the 8 $\text{fb}^{-1}$ of delivered luminosity.

The reason for all this in one word is: statistics. Three of the most recent and important D0 results with 1 $\text{fb}^{-1}$ of data and the $W$ boson helicity measurement with 0.37 $\text{fb}^{-1}$ of data will be described in this note and will be used to illustrate the importance of statistics in different measurements. The top quark mass statistical error is the smallest because three jet invariant masses are peaked. The statistical error in the cross section is next in line because the backgrounds are manageable. Single top quark production is hard to measure because the backgrounds are very large. And all measurements involving angles need very large statistics because angular distributions cover all the allowed range and are fairly flat. For detail information on all D0 top quark physics measurements see Reference [1].

## 2. TOP QUARK MEASUREMENTS

Top quarks at the Tevatron are produced either through the strong or weak interactions. The strong interaction produces top anti-top quark pairs when quarks and gluons inside

CP917, *Particles and Fields*, edited by H. Castilla Valdez, J. C. D'Olivo, and M. A. Perez

the protons and anti-protons collide with each other. The weak interaction produces single top quarks via two different channels. In the s-channel, a beam quark and a beam anti-quark annihilate into a virtual $W$ boson which in turn decays into a $b$ and a top quark. In the t-channel, a beam quark produces a real quark and a virtual $W$ boson and a gluon from the opposite beam produces a $b\bar{b}$ pair, with one real and one virtual $b$ quark; the virtual $b$ and $W$ then interact to produce a top quark. Even though the (strong) $t\bar{t}$ cross section is only about a factor of two larger than the (weak) single top quark cross section, evidence for the production of single top was reported for the first time only recently [3]. The reason being that single top quark production offers fewer constraints and since the final state has fewer jets the backgrounds are much larger.

In the SM the top quark decays almost 100% of the time to a $b$ quark and a $W$ boson. The $W$ in turns decays equally to a pair of leptons or a pair of quarks (e.g. $W^+ \to \bar{e}\nu_e, \bar{\mu}\nu_\mu, u\bar{d}, c\bar{s}$). Since there are three colors for every quark the $W$ boson decay to two jets is three times more probable than its decay to an electron or a muon. The $t\bar{t}$ final states can then be classified according to the $W$ boson decay into all jets, lepton+jets, or dileptons, which are produced in a ratio of 9:3:1. The most accurate measurements have been done in the lepton+jets channel because it has more statistics that the dilepton channel and fewer background than the all jets one.

In the $W$ boson center of mass (CM) the decay products have a momentum of 40 GeV/c, and in the top quark CM the $b$ and the $W$ have a momentum of 66 GeV/c (for a top quark mass of 170 GeV/c$^2$). So the main objects in top quark analysis are a 40 GeV/c electron, muon or neutrino, a 40 GeV/c light jet, and a 66 GeV/c b-jet. Lorentz boosts increase or decrease the particle's momenta depending on the direction of the boost relative to the particles. This has the effect of widening the momentum distributions keeping the peaks at the CM momenta. Therefore most event selection criteria require that jets, leptons and missing (neutrino) transverse momentum be bigger that 15-20 GeV/c, that the pseudorapidity $\eta$ of these objects be within the active part of the detector (typically $|\eta|<2$), and that the leptons are isolated from the jets to avoid confusion with leptons that can be generated from the decay of particles inside a jet.

### *2.0.1. $t\bar{t}$ cross section*

A cross section measurement is performed by counting events. In the simplest case there is a number $N$ of data events that passes all cuts and a prediction $B$ of background events. The number of signal events is $S = N - B$ and the cross section $\sigma$ is proportional to $S$, $\sigma = S/(\varepsilon \cdot L \cdot Br)$, where $\varepsilon$ is the selection efficiency, $L$ the integrated luminosity and $Br$ is the decay branching ratio for the final state under study. If the experiment is performed many times the distribution of the number of data events $N$ will have an rms equal to $\sqrt{N}$ (the rms of a Poisson distribution). The fluctuations in $N$ will be due to both the signal and background events, but $\sigma$ is only proportional to $S$, therefore the relative error in the cross section will be $\sqrt{N}/S$.

Cross section measurements are classified as b-tagged or un-tagged when they are performed with or without b-jet identification. In the un-tagged analysis a probability $P_S$ ($P_B$) for an event being signal (background) is calculated for every event. These proba-

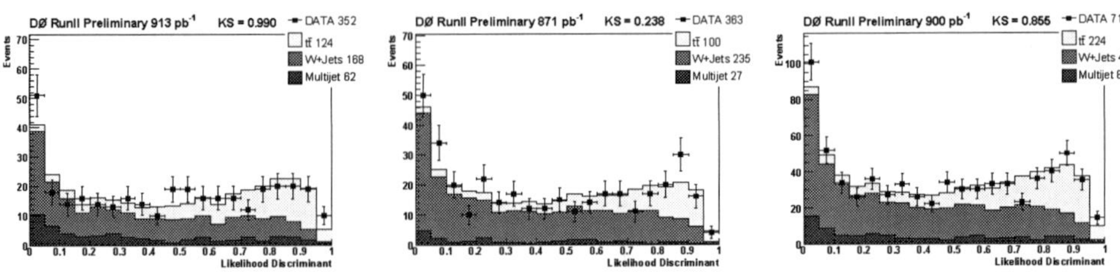


**FIGURE 1.** Likelihood discriminant distribution for data overlaid with the results from the fit to $t\bar{t}$ pair production, $W$+jets and multijet backgrounds, for $e$+jets (left), $\mu$+jets (center) and combined (right).

bilities are calculated multiplying properly normalized one dimensional histograms obtained from MC. The histograms are defined using variables like Aplanarity, Sphericity, Centrality, $H_T$, $\Delta\phi$ and $K_T^{min'}$ [4]. A discriminant $D = P_S/(P_S + P_B)$ is then used to separate signal from background. Figure 1 shows the value of the discriminant for $e$+jets, $\mu$+jets and both channels combined. In each histogram the discriminant is shown for data, $t\bar{t}$ signal and for the two main backgrounds $W$+jets and multijets. The signal is extracted by fitting the discriminant distributions allowing the number of signal and background events to float. The number of events resulting from the fit are shown in the upper right hand corner in the discriminant plots. The statistical error can be estimated in the following way. Two bins can be formed by splitting the combined discriminant histogram at $D \approx 0.3$. The lower bin contains mostly background, the upper bin most of the signal and the background is equally distributed in both bins. The extraction of the signal ($S = N - B$) from the second bin will be subject to fluctuations $\sqrt{S + B/2}$ in the number of events $N$ and $\sqrt{B/2}$ in the background prediction $B$. This last fluctuation comes from using the first bin to determine the number of background events. Adding the two errors in quadratures gives a total error for $S$ of $\sqrt{S+B}$ which implies a relative error in the cross section of $\sqrt{S+B}/S = \sqrt{715}/224 = 12$ %. This error agrees very well with the 13 % statistical error quoted in the final measurement given in Eq. 1 below. A relative error of 12% corresponds to 70 pure signal events as opposed to the 224 $t\bar{t}$ events measured in the un-tagged sample. This means that the presence of background produces most of the fluctuations observed in the statistical error and that this error could be reduced by reducing the background even at the expense of cutting a substantial amount of signal.

One way to reduce the background is by looking for the presence of $B$ mesons in b-jets. Due to their long lifetime the decay of $B$ mesons is observed as a vertex separated from the primary vertex of the interaction. If a secondary vertex is found inside a jet that jet is "b-tagged" and it is considered as a jet produced by a b quark. Given that there are two b-jets in each $t\bar{t}$ event and that only a small part of the $W$+jets background contains b-jets, using b-tagging keeps most of the signal while greatly reducing the backgrounds.

In the b-tagged cross section analysis the data is split into electrons and muons, three and four or more jets, and one or two or more b-tags. The number of expected signal and observed data events for these eight groups is shown in Table 1. If the experiment is performed many times the cross sections measured in each group will be distributed as a Gaussian with an rms given by the cross section $\sigma$ times the relative error in

**TABLE 1.** Event yields for signal (assuming $\sigma_{t\bar{t}}$= 8.3 pb) and observed event yield for the b-tagged cross section analysis.

| | | e+3 jets | e+≥4 jets | $\mu$+3 jets | $\mu$+≥4 jets |
|---|---|---|---|---|---|
| 1 b-tag | signal | 85.6 | 71.5 | 55.4 | 50.2 |
| | data | 172 | 88 | 135 | 92 |
| ≥ 2b-tags | signal | 32.0 | 35.3 | 22.1 | 26.2 |
| | data | 41 | 26 | 36 | 30 |

that group $r_i = \sqrt{N_i}/S_i$, where $N_i$ $(S_i)$ is the number of data (signal) events in group $i$. Since the measurements in each group are independent the cross section's statistical error can be derived from the product of the Gaussian distributions in each group. This product has an rms $s$ given by $1/s^2 = \Sigma_i 1/(\sigma r_i)^2$, which leads to a total relative error of $(s/\sigma)^2 = (\Sigma_i 1/r_i^2)^{-1} = (\Sigma_i S_i^2/N_i)^{-1}$. An error $s/\sigma$=6% is obtained using the numbers in Table 1. This corresponds to an equivalent gain of a factor of four in data relative to the un-tagged measurement. This gain is almost exclusively due to the background rejection achieved by b-tagging. Figure 2 shows the predicted sample compositions for the $l$+jets sample with exactly one and with two or more b-tagged jets.

The final cross section results are

$$l\text{+jets un-tagged:} \quad \sigma_{p\bar{p}\to t\bar{t}+X} = 6.3^{+0.9}_{-0.8}(\text{stat}) \pm 0.7(\text{syst}) \pm 0.4(\text{lumi}) \text{ pb} \quad (1)$$

$$l\text{+jets b-tagged:} \quad \sigma_{p\bar{p}\to t\bar{t}+X} = 8.3^{+0.6}_{-0.5}(\text{stat})^{+0.9}_{-1.0}(\text{syst}) \pm 0.5(\text{lumi}) \text{ pb} \quad (2)$$

It is clear from the b-tagged result that the cross section measurements are already not dominated by statistics. So a future increase in the amount of data will reduce the total error only if it can be used to reduce the systematic errors. The two largest systematic errors in the un-tagged analysis are the modeling of $W$+jets and the discriminant template shapes. In the b-tagged analysis the two largest systematic errors are the large error in the $W + b\bar{b}$+jets cross section and the uncertainty in the b-tagging probability.

### *2.0.2. Top quark mass*

The most accurate top quark mass measurements are done in the lepton+jets channel. In this channel one top quark decays into three jets and the other top quark decays to a jet, a lepton and a neutrino. Because the neutrino can not be measured, the top quark mass measurement in the $l$+jets channel is dominated by the invariant mass of the three jets coming from the decay of a top quark. This invariant mass has a typical width of about 20 GeV/c$^2$ and its mean, which is directly related to the top quark mass, has an error of $20/\sqrt{N}$, where $N$ is the number of signal events. For 1 fb$^{-1}$ of data $N \approx 150$, which means that the expected statistical error in the top quark mass is $\approx 1.6$ GeV/c$^2$. The reason it is possible to achieve a 1% measurement in the top quark mass but only a 10% measurement in the cross section is that the three jet invariant mass is very localized in comparison to its allowed range (say from 50 to 450 GeV/c$^2$).

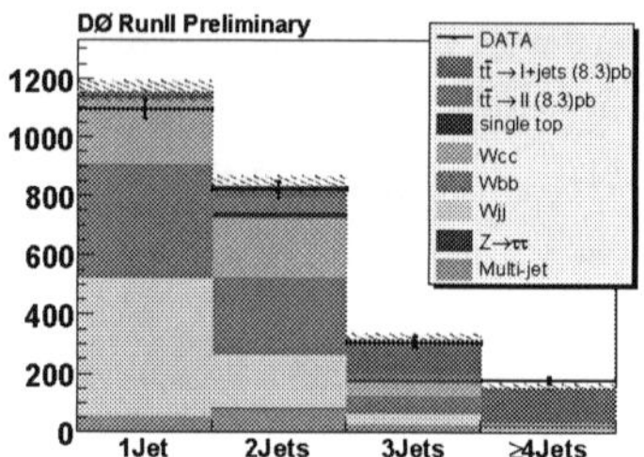

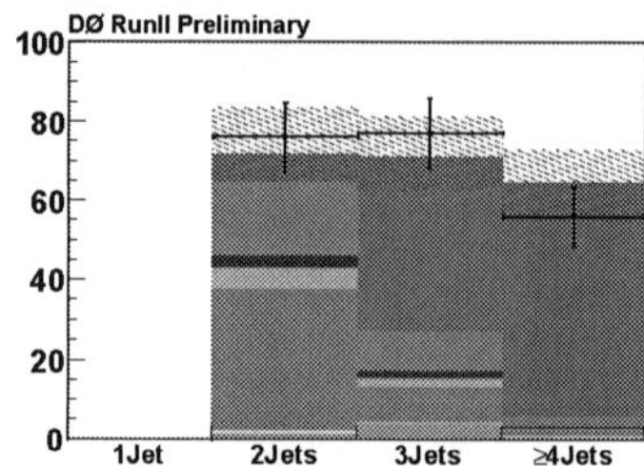


**FIGURE 2.** Predicted sample composition for the lepton+jets sample with exactly one b-tag (left) and two or more b-tags(right). The $t\bar{t}$ contribution assumes the measured cross section value of 8.3 pb.

Due to the presence of backgrounds and to the impossibility of knowing which jet comes from which parton (which gives rise to 24 possible combinations in the assignment between jets and partons), a fit to the three jet invariant mass is not the most effective way of extracting the top quark mass. In the analysis presented here a probability $P(x|\alpha)$ is calculated for every event. The four momenta of the lepton and jets are labeled by $x$ and the top quark mass $M_{top}$, Jet Energy Scale $JES$ and fraction of signal events $f$ are labeled by $\alpha = (M_{top}, JES, f)$. Given that all events are independent, the probability (or likelihood) for a sample of $N$ events is the product of the probabilities for each event $L = \prod_{i=1}^{N} P(x_i|\alpha)$. If $\alpha$ is known then $L$ will be very close to maximal, otherwise a different set of events would have been observed. Or if $\alpha$ is unknown it can be determined by maximizing $L$. A given event in the final sample has a certain probability $P_{t\bar{t}}(x|\alpha)$ of being signal and a probability $P_B(x)$ of being background. Therefore the probability for an individual event is $P(x|\alpha) = fP_{t\bar{t}}(x|\alpha) + (1-f)P_B(x)$.

A large sample of $\gamma$+jets events was used to perform an independent study of the Jet Energy Scale. A Gaussian likelihood $G(JES)$ with a mean $JES = 1$ and an rms of 0.037 is derived from this study. Since the $t\bar{t}$ and $\gamma$+jets samples are independent, the likelihood for the combination of both samples is just the product of $L$ and $G$. The top quark mass is extracted by projecting this product onto the $M_{top}$ axis:

$$L(M_{top}) = \int df\, dJES\; L(M_{top}, JES, f)\; G(JES) \tag{3}$$

The normalized likelihoods $L(M_{top})/L_{max}$ are shown in Figure 3. The figure on the left shows the likelihood for the combined $e$+jets and $\mu$+jets samples for the un-tagged analysis. The center (right) figure shows the likelihoods for the $e$+jets ($\mu$+jets) sample in the b-tagged analysis. The top quark masses derived from these plots are:

$$l\text{+jets un-tagged:}\quad m_{top} = 170.5 \pm 2.5(\text{stat+JES}) \pm 1.4(\text{syst})\quad \text{GeV}/c^2 \tag{4}$$

$$l\text{+jets b-tagged:}\quad m_{top} = 170.5 \pm 2.4(\text{stat+JES}) \pm 1.2(\text{syst})\quad \text{GeV}/c^2 \tag{5}$$

$$= 170.5 \pm 1.8(\text{stat}) \pm 1.6(\text{JES}) \pm 1.2(\text{syst})\quad \text{GeV}/c^2$$

With a weight of 39.7% measurement (5) has the largest single contribution to the world top quark mass combination [5].

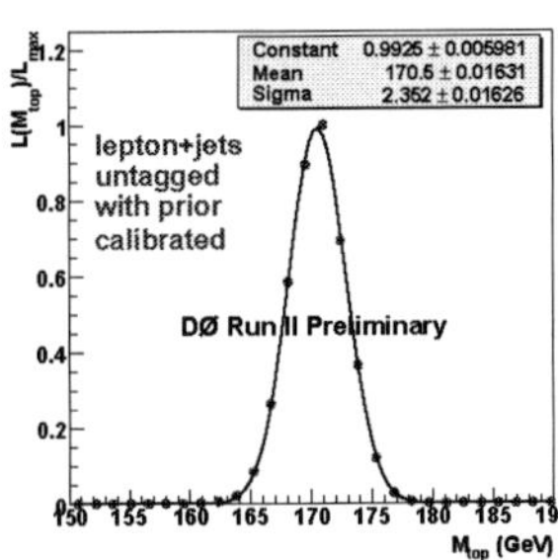


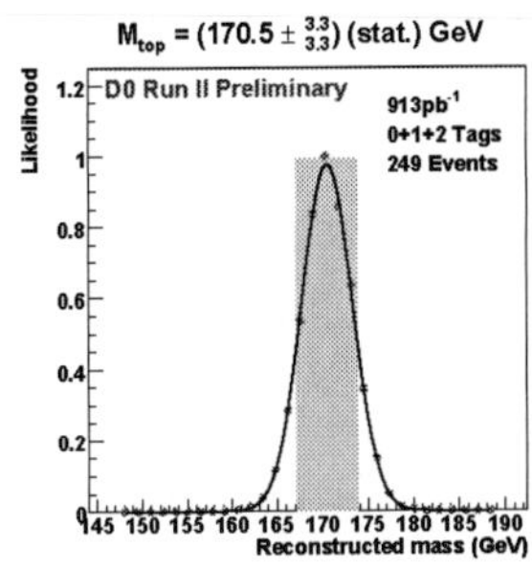


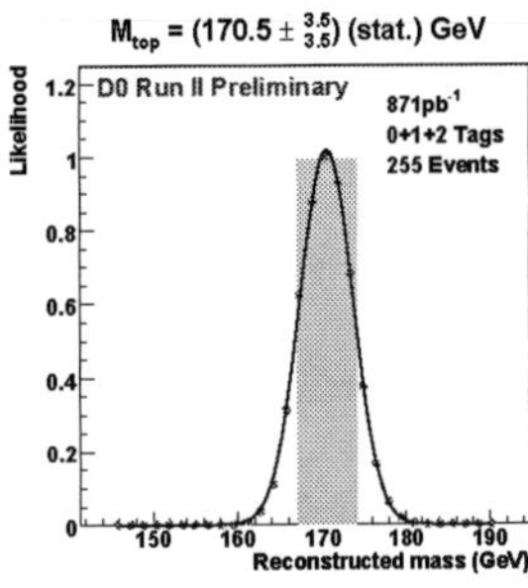


**FIGURE 3.** Likelihood plots (normalized to $L_{max}$) as a function of the top quark mass for the un-tagged analysis (left), and the b-tagged analysis in the e+jets channel (middle) and the $\mu$+jets channel (right).

### *2.0.3. W boson helicity*

In the $W^+$ boson rest frame, the general angular distribution of the down type decay product ($e^+$, $\mu^+$, $\bar{d}$ or $\bar{s}$) has the form $w(\theta^*) = |d^1_{01}(x)|^2 f_0 + |d^1_{-11}(x)|^2 f_- + |d^1_{11}(x)|^2 f_+$, where $x = cos\theta^*$, $d^J_{m,m'}$ are the rotation matrices, $f_\lambda$ is the probability of finding the $W^+$ boson in a state of helicity $\lambda = (0, -1, +1)$ and $f_0 + f_- + f_+ = 1$. In the $W^+$ rest frame the z-axis is defined in the $W^+$ momentum direction before the boost that brings the $W$ boson to rest and $\theta^*$ is the angle of the down type decay product with the z-axis. In the case of the $W^-$ decay all helicities are reversed and the angular distribution remains the same. Using the rotation matrices the angular distribution is $w(\theta^*) \propto 2(1-x^2)f_0 + (1-x)^2 f_- + (1+x)^2 f_+$.

In the SM $V-A$ current interaction the $W$ bosons originating from top quarks have well defined values of $f_0 = m_t^2/(m_t^2 + 2M_W^2 + m_b^2) \approx 0.70$ and $f_+ = m_b^2/(m_t^2 + 2M_W^2 + m_b^2) \approx 0$. For any linear combination of $V$ and $A$ currents the value of $f_0$ stays constant but $f_+$ changes. For example for a pure $V+A$ charge current interaction $f_+ = 0.3$.

In this analysis [6] the $cos\theta^*$ distribution is compared with templates for different values of $f_+$ obtained for constant $f_0$ =0.70. The data selection involves both the use of a discriminant and b-tagging. The different assignments between partons and jets are resolved by performing a constrained fit to each event and selecting the combination with the smallest $\chi^2$. Figure 4 shows the $cos\theta^*$ distribution for (a) $l$+jets and (b) dilepton events. The SM prediction is shown as a solid line, the dashed line shows a pure $V+A$ interaction. Templates were calculated for seven values of $f_+$ and a binned Poisson likelihood was used to extract $f_+$ from the $cos\theta^*$ distributions shown in Figure 4. The likelihood as a function of $f_+$ is shown in Figure 4 (right plot). The final result is [6]

$$f_+ = 0.056 \pm 0.080(\text{stat}) \pm 0.057(\text{syst}) \tag{6}$$

Since $f_0$ =0.70 the value of $f_+$ is bounded between 0 and 0.3. In the absence of prior knowledge or a measurement, all values of $f_+$ between 0 and 0.3 are equally probable (flat likelihood). Then without performing the experiment it is possible to say that $f_+$ has a 68% probability of being inside the interval 0.68*0.3=0.20 or 34% probability (1 sigma) of being inside a 0.10 interval. The comparison of this number

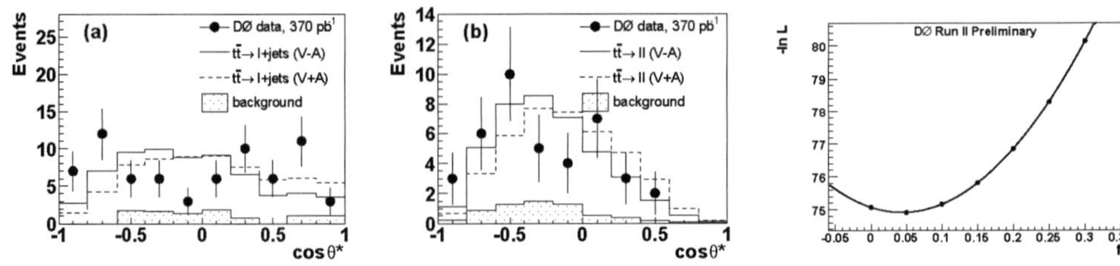


**FIGURE 4.** $cos\theta^*$ distribution in $l$+jets (left) and dileptons (center), and likelihood curve (right) as a function of $f_+$ for the combined dileptons and $l$+jets samples. The solid line on the $cos\theta^*$ plots shows the SM prediction, the dashed line shows the prediction for the pure $V+A$ interaction.

with the statistical error in Eq. 6 clearly indicates that not much has been gained yet by performing the measurement. The reason for this is that contrary to, for example three jet invariant masses, angular distributions cover the entire allowed range and are usually fairly flat. This means that larger statistics are needed in order to achieve the same precision as that found in measurements like the top quark mass. Therefore analyses that involve measurements of angular distributions will be the main beneficiaries of the future increase in statistics.

### *2.0.4. Single Top Quark Production*

Discovering single top quark production essentially involves a cross section measurement like the ones described in Subsection 2.0.1. The main difference is that the number of background events is much larger than the signal ones and therefore the use of both b-tagging and powerful discriminants is needed. D0 has recently announced for the first time evidence for single top quark production [3]. Three different algorithms were used in the analysis, Boosted Decision Trees (BDT), Matrix Element and Bayesian Neural Networks. Only the analysis with the largest significance (BDT) will be described here.

A BDT algorithm builds a sequence of "trees". In each tree a group of events starting in the tree "trunk" will, after a series of decisions, temporarily occupy different "branches" until they all end up in different "leaves." The decisions made to bifurcate a branch are optimized using signal and background MC events. Once optimized (or trained) the tree can be used to classify events. After a data or MC event that is run through the tree reaches a leaf a "leaf purity" $T_n(i)$ is assigned to it. $T_n(i)$, the output or leaf purity of tree $n$ for event $i$, is defined as $s/(s+b)$ where $s$ and $b$ are the sum of signal and background weights for that leaf. An event is called signal if $T_n(i)$ is greater than a specified number (usually 0.5) and background otherwise. During the training session an error $err_n$ is assigned to each tree based on the number of misclassified events, e.g. a MC signal event called background. The $n+1$ tree in the sequence is built weighting (or boosting) the misclassified events with a function of $err_n$. Until stability is reached, and as a consequence of weighting, each tree in the sequence will have a reduced number of misclassified events. The BDT output $T(i) = \sum_n \alpha_n T_n(i)$ for event $i$ is a linear combination of the outputs $T_n(i)$ of all the trees in the sequence. The coefficients $\alpha_n$ are functions

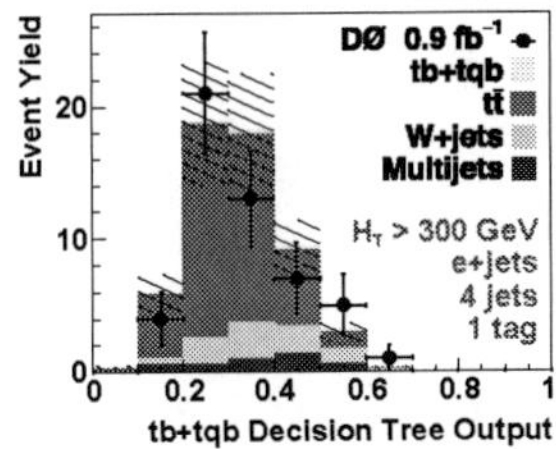

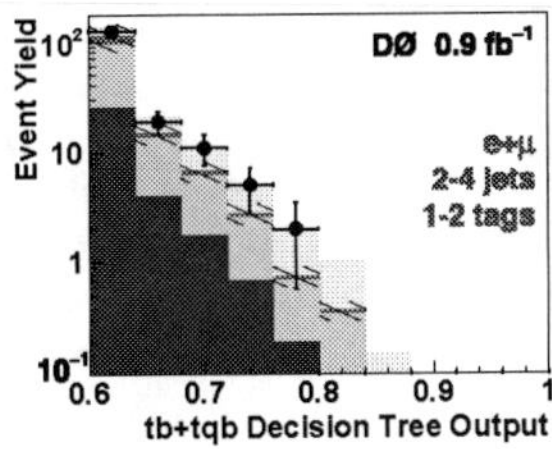

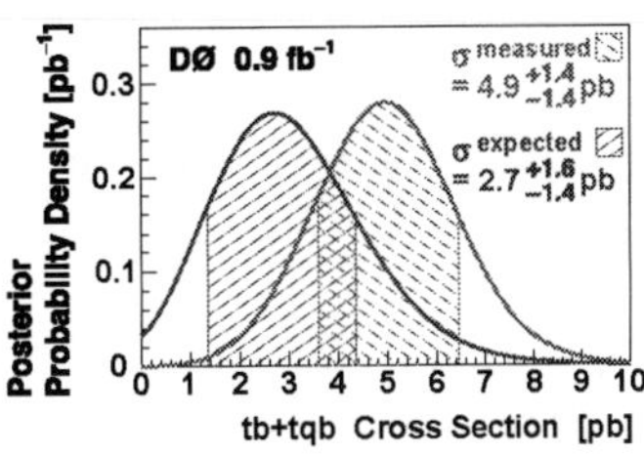


**FIGURE 5.** Boosted decision tree output for a $t\bar{t}$ dominated sample (left) and for the final selection (center). Bayesian posterior (right) for SM (blue line) and data (red line). The blue and red shaded regions indicate 68% confidence intervals.

of $err_n$.

Figure 5 shows the BDT output for a $t\bar{t}$ enriched sample (left plot) and for the final sample used in the analysis (center plot). The $tb$ and $tqb$ labels in the plots stand for s-channel and t-channel production of single top quarks which, as pointed out in Section 2, are produced together with a b-quark in the s-channel and with a b- and a light quark in the t-channel. A binned likelihood fit over the decision tree output or discriminant is used to extract the single top quark signal. The right plot on Figure 5 shows the posterior probability density obtained from the likelihood assuming a flat nonnegative prior probability. The cross section extracted from this plot is

$$\sigma(p\bar{p} \to tb+X, tqb+X) = 4.9 \pm 1.4 \text{ (stat+syst) pb} \quad (7)$$

Using ensemble test studies, the probability that the background would fluctuate up to produce a measured cross section of 4.9 pb or greater was calculated to be 0.035%, which corresponds to a 3.4 $\sigma$ significance of this result.

## ACKNOWLEDGMENTS

I would like to thank the organizers for an exciting and stimulating conference. And I would also like to express my sincere thanks to Linda Stutte, Lisa Shabalina and Yann Coadou for their corrections and suggestions after carefully reading the manuscript.

## REFERENCES

1. D0 public results. http://www-d0.fnal.gov/Run2Physics/top/top_public_web_pages/top_public.html
2. CDF public results. http://www-cdf.fnal.gov/physics/new/top/top.html
3. The D0 Collaboration. FERMILAB-PUB-06/475-E. To be published in PRL. More information can be found in http://www-d0.fnal.gov/Run2Physics/top/public/fall06/singletop/
4. The D0 Collaboration. Phys. Lett. B626, 45 (2005).
5. A Combination of CDF and D0 results on the mass of the top quark. hep-ex/0703034, http://www.slac.stanford.edu/spires/find/hep/www?r=fermilab-tm-2380-e
6. The D0 Collaboration. Phys. Rev. D75, 031102(R)(2007)

# New physics effects in Top quark interactions

F. Larios*, M.A. Pérez† and R. Martínez**

*Departamento de Física Aplicada, CINVESTAV-Mérida, A.P. 73, 97310 Mérida, Yucatán, México
†Departamento de Física, CINVESTAV, A.P. 14-740, 07000, México D.F., México
**Departamento de Física, Universidad Nacional, Apartado aéreo 14490, Bogotá, Colombia

**Abstract.** The Top quark stands out as the heaviest elementary particle known today and it is expected to be sensitive to the mechanism that breaks the electroweak symmetry. In particular, the study of its couplings to the electroweak gauge bosons is important as it a way to test different models beyond the SM. We review the status of Top quark couplings with the $W$ and $Z$ bosons, as well as FCNC couplings.

**Keywords:** top quark
**PACS:** 12.60.-i, 12.15.-y, 11.15.Ex, 14.65.Ha, 14.70.Fm

## INTRODUCTION

Although the top quark was discovered ten years ago [1, 2], its couplings to the gauge bosons $\gamma, g, W$ and $Z$ have not been yet measured directly [3]. Current data provide only weak indirect limits on the $tbW$ and $ttV$ couplings, with $V = \gamma, g, Z$. On the other hand, the Top quark mass is significantly heavy, in fact, it is of the order of the electroweak scale ($m_t \simeq 246/\sqrt{2}$GeV). Therefore, this particle is likely to play an essential role in the mechanism of electroweak symmetry breaking.

We will make a summary of our present knowledge of the couplings of the Top quark, what has been measured (directly and indirectly) so far, and what is the prospect for future colliders. We will start by considering the indirect limits on $ttZ$ and $tbW$ couplings from the low energy data, then we will focus on the $tbW$ coupling. This coupling can be studied by looking at the rate of the $t \to bW$ decay for different helicities of the $W$ boson, as well as by looking at the single Top production rates of the different channels. We will also consider the FCNC couplings of Top which are very suppresed in the SM due to the GIM mechanism. Therefore, new physics effects could be remarkably high for these couplings.

## DEVIATIONS FROM TOP-QUARK SM COUPLINGS

It is possible to parameterize possible deviations from the SM predictions for the $tbW$ and $ttZ$ couplings in terms of only four coefficients $\kappa_{L,R}^{NC}$ and $\kappa_{L,R}^{CC}$ defined as follows [4]:

$$\begin{aligned}\mathscr{L} &= \frac{g}{2c_W}\left(1-\frac{4s_W^2}{3}+\kappa_L^{NC}\right)\bar{t}_L\gamma^\mu t_L Z_\mu + \frac{g}{2c_W}\left(\frac{-4s_W^2}{3}+\kappa_R^{NC}\right)\bar{t}_R\gamma^\mu t_R Z_\mu \\ &+ \frac{g}{\sqrt{2}}\left(1+\kappa_L^{CC}\right)\bar{t}_L\gamma^\mu b_L W_\mu^+ + \frac{g}{\sqrt{2}}\left(1+\kappa_L^{CC\dagger}\right)\bar{b}_L\gamma^\mu t_L W_\mu^-\end{aligned}$$

CP917, *Particles and Fields*, edited by H. Castilla Valdez, J. C. D'Olivo, and M. A. Perez

$$+ \frac{g}{\sqrt{2}}\kappa_R^{CC}\bar{t}_R\gamma^\mu b_R W_\mu^+ + \frac{g}{\sqrt{2}}\kappa_R^{CC\dagger}\bar{b}_R\gamma^\mu t_R W_\mu^- \quad (1)$$

where $t_L$ denotes a top quark with left-handed chirality, etc. While the $ttZ$ vector and axial-vector couplings are tightly constrained by the LEP data [4, 5], the right handed $tbW$ coupling is severely bounded by the observed $b \to s\gamma$ rate [6] at the $2\sigma$ level,

$$\begin{aligned} |Re(\kappa_R^{CC})| &\leq 0.4\times 10^{-2} \\ -0.0035 &\leq Re(\kappa_R^{CC}) + 20|\kappa_R^{CC}|^2 \leq 0.0039. \end{aligned} \quad (2)$$

On the other hand, LEP/SLC data also constrains the other top-quark couplings included in Eq.(4). Even though these data do not restrict all the anomalous $\kappa$ terms, they induce the following inequalities

$$\begin{aligned} -0.019 &\leq (\kappa_R^{NC} - \kappa_L^{NC}) - (\kappa_R^{NC} - \kappa_L^{NC})^2 + \kappa_L^{CC} + \kappa_L^{CC\ 2} \leq 0.0013 \\ -0.33 &\leq (\kappa_R^{NC} - 4\kappa_L^{NC})(1 + 2\kappa_L^{CC}) \leq 0.1 \\ \kappa_L^{CC} &\sim \kappa_L^{NC} - \kappa_R^{NC} \end{aligned} \quad (3)$$

These relations impose in turn strong correlations on the $\kappa$ couplings so that if only one coupling, $\kappa_L^{CC}$ for instance, is not zero, the others are forced to be about the same order of magnitude [6].

On the other hand, at an $e^+e^-$ linear collider (ILC) with $\sqrt{s} = 500\ GeV$, and an integrated luminosity of $100-200\ fb^{-1}$, it will be possible to measure the $ttV$ couplings in $t\bar{t}$ production with a few percent precision [7]. In the LHC, with an integrated luminosity of $30 fb^{-1}$, it will be possible to probe the $tt\gamma$ coupling with a precision of $10-35\%$ per experiment [8]. The sensitivity limits on the $ttZ$ couplings will be significantly weaker than those expected for the $tt\gamma$ couplings. Thus, the ILC will be the best place to probe the $ttZ$ couplings at the few percent level.

## GENERAL $tbW$ COUPLINGS

Concerning the $W$ boson helicity in the $t \to bW$ decay, there are three modes depending on the polarization state of the $W$ boson. Each mode is associated with a fraction: $f_0$, $f_+$ or $f_-$ that corresponds to the longitudinal, right-handed or left-handed polarization, respectively. By definition, we have the restriction $f_0 + f_+ + f_- = 1$. Recent reports by the DØ and CDF collaborations at Fermilab give the following (95% C.L.)results for the longitudinal and right-handed fraction of $t \to bW$ in the $t\bar{t}$ pair events [9]:

$$\begin{aligned} f_0 &= 0.91 \pm 0.38\,(\mathrm{CDF}), & f_0 &= 0.56 \pm 0.32\,(\mathrm{D0}), \\ f_+ &\leq 0.18\,(\mathrm{CDF}), & f_+ &\leq 0.24\,(\mathrm{D0}). \end{aligned}$$

Therefore, the effects of our general effective Lagrangian to the processes considered here can be completely described by the following $tbW$ vertex:

$$\begin{aligned} \mathscr{L}_{tbW} &= \frac{g}{\sqrt{2}} W_\mu^- \bar{b}\gamma^\mu \left(f_1^L P_L + f_1^R P_R\right) t \\ &- \frac{g}{\sqrt{2}M_W} \partial_\nu W_\mu^- \bar{b}\sigma^{\mu\nu} \left(f_2^L P_L + f_2^R P_R\right) t \ + h.c., \end{aligned} \quad (4)$$

where we have changed the mass scale $\Lambda$ to $m_W$ to keep the same notation used in the literature [10, 12].

In the SM the values of the form factors are $f_1^L = V_{tb} \simeq 1$, $f_1^R = f_2^L = f_2^R = 0$. To focus on deviations from SM values, let us define $f_1^L \equiv 1 + \varepsilon_L$.

It is well known that $b \to s\gamma$ can impose a strong constraint on $f_1^R$ and $f_2^L$ to be less than 0.004 [6, 13]. Also, $b \to sl^+l^-$ can be sensitive to $f_2^R$ and impose constraint of order 0.03 [13]. For $\varepsilon_L$, the LEP precision data imposes some constraint: assuming no deviations from the SM $ttZ$ vertex we would have that $\varepsilon_L \leq 0.02$ [6]. These constraints do not use the combined effects of other couplings. For the dimension 5 couplings it is required an upper bound of about 0.5 in order to satisfy the unitarity condition [14].

Information on the helicity of the $W$ boson in $t \to bW$ can be obtained by measuring a forward-backward asymmetry ($A_{FB}$) based on the angle between the charged lepton and the b-jet of the observed decay process [15]. Preliminary studies show that if $A_{FB}$ is measured with 20% accuracy at the Tevatron, it may be sensitive to values of order $f_2^{L,R} \sim 0.3$; similarly, if $A_{FB}$ is measured with 1% accuracy at the LHC this may be translated to a sensitivity of order $f_2^L \sim 0.03$ and $f_2^R \sim 0.003$ [12]. We would like to point out that, since the observable $A_{FB}$ is only proportional to the difference between $f_+$ and $f_-$ [15], it is clear that it does not provide any more information than the separate measurements of (two of) the ratios $f_0$, $f_-$ and $f_+$.

In Table 1 we show the leading order (LO) and the next-to-leading order (NLO) SM predictions for $\sigma_t$ and $\sigma_s$ at the Tevatron and at the LHC [16]. For the LO predictions the CTEQ6L1 parton distribution function (PDF) has been used [17]. For the NLO predictions the CTEQ6M PDF has been used [16]. (We are taking $m_t = 178\,\text{GeV}$ and $m_W = 80.4\,\text{GeV}$.)

**TABLE 1.** SM single top production cross section predictions in units of pb [16]. ($m_t = 178\,\text{GeV}$.)

| Channel | Tevatron ($t$ LO) | ($t$ NLO) | LHC ($t$ LO) | ($t$ NLO) | LHC ($\bar{t}$ LO) | ($\bar{t}$ NLO) |
|---|---|---|---|---|---|---|
| t-channel | 0.827 | 0.924 | 146.0 | 150.0 | 84.9 | 88.5 |
| s-channel | 0.27 | 0.405 | 4.26 | 6.06 | 2.59 | 3.76 |

Neglecting terms proportional to the bottom mass, the Born level values of the top quark width and its $W$-polarization ratios are $\Gamma_t = 1.65\,\text{GeV}$, $f_0 = 0.71$, $f_- = 0.29$ and $f_+ = 0$. In the SM, including terms proportional to $m_b$, order $\alpha_s^2$ QCD, electroweak, and finite $W$ width corrections produce a 10% decrease in the top's width ($\Gamma_t = 1.49$) and a small $\sim$ 1% variation for decay ratios ($f_0 = 0.701$, $f_- = 0.297$ and $f_+ = 0.002$) [15].

Let us consider the deviations from the SM values (up to the NLO) that come from the effects of the anomalous $\varepsilon_L$, $f_1^R$, $f_2^L$ and $f_2^R$ couplings cf. Eq. (4), induced by heavy new physics effects. We will write down the Born level contributions of these couplings on the observables $f_0$, $f_+$, $f_-$, $\sigma_t$ and $\sigma_s$.

The tree level $t \to bW$ decay width of the top quark with the general $tbW$ vertex can be easily obtained with the helicity amplitude method, and it is given by [10]:

$$\begin{aligned}\Gamma_t &= \Gamma_0 + \Gamma_- + \Gamma_+ \\ &= \frac{g^2 m_t}{64\pi} \frac{(a_t^2 - 1)^2}{a_t^4} \left(a_t^2(1 + x_0) + 2(1 + x_m) + 2x_p\right),\end{aligned}$$

$$
\begin{aligned}
x_0 &\equiv (f_1^L + f_2^R/a_t)^2 + (f_1^R + f_2^L/a_t)^2 - 1\,, \\
x_m &\equiv (f_1^L + a_t f_2^R)^2 - 1\,, \\
x_p &\equiv (f_1^R + a_t f_2^L)^2\,, \\
a_t &\equiv \frac{m_t}{m_W}\,.
\end{aligned} \tag{5}
$$

As the notation suggests, $x_0$, $x_m$ and $x_p$ are the effective terms that originate the contribution to $f_0$, $f_-$ and $f_+$, respectively. Below, we will write down the explicit expressions for these decay ratios.

The contributions of the effective $tbW$ couplings to the observables of interest are [11]:

$$
f_0 = \frac{a_t^2(1+x_0)}{a_t^2(1+x_0)+2(1+x_m+x_p)}\,, \tag{6}
$$

$$
f_+ = \frac{2x_p}{a_t^2(1+x_0)+2(1+x_m+x_p)}\,, \tag{7}
$$

$$
f_- = \frac{2(1+x_m)}{a_t^2(1+x_0)+2(1+x_m+x_p)}\,,
$$

$$
\Delta\sigma_t = a_0 x_0 + a_m x_m + a_p x_p + a_5 x_5\,, \tag{8}
$$

$$
\Delta\sigma_s = b_0 x_0 + b_m x_m + b_p x_p + b_5 x_5\,, \tag{9}
$$

$$
x_5 \equiv a_t^2({f_2^L}^2 + {f_2^R}^2)\,,
$$

where $\Delta\sigma$ stands for the variation from the SM NLO prediction. The numerical values of the $a_i$ and $b_i$ coefficients are given in Table 2 for the Tevatron and the LHC. They have been obtained by integrating over the parton luminosities which are evaluated using the PDF CTEQ6L1 [17].

**TABLE 2.** The single top production cross section coefficients of Eqs. (8-9). In units of pb.

| t-channel: | $a_0$ | $a_m$ | $a_p$ | $a_5$ |
|---|---|---|---|---|
| Tevatron | 0.896 | -0.069 | -0.153 | 0.247 |
| LHC ($t$) | 165.2 | -19.1 | -34.2 | 62.5 |
| LHC ($\bar{t}$) | 105.8 | -20.9 | -12.5 | 38.6 |
| **s-channel:** | $b_0$ | $b_m$ | $b_p$ | $b_5$ |
| Tevatron | -0.081 | 0.352 | 0.352 | 0.230 |
| LHC ($t$) | -1.41 | 5.67 | 5.67 | 6.34 |
| LHC ($\bar{t}$) | -0.836 | 3.43 | 3.43 | 3.38 |

Eqs. (6)-(9) can be used to make a general analysis of the effective $tbW$ vertex. We note that in case a new light resonance is found, like a scalar or vector boson, the s-channel process could be significantly enhanced and its production rate may not be dominated by a virtual $W$-boson s-channel diagram [18]. We do not include possible new production channels for the s-channel single top events. We want to emphasize that in general all four observables of Eqs. (6)-(9) are needed to determine the four couplings of the $tbW$ vertex and to make a complete analysis that could test the different models of EWSB. Other studies of $f_2^{L,R}$ in connection with the single top quark production at hadron colliders have shown that a sensitivity of order 0.2 (0.05) might be achieved at the Tevatron (LHC) [19].

## TOP QUARK FCNC COUPLINGS.

The weak neutral current (WNC) was the primary prediction to be tested in the electroweak $SU(2)\times U(1)$ standard model (SM). The flavor-conserving structure of the WNC has been verified with high precision in many processes [20]. In the SM, there are no flavor changing neutral couplings (FCNC) mediated by the $Z$, $\gamma$, $g$ gauge bosons nor the Higgs boson $H$ at tree level because the fermions are rotated from gauge to mass eigenstates by unitary diagonalization matrices [21]. Furthermore, the top-quark FCNC induced by radiative effects are also highly suppressed [22, 23, 24]: the higher order contributions induced by the charged currents are proportional to $(m_i^2-m_j^2)/M_W^2$, where $m_{i,j}$ are the masses of the quarks circulating in the loop and $M_W$ is the $W$ gauge boson mass. As a consequence, in the SM all top-quark FCNC transitions $t\to qV, qH$ , with $V=Z,\gamma,g$, which involve down-type quarks in the loops, are suppressed far below the observable level at existing or upcoming high energy colliders [22, 23, 24]. For example, in the $t\to cV$ transitions the scale of the respective partial widths is set by the $b$ quark mass [22, 23],

$$\Gamma(t\to V_i c) \;=\; |V_{bc}|^2\alpha\alpha_i\, m_t\,(\frac{m_b}{M_W})^4(1-\frac{m_{V_i}^2}{m_t^2}) \tag{10}$$

where $\alpha_i$ is the respective coupling for each gauge boson $V_i$. From the above result, it follows the approximated branching ratios $BR(t\to\gamma,Z)\sim 10^{-13}$ and $BR(t\to cg)\sim 10^{-11}$. In contrast, in the $b\to s\gamma$ transitions the leading contribution is proportional to $m_t^4/M_W^4$ and thus the GIM mechanism [21] induces in this case an enhancement factor.

The most general effective Lagrangian describing the FCNC top-quark interactions with a light quark $q'=u,c$, containing terms up to dimension five, can be written as [25]

$$\begin{aligned}\mathscr{L} \;=\;& \bar{t}\{\frac{ie}{2m_t}(\kappa_{tq'\gamma}+i\tilde{\kappa}_{tq'\gamma}\gamma_5)\sigma_{\mu\nu}F^{\mu\nu}\\ &+\;\bar{t}\{\frac{ig_s}{2m_t}(\kappa_{tq'g}+i\tilde{\kappa}_{tq'g}\gamma_5)\sigma_{\mu\nu}\frac{\lambda^a}{2}G_a^{\mu\nu}\\ &+\;\frac{i}{2m_t}(\kappa_{tq'Z}+i\tilde{\kappa}_{tq'Z}\gamma_5)\sigma_{\mu\nu}Z^{\mu\nu}\\ &+\;\frac{g}{2c_w}\gamma_\mu(v_{tq'Z}+a_{tq'Z}\gamma_5)Z^\mu\\ &+\;\frac{g}{2\sqrt{2}}(h_{tq'H}+i\tilde{h}_{tq'H}\gamma_5)H\}q'.\end{aligned} \tag{11}$$

where we have assumed also that the top quark and the neutral bosons are on shell or coupled effectively to massless fermions. In terms of these coupling constants, the respective partial widths for FCNC decays are given by [26]

$$
\begin{aligned}
\Gamma(t \to qZ)_\gamma &= \frac{\alpha}{32 s_W^2 c_W^2}\left(|\,\kappa_{tqZ}\,|^2 + |\,\tilde{\kappa}_{tqZ}\,|^2\right)\frac{m_t^3}{M_Z^2}\left[1-\frac{M_Z^2}{m_t^2}\right]^2\left[1+2\frac{M_Z^2}{m_t^2}\right] \\
\Gamma(t \to qZ)_\sigma &= \frac{\alpha}{16 s_W^2 c_W^2}\left(|\,v_{tqZ}\,|^2 + |\,a_{tqZ}\,|^2\right) m_t\left[1-\frac{M_Z^2}{m_t^2}\right]^2\left[2+\frac{M_Z^2}{m_t^2}\right] \\
\Gamma(t \to q\gamma) &= \frac{\alpha}{2}\left(|\,\kappa_{tq\gamma}\,|^2 + |\,\tilde{\kappa}_{tq\gamma}\,|^2\right) m_t \\
\Gamma(t \to qg) &= \frac{2\alpha_s}{3}\left(|\,\kappa_{tqg}\,|^2 + |\,\tilde{\kappa}_{tqg}\,|^2\right) m_t \\
\Gamma(t \to qH) &= \frac{\alpha}{32 s_W^2}\left(|\,h_{tqH}\,|^2 + |\,\tilde{h}_{tqH}\,|^2\right) m_t\left[1-\frac{M_H^2}{m_t^2}\right]^2
\end{aligned}
\tag{12}
$$

If we use $m_t = 178.0 \pm 4.3$ GeV, $\alpha(m_t) = 1/128.921$, $s_W^2 = 0.2342$, $\alpha_S(mt) = 0.108$, $m_H = 115\ GeV$ and the tree level prediction for the leading $t \to bW$ decay [20]

$$
\Gamma(t \to bW) = \frac{\alpha}{16 s_W^2}\,|\,V_{tb}\,|^2\,\frac{m_t^3}{M_W^2}\left[1-3\frac{M_W^4}{m_t^4}+2\frac{M_W^6}{m_t^6}\right], \tag{13}
$$

an update of the original SM calculations [22, 23, 24] for the FCNC top-quark branching ratios gives thus the following results [26]

$$
\begin{aligned}
BR(t \to q\gamma) &= (4.6^{+1.2}_{-1.0} \pm 0.2 \pm 0.4^{+1.6}_{-0.5}) \times 10^{-14} \\
BR(t \to qg) &= (4.6^{+1.1}_{-0.9} \pm 0.2 \pm 0.4^{+2.1}_{-0.7}) \times 10^{-12} \\
BR(t \to qZ) &\approx 1 \times 10^{-14} \\
BR(t \to qH) &\approx 3 \times 10^{-15}
\end{aligned}
\tag{14}
$$

where the uncertainties shown in the $t \to c\gamma, cg$ branching ratios are associated to the top and bottom quark masses, the CKM matrix elements and the renormalization scale. These updated results are about one order of magnitude smaller than the ones previously obtained [22, 23, 24]. For the decays involving the $u$ quark, the respective BR are a factor $|Vub/Vcb|^2 \sim 0.0079$ smaller than those shown in (14).

It has been realized that some top-quark FCNC decay modes can be enhanced by several orders of magnitude in scenarios beyond the SM, and some of them falling within the LHC's reach. In this case, the enhancement arises either from a large virtual mass or from the couplings involved in the loop. Top-quark FCNC processes may thus serve as a window for probing effects induced by new physics.

The absence of the vertex Htc at tree-level in the SM can be traced down to the presence of only one Higgs doublet. The process involved in the diagonalization of the fermion masses induces simultaneously diagonal Yukawa couplings for the physical Higgs boson. In models with more than one Higgs doublet, additional conditions have to be imposed to ensure that no FCNC arise at tree level. In particular, a discrete symmetry that makes quarks of same charge to interact with only one of the two (or more) Higgs doublets will, by the Glashow-Weinberg mechanism, cause all the Yukawa couplings involving physical neutral Higgs boson states become diagonal[27]. On the other hand, without any FCNC suppression mechanism these type of models may produce $tqH$ couplings at tree level, which in turn may induce large enhancements of the FCNC $tqV$ by radiative effects [28]. In Table 3 we make a summary of the predictions from several models beyond the SM.

**TABLE 3.** Predictions on the FCNC $t \to qV$ decays from different models.

| BR | THDM II | THDM III | MSSM | R-MSSM | TC2 | LR-SUSY | Extra q |
|---|---|---|---|---|---|---|---|
| $t \to qZ$ | $10^{-8}$ | $10^{-6}$ | $10^{-6}$ | $10^{-4}$ | $10^{-5}$ | $10^{-4}$ | $10^{-4}$ |
| $t \to q\gamma$ | $10^{-7}$ | $10^{-7}$ | $10^{-6}$ | $10^{-5}$ | $10^{-7}$ | $10^{-6}$ | $10^{-8}$ |
| $t \to qg$ | $10^{-5}$ | $10^{-4}$ | $10^{-4}$ | $10^{-3}$ | $10^{-5}$ | $10^{-5}$ | $10^{-7}$ |
| $t \to qH$ | $10^{-4}$ | $10^{-3}$ | $10^{-4}$ | $10^{-5}$ | – | – | $10^{-5}$ |

The interest in FCNC top-quark physics is expected to increase since processes involving top and Higgs FCNC will be examined with significant precision at both the LHC and ILC. In the first case, with a LHC luminosity of $100 fb^{-1}$, 80 million of $t\bar{t}$ pairs per year will make it possible to reach the following limits [26]:

$$\begin{aligned} BR(t \to cH) &< 6 \times 10^{-5}\,, \\ BR(t \to c\gamma) &< 1 \times 10^{-5}\,, \\ BR(t \to cZ) &< 4 \times 10^{-5}\,, \\ BR(t \to cg) &< 2 \times 10^{-5}\,, \end{aligned} \tag{15}$$

while at the ILC, with an integrated luminosity of $100 - 200 fb^{-1}$ one can hope to reach the sensibilities [26, 7]:

$$\begin{aligned} BR(t \to cH) &< 4.5 \times 10^{-5} \\ BR(t \to c\gamma) &< 7.7 \times 10^{-6}\,. \end{aligned} \tag{16}$$

The physics associated to the production mechanisms of the processes induced by the top-quark FCNC at future accelerators have been surveyed in several reviews [3].

**TABLE 4.** Limits on the FCNC $t \to qV$ decays from collider data.

| BR | Indirect | LEP2+Tevatron | LHC ($t \to qV$) | LHC ($tV$) | LHC ($tt$) |
|---|---|---|---|---|---|
| $t \to qZ$ | $10^{-2}$ | $2 \times 10^{-1}$ | $2 \times 10^{-4}$ | $1 \times 10^{-4}$ | $0.7 \times 10^{-1}$ |
| $t \to q\gamma$ | $10^{-3}$ | $3 \times 10^{-2}$ | $4 \times 10^{-5}$ | $0.4 \times 10^{-5}$ | $1 \times 10^{-3}$ |
| $t \to qg$ | $10^{-1}$ | $2 \times 10^{-1}$ | had. backg. | $0.3 \times 10^{-5}$ | $0.5 \times 10^{-2}$ |
| $t \to qH$ | $10^{-3}$ | – | – | – | $10^{-1}$ |

## ACKNOWLEDGMENTS

The part of the talk that deals with the $ttZ$ and $tbW$ couplings has been the result of work in collaboration with C.-P. Yuan and Chuan-Ren Chen. We thank Conacyt for support.

## REFERENCES


1. S. Abachi et al. (D0 Collaboration), *Phys. Rev. Lett.* **74**, 2632 (1995).
2. F. Abe et al. (CDF Collaboration), *Phys. Rev. Lett.* **74**, 2626 (1995).
3. W. Wagner, *Rept. Prog. Phys.* **68**, 2409 (2005); A. Juste et al., Report of the *2005 Snowmass Top/QCD Working Group*, hep-ph/0601112; J.M. Yang, *Ann. Phys. (N.Y.)* **316**, 529 (2005); D. Chakraborty, J. Konigsberg, D. Rainwater, *Ann. Rev. Nucl. Part. Sci.* **53**, 301 (2003); M. Beneke et al., Report of the *1999 CERN Workshop on SM physics (and more) at the LHC*, Geneva, hep-ph/0003033.
4. E. Malkawi and C.-P. Yuan, *Phys. Rev.* **D50**, 4462 (1994); ibid. **D52**, 472 (1995); F. Larios, E. Malkawi and C.-P. Yuan, hep-ph/9704288.
5. F. Larios and C.-P- Yuan, *Phys. Rev.* **D55**, 7218 (1997); O.J.P. Eboli, M.C. González-García and S.F. Novaes, *Phys. Lett.* **B415**, 75 (1997).
6. F. Larios, M.A. Pérez and C.-P. Yuan, *Phys. Lett. B457*, 334 (1999).
7. T. Abe et al., *Linear Collider Physics Resource Book for Snowmass 2001 - Part 3.*, hep-ex/0106057.
8. U. Baur et al., *Phys. Rev.* **D71**, 054013 (2005); U. Baur, hep-ph/0508151.
9. By the DØ collaboration, hep-ex/0404040 and references there in; by the CDF collaboration, hep-ex/0411070 and references therein.
10. G.L. Kane, G.A. Ladinsky and C.-P. Yuan, Phys. Rev. D**45** (1992), 124.
11. Chuan-Ren Chen, F. Larios and C.-P. Yuan, Phys. Lett. B**631** (2005), 126.
12. F. del Aguila and J.A. Aguilar-Saavedra, Phys. Rev. D**67** (2003), 014009.
13. G. Burdman, M.C. Gonzalez-Garcia and S.F. Novaes, Phys. Rev. D**61** (2000), 114016.
14. G.J. Gounaris, F.M. Renard and C. Verzegnassi, Phys. Rev. D**52** (1995), 451.
15. H.S. Do, S. Groote, J.G. Korner and M.C. Mauser, Phys. Rev. D**67** (2003), 091501. and references therein.
16. Z. Sullivan, Phys. Rev. D**70** (2004), 114012. and references therein; J. Campbell, R.K. Ellis and F. Tramontino, Phys. Rev. D**70**(2004), 094012.
17. J. Pumplin, D.R. Stump, J. Huston, H.L. Lai, P. Nadolsky and W.K. Tung, JHEP **0207** (2002) 012.
18. H.-J. He, L. Diaz-Cruz and C.-P. Yuan, Phys. Lett. B**530** (2002), 179.
19. E. Boos, L. Dudko and T. Ohl, Eur. Phys. J. C**11** (1999) 473.
20. S. Eidelman et al., Particle Data Group *Phys. Lett.* **B592,** 1 (2004).
21. S.L. Glashow, J. Iliopoulos, L. Maiani, *Phys. Rev.* **D2,** 1285 (1970).
22. J.L. Díaz-Cruz, R. Martínez, M.A. Pérez, A. Rosado, *Phys. Rev.* **D41,** 891 (1990).
23. G. Eilam, J.L. Hewett, A. Soni, *Phys. Rev.* **D44,** 1473 (1991); Erratum-ibid. **D59,** 039901 (1999).
24. B. Mele, S. Petrarca, A. Soddu, *Phys. Lett.* **B435,** 401 (1998); H. Fristch, *Phys. Lett.* **B224,** 179 (1989).
25. T. Han and J.L. Hewett, *Phys. Rev.* **D60,** 074015 (1999).
26. Aguilar-Saavedra, *Acta Phys. Pol.* **B35,** 2695 (2004); *Phys. Rev.* **D67,** 035003 (2003); ibid. **D69,** 099901 (2004); J.A. Aguilar-Saavedra and B.M. Nobre, *Phys. Lett.* **B553,** 251 (2003); J.A. Aguilar-Saavedra and G.C. Branco, *Phys. Lett.* **B495,** 347 (2000); J.A. Aguilar-Saavedra, *Phys. Lett.* **B502,** 115 (2001).
27. S.L. Glashow and S. Weinberg, *Phys. Rev.* **D15,** 1958 (1977).
28. A. Cordero-Cid, M.A. Pérez, J.J. Toscano and G. Tavares-Velasco, *Phys. Rev.* **D70,** 074003 (2004).

# POSTER SESSION

# Constraining the mSUGRA parameter space through entropy and abundance criteria [1]

Luis G. Cabral-Rosetti*, Myriam Mondragón†, Dario Núñez**, Roberto A. Sussman**, Jesus Zavala** and Lukas Nellen**

*Departamento de Posgrado,
Centro Interdisciplinario de Investigación y Docencia en Educación Técnica (CIIDET),
Av. Universidad 282 Pte., Col. Centro, A. Postal 752, C. P. 76000,
Santiago de Querétaro, Qro., México.
†Instituto de Física,
Universidad Nacional Autónoma de México (IF-UNAM),
Apdo. Postal 20-364, 01000 México D.F., México.
**Instituto de Ciencias Nucleares,
Departamento de Gravitación y Teoría de Campos,
Universidad Nacional Autónoma de México (ICN-UNAM).
A. Postal 70-543, 04510 México, D.F., México.

**Abstract.** We explore the use of two criteria to constrain the allowed parameter space in mSUGRA models; both criteria are based in the calculation of the present density of neutralinos $\chi_0$ as Dark Matter in the Universe. The first one is the usual "abundance" criterion that requieres that present neutralino relic density complies with $0.0945 < \Omega_{CDM}h^2 < 0.1287$, which are the $2\sigma$ bounds according to WMAP [1]. To calculate the relic density we use the public numerical code micrOMEGAS [2]. The second criterion is the original idea presented in [3] that basically applies the microcanonical definition of entropy to a weakly interacting and self-gravitating gas, and then evaluate the change in entropy per particle of this gas between the freeze-out era and present day virialized structures. An *"entropy consistency"* criterion emerges by comparing theoretical and empirical estimates of this entropy. One of the objetives of the work is to analyze the joint application of both criteria, already done in [3], to see if their results, using approximations for the calculations of the relic density, agree with the results coming from the exact numerical results of micrOMEGAS. The main objetive of the work is to use this method to constrain the parameter space in mSUGRA models that are inputs for the calculations of micrOMEGAS, and thus to get some bounds on the predictions for the SUSY spectra.

**Keywords:** dark matter, supersymmetry

## PRELIMINARY RESULTS

The results obtained in [3] used several approximations, some of them connected to a restricted region in the parameter space of an mSUGRA model; one of them is the hypothesis that $m_R$ is similar to $m_\chi$ ($m_R$ is the mass of the right-handed s-lepton and $m_\chi$ is the mass of the lightest neutralino $\chi_0$). The choice of an annihilitation channel also restricts the parameter space. Using these hypothesis for the B-ino annihilation

[1] To appear in the **Proceedings of XII Mexican School on Particles and Fields and VI Latin American Symposium on High Energy Physics**, Puerto Vallarta, México, November 1-8, 2006.

CP917, *Particles and Fields,* edited by H. Castilla Valdez, J. C. D'Olivo, and M. A. Perez

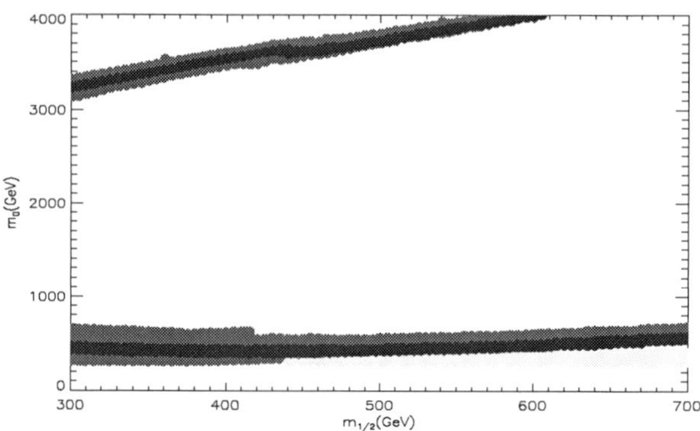


**FIGURE 1.** Comparison of criteria in the $m_0 - m_{1/2}$ plane for $A_0 = 0$, $\tan\beta = 10$ and $\mu > 0$. The blue and red regions show the allowed zones according to the WMAP bound for $\Omega_{CDM}h^2$ for the abundance and entropy criteria respectively. The yellow region shows the region where the stau is the LSP. The upper and lower zones in the figure correspond to the so called Focus Point and Coannihilation regions respectively.

channel we found that our results disagree with those presented in [3], see figures 4 and 5. In figure 1 we see the region of the parameter space taken to make the analysis. The graph shows the plane $m_0$ vs. $m_{1/2}$, $\tan\beta = 2$ and $sgn\mu = +$; the yellow region is where the s-tau is the LSP (ligtest supersymmetric particle), the red region is for $m_R \sim m_\chi$, and the blue region corresponds to the allowed region according to WMAP, for all this region the neutralinos are pure B-ino. Figure 4 is a logarithmic plot of $\Omega_{CDM}h^2$ vs. $M_\chi$, we see that there is no intersection between the two criteria: the black region, which corresponds to the *"abundance"* criterion, and all the space between the blue regions, which corresponds to the *"entropy"* criterion. Comparing figures 4 and 5, that come from the work [3], we see that with the use of micrOMEGAS we are not reproducing their results, concluding then that some of the approximations made by the authors, particularly in the calculation of $x_f = \frac{m_\chi}{T_f}$ ($T_f$ is the temperature of the gas at the freeze-out era), which is a key quantity for the *"entropy"* criterion, are not good in the application of this method.

Finding theoretical and phenomenological ways of constraining the supersymmetry parameters is one of the main challenges we face to discriminate among the many susy models, and to reconstruct, if possible, the fundamental parameters and understand the mechanism of susy breaking [5]. In figures 2 and 3 we show the allowed values for the LSP and the Higgs mass after constraining the parameter space with the abundance and entropy criteria. As can be seen from Fig. 3, the current limit for the Higgs mass favours a large value of $\tan\beta$. This also puts a constraint on the allowed susy spectra, with an LSP that is around 168 GeV for $\tan\beta 10$, and an LSP that starts from 150 GeV for large $\tan\beta$, and increases with increasing $m_{1/2}$.

We acknowledge partial support by CONACyT México, under grants 32138-E, 34407-E and 42026-F, and PAPIIT-UNAM IN-122002, IN117803 and IN115207 grants. JZ acknowledges support from DGEP-UNAM and CONACyT scholarships.

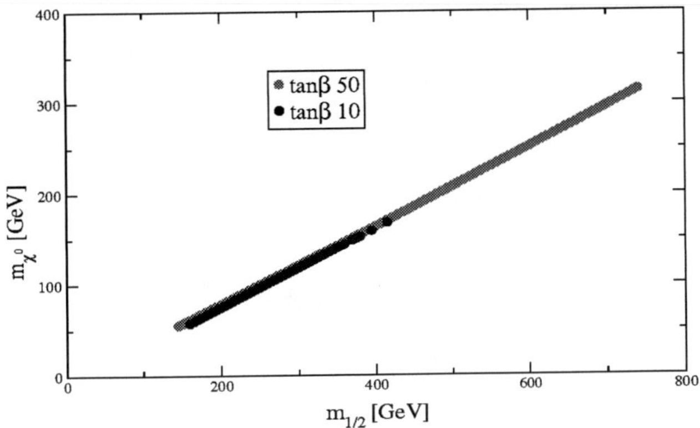


**FIGURE 2.** Allowed values for the neutralino mass $m\chi^0$ as fuction of the unified gaugino mass $m_{1/2}$.

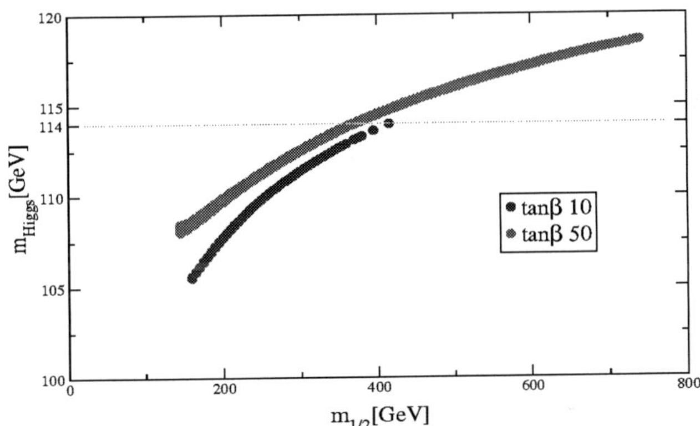


**FIGURE 3.** Allowed values for $M_{Higgs}$ as function of $m_{1/2}$. As can be seen from the figure, the actual bound on the Higgs mass favours a large $\tan\beta$.

# REFERENCES


1. G. Belanger, S. Kraml and A. Pukhov, Phys. Rev. D **72** (2005) 015003 [arXiv:hep-ph/0502079].
2. G. Belanger, F. Boudjema, A. Pukhov and A. Semenov, Comput. Phys. Commun. **174** (2006) 577 [arXiv:hep-ph/0405253].
3. L. G. Cabral-Rosetti, X. Hernandez and R. A. Sussman, Phys. Rev. D **69** (2004) 123006.
4. D. Nuñez, R. A. Sussman, J. Zavala, L. Nellen, L. G. Cabral-Rosetti and M. Mondragón, AIP Conf. Proc. **857** (2006) 321 [arXiv:astro-ph/0604127].
5. J. A. Aguilar-Saavedra *et al.*, Eur. Phys. J. C **46** (2006) 43 [arXiv:hep-ph/0511344].

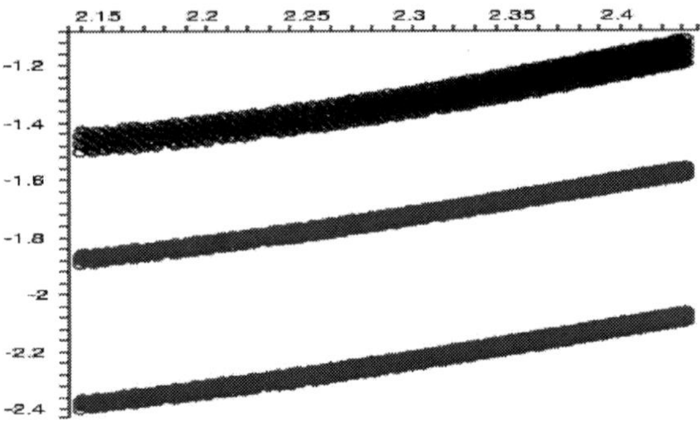


**FIGURE 4.** Logarithmic plot, $\Omega_{CDM}h^2$ vs. $m_\chi$. Abundance criterion (upper black zone), entropy criterion (space between the lower blue regions).

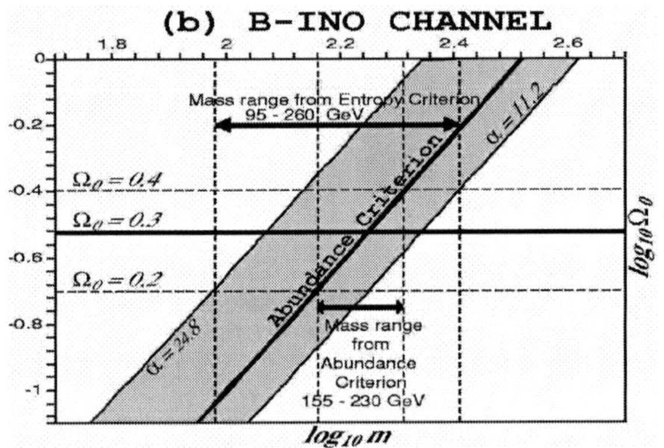


**FIGURE 5.** The same as in figure 4 with $h = 0.65$ according to the work [3].

# Neutrino masses and mixings in a Minimal $S_3$-invariant Extension of the Standard Model.

O. Félix, A. Mondragón, M. Mondragón and E. Peinado

*Departamento de Física Teórica,*
*Instituto de Física, Universidad Nacional Autónoma de México ,*
*Apdo. Postal 20-364, 01000 México D.F., México.*

**Abstract.** The mass matrices of the charged leptons and neutrinos, that had been derived in the framework of a Minimal $S_3$-invariant Extension of the Standard Model [1], are here reparametrized in terms of their eigenvalues. The neutrino mixing matrix, $V_{PMNS}$, is then computed and exact, explicit analytical expressions for the neutrino mixing angles as functions of the masses of the neutrinos and charged leptons are obtained. The reactor, $\theta_{13}$, and the atmosferic, $\theta_{23}$, mixing angles are found to be functions only of the masses of the charged leptons. The numerical values of $\theta_{13}^{th}$ and $\theta_{23}^{th}$ computed from our theoretical expressions are found to be in excellent agreement with the latest experimental determinations. The solar mixing angle, $\theta_{12}^{th}$, is found to be a function of both, the charged lepton and neutrino masses, as well as of a Majorana phase $\phi_\nu$. A comparison of our theoretical expression for the solar angle $\theta_{12}^{th}$ with the latest experimental value $\theta_{12}^{exp} \approx 34°$ allowed us to fix the scale and origin of the neutrino mass spectrum and obtain the mass values $|m_{\nu_2}| = 0.0507eV$, $|m_{\nu_1}| = 0.0499eV$ and $|m_{\nu_3}| = 0.0193eV$, in very good agreement with the observations of neutrino oscillations, the bounds extracted from neutrinoless double beta decay and the precision cosmological measurements of the CMB.

**Keywords:** Flavour symmetries, Quark and lepton masses and mixings, Neutrino masses and mixings

## INTRODUCTION

Recent neutrino oscillation observations [2, 3, 4, 5, 6, 7] and experiments [8, 9] have allowed the determination of the differences of the neutrino mass squared and the flavour mixing angles in the leptonic sector. Upper bounds on neutrino masses have been provided by the searches that probe the neutrino mass value at rest: beta decay [10], neutrinoless double beta decay [11] and precision cosmology [12].

In this short note, we will present the results of an analysis of neutrino masses and mixings in the Minimal $S_3$-Invariant Extension of the Standard Model [1], without going into the details of the derivation of these results. A detailed account of this analysis may be found in another recent paper [13].

## A MINIMAL $S_3$ INVARIANT EXTENSION OF THE STANDARD MODEL.

A Minimal $S_3$ Invariant Extension of the Standard Model [1] is formulated by introducing in the theory three $SU_L(2)$ Higgs doublet fields and extending the concept of flavour and generations to the Higgs sector. In this way, all the matter fields, that is, quarks,

CP917, *Particles and Fields,* edited by H. Castilla Valdez, J. C. D'Olivo, and M. A. Perez

leptons and Higgs fields can be accommodated in the three dimensional, real representation $\mathbf{1}_s \oplus 2$ of the flavour permutational group $S_3$. Neither, the fermionic nor the gauge couplings sectors of the theory are modified, but we add to the Lagrangian a Majorana mass term for the right handed neutrinos.

## EXTENDED HIGGS SECTOR, YUKAWA COUPLINGS AND FERMIONIC MASS MATRICES.

In a flavour symmetry adapted weak basis, the Higgs fields $H$ is the direct sum of a singlet $H_s$ and doublet $(H_1, H_2)^T$ representations of $S_3$. The presence of three Higgs fields in the Minimal $S_3$ Invariant Extension of the Standard Model, makes the Higgs potential somewhat more complicated than that of the Standard Model. But we may still assume that all the vaccumm expectation values are real and that $\langle H_1 \rangle = \langle H_2 \rangle$. They also satisfy the constraint $\langle H_s \rangle^2 + \langle H_1 \rangle^2 + \langle H_2 \rangle^2 = (246GeV)^2/2$.

The most general renormalizable Yukawa interactions are obtained by forming all the $S_3$ invariant Yukawa couplings of the Higgs fields and the fermion fields: $\mathscr{L}_Y = \mathscr{L}_{Y_D} + \mathscr{L}_{Y_u} + \mathscr{L}_{Y_E} + \mathscr{L}_{Y_\nu}$ where

$$\begin{aligned} \mathscr{L}_{Y_{D,E}} &= -Y_1^{d,e}\overline{G}_I H_S w_{IR} - Y_3^{d,e}\overline{G}_3 H_S w_{3R} \\ &- Y_2^{d,e}[\,\overline{G}_I \kappa_{IJ} H_1 w_{JR} + \overline{G}_I \eta_{IJ} H_2 w_{JR}\,] \\ &- Y_4^{d,e}\overline{G}_3 H_I w_{IR} - Y_5^{d,e}\overline{G}_I H_I w_{3R} + h.c \end{aligned} \tag{1}$$

and

$$\begin{aligned} \mathscr{L}_{Y_{U,\nu}} &= -Y_1^{u,\nu}\overline{G}_I (i\sigma_2) H_S^* v_{IR} - Y_3^{u,\nu}\overline{G}_3 (i\sigma_2) H_S^* v_{3R} \\ &- Y_2^{u}[\,\overline{G}_I \kappa_{IJ} (i\sigma_2) H_1^* v_{JR} + \eta \overline{G}_I \eta_{IJ} (i\sigma_2) H_2^* v_{JR}\,] \\ &- Y_4^{u,\nu}\overline{G}_3 (i\sigma_2) H_I^* v_{IR} - Y_5^{u,\nu}\overline{G}_I (i\sigma_2) H_I^* v_{3R} + h.c. \end{aligned} \tag{2}$$

and $G = Q, L$, $w = d, e$, $v = u, \nu$ and $I, J = 1, 2$ $\kappa = \begin{pmatrix} 0 & 1 \\ 1 & 0 \end{pmatrix}$ and $\eta = \begin{pmatrix} 1 & 0 \\ 0 & 1 \end{pmatrix}$.

After the spontaneous breaking of the gauge symmetry, the Yukawa interactions generate mass matrices for the Dirac fermions of the generic form

$$\mathbf{M} = \begin{pmatrix} \mu_1 + \mu_2 & \mu_2 & \mu_5 \\ \mu_2 & \mu_1 - \mu_2 & \mu_5 \\ \mu_4 & \mu_4 & \mu_3 \end{pmatrix} \tag{3}$$

The masses of the left handed Majorana neutrinos $\nu_L$ are obtained from the see-saw mechanism [14, 15, 16] and the corresponding mass matrix is given by $M_\nu = M_{\nu D} \tilde{\mathrm{M}}_R^{-1} (M_{\nu D})^T$ with $\tilde{\mathrm{M}}_R = diag(M_1, M_1, M_3)$. All the entries in the matrices can be complex, there are no restrictions coming from $S_3$.

### *$Z_2$ symmetry and the lepton mass matrices*

A further reduction of the number of free parameters is achieved if we introduce a $Z_2$ symmetry in the leptonic sector according with the $Z_2$ charge assignments: $(-)$ to the $H_I$ and $\nu_{3R}$ fields, $(+)$ for $H_S$, $L_3$, $L_I$, $e_{3R}$, $e_{IR}$ and $\nu_{IR}$ with $I = 1, 2$. This symmetry forbids certain Yukawa couplings: $Y_1^e = Y_3^e = Y_1^\nu = Y_5^\nu = 0$. The corresponding entries in the mass matrices vanish: $\mu_1^e = \mu_3^e = 0$ and $\mu_1^\nu = \mu_5^\nu = 0$.

The resulting mass matrix for the charged leptons, reparametrized in terms of its eigenvalues is

$$M_e \approx m_\tau \begin{pmatrix} \frac{1}{\sqrt{2}}\frac{\tilde{m}_\mu}{\sqrt{1+x^2}} & \frac{1}{\sqrt{2}}\frac{\tilde{m}_\mu}{\sqrt{1+x^2}} & \frac{1}{\sqrt{2}}\sqrt{\frac{1+x^2-\tilde{m}_\mu^2}{1+x^2}} \\ \frac{1}{\sqrt{2}}\frac{\tilde{m}_\mu}{\sqrt{1+x^2}} & -\frac{1}{\sqrt{2}}\frac{\tilde{m}_\mu}{\sqrt{1+x^2}} & \frac{1}{\sqrt{2}}\sqrt{\frac{1+x^2-\tilde{m}_\mu^2}{1+x^2}} \\ \frac{\tilde{m}_e(1+x^2)}{\sqrt{1+x^2-\tilde{m}_\mu^2}}e^{i\delta_e} & \frac{\tilde{m}_e(1+x^2)}{\sqrt{1+x^2-\tilde{m}_\mu^2}}e^{i\delta_e} & 0 \end{pmatrix}. \tag{4}$$

In this expression, the mass ratios $\tilde{m}_\mu = m_\mu/m_\tau$ and $x = m_e/m_\mu$ are real. Similarly, the mass matrix of the left handed Majorana neutrinos, generated by the see saw mechanism and reparametrized in terms of its complex mass eigenvalues is

$$M_\nu = \begin{pmatrix} m_{\nu_3} & 0 & \sqrt{(m_{\nu_3}-m_{\nu_1})(m_{\nu_2}-m_{\nu_3})}e^{-i\delta_\nu} \\ 0 & m_{\nu_3} & 0 \\ \sqrt{(m_{\nu_3}-m_{\nu_1})(m_{\nu_2}-m_{\nu_3})}e^{-i\delta_\nu} & 0 & m_{\nu_1}+m_{\nu_2}-m_{\nu_3} \end{pmatrix}. \tag{5}$$

$M_\nu$ is a complex symmetric matrix, which is diagonalized as $U_\nu^T M_\nu U_\nu = \left(|m_{\nu_1}|e^{i\phi_1}, |m_{\nu_2}|e^{i\phi_2}, |m_{\nu_3}|e^{i\phi_\nu}\right)$.

## THE NEUTRINO MIXING MATRIX $V_{PMNS}$

The neutrino mixing matrix is, by definition,

$$V_{PMNS} = U_{eL}^\dagger U_\nu K, \tag{6}$$

where $U_{eL}$ and $U_\nu$ are the unitary matrices that diagonalize the mass matrices and $K = diag\left(1, e^{i\alpha}, e^{i\beta}\right)$ is the diagonal matrix of the Majorana phases. Since the mass matrices have already been parametrized in terms of the physical masses, we can compute and express $U_{eL}$ and $U_\nu$ also in terms of the physical masses The unitary matrix $U_{eL}$ that diagonalizes $M_e M_e^\dagger$ and enters in the definition of the neutrino mixing matrix $V_{PMNS}$, eq. (6), is

$$U_{eL} = \begin{pmatrix} 1 & 0 & 0 \\ 0 & 1 & 0 \\ 0 & 0 & e^{i\delta_e} \end{pmatrix} \begin{pmatrix} O_{11} & -O_{12} & O_{13} \\ -O_{21} & O_{22} & O_{23} \\ -O_{31} & -O_{32} & O_{33} \end{pmatrix} \tag{7}$$

where

$$\begin{pmatrix} O_{11} & -O_{12} & O_{13} \\ -O_{21} & O_{22} & O_{23} \\ -O_{31} & -O_{32} & O_{33} \end{pmatrix} =$$

$$\begin{pmatrix} \frac{1}{\sqrt{2}}x\frac{(1+2\tilde{m}_\mu^2+4x^2+\tilde{m}_\mu^4+2\tilde{m}_e^2)}{\sqrt{1+\tilde{m}_\mu^2+5x^2-\tilde{m}_\mu^4-\tilde{m}_\mu^6+\tilde{m}_e^2+12x^4}} & -\frac{1}{\sqrt{2}}\frac{(1-2\tilde{m}_\mu^2+\tilde{m}_\mu^4-2\tilde{m}_e^2)}{\sqrt{1-4\tilde{m}_\mu^2+x^2+6\tilde{m}_\mu^4-4\tilde{m}_\mu^6-5\tilde{m}_e^2}} & \frac{1}{\sqrt{2}} \\ -\frac{1}{\sqrt{2}}x\frac{(1+4x^2-\tilde{m}_\mu^4-2\tilde{m}_e^2)}{\sqrt{1+\tilde{m}_\mu^2+5x^2-\tilde{m}_\mu^4-\tilde{m}_\mu^6+\tilde{m}_e^2+12x^4}} & \frac{1}{\sqrt{2}}\frac{(1-2\tilde{m}_\mu^2+\tilde{m}_\mu^4)}{\sqrt{1-4\tilde{m}_\mu^2+x^2+6\tilde{m}_\mu^4-4\tilde{m}_\mu^6-5\tilde{m}_e^2}} & \frac{1}{\sqrt{2}} \\ -\frac{\sqrt{1+2x^2-\tilde{m}_\mu^2-\tilde{m}_e^2}(1+\tilde{m}_\mu^2+x^2-2\tilde{m}_e^2)}{\sqrt{1+\tilde{m}_\mu^2+5x^2-\tilde{m}_\mu^4-\tilde{m}_\mu^6+\tilde{m}_e^2+12x^4}} & -x\frac{(1+x^2-\tilde{m}_\mu^2-2\tilde{m}_e^2)\sqrt{1+2x^2-\tilde{m}_\mu^2-\tilde{m}_e^2}}{\sqrt{1-4\tilde{m}_\mu^2+x^2+6\tilde{m}_\mu^4-4\tilde{m}_\mu^6-5\tilde{m}_e^2}} & \tilde{m}_e\tilde{m}_\mu\frac{\sqrt{1+x^2}}{\sqrt{1+x^2-\tilde{m}_\mu^2}} \end{pmatrix} \tag{8}$$

the mass ratios $\tilde{m}_e = m_e/m_\tau$ and $x = m_e/m_\mu$ are real.

Similarly, the unitary matrix $U_\nu$ that diagonalizes $M_\nu$ is

$$U_\nu = \begin{pmatrix} \sqrt{\frac{m_{\nu_2}-m_{\nu_3}}{m_{\nu_2}-m_{\nu_1}}} & \sqrt{\frac{m_{\nu_3}-m_{\nu_1}}{m_{\nu_2}-m_{\nu_1}}} & 0 \\ 0 & 0 & 1 \\ -\sqrt{\frac{m_{\nu_3}-m_{\nu_1}}{m_{\nu_2}-m_{\nu_1}}}e^{i\delta_\nu} & \sqrt{\frac{m_{\nu_2}-m_{\nu_3}}{m_{\nu_2}-m_{\nu_1}}}e^{i\delta_\nu} & 0 \end{pmatrix} \tag{9}$$

The unitarity of $U_\nu$ constrains

$$\sin^2\eta = \frac{m_{\nu_3}-m_{\nu_1}}{m_{\nu_2}-m_{\nu_1}}, \tag{10}$$

to be real and $|\sin\eta| \leq 1$. This condition fixes the phases $\phi_1$ and $\phi_2$ as

$$|m_{\nu_1}|\sin\phi_1 = |m_{\nu_2}|\sin\phi_2 = |m_{\nu_3}|\sin\phi_\nu. \tag{11}$$

The only free parameters in $U_e$ and $U_\nu$ other than the physical masses, are the phase $\phi_\nu$ implicit in $m_{\nu_1}$, $m_{\nu_2}$ and $m_{\nu_3}$ and the two Dirac phases $\delta_e$ and $\delta_\nu$.

The theoretical expression for the neutrino mixing matrix $V_{PMNS}$, is obtained by taking the product $U_{eL}^\dagger U_\nu K$ and making an appropriate transformation of phases

$$V_{PMNS} \approx \begin{pmatrix} O_{11}\cos\eta + O_{31}\sin\eta & O_{11}\sin\eta - O_{31}\cos\eta & -O_{21}e^{-i\delta} \\ -O_{12}\cos\eta + O_{32}\sin\eta e^{i\delta} & -O_{12}\sin\eta - O_{32}\cos\eta e^{i\delta} & O_{22} \\ O_{13}\cos\eta - O_{33}\sin\eta e^{i\delta} & O_{13}\sin\eta + O_{33}\cos\eta e^{i\delta} & O_{23} \end{pmatrix} \times K, \tag{12}$$

where $\cos\eta$ and $\sin\eta$ are given in eq (10), $O_{ij}$ are given in eq (8) and $\delta = \delta_\nu - \delta_e$.

A comparison of this expression with the standard parametrization advocated by the *PDG* [17], allowed us to derive expressions for the mixing angles in terms of the charged lepton and neutrino masses. Keeping only terms of order $(m_e^2/m_\mu^2)$ and $(m_\mu/m_\tau)^4$, we get

$$\sin\theta_{13} \approx -\frac{1}{\sqrt{2}} x \frac{(1+4x^2-\tilde{m}_\mu^4)}{\sqrt{1+\tilde{m}_\mu^2+5x^2-\tilde{m}_\mu^4}}, \quad \sin\theta_{23} \approx \frac{1}{\sqrt{2}} \frac{1-2\tilde{m}_\mu^2+\tilde{m}_\mu^4}{\sqrt{1-4\tilde{m}_\mu^2+x^2+6\tilde{m}_\mu^4}} \tag{13}$$

and

$$\tan\theta_{12} = -\sqrt{\frac{m_{\nu_2}-m_{\nu_3}}{m_{\nu_3}-m_{\nu_1}}} \times \left( \frac{\sqrt{1+2x^2-\tilde{m}_\mu^2(1+\tilde{m}_\mu^2+x^2)} - \frac{1}{\sqrt{2}} x(1+2\tilde{m}_\mu^2+4x^2) \sqrt{\frac{m_{\nu_3}-m_{\nu_1}}{m_{\nu_2}-m_{\nu_3}}}}{\sqrt{1+2x^2-\tilde{m}_\mu^2(1+\tilde{m}_\mu^2+x^2)} + \frac{1}{\sqrt{2}} x(1+2\tilde{m}_\mu^2+4x^2) \sqrt{\frac{m_{\nu_2}-m_{\nu_3}}{m_{\nu_3}-m_{\nu_1}}}} \right). \tag{14}$$

where

$$\frac{m_{\nu_2}-m_{\nu_3}}{m_{\nu_3}-m_{\nu_1}} = \frac{(|m_{\nu_2}|^2 - |m_{\nu_3}|^2 \sin^2\phi_\nu)^{1/2} - |m_{\nu_3}||\cos\phi_\nu|}{(|m_{\nu_1}|^2 - |m_{\nu_3}|^2 \sin^2\phi_\nu)^{1/2} + |m_{\nu_3}||\cos\phi_\nu|}. \tag{15}$$

Similarly, the Majorana phases are given by

$$\sin 2\alpha = \sin(\phi_1 - \phi_2) = \frac{|m_{\nu_3}|\sin\phi_\nu}{|m_{\nu_1}||m_{\nu_2}|} \times \left( \sqrt{|m_{\nu_2}|^2 - |m_{\nu_3}|^2 \sin^2\phi_\nu} + \sqrt{|m_{\nu_1}|^2 - |m_{\nu_3}|^2 \sin^2\phi_\nu} \right),$$

$$\sin 2\beta = \sin(\phi_1 - \phi_\nu) = \frac{\sin\phi_\nu}{|m_{\nu_1}|} \left( |m_{\nu_3}|\sqrt{1-\sin^2\phi_\nu} + \sqrt{|m_{\nu_1}|^2 - |m_{\nu_3}|^2 \sin^2\phi_\nu} \right). \tag{16}$$

A more complete and detailed discussion of the Majorana phases in the neutrino mixing matrix $V_{PMNS}$ obtained in our model is given by J. Kubo [18].

## NEUTRINO MASSES AND MIXINGS

In the present model, $\sin^2\theta_{13}$ and $\sin^2\theta_{23}$ are determined by the masses of the charged leptons in very good agreement with the experimental values [19, 20, 21],

$$(\sin^2\theta_{13})^{th} = 1.1 \times 10^{-5}, \quad (\sin^2\theta_{13})^{exp} \leq 0.046,$$

and

$$(\sin^2\theta_{23})^{th} = 0.499, \quad (\sin^2\theta_{23})^{exp} = 0.5^{+0.06}_{-0.05}.$$

The experimental restriction $|\Delta m_{12}^2| < |\Delta m_{13}^2|$ implies an inverted neutrino mass spectrum, $|m_{\nu_3}| < |m_{\nu_1}| < |m_{\nu_2}|$ [1].

From eqs. (14) and (15), the solar mixing angle is sensitive to the neutrino mass differences and the phase $\phi_\nu$ but is only very weakly sensitive to the charged lepton masses. Writing the neutrino mass differences $m_{\nu_i} - m_{\nu_j}$ in terms of the differences of the mass squared and one of the neutrino masses, say $|m_{\nu_2}|$, from (14) and (15), we obtain

$$\begin{aligned}\frac{m_{\nu_2}^2}{\Delta m_{13}^2} &= \frac{1+2t_{12}^2+t_{12}^4-rt_{12}^4}{4t_{12}^2(1+t_{12}^2)(1+t_{12}^2-rt_{12}^2)cos^2\phi_\nu} - \tan^2\phi_\nu + O\left(\frac{m_e^2}{m_\mu^2}\right) \\ &\approx \frac{1}{\sin^2 2\theta_{12}\cos^2\phi_\nu} - \tan^2\phi_\nu \text{ for } r << 1,\end{aligned} \tag{17}$$

where $t_{12} = \tan\theta_{12}$ and $r = \Delta m_{21}^2/\Delta m_{13}^2$.

The mass $|m_{\nu_2}|$ assumes its minimal value when $\sin\phi_\nu$ vanishes, then

$$|m_{\nu_2}| \approx \frac{\sqrt{\Delta m_{13}^2}}{\sin 2\theta_{12}}. \tag{18}$$

Hence, we find

$$|m_{\nu_2}| \approx 0.0507eV, \quad |m_{\nu_1}| \approx 0.0499eV, \quad |m_{\nu_3}| \approx 0.0193eV, \tag{19}$$

where we used the values $\Delta m_{13}^2 = 2.2^{+0.37}_{-0.27} \times 10^{-3}eV^2$ and $\sin^2\theta_{12} = 0.31^{+0.02}_{-0.03}$ taken from M. Maltoni et al. [19], T. Schwetz [20] and G. L. Fogli et al. [21].

With those values for the neutrino masses we compute the effective electron neutrino mass $m_\beta$

$$m_\beta = \left[\sum_i |U_{ei}|^2 m_{\nu_i}^2\right]^{\frac{1}{2}} = 0.0502eV, \tag{20}$$

well below the upper bound $m_\beta < 1.8eV$ coming from the tritium $\beta$-decay experiments [10, 21, 22].

## SUMMARY AND CONCLUSIONS

By introducing three Higgs fields that are $SU(3)_L$ doublets in the theory, the concepts of flavour and generations were extended to the Higgs sector. Thus, all the matter fields, quarks, leptons and Higgs fields were accommodated in the three dimensional, real representation of the flavour permutational group $S_3$. In this way a Minimal $S_3$-Invariant Extension of the Standard Model was formulated which includes in its Lagrangian a Majorana mass term for the right handed neutrinos. A well defined structure of the Yukawa couplings is obtained which permits the calculation of mass and mixing matrices for quarks and leptons in a unified way. A further reduction of redundant parameters is achieved by introducing a $Z_2$ symmetry in the leptonic sector. The Majorana neutrinos acquire mass via the see saw mechanism. The mass matrices of the charged leptons and Majorana neutrinos were reparametrized in terms of their eigenvalues. This allowed us to compute the neutrino mixing matrix explicitly in terms of the masses of the charged leptons and neutrinos. In this model, the magnitudes of the three neutrino mixing angles are determined by the interplay of the flavour $S_3 \times Z_2$ symmetry, the see saw mechanism and the charged lepton mass hierarchy. We also found that $V_{PMNS}$ has three CP violating phases, one Dirac phase $\delta = \delta_\nu - \delta_e$ and two Majorana phases $\alpha$ and $\beta$, which are functions of the neutrino masses and another phase $\phi_\nu$ which is independent of the Dirac phase. The numerical values of the reactor $\theta_{13}$, and atmospheric, $\theta_{23}$, mixing angles are determined by the masses of the charged leptons only, in very good agreement with the latest experimental values. The solar mixing angle $\theta_{12}$ is almost insensitive to the values of the charged leptons, but its experimental value allowed us to fix the scale and origin of the neutrino mass spectrum, which has an inverted hierarchy with the mass values $|m_{\nu_1}| = 0.0499eV$ and $|m_{\nu_3}| = 0.0193eV$.

## REFERENCES


1. Kubo J, Mondragón A, Mondragón M, Rodríguez-Jáuregui E. *Prog. Theor. Phys.* **109**, (2003), 795.
2. M. Altmann *et al.* [GNO collaboration], *Phys. Lett.* B **616**, (2005), 174.
3. M. B. Smy *et al.* [SK collaboration], *Phys. Rev.* D **69** (2004), 011104.
4. Q.R. Ahmad *et al.* [SNO collaboration], *Phys. Rev. Lett.* **89** (2002) 011301.
5. B. Aharmim *et al.* [SNO collaboration], *Phys. Rev.* **C72** (2005), 055502. arxiv: nucl-ex/0502021.
6. S. Fukuda *et al.* ,[SK collaboration] *Phys. Lett.* **B539** (2002) 179.
7. Y. Ashie *et al.*, *Phys. Rev. Lett.* **93** (2004) 101801; [hep-ex/0501064].
8. C. Bemporad, G. Gratta and P. Vogel, *Rev. Mod. Phys.*, **74** (2002) 297.
9. T. Araki *et al.* (KamLAND collaboration), *Phys. Rev. Lett.* **94**, (2005) 081801.
10. K. Eitel in "Neutrino 2004", 21st International Conference on Neutrino Physics and Astrophysics (Paris, France 2004) Ed. J. Dumarchey, Th. Patyak and F. Vanuucci. *Nucl. Phys.* B (*Proc Suppl.*) **143**, (2005) 3.
11. S. R. Eliot and J. Engel, *J. Phys.* G **30** (2004) R183.
12. O. Elgaroy, and O. Lahav, *New J. Phys.* **7** (2005) 61.
13. O. Felix, A. Mondragon, M. Mondragon and E. Peinado, *Rev. Mex. Fís.* **S 52**, (2006), 67-73. arxiv: hep-ph/0610061.
14. P. Minkowski, Phys. Lett. B **67**, 421 (1977).
15. T. Yanagida, in Proceedings of the Workshop on Unified Theory and Baryon Number of the Universe, eds. O. Sawada and A. Sugamoto (KEK, 1979) p.95
16. M. Gell-Mann, P. Ramond, and R. Slansky, in Supergravity, edited by P. Van Nieuwenhuizen and D. Z. Freedman (North-Holland, Amsterdam, 1979).
17. W-M Yao et al 2006 *J. Phys. G: Nucl. Part. Phys.* **33** 1.
18. J. Kubo, *Phys. Lett.* **B578**, (2004), 156; Erratum: ibid **619** (2005) 387.
19. M. Maltoni, T. Schwetz, M.A. Tórtola and J.W.F. Valle, *New J. Phys.* **6** (2004) 122.
20. T. Schwetz, "Neutrino oscillations: Current status and prospects", *Acta Phys. Polon.* **B36** (2005) 3203. arxiv: hep-ph/0510331.
21. G. L. Fogli, E. Lisi, A. Marrone and A. Palazzo, *Prog. Part. Nucl. Phys.* **57** (2006) 742-795. arxiv: hep-ph/0506083.
22. G. L. Fogli, E. Lisi, A. Marrone, A. Melchiorri, A. Palazzo, P. Serra and J. Silk, *Phys. Rev.* **D70** (2004), 113003.

# Stability of the Tree-Level Vacuum in a Minimal $S_3$ Extension of the Standard Model

D. Emmanuel-Costa*, O. Félix-Beltrán†, M. Mondragón** and E. Rodríguez-Jáuregui‡

*Departamento de Física and Centro de Física Teórica (CFTP), Instituto Superior Técnico, Lisboa, Portugal
†Facultad de Cs. de la Electrónica, BUAP, Puebla, Pue. México
**Instituto de Física, UNAM, México, D.F.
‡Depto. de Física UNISON, Sonora, México

**Abstract.** In this work we analyzed the stability of the minimal $S_3$ invariant extension of the Higgs potential. In the $S_3$ invariant minimal standard model it is assumed that the Higgs fields belong to the three-dimensional reducible representation of the permutation group $S_3$. We show that in the three $S_3$ Higgs doublet model at tree-level the potential minimum preserving electric charge and CP symmetries, when it exists, is the global one.

**Keywords:** Extension SM, Higgs Potential, $S_3$ symmetry
**PACS:** 14.80.Cp, 14.80.Bn, ,12.60.Fr ,11.30.Qc ,11.30.Ŭj ,11.15.Ex.

## INTRODUCTION

In the Standard Model each family of fermions enters independently. In order to understand the replication of family fermion generation and reduce the number of free parameters, one is lead to introduce a flavour family fermion group. Recently succesfuly work has been done using a flavour family discrete $S(3)$ group. Although their existence is a fundamental piece of the theory, the least understood part of the Standard Model is the property of Higgs boson. With exact $S(3)$ flavour symmetry the fermion masses are degenerated, and a flavoured Higgs boson is needed to avoid this problem. In this paper we extend the Standard Model Higgs Sector using a flavour discrete symmetry $S(3)$ assuming that Higgs fields belong to the three-dimensional reducible representation of the flavour permutational group $S(3)$. Since the potential and its minimization play a vital part in the successful construction of the model we analyze the $S(3)$-invariant Higgs boson potential and found the conditions under which the potential minimum preserving electric charge and CP symmetries, is the global one.

## EXTENSION SM$\otimes S_3$

The most general Higgs potential invariant under exact $SU(2)_L \times U(1)_Y \times S_3$ can be expressed as [1]:

$$\begin{aligned} V &= \mu_1^2(x_1+x_2)+\mu_0^2 x_3+a x_3^2+b x_3(x_1+x_2)+c(x_1+x_2)^2-4d x_5^2+g\left[(x_1-x_2)^2+4x_4^2\right] \\ &\quad +2e\left[2x_4x_6+x_8(x_1-x_2)\right]+f\left(x_6^2+x_7^2+x_8^2+x_9^2\right)+2h\left(x_6^2-x_7^2+x_8^2-x_9^2\right), \end{aligned}$$

CP917, *Particles and Fields,* edited by H. Castilla Valdez, J. C. D'Olivo, and M. A. Perez

where $x_i$ are the components of a vector $\mathbf{X}$ given by $\mathbf{X}^T = (x_1, x_2, x_3, x_4, x_5, x_6, x_7, x_8, x_9)$, following the notation for the Higgs potential given in Ref.[2]:

$$\begin{array}{lll} x_1 = \Phi_1^\dagger \Phi_1 & x_2 = \Phi_2^\dagger \Phi_2 & x_3 = \Phi_S^\dagger \Phi_S \\ x_4 = \mathscr{R}\left(\Phi_1^\dagger \Phi_2\right) & x_6 = \mathscr{R}\left(\Phi_1^\dagger \Phi_S\right) & x_8 = \mathscr{R}\left(\Phi_2^\dagger \Phi_S\right) \\ x_5 = \mathscr{I}\left(\Phi_1^\dagger \Phi_2\right) & x_7 = \mathscr{I}\left(\Phi_1^\dagger \Phi_S\right) & x_9 = \mathscr{I}\left(\Phi_2^\dagger \Phi_S\right) \end{array}$$

Here, the $SU(2)_L$ doublet Higgs fields, $\Phi_1$ and $\Phi_2$ belongs to the $S(3)$ doublet $H_D$ and $\Phi_S$ is the $S(3)$ singlet:

$$\Phi_1 = \begin{pmatrix} \phi_1 + i\phi_2 \\ \phi_7 + i\phi_{10} \end{pmatrix}, \; \Phi_2 = \begin{pmatrix} \phi_3 + i\phi_4 \\ \phi_8 + i\phi_{11} \end{pmatrix}, \; \Phi_S = \begin{pmatrix} \phi_5 + i\phi_6 \\ \phi_9 + i\phi_{12} \end{pmatrix}.$$

Defining a vector $\mathbf{A}$ and the square symmetric matrix $\mathbf{B}$ as

$$\mathbf{A} = \begin{pmatrix} \mu_1 \\ \mu_1 \\ \mu_0 \\ 0 \\ 0 \\ 0 \\ 0 \\ 0 \\ 0 \end{pmatrix}, \; \mathbf{B} = \begin{pmatrix} 2(c+g) & 2(c-g) & b & 0 & 0 & 0 & 0 & 2e & 0 \\ 2(c-g) & 2(c+g) & b & 0 & 0 & 0 & 0 & -2e & 0 \\ b & b & 2a & 0 & 0 & 0 & 0 & 0 & 0 \\ 0 & 0 & 0 & 8g & 0 & 4e & 0 & 0 & 0 \\ 0 & 0 & 0 & 0 & -8d & 0 & 0 & 0 & 0 \\ 0 & 0 & 0 & 4e & 0 & 2(f+2h) & 0 & 0 & 0 \\ 0 & 0 & 0 & 0 & 0 & 0 & 2(f-2h) & 0 & 0 \\ 2e & -2e & 0 & 0 & 0 & 0 & 0 & 2(f+2h) & 0 \\ 0 & 0 & 0 & 0 & 0 & 0 & 0 & 0 & 2(f-2h) \end{pmatrix}.$$

we can write the $S(3)$ Higgs potential as:

$$V = \mathbf{A^T X} + \frac{1}{2}\mathbf{X^T B X}.$$

This potential has three types of minimum or stationary points, that is: the normal minimum with the following field configuration: $\phi_7 = v_1, \phi_8 = v_2, \phi_9 = v_3$ and $\phi_i = 0\,(i = 1,2,3,..,12)$; the electric charge breaking stationary point, where three of the charged fields $\phi_i$ has a non zero vev: $\phi_7 = v_1', \phi_8 = v_2', \phi_9 = v_3', \phi_1 = \alpha, \phi_3 = \beta, \phi_5 \neq 0$; and the CP-symmetry breaking stationary point where three of the imaginary neutral component fields $\phi_i$ gets a non zero vev: $\phi_7 = v_1'', \phi_8 = v_2'', \phi_9 = v_3'', \phi_{10} = \delta, \phi_{11} = \gamma, \phi_{12} \neq 0$.

## THE NORMAL, CB AND CP BREAKING MINIMUMS

**(A)** Normal Minimum: From the definitions given above, the normal minimum corresponds to the following field configuration $x_i = v_i^2$ for $i = 1,2,3$, $x_4 = v_1 v_2$, $x_6 = v_1 v_3$, $x_8 = v_2 v_3$, and $x_5 = x_7 = x_9 = 0$, with this notation, the minimization conditions can be written as

$$\frac{\partial V}{\partial v_i} = 0 \leftrightarrow \frac{\partial V}{\partial x_j}\frac{\partial x_j}{\partial v_i} = 0 \quad i = 1,2,3 \quad j = 1,2,..,9$$

The solution to this coupled equations implies $e = 0$ or ${v_1}^2 = 3{v_2}^2$. Defining the vectors $\mathbf{V}' = \frac{\partial V}{\partial X_i}\,|_{min}$, and $\mathbf{X}_N = \mathbf{X}\,|_{normal\,min}$ we obtain that in the normal minimum

$$\mathbf{V}' = \mathbf{A} + \mathbf{B}\mathbf{X}_N \qquad \mathbf{X}_N^T \mathbf{V}' = 0. \tag{1}$$

In this notation the potential in the normal minimum can be written as

$$V_N = -\frac{1}{2}\mathbf{X}_N^T\mathbf{B}\mathbf{X}_N = \frac{1}{2}\mathbf{A}^T\mathbf{X}_N. \tag{2}$$

(**B**) *Charge Breaking*: We consider the case where the charge breaking comes from the $S_3$ Higgs doublet $H_D$ for this configuration the fields with vev's different from zero are $\phi_7 = v'_1$, $\phi_8 = v'_2$, $\phi_9 = v'_3$ and $\phi_1 = \alpha$, $\phi_3 = \beta$. These vevs break charge conservation and would give mass to the photon, but in the CB minimum we can always have a solution with $\alpha = \beta = 0$. In the CB stationary point the vector $\mathbf{X}_{CB} = \mathbf{X}\,|_{CB\,min}$ is:

$$\mathbf{X}_{CB}^T = \left(\alpha^2 + v_1'^2, \beta^2 + v_2'^2, v_3'^2, \alpha\beta + v_1'v_2', 0, v_1'v_3', 0, v_2'v_3', 0\right)$$

and the stationarity conditions can be expressed in vectorial form as follows:

$$\frac{\partial \mathbf{V}}{\partial \mathbf{X}}\,|_{\mathbf{X}=\mathbf{X}_{CB}} = 0 \Leftrightarrow \mathbf{A} + \mathbf{B}\mathbf{X}_{CB} = 0. \tag{3}$$

This is a linear equation, then, if it has a solution for $\mathbf{X}_{CB}$ this is unique. The potential evaluated in CB is:

$$\mathbf{V}_{CB} = -\frac{1}{2}\mathbf{X}_{CB}^T\mathbf{B}\mathbf{X}_{CB}\,. \tag{4}$$

From this equation and equations (1)-(3), we can write

$$\mathbf{V}_{CB} - \mathbf{V}_N = \frac{1}{2}\mathbf{X}_{CB}^T\mathbf{V}'. \tag{5}$$

If the right side is positive the normal minimum is always deeper then the CB minimum.

(**C**) *CP Breaking*: We assume that CP breaking comes from the $S_3$ Higgs doublet $H_D$, in this case, the CP breaking stationary point is given by the non-zero vevs $\phi_7 = v'_1$, $\phi_8 = v'_2$, $\phi_9 = v'_3$, $\phi_{10} = \delta$ and $\phi_{11} = \gamma$. Then,defining the vectors $\mathbf{V}'_{CP} = \frac{\partial V}{\partial X_i}\,|_{CP\,min}$, and $\mathbf{X}_{CP} = \mathbf{X}\,|_{CP\,min}$, we obtain that

$$\mathbf{X}_{CP}^T\mathbf{V}'_{CP} = 0 \qquad \text{and} \qquad \mathbf{V}_{CP} = \frac{1}{2}\mathbf{A}^T\mathbf{X}_{CP}.$$

Now, we have all the elements to show that the normal minimum is deeper than CP minimum. As in the previous case, we obtain the expression

$$\mathbf{V}_{CP} - \mathbf{V}_N = \frac{1}{2}\mathbf{X}_{CP}^T\mathbf{V}'. \tag{6}$$

Also, the normal minimum is deeper if $\mathbf{X}_{CP}^T\mathbf{V}'$ is positive.

## MASS MATRICES

The second derivatives of the Higgs potential are necessary to know the nature of the stationary points. These derivatives are given by:

$$\frac{\partial^2 V}{\partial \phi_i \partial \phi_j} = \frac{\partial V}{\partial x_l}\frac{\partial^2 x_l}{\partial \phi_i \partial \phi_j} + \frac{\partial^2 V}{\partial x_l \partial x_m}\frac{\partial x_l}{\partial \phi_i}\frac{\partial x_m}{\partial \phi_j}. \tag{7}$$

The corresponding squared Higgs mass matrix has the form

$$[\mathbf{M}^2] = \frac{1}{2}\left([\mathbf{M}_I^2] + \mathbf{C}^T\mathbf{B}\mathbf{C}\right). \tag{8}$$

The derivatives $\partial x_i/\partial\phi_j$ form a $9\times 12$ matrix, which we call $\mathbf{C}$ and the second derivatives in the second term of equation (7) are clearly the matrix elements $B_{lm}$ of $\mathbf{B}$ matrix. Evaluated at each of the different stationary points only the fields $\phi_7,\phi_8,\phi_9$, $\phi_1,\phi_3,\phi_5$ and $\phi_{12}$ appear, and the rest vev's are always zero at the stationary points. In the normal minimum $[\mathbf{M}_I^2] = diag(\mathbf{M}_{11}^2,\mathbf{M}_{12}^2)$ where $\mathbf{M}_{11}^2$ and $\mathbf{M}_{12}^2$ are $6\times 6$ matrices. Then, the squared mass matrix is computed and take the form $diag(\mathbf{M}_C^2,\mathbf{M}_C^2,\mathbf{M}_S^2,\mathbf{M}_P^2)$. In particular, the physics charged Higgs masses can be expressed as

$$m_{H^\pm}^2 = V_1' + V_8' \pm \left(4V'^2_1 - 8V_1'V_8' + 8V_4'V_6' + 16V_3'V_2'\right)^{1/2}.$$

$\mathbf{M}_S^2$ and $\mathbf{M}_P^2$ are block diagonal symmetric scalar and pseudoscalar Higgs mass matrices, respectively.

## CONCLUSIONS

We analyzed the $S_3$ extended Higgs potential of the SM including a Higgs doublet $H_D$ and a Higgs singlet $\Phi_S$ of the non-abelian discrete symmetry $S_3$. In particular, we studied the Higgs potential stationary points: the normal one, charge breaking and CP breaking. We found the conditions in eq. (5)and (6) showing under which conditions the normal minimum, is a global and stable one. A complete analisis of the parameters space for the stationary points will be published else where.

## ACKNOWLEDGMENTS

We acknowledge support by CONACyT México, under grant 42026-F, and by PAPIIT-UNAM IN115207. The work of E.R.J. was suported by a SESIC-PROMEP(México) under F-PROMEP-39 grant. We have enjoyed fruitful and enlighten discussions with Alfonso Mondragón B.

## REFERENCES

1. S. Pakvasa and H. Sugawara, *Phys. Lett.* **B**73, 61 (1978); S. Pakvasa and H. Sugawara, *Phys. Lett.* **B**82, 105 (1979); J. Kubo, A. Mondragón, M. Mondragón, E. Rodríguez-Jáuregui, O. Félix-Beltrán and E. Peinado, J. *Phys.: Conf. Ser.* 18, 380-384 (2005); J. Kubo, H. Okada and F. Sakamaki, *Phys. Rev.* **D**70, 036007 (2004) Y. Koide, *Phys. Rev.* **D** 73, 057901 (2006)
2. P. M. Ferreira, R. Santos and A. Barroso, *Phys. Lett.* **B**603, 219 (2004) [Erratum-ibid. **B** 629, 114 (2005)] A. Barroso, P. M. Ferreira and R. Santos, arXiv:hep-ph/0507329; A. Barroso, P. M. Ferreira and R. Santos, *PoS HEP2005*, 337 (2006).

# Upper Bound on the Hadronic Light-by-Light Contribution to the Muon $g-2$

Jens Erler and Genaro Toledo Sánchez

*Departamento de Física Teórica, Instituto de Física, Universidad Nacional Autónoma de México, México D.F. 04510, México*

**Abstract.** There are indications that hadronic loops in some electroweak observables are almost saturated by parton level effects. Taking this as the hypothesis for this work, we propose a genuine parton level estimate of the hadronic light-by-light contribution to the anomalous magnetic moment of the muon, $a_\mu^{\rm LBL}(\text{had})$. For infinitely heavy quarks our treatment is exact, while for asymptotically small quark masses $a_\mu^{\rm LBL}(\text{had})$ is overestimated. Interpolating, this suggests to quote an upper bound. We obtain $a_\mu^{\rm LBL}(\text{had}) < 1.59 \times 10^{-9}$ (95% CL).

**Keywords:** Muon, magnetic moment.
**PACS:** 13.40.Em, 14.60.Ef, 12.20.Ds, 12.38.Bx.

The E–821 Collaboration at BNL [1] has measured the anomalous magnetic moment of the muon, $a_\mu \equiv (g_\mu - 2)/2$, with an uncertainty of $\Delta a_\mu = \pm 0.63 \times 10^{-9}$. This would provide sensitivity to new physics scales, $\Lambda$, as large as $\Lambda \sim m_\mu/\sqrt{\Delta a_\mu} =$ 4.2 TeV, where $m_\mu$ is the mass of the muon [2]. Unfortunately, the interpretation of $a_\mu$ is compromised by large theoretical uncertainties introduced by hadronic effects diluting the new physics sensitivity. These mainly arise from two-loop vacuum polarization (VP) effects, $a_\mu^{\rm VP}(2,\text{had})$, and from the three-loop contribution of light-by-light type, $a_\mu^{\rm LBL}(\text{had})$. The calculations of $a_\mu^{\rm LBL}(\text{had})$ based on chiral perturbation theory [3, 4] are the only ones to date solidly based on a systematic expansion. However, the estimated uncertainty, $\gtrsim \pm 10^{-9}$, is significantly larger [4] than the measurement error. There is a variety of model estimates, all supplementing the dominant $\pi^0$-exchange contribution with other resonance exchanges and $\pi^\pm$-loops. Current analyzes [5, 6] agree reasonably well on the magnitude (where residual differences are largely understood) and the sign of $a_\mu^{\rm LBL}(\text{had})$, but the error estimates remain rough guesses.

In this work we offer a complementary way to estimate $a_\mu^{\rm LBL}(\text{had})$, with no need to commit to the dominance of any particular type of contribution. It is a naïve parton level estimate which is solid in the heavy quark limit where it matches onto perturbative QCD, but overestimates the light-by-light contribution in the chiral limit. Therefore, our approach naturally implies an *upper bound* for $a_\mu^{\rm LBL}(\text{had})$, which is very useful in view of the $a_\mu$ measurement lying above the Standard Model prediction. It can also be applied to $a_\mu^{\rm VP}(\text{had})$, which serves as a reference case and allows for a *transparent* error estimate.

Attempting to obtain $a_\mu^{\rm LBL}(\text{had})$ directly at the parton level seems hopeless at first since perturbative QCD cannot be applied except for the heavy $c$ and $b$ quarks. Moreover, in the chiral limit, in which the bulk of the effect arises from light, charged, and essentially point-like pseudo-Goldstone bosons, one has $a_\mu^{\rm LBL}(\text{had}) < 0$ from scalar

CP917, *Particles and Fields,* edited by H. Castilla Valdez, J. C. D'Olivo, and M. A. Perez

QED [7], while leptons and heavy quarks contribute positively [8]. However, as shown in Ref. [6], the $\pi^{\pm}$-loops receive large chiral symmetry breaking corrections, rendering the net $\pi^{\pm}$-contribution almost negligible. Thus, at least as far as $a_\mu^{\rm LBL}$(had) is concerned, we are dealing with a kinematic regime that conceivably admits an alternative description in terms of quark in place of chiral degrees of freedom. Indeed, an estimate [6] of the typical momenta, $\mu$, in the $\pi^{\pm}$-loops yields $\mu \sim 4.25\, m_\pi \approx 0.6$ GeV, *i.e.* large enough to feel individual partons even within pions.

The main problem with any parton approach is to find a consistent set of quark mass definitions because the strong coupling constant $\hat{\alpha}_s$ and consequently the anomalous quark dimensions formally diverge. In a similar context, however, it was possible to proceed phenomenologically. Namely, the hadronic contribution to the renormalization group evolution (RGE) of the QED coupling (the running), $\hat{\alpha}(\mu)$, can be mimicked [9] by effective quark masses. These have been defined more precisely in Ref. [10] as *threshold quark masses*, $\bar{m}_q$, and correspond to the scale at which the theory is changed to include or exclude the quark $q$ and where the one-loop QCD $\beta$ function coefficient changes correspondingly. Since this is *not* a perturbative description, higher order QCD effects are assumed to be absorbed in the threshold masses, leading to [10],

$$\bar{m}_u = \bar{m}_d = 176 \pm 9 \text{ MeV}, \tag{1}$$

$$\bar{m}_s = 305^{-65}_{+82} \text{ MeV}, \tag{2}$$

$$\bar{m}_c = 1.18 \pm 0.04 \text{ GeV}, \tag{3}$$

$$\bar{m}_b = 4.00 \pm 0.03 \text{ GeV}, \tag{4}$$

where isospin symmetry, $\hat{m}_u(\hat{m}_u) = \hat{m}_d(\hat{m}_d)$, was imposed. The uncertainties in the first (last) two lines have a correlation of $-100\%$ (+29%).

In Ref. [10] the threshold masses (1–4) were used for the RGE of the weak mixing angle, $\sin^2\hat{\theta}(\mu)$, which is rigorously justified as the same vector current correlator convoluted by the same weight function enters in both, $\hat{\alpha}(\mu)$ and $\sin^2\hat{\theta}(\mu)$. What we are proposing here, is that $a_\mu^{\rm LBL}$(had) can be *modeled* by treating quarks like massive leptons and using the values (1–4). Notice, that these values are systematically lower than typical constituent masses, which is reasonable since we do not add an independent contribution from pion degrees of freedom.

Although our treatment clearly improves with growing $\bar{m}_q$ and becomes exact for $\bar{m}_q \to \infty$, it is important to establish *quantitative* tests and to apply our approach to physical quantities for which the answer is known with sufficient precision. For this we note that $a_\mu^{\rm VP}(2,\text{had})$ is described by the same vector current correlator as the RGE of $\hat{\alpha}(\mu)$ but is convoluted with a different weight function, $K^{(1)}(s)$ [11], so that just as in the case of $a_\mu^{\rm LBL}$(had) it is *a priori* not obvious that our approach and quark mass values will generate the correct answer,

$$a_\mu^{\rm VP}(2,\text{had}) \times 10^9 = 70.30 \pm 0.43 \pm 0.88 = 70.3 \pm 1.0 \tag{5}$$

To arrive at Eq. (5) we averaged the results [12] based on $e^+e^-$ annihilation data and $\tau$ decay data. The second uncertainty accounts for the discrepancy between the two data sets. We compare this result with the analytical expression [13, 14],

$$a_\mu^{\rm VP}(2,{\rm had}) = \frac{\alpha^2}{\pi^2} N_C \sum_q Q_q^2 \sum_{n=0}^{\infty} \left(\frac{m_\mu}{\bar{m}_q}\right)^{2n+2} \times \tag{6}$$

$$\left[\frac{n \ln(m_\mu^2/\bar{m}_q^2)}{(n+3)(2n+3)(2n+5)} - \frac{8n^3+28n^2-45}{[(n+3)(2n+3)(2n+5)]^2}\right] = 65.1^{-2.6}_{+4.2} \mp 0.1 \times 10^{-9}.$$

The first and second uncertainty are induced, respectively, from Eqs. (1–2) and Eqs. (3–4). Effects due to $\bar{m}_d \neq \bar{m}_u$ cancel to first order in $a_\mu^{\rm VP}(2,{\rm had})$ (but not in $a_\mu^{\rm LBL}({\rm had})$ below) and can be neglected. Notice, that the estimate (6) is in reasonable agreement with Eq. (5) and that our central value is *below* it. This is consistent with the chiral limit, $m_\pi \to 0$ with $m_\mu/m_\pi$ fixed, underestimating $a_\mu^{\rm VP}(2,{\rm had})$. Indeed, in this limit,

$$m_\pi \to 0 \Longrightarrow \frac{\bar{m}_q}{\mu} \to \left(\frac{m_\pi}{\mu}\right)^{\sqrt{\frac{1}{4N_C(Q_u^2+Q_d^2)}}} \approx \left(\frac{m_\pi}{\mu}\right)^{0.39},$$

(where $\mu$ is a fixed reference scale) so that $m_\mu/\bar{m}_q \to 0$. Combined with $K(s) > 0$ this implies a systematic *underestimate*. The fact that the estimate (6) reproduces the experimental result within about 10% provides phenomenological support for our approach and its use of quark degrees of freedom.

The success of the test in the previous paragraph may be coincidental. Therefore, we performed three similar tests on various three-loop VP effects, $a_\mu^{\rm VP}(3{\rm i},{\rm had})$, each with a different kernel function, $K^{(2i)}(s)$ [15, 16]. $i = a$ contains an additional photonic correction or muon-loop relative to $a_\mu^{\rm VP}(2,{\rm had})$, while $i = b$ ($c$) corresponds to an additional electron (hadron) loop insertion. Our method (in parentheses) reproduces the central values of Ref. [17],

$$\begin{aligned} a_\mu^{\rm VP}(3{\rm a},{\rm had}) &= \; -2.11 \times 10^{-9} \quad (-1.82 \times 10^{-9}), \\ a_\mu^{\rm VP}(3{\rm b},{\rm had}) &= \; +1.07 \times 10^{-9} \quad (+0.99 \times 10^{-9}), \\ a_\mu^{\rm VP}(3{\rm c},{\rm had}) &= \; +2.7 \times 10^{-11} \quad (+2.8 \times 10^{-11}), \end{aligned} \tag{7}$$

within 14%, 7%, and 4%, respectively. Note, that these represent genuine "hit or miss" tests involving no adjustable parameters, and all four passed at the 10% level.

Now we apply our approach to $a_\mu^{\rm LBL}({\rm had})$, using the formula for heavy leptons [8],

$$a_\mu^{\rm LBL}({\rm had}) = \frac{\alpha^3}{\pi^3} N_C \sum_q Q_q^4 \sum_{n=1}^{\infty} \frac{m_\mu^{2n}}{\bar{m}_q^{2n}} \sum_{k=0}^{2} C_{n,k} \ln^k \frac{m_\mu^2}{\bar{m}_q^2},$$

where $C_{1,0} = 3/2\zeta_3 - 19/16$, $C_{1,1} = C_{1,2} = 0$, and where the other $C_{n,k}$ up to $n = 5$ can be found in Ref. [7]. The central values in Eqs. (1–4) then imply $a_\mu^{\rm LBL}({\rm had}) = 1.36 \pm 0.13 \times 10^{-9}$, where the uncertainty is the model error, determined by the 9.5% difference between the upper end of the range (5) from the central value in Eq. (6). Similarly, the upper and lower error in Eqs. (1–2) and (6) correspond to $a_\mu^{\rm LBL}({\rm had}) = 1.27 \pm 0.18 \times 10^{-9}$ and $a_\mu^{\rm LBL}({\rm had}) = 1.47 \pm 0.04 \times 10^{-9}$, *i.e.* model errors of 14.1% and

2.9%, respectively. As expected, $b$ quark effects are negligible, but the contribution from $c$ quarks (usually ignored) turns out to be $0.04 \times 10^{-9}$. To account for isospin breaking we shift $\bar{m}_d$ by +2 MeV and $\bar{m}_u$ by $-1$ MeV, increasing all results by $0.01 \times 10^{-9}$ to

$$a_\mu^{\mathrm{LBL}}(\mathrm{had}) = 1.37^{-0.27}_{+0.15} \times 10^{-9}, \tag{8}$$

where the error covers the three ranges above and is constructed to include the intrinsic model uncertainty. The result, $1.36 \pm 0.25 \times 10^{-9}$, of Ref. [6] is the first to take short-distance QCD constraints explicitly into account. This is a feature of our approach, as well, so it is gratifying that the estimate (8) turns out to agree perfectly. Our central value is *higher* than previous ones [5], which is expected given that in the chiral limit, $m_\pi \to 0$ with $m_\mu/m_\pi$ fixed we *overestimate* $a_\mu^{\mathrm{LBL}}(\mathrm{had})$ since $\pi^\pm$-loops contribute negatively.

Now we repeat the above analysis with the model errors multiplied by 1.645 to shift from uncertainties estimated in a "$1\sigma$ spirit" to 90% ranges, and find $a_\mu^{\mathrm{LBL}}(\mathrm{had}) = 1.37 \pm 0.21 \times 10^{-9}$, $a_\mu^{\mathrm{LBL}}(\mathrm{had}) = 1.28 \pm 0.30 \times 10^{-9}$, and $a_\mu^{\mathrm{LBL}}(\mathrm{had}) = 1.48 \pm 0.07 \times 10^{-9}$, respectively. The upper errors have converged, and since as mentioned our approach tends to overestimate $a_\mu^{\mathrm{LBL}}(\mathrm{had})$, we quote as our final result the 95% CL upper bound,

$$a_\mu^{\mathrm{LBL}}(\mathrm{had}) < 1.59 \times 10^{-9}. \tag{9}$$

For more details and further references, see Ref. [18].

## ACKNOWLEDGMENTS

This work was supported by contracts CONACyT (México) 42026–F and DGAPA–UNAM PAPIIT IN112902.

## REFERENCES

1. BNL–E–821 Collaboration: G.W. Bennett *et al.*, *Phys. Rev. Lett.* **92**, 161802 (2004).
2. For a review, see A. Czarnecki and W.J. Marciano, *Phys. Rev.* **D64**, 013014 (2001).
3. M. Knecht, A. Nyffeler, M. Perrottet and E. de Rafael, *Phys. Rev. Lett.* **88**, 071802 (2002).
4. M. Ramsey-Musolf and M.B. Wise, *Phys. Rev. Lett.* **89**, 041601 (2002).
5. M. Knecht and A. Nyffeler, *Phys. Rev.* **D65**, 073034 (2002); J. Bijnens, E. Pallante and J. Prades, *Nucl. Phys.* **B626**, 410 (2002); M. Hayakawa and T. Kinoshita, *Phys. Rev.* **D57**, 465 (1998).
6. K. Melnikov and A. Vainshtein, *Phys. Rev.* **D70**, 113006 (2004).
7. J.H. Kühn, A.I. Onishchenko, A.A. Pivovarov and O.L. Veretin, *Phys. Rev.* **D68**, 033018 (2003).
8. S. Laporta and E. Remiddi, *Phys. Lett.* **301B**, 440 (1993).
9. W.J. Marciano and A. Sirlin, *Phys. Rev.* **D22**, 2695 (1980).
10. J. Erler and M.J. Ramsey-Musolf, *Phys. Rev.* **D72**, 073003 (2005).
11. S.J. Brodsky and E. De Rafael, *Phys. Rev.* **168**, 1620 (1968).
12. M. Davier, A. Höcker and Z. Zhang, hep-ph/0507078.
13. H.H. Elend, *Phys. Lett.* **20**, 682 (1966).
14. M.A. Samuel and G. Li, *Phys. Rev.* **D44**, 3935 (1991).
15. R. Barbieri and E. Remiddi, *Nucl. Phys.* **B90**, 233 (1975).
16. J. Calmet, S. Narison, M. Perrottet and E. de Rafael, *Phys. Lett.* **61B**, 283 (1976).
17. B. Krause, *Phys. Lett.* **390B**, 392 (1997).
18. J. Erler and G. Toledo Sanchez, *Phys. Rev. Lett.* **97**, 161801 (2006).

# B meson decays into charmless pseudoscalar scalar mesons

D. Delepine*, J. L. Lucio M.*, Carlos A. Ramírez† and J. A. Mendoza S.†

*Instituto de Física, Universidad de Guanajuato
Loma del Bosque # 103, Lomas del Campestre,
37150 León, Guanajuato; México
†Depto. de Física, Universidad Industrial de Santander,
A.A. 678, Bucaramanga, Colombia
Depto. de Fsica-matemáticas, Universidad de Pamplona
Pamplona, Norte de Santander, Colombia.

**Abstract.** The nonleptonic weak decays of meson $B$ into a scalar and pseudoscalar meson are studied. The scalar mesons under consideration are $\sigma$ ( or $f_0(600)$), $f_0(980)$, $a_0(980)$ and $K_0^*(1430)$. We calculate the Branching ratios in the Naive Factorization approximation. Scalars are assumed to be $q\bar{q}$ bounded sates, but an estimation can be obtained in the case they are four bounded states.

## 1. INTRODUCTION

There are several reasons to study the $B \to PS$ (P and S are pseudoscalar and scalar mesons) decays, let us mention several ones. Scalars have been an issue for long tine [1] and $B \to PS$ is another window to study their properties [2]. CP asymmetries of these decays provide another way to mesure the CKM angles $\beta$ and maybe $\alpha$ [3]. Additionally $B \to PS$ decays have to be taken into account in order to analyze the $B \to 3P$ decays in the different channels [4] and perhaps these decays can be used to study mew physics [5].

From the experimental point of view several results have been already obtained [6] and future experiments like LHCB [7] are expected to improve them. It is remarkable how the $f_0$ properties measured in the $B \to f_0 K$ decay are in total agreement width those obtained at lower energies [6]. Similarly the $\beta$ angle also agrees with more accurate measurements like $B \to J/\Psi K$ decays. Recently, BABAR has searched for the decays $B \to a_0\pi$ and $B \to a_0 K$ that may allow us to discriminate between 2 and 4 quarks scalars [2], unfortunately the experiment is still not sensible to these decays.

Given the actual experimental an theoretical accuracies we work in the Naive Factorization Approximation (NFA) [8, 9]. This approximation has given a reasonable fit for $B \to PP$ decays, except for these decays dominated by the so called 'color suppressed 'diagrams and for the strong CP phases. Naturally when our knowledge of $B \to PS$ decays improves one has to go use more accurate formalisms

CP917, *Particles and Fields*, edited by H. Castilla Valdez, J. C. D'Olivo, and M. A. Perez

like QCDF, etc [9].

## 2. NAIVE FACTORIZATION

The effective Hamiltonian for this case is [8, 9, 10] :

$$\mathcal{H}_{\text{eff}} = \frac{G_F}{\sqrt{2}}\left[V_{ub}V_{uq}^*(C_1O_1^u+C_2O_2^u)-V_{tb}V_{tq}^*\left(\sum_{i=3}^{10}C_iO_i+C_gO_g+C_\gamma O_\gamma\right)\right]+\text{h.c.}$$

where $\lambda_{q'q}=V_{q'b}V_{q'q}^*, q=d,s,q'=u,c,t, V_{ij}$ are the Cabibbo-Kobayashi-Maskawa (CKM) matrix elements. Whit $O_i$ is the four fermion operators:
At tree level

$$\mathcal{O}_1=(\bar{q}u)_{V-A}(\bar{u}b)_{V-A}; \qquad \mathcal{O}_2=(\bar{u}_\alpha b_\beta)_{V-A}(\bar{q}_\beta u_\alpha)_{V-A}$$

The QCD penguin contributions

$$\mathcal{O}_3 = (\bar{q}b)_{V-A}\sum_{q'}(\bar{q}'q')_{V-A}; \quad \mathcal{O}_4=(\bar{q}_\alpha b_\beta)_{V-A}\sum_{q'}(\bar{q}_{\beta}{}'q_{\alpha}{}')_{V-A}$$

and similarly the operators $\mathcal{O}_5,\mathcal{O}_6$ [8]. The electroweak penguin contributions

$$\mathcal{O}_7 = \frac{3}{2}(\bar{q}b)_{V-A}\sum_{q'}e_{q'}(\bar{q}'q')_{V+A} \quad \mathcal{O}_8=\frac{3}{2}(\bar{q}_\alpha b_\beta)_{V-A}\sum_{q'}e_{q'}(\bar{q}_{\beta}{}'q_{\alpha}{}')_{V+A}$$

similarly the operators $\mathcal{O}_9,\mathcal{O}_{10}$ [8].The $C_i$ is the Coefficient of Wilson [8], and $a_{2i-1}=C_{2i-1}+C_{2i}/N$, $a_{2i}=C_{2i}+C_{2i-1}/N$. The numerical values of the $a_i$ coefficients are taken of [8, 10].
In the NFA one can reduce the matrix elements to decay constants and form factors as (e.g. takin the operator $\mathcal{O}_1$ ):

$$\begin{aligned}\langle a_0^+\pi^-|\mathcal{O}_1|\bar{B}^0\rangle &= \langle a_0^+\pi^-|(\bar{d}u)_L(\bar{u}b)_L|\bar{B}^0\rangle \\ &= \langle\pi^-|(\bar{d}u)_L|0\rangle\langle a_0^+|(\bar{u}b)_L|\bar{B}^0\rangle+\frac{1}{N}\langle\pi^-a_0^+|(\bar{u}u)_L|0\rangle\langle 0|(\bar{d}b)_L|\bar{B}^0\rangle \\ &= X_{\bar{B}^0a_0^+}^{\pi^-}+\frac{1}{N}X_{a_0^+\pi^-}^{\bar{B}^0}\end{aligned}$$

Where the Fiertz rotations, $(\bar{\psi}_1\psi_2)_L(\bar{\psi}_3\psi_4)_L=(\bar{\psi}_1\psi_4)_L(\bar{\psi}_3\psi_2)_L$; have been used. Similarly $2(T_i)_{\alpha\beta}(T_i)_{\gamma\delta}=\delta_{\alpha\delta}\delta_{\beta\gamma}-(1/N)\delta_{\alpha\gamma}\delta_{\beta\delta}$. for the SU(N) matrices. The X are obtained in terms of decay constants and form factors. They can be found in ref. [2]. Form factors are defined as

$$\begin{aligned}\langle M_2(p_2)|L_\mu|M_1(p_1)\rangle &= f_+(q^2)(p_1+p_2)_\mu+f_-(q^2)q_\mu \\ &= \left(p_1+p_2-\frac{m_1^2-m_2^2}{q^2}q\right)_\mu F_+^{M_1M_2}+\frac{m_1^2-m_2^2}{q^2}q_\mu F_0^{M_1M_2}(q^2),\end{aligned}$$

width $q=(p_1-p_2)$ and an additional prefactor of $-i$ has to be added to the form factors. The decay constants are defined as

$$\langle 0|A_\mu|P(q)\rangle = if_P q_\mu = -\langle P(q)|A_\mu|0\rangle; \qquad \langle 0|V_\mu|S(q)\rangle = f_s q_\mu$$

For the neutral scalars like $\sigma, f_0$ and $a_0^0$, the decay constant must be zero due to charge conjugation invariance or conservation of vector current $f_\sigma = f_{f_0} = f_{a_0^0} = 0$. For other scalars the decay constant of the scalar meson are taken to be $f_{a_0^\pm} = 1.1$ MeV, $f_{K_0^*} = 42$ Mev [11]. The corresponding form factors are taken from literature as well as for the decay constants [2, 6, 8]. In the case of four quarks scalar mesons the form factors are assumed to be equal to the $q\bar{q}$ case, except for the annihilation ones that are assumed to be zero [2].

## 3. $B \to PS$ AMPLITUDES

Next we present several amplitudes for several decays.

$$\begin{aligned}A_{\bar{B}^0\to\pi^-a_0^+} &= \lambda_{ud}(a_1X^{\pi^-}_{\bar{B}^0a_0^+}+a_2X^{\bar{B}^0}_{\pi^-a_0^+})-\lambda_{td}\left[\left(a_4+a_{10}-\frac{(a_6+a_8)m_{\pi^2}}{\hat{m}(m_b+m_u)}\right)X^{\pi^-}_{\bar{B}^0a_0^+}\right. \\ &\left.+\left(2(a_3-a_5)+a_4+\frac{a_9-a_7-a_{10}}{2}-\frac{(a_6-a_8/2)m_B^2}{m_u(m_b+m_d)}\right)X^{\bar{B}^0}_{\pi^-a_0^+}\right]\end{aligned}$$

$$\begin{aligned}A_{B^-\to\pi^0a_0^-} &= \lambda_{ud}\left[a_1\left(X^{a_0^-}_{B^-\pi^0}+X^{B^-}_{a_0^-\pi^0}\right)+a_2X^{\pi^0_u}_{B^-a_0^-}\right]-\lambda_{td}\left[(a_4+a_{10})X^{a_0^-}_{B^-\pi^0}\right. \\ &-2(a_6+a_8)\widetilde{X}^{a_0^-}_{B^-\pi^0}-\left(a_4-\frac{3}{2}(a_9-a_7)-\frac{1}{2}a_{10}+\frac{(a_6+a_8)m_\pi^2}{m_u(m_b+m_d)}\right)X^{\pi^0_u}_{B^-a_0^-} \\ &\left.+\left(a_4+a_{10}+\frac{(a_6+a_8)m_B^2}{\hat{m}(m_b+m_u)}\right)X^{B^-}_{a_0^-\pi^0}\right]\end{aligned}$$

$$\begin{aligned}A_{B^-\to\pi^-\sigma} &= \lambda_{ud}a_1\left(X^{\pi^-}_{B^-\sigma}+X^{B^-}_{\sigma\pi^-}\right)-\lambda_{td}\left[\left(a_4+a_{10}-\frac{(a_6+a_8)m_\pi^2}{\hat{m}(m_b+m_u)}\right)X^{\pi^-}_{B^-\sigma}\right. \\ &\left.+\left(a_4+a_{10}-\frac{(a_6+a_8)m_B^2}{\hat{m}(m_b+m_u)}\right)X^{B^-}_{\sigma\pi^-}+(a_8-2a_6)\widetilde{X}^{\sigma_d}_{B^-\pi^-}\right]\end{aligned}$$

$$A_{B^-\to\pi^- f_0} = \lambda_{ud}a_1\left(X^{\pi^-}_{B^- f_0}+X^{B^-}_{f_0\pi^-}\right)-\lambda_{td}\left[\left(a_4+a_{10}-\frac{(a_6+a_8)m_\pi^2}{\hat{m}(m_b+m_u)}\right)X^{\pi^-}_{B^- f_0}\right.$$
$$\left.+\left(a_4+a_{10}-\frac{(a_6+a_8)m_B^2}{\hat{m}(m_b+m_u)}\right)X^{B^-}_{f_0\pi^-}+(a_8-2a_6)\widetilde{X}^{f_0}_{B^-\pi^-}\right]$$

$$A_{(\bar{B}^0\to f_0\bar{K}^0)} \simeq -\lambda_{ts}\left[\left(a_4-\frac{a_{10}}{2}-\frac{2(a_6-a_8)m_B^2}{(m_b+m_d)(m_s+m_d)}\right)X^{\bar{B}^0}_{\bar{K}^0 f_0}\right.$$
$$\left.+\left(a_4-\frac{a_{10}}{2}-\frac{2(a_6-a_8)m_K^2}{(m_s+m_d)(m_b+m_d)}\right)X^{K^0}_{B^0 f_0}-(2a_6-a_8-)\widetilde{X}^{f_0}_{B^0K^0}\right]$$

$$A_{(B^-\to\bar{K}^0_{0*}\pi^-)} \simeq \lambda_{us}a_1X^{B^-}_{\bar{K}^0_{0*}\pi^-}-\lambda_{ts}\left[\left(a_4-\frac{a_{10}}{2}-\frac{(2a_6-a_8)m^2_{K_{0*}}}{(m_s-m_d)(m_b-m_d)}\right)\cdot\right.$$
$$\left.X^{\bar{K}^0_{0*}}_{B^-\pi^-}+\left(a_4+a_{10}-\frac{(2a_6+a_8)m_B^2}{(m_b+m_u)(m_s+m_u)}\right)X^{B^-}_{\bar{K}^0_{0*}\pi^-}\right]$$

## 4. NUMERICAL RESULTS

Table 1 summarizes the NFA predictions, for the $2q$ and $4q$ scalars compared with the experimental results available [6].

## 5. CONCLUSIONS

1. There is an overall agreement with the experimental information, except for the case $\bar{B}^0\to\pi^-K_0^*(1430)^+$ where the experimental situation is not very clear. However our NFA result favors the small branching ratio.
2. Given that several branchings are sensitive to annihilation contribution, therefore to the $2q$ or $4q$ scalars nature one way in principe use the experimental results to distinguish between the two models. Unfortunately the experimental resolution is too low at the moment [2], but this is expected to improve in the future.

## REFERENCES


1. F. E. Close and N. A. Tornqvist, J. Phys. G **28**. R249 (2002)[arXiv:hep-ph/0204205]
   C. McNeile and C. Michel [UKQCD Collaboration], Phys. Rev **D74**, 014508 (2006) [arXiv:hep-lat/0604009]
2. D. Delepine, J. L. Lucio M. and C. A. Ramirez, Eur. Phys. J. C **45**, 693 (2006) [arXiv:hep-ph/0501022].
   H. Y. Cheng, C. K. Chua, and K. C. Yang, Phys Rev. D **73**, 014017 (2006).
   BABAR Collaboration, [arXiv:hep-ex/0607064].

**TABLE 1.** Branching ratios of the decays $B \to PS$. (in units of $10^{-6}$)

| Decaimiento | BELLE | BABAR | PROM | $B_{exp}$ | $NFA_{(q\bar{q})}$ | $NFA_{(4q)}$ |
|---|---|---|---|---|---|---|
| $\pi^- a_0^+$ | | $2.8^{+1.5}_{-1.3} \pm 0.7$ | | | 1.2 | 1.2 |
| $\pi^+ a_0^-$ | | | | | 0.04 | 0.2 |
| $\pi^0 a_0^-$ | | $2.6^{+2.0}_{-1.7} \pm 1$ | | | 0.01 | 0.07 |
| $\pi^- a_0^0$ | | $2.6^{+2.0}_{-1.7} \pm 1$ | | | 0.6 | 0.6 |
| $\pi^- \sigma(\pi^+\pi^-)$ | | $< 4.1$ | | | 0.7 | 0.7 |
| $\pi^- f_0(\pi^+\pi^-)$ | | $1.2 \pm 0.6 = 0.1 \pm 0.4$ | | | 0.2 | 0.2 |
| $K^- f_0(\pi^+\pi^-)$ | $8.78 \pm 0.82 \pm 0.65^{+0.55}_{-1.64}$ | $9.47 \pm 0.97 \pm 0.46^{+0.42}_{-0.75}$ | 9.1(6)* | 17(3) | 15 | 16.4 |
| $K^0 f_0(\pi^+\pi^-)$ | $7.60 \pm 1.66 \pm 0.59^{+0.48}_{-0.67}$ | $5.5 \pm 0.7 \pm 0.6 \pm 0.3$ | 5.8(6)* | 11(2) | 15 | 16 |
| $\pi^+ K_0^*(1430)^0 (K^+\pi^-)$ | $32.0 \pm 1.0 \pm 2.4^{+1.1}_{-1.9}$ | $25.1 \pm 2 \pm 2.9^{+9.4}_{-0.5}$ | 30.6(9)* | 49(6) | 11 | 13 |
| $\pi^- K_0^*(1430)^+ (K^+\pi^0)$ | $5.1 \pm 1.5^{+0.6}_{-0.7}$ | $11 \pm 1.5 \pm 0.5 \pm 0.4$ | 8(1) | 8(1) | 11 | 13 |
| $\pi^0 K_0^*(1430)^0$ | $6.1^{+1.6+0.5}_{-1.5-0.6}$ | $7.9 \pm 1.5 \pm 2.7$ | 7(1) | 7(1) | 6.7 | 6 |


3. G. Sciolla [BABAR Collaboration], Nucl. Phys. Proc. Suppl. **156**, 16 (2006) [arXiv:hep-ex/0509022]
S. Laplace and V. Shelkov, Eur. Phys. J. C **22**, 431 (2001) [arXiv:hep-ph/0105252]
M. Suzuki, Phys. Rev. **D 65**, 097501 (2002) [arXiv:hep-ph/0202222]
A. S. Dighe and C. S. Kim, Phys Rev. **D 62** 111302 (2000) [arXiv:hep-ph/0004244]
4. B. El-Bennich, A. Furman, R. Kaminski, L. Lesniak and B. Loiseau, [arXiv:hep-ph/0608205]
5. A. K. Giri, B. Mawlong and R. Mohanta, [arXiv:hep-ph/0608088]

6. E. Barbeiro *et al.* [Heavy Flavor Averaging Group (HFAG)], [arXiv:hep-ex/0603003].
A. Garmash [Belle Collaboration], AIP Conf. Proc. **814**, 680 (2006) [arXiv:hep-ex/0510059].
B. Aubert *et al* [BABAR Collaboration], Phys. Rev. D **73**. 031101 (2006) [arXiv:hep-ex/0508013].
W. M. Yao *et al.* [Particle Data Group], J. Phys. G. **33** (2006) 1. pdgl.lbl.gov.
7. P. Ball *et al.* [arXiv:hep-ph/0003228]. lhcb.web.cern.ch/lhcb/.
A. G. Akeroyd *et al.* [arXiv:hep-ex/0406071].
8. A. Ali, G. Kramer and C. D. Lu, Phys. Rev. D **58**, 094009 (1998) [arXiv:hep-ph/9804363].
Y. H. Chen, H. Y. Cheng, B. Tseng and K. C. Yang, Phys. Rev. D. **60**, 094014 (1999). [arXiv:hep-ph/9903453].
G. Buchala, A. J. Buras and M. E. Lautenbacher, Rev. Mod. Phys. **68**, 1125 (1996). [arXiv:hep-ph/9512380].
9. M. Beneke and M. Neubert, Nucl. Phys. B **675**, 333 (2003). [arXiv:hep-ph/308039].
M. Beneke, G. Buchala, M. Neubert and C. T. Sachrajda, Phys. Rev. Lett. **83**, 1914 (1999). [arXiv:hep-ph/9905312].
10. Y. Y. Keum and A. I. Sanda, Phys. Rev. D **67** 054009 (2003). [arXiv:hep-ph/0209014].
Y. Y. Keum and H. N. Li, Phys. Rev. D. **63**, 074006 (2001) [arXiv:hep-ph/0006001].
11. K. Maltman, Phys. Lett. **B462**, 14 (1999).

# Family Dependence in 331 Models

R. Martinez and Fredy Ochoa

*Depto de Física, Universidad Nacional// Bogotá, Colombia*

**Abstract.** Using experimental results at the Z-pole and atomic parity violation, we perform a $\chi^2$ fit at 95% CL to obtain family-dependent bounds to $Z'$ mass and Z-Z' mixing angle in the framework of the main versions of 331 models. The allowed regions depend on the assignment of the quark families into different representations that cancel anomalies.

**Keywords:** Electroweak theories, Unified theories
**PACS:** 12.60.Cn, 12.15.Ff

A very common alternative to solve some of the problems of the standard model (SM) consists to enlarging the gauge symmetry. For example, a very interesting alternative to explain the origin of the fermion families comes from the cancellation of chiral anomalies. In particular, models with gauge symmetry $SU(3)_c \otimes SU(3)_L \otimes U(1)_X$, also called 331 models, arise as a possible solution to this puzzle, where three families are required in order to cancel chiral anomalies [1, 2, 3]. In general, the electric charge is defined as a linear combination of the diagonal generators of the group

$$Q = T_3 + \beta T_8 + XI, \tag{1}$$

where $\beta$ allow classify the different 331 models. The two main versions corresponds to $\beta = -\sqrt{3}$ [1] and $\beta = -\frac{1}{\sqrt{3}}$ [2]. On the other hand, the group structure of these models leads, along with the SM-neutral boson $Z$, to the prediction of an additional current associated with a new massive and neutral gauge bosons, usually called $Z'$. Unlike the $Z-$boson whose couplings are family independent and the the weak interaction at low energy are of universal character, the couplings of $Z'$ are different for the three families due to the $U(1)_X$ values to each of them. In the quark sector, each 331-family can be assigned in 3 different ways. Therefore, in a phenomenological analysis, the allowed region associated with the $Z-Z'$ mixing angle and the physical mass $M_{Z'}$ of $Z'$ will depend on the family assignment. In this work we report a phenomenological study through a $\chi^2$ fit at the $Z-$pole to find the allowed region for the $Z-Z'$ mixing angle and the mass of the $Z'$ boson at 95% CL for 3 different assignments of the quark families. We take into account the two main versions of the 331 models.

We recognize 3 different possibilities to assign the physical quarks in each family representation as shown in the table below. At low energy, the three models from the table are equivalent and there is not any phenomenological feature that allows us to detect differences between them. However, through the couplings of the $Z'$ boson and the introduction of a mixing angle between $Z$ and $Z'$ boson, it is possible to recognize differences among the three models. In the neutral gauge spectrum associated with the

CP917, *Particles and Fields,* edited by H. Castilla Valdez, J. C. D'Olivo, and M. A. Perez

**TABLE 1.** Different family assigments

| Representation $A$ | Representation $B$ | Representation $C$ |
|---|---|---|
| $q_{mL} = \begin{pmatrix} d,s \\ -u,-c \\ J_1,J_2 \end{pmatrix}_L$ | $q_{mL} = \begin{pmatrix} d,b \\ -u,-t \\ J_1,J_3 \end{pmatrix}_L$ | $q_{mL} = \begin{pmatrix} s,b \\ -c,-t \\ J_2,J_3 \end{pmatrix}_L$ |
| $q_{3L} = \begin{pmatrix} t \\ b \\ J_3 \end{pmatrix}_L$ | $q_{3L} = \begin{pmatrix} c \\ s \\ J_2 \end{pmatrix}_L$ | $q_{3L} = \begin{pmatrix} u \\ d \\ J_1 \end{pmatrix}_L$ |

group $SU(3)_L \otimes U(1)_X$, we obtain the following mass eigenstates

$$Z_{1\mu} = Z_\mu C_\theta + Z'_\mu S_\theta; \quad Z_{2\mu} = -Z_\mu S_\theta + Z'_\mu C_\theta, \tag{2}$$

where a small mixing angle $\theta$ between the neutral currents $Z_\mu$ and $Z'_\mu$ appears, with

$$\begin{aligned} Z_\mu &= C_W W^3_\mu - S_W \left( \beta T_W W^8_\mu + \sqrt{1-\beta^2 T^2_W} B_\mu \right) \\ Z'_\mu &= -\sqrt{1-\beta^2 T^2_W} W^8_\mu + \beta T_W B_\mu, \end{aligned} \tag{3}$$

where the Weinberg angle is defined as $S_W = \frac{g'}{\sqrt{g^2+(1+\beta^2)g'^2}}$ and $g$, $g'$ correspond to the coupling constants of the groups $SU(3)_L$ and $U(1)_X$, respectively.

The neutral Lagrangian associated to the SM-boson $Z_{1\mu}$ in the weak basis of the SM quarks $U^0 = (u^0, c^0, t^0)^T$ and $D^0 = (d^0, s^0, b^0)^T$, is

$$\mathscr{L}^{NC} = \frac{g}{2C_W} \left\{ \overline{U^0} \gamma_\mu \left[ G_V^{U(r)} - G_A^{U(r)} \gamma_5 \right] U^0 + \overline{D^0} \gamma_\mu \left[ G_V^{D(r)} - G_A^{D(r)} \gamma_5 \right] D^0 \right\} Z_1^\mu . \tag{4}$$

The couplings of the $Z_{1\mu}$ have the same form as the SM couplings, where the usual vector and axial vector couplings $g^{SM}_{V,A}$ are replaced by $G^{(r)}_{V,A} = g^{SM}_{V,A} I + \delta g^{(r)}_{V,A}$, with $\delta g^{(r)}_{V,A} = \widetilde{g}^{(r)}_{V,A} S_\theta$ the correction due to the small $Z_\mu - Z'_\mu$ mixing angle $\theta$, and $r = (A,B,C)$. We will consider that the three realization in the table are in the mass basis. Thus, all the analytical parameters ($O$) at the Z pole have the same SM-form but with small correction factors ($\delta O$) due to the factor $\delta g^{(r)}_{V,A}$ that depend on the family assignment, where each observable takes the form $O = O_{SM}(1+\delta O)$. With the expressions for the Z-pole observables and the experimental data from the LEP [4], we perform a $\chi^2$ fit for each representation $A, B$ and $C$ at 95% CL, which will allow us to display the family dependence in the models. We find the best allowed region in the plane $M_{Z'} - S_\theta$ for both models, i.e. $\beta = -\sqrt{3}$ and $\beta = -\frac{1}{\sqrt{3}}$ as shown by the figures, respectively. We can see that A and B spectrum display teh same alloed region, however they yield broader allowed regions for the mixing angle than model C. It is also noted that the lowest bound for the mass of $Z'$ for A-B cases are about 4 TeV and 1.5 TeV for each model $\beta = -\sqrt{3}$ and $\beta = -\frac{1}{\sqrt{3}}$, respectively, while for C case are about 8 TeV and 2 TeV for each 331 model, respectively.

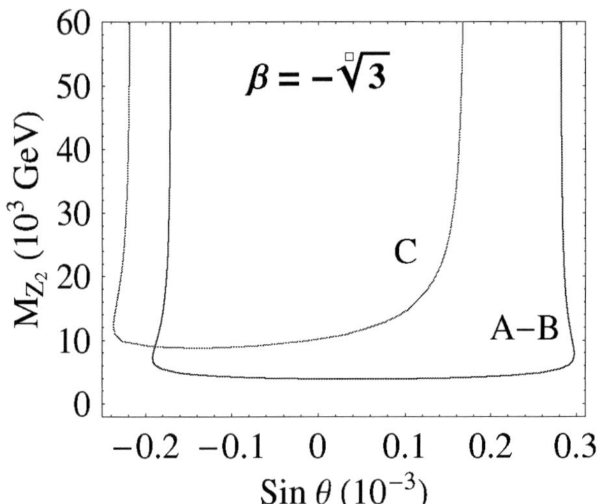


**FIGURE 1.** $M_{Z_2}$ as a function of $\sin\theta$ for $\beta = -\sqrt{3}$

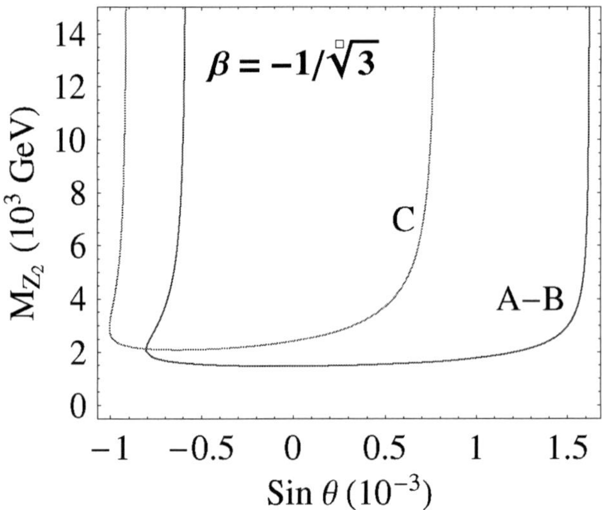


**FIGURE 2.** $M_{Z_2}$ as a function of $\sin\theta$ for $\beta = -1/\sqrt{3}$

## ACKNOWLEDGMENTS


We acknowledge the financial support from COLCIENCIAS.

# REFERENCES


1. F. Pisano and V. Pleitez, Phys. Rev. **D46**, 410 (1992); P.H. Frampton, Phys. Rev. Lett. **69**, 2889 (1992).
2. R. Foot, H.N. Long and T.A. Tran, Phys. Rev. **D50**, R34 (1994).
3. L.A. Sánchez, W.A. Ponce, and R. Martínez, Phys. Rev. **D64**, 075013 (2001);Rodolfo A. Diaz, R. Martinez, F. Ochoa, Phys. Rev. **D69**, 095009 (2004); Rodolfo A. Diaz, R. Martinez, F. Ochoa, Phys. Rev. **D72,** 035018 (2005).
4. S. Eidelman et. al. Particle Data Group, Phys. Lett. **B592**, (2004), pp. 120-121; S. Schael et. al. arXiv: hep-ph/0509008 (submitted to Physics Reports)

# Effects of charged Higgs bosons in the $\nu_\tau \mathcal{N}$ deep inelastic scattering within the 2HDM(II) and the UHE $\nu_\tau$ detection

M.I. Pedraza-Morales[(a)], A. Rosado[(a)] and H. Salazar[(b)]

[(a)] *Instituto de Física, BUAP. Apdo. Postal J-48, C.P. 72570 Puebla, Pue., México,*
[(b)] *Fac. de Cs. Físico-Matemáticas, BUAP. Apdo. Postal 1364, C.P. 72000 Puebla, Pue., México*

**Abstract.** We study the deep inelastic process $\nu_\tau + \mathcal{N} \to \tau^- + X$ (with $\mathcal{N} \equiv (n+p)/2$ an isoscalar nucleon), in the context of the two Higgs doublet model type II (2HDM(II)). We discuss the contribution to the total cross section of diagrams, in which a charged Higgs boson is exchanged. The results show the strong dependence of such contribution on $\tan\beta$ and $M_{H^\pm}$. In the region $50 \leq \tan\beta \leq 200$ and 90 GeV $\leq M_{H^\pm} \leq$ 600 GeV with the additional experimental constraint on the involved model parameters $M_{H^\pm} \geq 1.5 \times \tan\beta$ GeV, the contribution of the charged Higgs boson exchange diagrams to the cross section of the charged current inclusive $\nu_\tau \mathcal{N}$ collision can become important. The contribution for an inclusive dispersion generated through the collision of an ultrahigh energy tau-neutrino with $E_\nu \approx 10^{20}$ eV on a target nucleon can be larger than the value of the contribution of the $W^\pm$ exchange diagrams, provided that $M_{H^\pm} \approx 300$ GeV and $\tan\beta \approx 200$. Such enhancement and the induced variation on the mean inelasticity $\langle y \rangle^{CC}$ could lead to sizeable effects in the acceptance of cosmic tau-neutrino detectors.

**Keywords:** Charged Higgs boson, $\nu_\tau \mathcal{N}$ deep inelastic scattering, Cosmic neutrino detection
**PACS:** 13.15.+g, 13.85.Tp, 14.60.Fg, 14.80.Cp, 95.55.Vj

## INTRODUCTION.

In the SM, the Higgs sector consists of a single $SU(2)$ doublet, and after spontaneous symmetry breaking (SSB) it remains a physical state, the Higgs boson ($h^0_{sm}$), whose mass is not predicted in the theory. On the other hand, the SM is not expected to be the ultimate theoretical structure responsible for electroweak symmetry breaking (EWSB) [1, 2]. One of the most simple extensions of the SM is the so called two Higgs doublet model (2HDM). There are four classes of 2HDM which naturally avoid tree-level flavor-changing neutral currents that can be induced by Higgs boson exchange [3]. These models include a Higgs sector with two scalar doublets, which give masses to the up and down-type fermions as well as the gauge bosons. Model II is particularly interesting, where one of the Higgs scalar doublet couples to the up-components of isodoublets while the other to the down-components. Model II is that utilized in SUSY theories. The 2HDM(II) has a rich structure and predicts interesting phenomenology [1]. The physical spectrum consists of two neutral CP-even states ($h^0, H^0$) and one CP-odd ($A^0$), as well as a pair of charged scalar particles ($H^\pm$). The advantage of such model is the fact that any Higgs sector built only upon doublets preserves naturally the lowest-order electroweak

CP917, *Particles and Fields,* edited by H. Castilla Valdez, J. C. D'Olivo, and M. A. Perez

relation $\rho = 1$, with $\rho = M_{W^\pm}^2/(M_Z^2 \cos^2\theta)$, which has been tested with a good accuracy. On the phenomenological side, an important aspect of the 2HDM is that the Higgs sector may provide an additional source of CP violation [4].

Several experimental lower limits on the charged Higgs boson mass in this model have been reported in the literature:

$$M_{H^\pm} > 79.3\,\text{GeV} \qquad (95\%\ \text{C.L.})\ [5], \tag{1}$$

$$M_{H^\pm} > (0.97, 1.28, 1.89) \times \tan\beta\,\text{GeV} \qquad (95\%\ \text{C.L.})\ [6], [7], [8], \tag{2}$$

$$M_{H^\pm} > (0.97, 1.5, 1.9, 2.5, 2.6) \times \tan\beta\,\text{GeV} \qquad (90\%\ \text{C.L.})\ [9], [10], [11], [12], [13]. \tag{3}$$

Further, based on the discussions on $\tan\beta$ given in Refs. [14] and [15], we restrict ourselves by taking the following upper limit on $\tan\beta$

$$\tan\beta \leq 200. \tag{4}$$

## RESULTS FOR DEEP INELASTIC $\nu_\tau\mathcal{N}$ IN THE SM AND THE 2HDM(II)

The differential cross section for the inclusive reaction

$$\nu_\tau(p) + \mathcal{N}(P_\mathcal{N}) \to \tau^-(p') + X, \tag{5}$$

where $\mathcal{N} \equiv (n+p)/2$ is an isoscalar nucleon, at the lowest order in $\alpha$ in the frame of the SM can be found in Ref. [16]. The differential cross section for the inclusive reaction (5), at the lowest order in $\alpha$ in the frame of the 2HDM(II), is given in Ref. [17]:

We take $10^{14}$ eV $\leq E_\nu$ which leads to a condition where all fermion masses become negligible with respect to the total energy $s = 2M_\mathcal{N}E_\nu$, even the top quark mass. We take cuts of $\sim 2$ GeV$^2$ and 10 GeV$^2$ for $Q^2$ and the invariant mass $W$, respectively. These values for the cuts are suited for the parton distribution functions of J. Pumplin *et al.* [18] which we will use in our calculations. We have checked numerically that the total cross section rates do not depend on the choice of the cuts on the momentum transfer square $Q^2$, when they take on values of a few GeV$^2$. Taking into account the constraints on $M_{H^\pm}/\tan\beta$ given by (2) and (3), we presented in Ref.[19] results for the cases: $M_{H^\pm} = (1.5, 2.0, 3.0)\tan\beta$ GeV. Further, in the numerical performances the conditions (1) and (4) have been fulfilled.

## CHARGED HIGGS BOSON EFFECTS IN THE POSSIBILITY OF DETECTING ULTRA-HIGH ENERGY TAU-NEUTRINOS

Cosmic neutrino fluxes can initiate air-showers through interaction in the atmosphere, or in the Earth. Neutrino trajectories will be going down in a nearly horizontal way in the former case, whereas in a Earth-skimming manner in the latter case. Thus, it is important to know the acceptance (event rate/flux) of proposed air shower experiments for detecting both types of neutrino -initiated events [Horizontal air-showers (HAS) and Upgoing (Earth-skimming) air-showers (UAS)]. These acceptances for fluorescence detectors have been calculated in [20] for experiments as High Resolution Fly's Eye Detector (HiRes) [21], Pierre Auger Observatory (PAO) [22], and the Cosmic Ray Tau Neutrino Telescopes (CRTNT) [23], which are anchored to the ground, and at experiments as Extreme Universe Space Observatory (EUSO) [24] and Orbiting wide angle light-collectors (OWL) [25], which are proposed to orbit around the Earth. In Ref.[20], it is stated that the very different dependence on the cross section of the HAS (linear) and UAS (non linear) acceptances will provide a practical method to measure the charged current cross section of the neutrino nucleon scattering $\sigma^{CC}_{\nu\mathcal{N}}$. One simply has to exploit the ratio of UAS to HAS event rates. However, one assumption made in Ref.[20] was taking that $\sigma^{CC}_{\nu_e\mathcal{N}} = \sigma^{CC}_{\nu_\mu\mathcal{N}} = \sigma^{CC}_{\nu_\tau\mathcal{N}}$. This is valid in the SM, but it is no longer valid in the 2HDM(II), because the $\sigma^{CC}_{\nu_\tau\mathcal{N}}$ could be twice larger than $\sigma^{CC}_{\nu_e\mathcal{N}} = \sigma^{CC}_{\nu_\mu\mathcal{N}}$ in this model. Hence, the neutrino charged current interaction mean-free path (introduced in Ref.[20]) could be for $\nu_\tau$ one half of that for $\nu_e$ and $\nu_\mu$, in the mentioned model. Therefore, the existence of charged Higgs bosons could imply a large deviation from the results for the acceptances for space based (or ground based) tau-neutrino detectors presented in Fig. 7 of Ref.[20].

Further, it was already pointed out in Ref.[26], that besides the obvious role of the neutrino cross section in the actual interaction that leads to the possible neutrino detection there are other more subtle effects. The $y$ distribution of the cross section also has an impact on the detection rate [26, 27, 28]. In fact, in Ref.[19] we showed how the mean inelasticity $\langle y\rangle^{CC}$ would change in the THDM(II). Our results obtained for $\langle y\rangle^{CC}$ in the frame of the SM are in good agreement with those reported in Ref.[29]. We showed in Ref.[19] that $\langle y\rangle^{CC}_{2hdm} \cong 2.5 \times \langle y\rangle^{CC}_{sm}$ for $M_{H^\pm} = 300$ GeV, $\tan\beta = 200$ and $E_\nu = 10^{20}$ eV. Such a drastic variation in $\langle y\rangle^{CC}$ could induce a large variation in the possibility of detecting ultra-high energy tau-neutrinos through inclined showers [26].

## CONCLUSIONS

We showed that the contribution of the charged Higgs boson exchange diagrams can lead to a sizeable enhancement with respect to the SM cross section rates for the charged current $\nu_\tau\mathcal{N}$ deep inelastic scattering. For the case of an ultrahigh energy tau-neutrino with $E_\nu = 10^{20}$ eV colliding on a target nucleon such enhancement reached 110% provided that $\tan\beta = 200$ and $M_{H^\pm} = 300$ GeV. Besides, we obtain $\langle y\rangle^{CC}_{2hdm} \approx 2.5 \times \langle y\rangle^{CC}_{sm}$ for the same values of $E_\nu$, $\tan\beta$ and $M_{H^\pm}$. This enhancement and the induced variation in the mean inelasticity $\langle y\rangle^{CC}$ could lead to sizeable effects in the acceptance of the cosmic tau-neutrino detectors at space based experiments such as the EUSO and OWL proposals and at ground-based experiments such as PAO, HiRes and the CRTNT.

## ACKNOWLEDGMENTS

The authors thank the *Sistema Nacional de Investigadores* and *CONACyT* (México) for financial support.

## REFERENCES

1. J. Gunion, H. Haber, G. Kane and S. Dawson, *The Higgs Hunter's Guide*, Addison-Wesley Publishig Company, Reading, MA, 1990.
2. H. E. Haber, in *Testing the Standard Model*, Proceedings of the 1990 Theoretical Advanced Study Institute in Elementary Particle Physics, edited by M. Cvetic and P. Langacker (World Scientific, Singapore, 1991) p. 340-475.
3. V. D. Barger, J. L. Hewett and R. J. N. Phillips, Phys. Rev. D **41**, 3421 (1990).
4. T. D. Lee, Phys. Rev. D **8**, 1226 (1973); T. D. Lee, Phys. Rept. **9**, 143 (1974); S. Weinberg, Phys. Rev. Lett. **37**, 657 (1976); Y. L. Wu and L. Wolfenstein, Phys. Rev. Lett. **73**, 1762 (1994) [arXiv:hep-ph/9409421]; J. Liu and L. Wolfenstein, Nucl. Phys. B **289**, 1 (1987).
5. A. Heister *et al.* [ALEPH Collaboration], Phys. Lett. B **543**, 1 (2002).
6. K. Ackerstaff *et al.* [OPAL Collaboration], Eur. Phys. J. C **8**, 3 (1999).
7. G. Abbiendi *et al.* [OPAL Collaboration], Phys. Lett. B **551**, 35 (2003).
8. G. Abbiendi *et al.* [OPAL Collaboration], Phys. Lett. B **520**, 1 (2001).
9. R. Ammar *et al.* [CLEO Collaboration], Phys. Rev. Lett. **78**, 4686 (1997).
10. A. Stahl and H. Voss, Z. Phys. C **74**, 73 (1997).
11. D. Buskulic *et al.* [ALEPH Collaboration], Phys. Lett. B **343**, 444 (1995).
12. R. Barate *et al.* [ALEPH Collaboration], Eur. Phys. J. C **19**, 213 (2001).
13. M. Acciarri *et al.* [L3 Collaboration], Phys. Lett. B **396**, 327 (1997).
14. D. P. Roy, AIP Conf. Proc. **805**, 110 (2006).
15. H. Baer, J. Ferrandis and X. Tata, Phys. Lett. B **561**, 145 (2003).
16. M. H. Reno and C. Quigg, Phys. Rev. D **37**, 657 (1988); C. Quigg, M. H. Reno and T. P. Walker, Phys. Rev. Lett. **57**, 774 (1986).
17. A. Rosado, Phys. Rev. D **74**, 057301 (2006).
18. J. Pumplin *et al.*, JHEP **207**, 12 (2002); D. Stump *et al.*, e-Print Archive: hep-ph/0303013.
19. A. Rosado, arXiv:hep-ph/0610220.
20. S. Palomares-Ruiz, A. Irimia and T. J. Weiler, Phys. Rev. D **73**, 083003 (2006).
21. P. Sokolsky and J. Belz [the HiRes Collaboration], "Comparison of UHE composition measurements by Fly's Eye, HiRes-prototype / MIA and stereo HiRes experiments," arXiv:astro-ph/0507485.
22. H. Blumer *et al.* [Auger Collaboration], "The Auger fluorescence detector prototype telescope," FZKA-6345P *Prepared for 26th International Cosmic Ray Conference (ICRC 99), Salt Lake City, UT, 17-25 Aug 1999*; J. Abraham *et al.* [Pierre Auger Collaboration], Nucl. Instrum. Meth. A **523**, 50 (2004); G. Miele, S. Pastor and O. Pisanti, Phys. Lett. B **634**, 137 (2006); D. Gora, M. Roth and A. Tamburro, arXiv:astro-ph/0610504.
23. Z. Cao, Nucl. Phys. Proc. Suppl. **151**, 287 (2006).
24. Ph. Gorodetzky [The EUSO Collaboration], Nucl. Phys. Proc. Suppl. **151**, 401 (2006); G. D'Ali Staiti [EUSO Collaboration], Nucl. Phys. Proc. Suppl. **136**, 415 (2004).
25. F. W. Stecker, J. F. Krizmanic, L. M. Barbier, E. Loh, J. W. Mitchell, P. Sokolsky and R. E. Streitmatter, Nucl. Phys. Proc. Suppl. **136C**, 433 (2004).
26. E. Zas, New J. Phys. **7**, 130 (2005).
27. R. Gandhi, C. Quigg, M. H. Reno and I. Sarcevic, Astropart. Phys. **5**, 81 (1996).
28. J. C. Arteaga-Velazquez and A. Zepeda, "Estimation of the detectable flux of astrophysical neutrinos at the Pierre Auger observatory by means of horizontal air showers," *Proceedings of 29th International Cosmic Ray Conference Pune* **Vol. 9**, *151-154 (2005)*.
29. C. Aramo, A. Insolia, A. Leonardi, G. Miele, L. Perrone, O. Pisanti and D. V. Semikoz, Astropart. Phys. **23**, 65 (2005).

# Signatures of First Cosmological Structures in the CH Rotational Lines

Julio C. Campos-García, Anton A. Lipovka and Julio C. Saucedo

*Departamento de Investigación en Física. Universidad de Sonora.*
*Apdo. Postal 5-88, C. P. 83190 Hermosillo, Sonora, México.*

**Abstract.** In the present work we discuss the possible signatures of first cosmological structures formed at the end of the dark-age epoch ($z$ ~ 40-20). These correspond to the spectral-spatial fluctuations in the cosmic microwave background radiation (CMBR), produced by elastic resonant scattering of CMBR photons by the rotational structure of the CH molecules formed in proto-structures. The chemical kinetic of CH molecule is calculated in the expanding homogeneous medium and applied to proto-structures in their linear stage of collapse, whose properties were estimated by using a top-hat spherical approach and Lambda-CDM cosmology. It is shown that the optical depth in the rotational structure of the CH molecule for some cases of non-standard BBN (NBBN) is rather high and could be observed with radio telescopes under construction. Possible future observations are discussed as a crucial test for NBBN models.

**Keywords:** Spectroscopy and spectrophotometry; Submillimeter; Cosmology
**PACS:** 95.75.Fg; 95.85.Fm; 98.80.-K

## OBSERVATIONAL ESTIMATES

In the range of $z$ = 20 - 40, for the case of a proto-object of size L = 2 x $10^{20}$ cm, one can obtain the fluctuations in the temperature at the order $\Delta T/T \sim 3 \times 10^{-10} - 10^{-9}$ at the third rotational transition (5/2-3/2) of the CH molecule which form the main contribution to the optical depth.

Let us estimate the integration time required to detect secondary anisotropies with modern mm telescopes. For noise temperature we assume $T_n$ ~ 40-80 K, and $\Delta\upsilon/\upsilon$ ~ 0.1 for the bandwidth. In this case the integration time for the already collapsed haloes (see [2] for details) is estimated to be about 10 hours, whereas for the $3\sigma$ - $6\sigma$ haloes it is still too large to be observed. But it must be stressed that these estimations were made for the primordial abundance [C]/[H] = $10^{-10}$, and in some models of NBBN this value is predicted to be as high as $10^{-8}$ - $10^{-7}$ [1]. In this case we can calculate the integration time for the $3\sigma$ - $6\sigma$ haloes to be about 200 h for [C]/[H]=$10^{-8}$ and about 2 hours for [C]/[H]=$10^{-7}$.

## REFERENCES

1. Rauscher T., Applegate J.H., Cowan J.J., Thielemann F.K., Wiescher H., *Ap.J.*, (1994), 429, 499-530.
2. Núñes-López R., Lipovka A. Avila-Reese V. (2006), *MNRAS*, 369, 2005

CP917, *Particles and Fields,* edited by H. Castilla Valdez, J. C. D'Olivo, and M. A. Perez

# $\Lambda^0$ Polarization In $pp \rightarrow p_f p_s \Lambda^0 \text{anti-}\Lambda^0$ At 800-GeV/c

J. Castorena,[1] J. Félix,[1] M.C. Berisso,[2] D.C. Christian,[3] A. Gara,[4] E.E. Gottschalk,[3] G. Gutiérrez,[3] E.P. Hartouni,[5] B.C. Knapp,[4] M.N. Kreisler,[2] S. Lee,[2] K. Markianos,[2] G. Moreno,[1] M. A. Reyes,[1] M.H.L.S. Wang,[2] A. Wehmann,[3] D. Wesson[2].

[1] *Universidad de Guanajuato, León, Guanajuato, México.*
[2] *University of Massachusetts, Amherst, Massachusetts, USA.*
[3] *Fermilab, Batavia, Illinois, USA.*
[4] *Columbia University, Nevis Laboratory, New York, New York, USA.*
[5] *Lawrence Livermore National Laboratory, Livermore, California, USA.*

**Abstract.** The preliminary results of a $\Lambda^0$ and anti-$\Lambda^0$ polarization study as functions of the $M_{(\Lambda\text{-}b\Lambda)}$, $X_F$ and their corresponding $P_T$ are reported in this work.

**Keywords:** Polarization, E690.
**PACS:** 13.88.+e

The polarization of $\Lambda^0$ or anti-$\Lambda^0$ is measured with respect to the normal vector to its corresponding production plane; it is extracted by fitting to a straight line the angular distribution ($\cos\theta$) of the proton decaying from $\Lambda^0$ or anti-$\Lambda^0$, $\cos\theta$ has the following dependence on the polarization ($P$): $dN/d\Omega = N_0(1-\alpha P\cos\theta)$.

$\Lambda^0$ polarization and anti-$\Lambda^0$ polarization depend on $M_{(\Lambda\text{-}b\Lambda)}$ ($\Lambda^0$anti-$\Lambda^0$ invariant mass), $P_T$ and $X_F$. Both $\Lambda^0$ and anti-$\Lambda^0$ polarizations have a similar dependence on $P_T$; and a symmetric dependence as functions of $X_F$. The dependence of $\Lambda^0$ polarization on $M_{(\Lambda\text{-}b\Lambda)}$ suggests that the origin of the polarization might be related to the production of the $\Lambda^0$ anti-$\Lambda^0$ system. The preliminary results are shown in Figure 1.

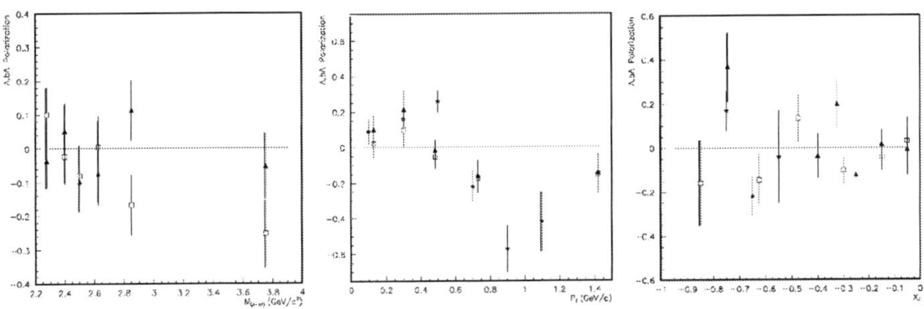

**FIGURE 1.** Polarization as a function of $M_{(\Lambda\text{-}b\Lambda)}$ (left), $X_F$ (center) and $P_T$ (right). The squares (□) correspond to the results for Lambda, triangles (▲) for anti-Lambda and stars for results reported in [1].

## REFERENCES

[1] J. Félix et al, *Phys. Rev. Lett.* 88, 061801 (2002).

[2] Michael H.L.S. Wang, Ph.D. *thesis*, University of Massachusetts, Amherst, 2000.

CP917, *Particles and Fields,* edited by H. Castilla Valdez, J. C. D'Olivo, and M. A. Perez

# Rare top quark decays in Alternative Left-Right Symmetric Models

R. Gaitán*, O. G. Miranda† and L. G. Cabral-Rosetti**

**Centro de Investigaciones Teóricas,*
*Facultad de Estudios Superiores – Cuautitlán,*
*Universidad Nacional Autónoma de México, (FESC-UNAM).*
*A. Postal 142,Cuautitlán-Izcalli, Estado de México, C. P. 54700, México.*
*†Departamento de Física,*
*Centro de Investigación y de Estudios Avanzados del IPN*
*A. Postal 14-740, México D. F. 07000, México.*
*** Departamento de Posgrado,*
*Centro Interdisciplinario de Investigación y Docencia en Educación Técnica (CIIDET),*
*Av. Universidad 282 Pte., Col. Centro, A. Postal 752,*
*C. P. 76000, Santiago de Queretaro, Qro., México.*

**Abstract.** We evaluate the flavor changing neutral currents (FCNC) decay $t \to H^0 + c$ in the context of Alternative Left-Right symmetric Models (ALRM) with extra isosinglet heavy fermions; the FCNC decays may place at tree level and are only supressed by the mixing between ordinary top and charm quarks. We also comment on the decay process $t \to c + \gamma$, which involves radiative corrections.

We evaluate de flavor changing neutral currents (FCNC) decays $t \to H^0 + c$ and $t \to \gamma + c$ in the context of Alternative Left-Right Symmetry Models (ALRM) with extra isosinglet heavy fermions; in the first case FCNC decay may take place at tree level, and in the second one at one loop level.

The ALRM formulation is based on the gauge group $SU(2)_L \otimes SU(2)_R \otimes U(1)_{B-L}$. It has been formulated in order to solve different problems such as the hierarchy of quark and lepton masses or the strong CP problem, different authors have enlarged the fermion content [1]

The partial width for the $t \to c + H$ can be present at tree-level in this model and it is given by [2]:

$$\Gamma(t \to H^0 + c) =$$

$$\frac{G_F\, \eta_{32}^2 \cos^2\alpha}{16\sqrt{2}\pi m_t}\left(m_t^2 + m_c^2 - M_H^2\right)\left[\left(m_t^2 - \left(M_H + m_c\right)^2\right)\left(m_t^2 - \left(M_H - m_c\right)^2\right)\right]^{\frac{1}{2}} \tag{1}$$

where $G_F$ is the Fermi constant, $m_t$ denotes the top mass, $m_c$ is the charm mass, and $M_H$ is the mass of the neutral Higgs boson. We can see from this formula that the branching ratio will be proportional to the product $\eta_{32} \cos\alpha$, of the top-quark mixing with the charm-quark, $\eta_{32}$ and the mixing of the SM Higgs boson mixing with the extra Higgs boson, $\cos\alpha$. Thanks to the possible combined effect of a big $\cos\alpha$ (null mixing between the SM Higgs boson and the additional Higgs bosons) and a big value of $\eta_{32}$

CP917, *Particles and Fields,* edited by H. Castilla Valdez, J. C. D'Olivo, and M. A. Perez

this branching ratio could be as high as $10^{-5}$ [2], for a Higgs mass of 117 *GeV* and a value $\eta_{32} = 0.009$, that could be considered as the most stringent constraint.

On the other hand, in the process $t \to \gamma + c$, the final state particles are on-shell and the photon has transverse polarization, therefore, the partial width can be written generality as

$$\Gamma(t \to c + \gamma) = \frac{1}{\pi} \left[ \frac{m_t^2 - m_c^2}{2m_t} \right] \left( |A|^2 + |B|^2 \right) . \tag{2}$$

with $A$ and $B$ are the vector and axial vector form factors.
In the limit $m_c = 0$ the vector and axial vector form factors are equal ($A = B$). For instance, the contribution of the diagram shown in Fig. (1) in this limit is:

$$\begin{aligned} A = B = \frac{eg\hat{U}^*_{tb} V^R_{c\hat{b}}}{192\pi^2 m_t^3} \Big\{ & m_t^2 + \left( m_b^2 - m_{\hat{b}}^2 \right) B_0 \left( 0; m_b^2, m_{\hat{b}}^2 \right) \\ & + \left( m_b^2 - M_{\hat{W}}^2 \right) B_0 \left( 0; m_b^2, M_{\hat{W}}^2 \right) - \left( 2m_t^2 + 3m_{\hat{b}}^2 - 3M_{\hat{W}}^2 \right) B_0 \left( 0; m_{\hat{b}}^2, M_{\hat{W}}^2 \right) \\ & +4 \left( m_{\hat{b}}^4 - m_b^2 m_{\hat{b}}^2 + m_t^2 m_{\hat{b}}^2 - M_{\hat{W}}^2 m_{\hat{b}}^2 - \frac{1}{2} m_b^2 m_t^2 + m_b^2 M_{\hat{W}}^2 \right) C_0 \left( 0, 0, m_t^2; m_b^2, m_{\hat{b}}^2, M_{\hat{W}}^2 \right) \Big\} \end{aligned} \tag{3}$$

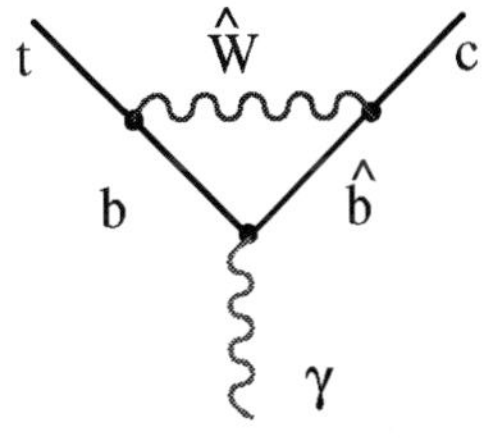

**FIGURE 1.** Example of Feynman diagram contributing to the $t \to c + \gamma$ decay amplitud in the ALRM

This is just and example of the kind of diagrams that contributes to this decay. The computation of all the diagrams in the ALRM is now under way and we expect to report final results on this process in the near future.

**Acknowledgments** This work has been supported by CONACYT, SNI and also by PAPIIT project No.IN113206.

## REFERENCES

1. Rabindra N. Mohapatra *Unification and Supersymmetry* Springer 2003 and references therein.
2. R. Gaitan, O. G. Miranda and L. G. Cabral-Rosetti, Phys. Rev. D **72**, 034018 (2005) [arXiv:hep-ph/0410268] and references there in.

# List of Participants PASI2006/VI-SILAFAE/XII-MSPF

| Name | Institution | Country |
|---|---|---|
| Aldana Segura Waleska | EFPEM USAC | Guatemala |
| Alfaro Jorge | PUC-Chile | Chile |
| Alvarez Ramírez Erika L. | IF-UNAM | Mexico |
| Anzo Hernández Andres | IF-UNAM | Mexico |
| Ayala Alejandro | ICN-UNAM | Mexico |
| Barranco Monarca Juan | Cinvestav | Mexico |
| Barrañón Armando | UAM-A | Mexico |
| Barros Pablo | U Campinas | Brazil |
| Bashir Adnan | IFM-UMSNH | Mexico |
| Bazo Alba Jose Luis | FNAL/PUC-Chile | USA |
| Bernabeu Jose | U Valencia | Spain |
| Bertou Xavier | U Bariloche | Argentina |
| Bifani Simone | U Torino/INFN | Italy |
| Blanco Ernesto | IF-UASLP | Mexico |
| Bojowald Martin | Penn State U | USA |
| Bolaños Carrera Azucena | Cinvestav | Mexico |
| Brockill Patrick | CBPF | Brazil |
| Bustamante R. Mauricio | Cinvestav/PUC-Peru | Peru |
| Caicedo José A. | ICN-UNAM | Colombia |
| Calderón Manuel | UC Davis | USA |
| Campos García Julio C. | U de Sonora | Mexico |
| Carena Marcela | Fermilab | USA |
| Casimiro Edgar | CBPF | Brazil |
| Castilla Heriberto | Cinvestav | Mexico |
| Castillo-Vallejo V. Manuel | CIO | Mexico |
| Castorena G. Jorge A. | IFUG | Mexico |
| Castromonte Cesar | CBPF | Brazil |
| Cázarez-Bush Federico | IM-UNAM | Mexico |
| Chernicoff Minsberg Mariano | ICN-UNAM | Mexico |
| Chevis Ben | U Tennessee | USA |
| Chikahiro Nishi Celso | IFT-UNESP | Brazil |
| Christian David | Fermilab | USA |
| Chung Daniel | U Wiscousin | USA |
| Compeán Jasso Cliffor B. | IF-UASLP | Mexico |
| Cortes Maldonado Ismael | FCFM-BUAP | Mexico |
| Cortes Rubio Manuel | UNAM | Mexico |
| Cortese Mombelli Ignacio | ICN-UNAM | Mexico |
| Cronin James W. | U Chicago | USA |
| Cuautle Eleazar | ICN-UNAM | Mexico |
| Cuestra Sánchez Vladimir | Cinvestav | Mexico |
| Cywiak Cordova David | IFUG | Mexico |
| D´Olivo Juan Carlos | ICN-UNAM | Mexico |
| De Bernardini Alex E. | U Estadual de Campinas | Brazil |
| De Florian Daniel | U Buenos Aires | Argentina |
| De Gouvea Andre | Northwestern U | USA |
| De la Macorra Axel | IF-UNAM | Mexico |
| De Souza Luna Emerson G. | IFT-UNESP | Brazil |
| Dehmelt Klaus | Florida IT | USA |
| Delbourgo Robert | U Tasmania | Australia |
| Delepine David | IFUG | Mexico |

| | | |
|---|---|---|
| Denigri Daniel | CEN Saclay/DAPNIA | France |
| Díaz Cruz J. Lorenzo | FCFM-BUAP | Mexico |
| Dodelson Scott | U Chicago | USA |
| Dorlan Tyler | Notre Dame U | USA |
| Dos Anjos João | CBPF | Brazil |
| Dvoeglazov Valeri | U de Zacatecas | Mexico |
| Elsen Eckhard | DESY | Germany |
| Engelfried Jurgen | IF-UASLP | Mexico |
| Erler Jens | IF-UNAM | Mexico |
| Ernest David I. | Vanderbilt U | USA |
| Espinosa José R. | CERN | Switzerland |
| Félix Beltran Olga G. | FCE-BUAP | Mexico |
| Félix Julián | IFUG | Mexico |
| Fernández García José N. | Cinvestav | Mexico |
| Fernández Telles Arturo | FCFM-BUAP | Mexico |
| Ferrari Sergio | U Nal de la Plata | Argentina |
| Flores Báez Fco. Vicente | Cinvestav | Mexico |
| Flores Tlalpa Alain | Cinvestav | Mexico |
| Franklin Allan | U Colorado | USA |
| Gaitán Lozano Ricardo | FES-Cuautitlan UNAM | Mexico |
| Galaviz Vilchis Pablo | ICN-UNAM | Mexico |
| García Augusto | Cinvestav | Mexico |
| García José Antonio | ICN-UNAM | Mexico |
| Garcia Solis Edmundo | UIC | USA |
| García-Guerra G. Alejandro | Cinvestav | Mexico |
| German Velarde Gabriel | CCF-UNAM | Mexico |
| Gill Mandeep | Ohio State U | USA |
| Göbel Carla | PUC-Rio | Brazil |
| Goldenzweig Pablo | U Cincinnati | USA |
| Gómez Izquierdo J. Carlos | Cinvestav | Mexico |
| González M. Vannia | IFUG | Mexico |
| González Ma Concepción | U Valencia | Spain |
| Goodman Jordan | U Maryland | USA |
| Guijosa Alberto | IF-UNAM | Mexico |
| Gutiérrez Gastón R. | Fermilab | USA |
| Gutiérrez-Rodríguez Alejando | U de Zacatecas | Mexico |
| Guzmán Ramírez Walberto | IFUG | Mexico |
| Herrera Gerardo | Cinvestav | Mexico |
| Hollik Wolfgang | Max Planck Institut | Germany |
| Kemp Ernesto | Unicamp | Brazil |
| Kim Young-Kee | Fermilab | USA |
| Konigsberg Jacobo | U Florida | USA |
| Kruczenski Martin | Pinceton U | USA |
| Kusenko Alex | UCLA | USA |
| Kyser Boris | Fermilab | USA |
| Larios Francisco | Cinvestav Mérida | Mexico |
| Laura Helena Gonzalez | FC-UNAM | Mexico |
| Lederman León M. | IIT/IMSA | USA |
| Lehnert Ralf | MIT | USA |
| León Ely | Florida A&M U | USA |
| León Vargas Hermes | IF-UNAM | Mexico |
| López Ricardo | Cinvestav | Mexico |
| López-Castro Gabriel | Cinvestav | Mexico |

| | | |
|---|---|---|
| Losada Marta | UA Nariño | Colombia |
| Loza José Antonio | ICN-UNAM | Mexico |
| Ma Ernest | UC Riverside | USA |
| Malbouisson Jorge | U F Bahia | Brazil |
| Martínez H Mario I. | ICN-UNAM | Mexico |
| Martínez Roberto | U Nal de Colombia | Colombia |
| Mateos Gisela | CEIICH-UNAM | Mexico |
| McDonald Arthur | Queen's U | Canada |
| McLerran Larry | BNL | USA |
| Medina Anibal | U Chicago | USA |
| Medina María C. | Lab. Tandar - CAC - CNEA | Argentina |
| Medina-Tanco Gustavo | ICN-UNAM | Mexico |
| Melfo Alejandra | U de los Andes | Venezuela |
| Méndez Héctor | U Puerto Rico | USA |
| Mendoza Suárez Jairo A. | U de Pamplona | Colombia |
| Metcalf William | LSU | USA |
| Miranda Omar | Cinvestav | Mexico |
| Mondragón Alfonso | IF-UNAM | Mexico |
| Mondragón Myriam | IF-UNAM | Mexico |
| Montaño Zetina Luis M. | Cinvestav | Mexico |
| Montemayor Rafael L. | CA Bariloche | Argentina |
| Montes de Oca Yemha Jose H. | ESFM-IPN | Mexico |
| Morales-Tecotl Hugo A | UAM-I | Mexico |
| Morfín Jorge | Fermilab | USA |
| Morselli Aldo | U Roma / INFN | Italy |
| Muñoz Ñungo José H. | U del Tolima | Colombia |
| Nardi Enrico | U Antioquia | Colombia |
| Navarro Estrada Jorge L. | ICN-UNAM | Mexico |
| Neubert Matthias | Cornell U | USA |
| Oddone Piermaria J. | Fermilab | USA |
| Ortiz González Carlos A. | IFUG | Mexico |
| Paic Guy | ICN-UNAM | Mexico |
| Paniagua López Pablo | Cinvestav | Mexico |
| Pastor Sergio | IFIC/CSIC-U Valencia | Spain |
| Paucar Acosta Manuel G. | JINR | Rusia |
| Peinado Rodriguez Eduardo | IF-UNAM | Mexico |
| Peña Castañeda Carlos A. | U Nal de Colombia | Colombia |
| Peñuñuri A. Francisco R. | Cinvestav Mérida | Mexico |
| Pérez Abdel | Cinvestav | Mexico |
| Pérez Adrían | U Maryland | USA |
| Pérez Angón Miguel | Cinvestav | Mexico |
| Pérez Lara Carlos E. | PUC-Peru | Peru |
| Pérez Vargas Felipe | U de Zacatecas | Mexico |
| Podesta Pedro L.M. | ICN-UNAM | Mexico |
| Ponton Eduardo | Columbia U | USA |
| Rashba Timur | Max Planck Institut | Germany |
| Raya Alfredo | IFM-UMSNH | Mexico |
| Rendon Carolina | FC-UNAM | Mexico |
| Reyes Martinez Carlos M. | ICN-UNAM | Mexico |
| Rivera Max | PUC-Chile | Chile |
| Rodrigo Teresa | U Cantrabria | Spain |
| Rodríguez Ezequiel | UniSon | Mexico |
| Rodríguez López Jairo A. | U Nal de Colombia | Colombia |

| | | |
|---|---|---|
| Rodríguez Tatiana | U Pennsylvania | USA |
| Rojas Eduardo | ICN-UNAM | Mexico |
| Román López Sergio | CIDS/IC-BUAP | Mexico |
| Rosado Alfonso | IF-BUAP | Mexico |
| Russ James S. | Carnegie Mellon | USA |
| Sabido Moreno Oscar M. | IFUG | Mexico |
| Sahu Sarira | ICN-UNAM | Mexico |
| Salazar Humberto | FCFM BUAP | Mexico |
| Sánchez Alberto | Cinvestav | Mexico |
| Sánchez Cecilio Angel | ICN-UNAM | Mexico |
| Sánchez Federico | ICN-UNAM | Mexico |
| Sánchez Rodrigo | Instituto Balseiro | Argentina |
| Sandoval Andres | ICN-UNAM | Mexico |
| Schlobohm Sarah | Notre Dame U | USA |
| Schmidt Ivan | UTFSM | Chile |
| Schubert Christian | IFM-UMSNH | Mexico |
| Shah Nausheen R. | U Chicago | USA |
| Shapiro Charles | U Chicago | USA |
| Sheaff Marleigh | U Wiscousin | USA |
| Szabo Richard | IF-UNAM | Mexico |
| Toledo Sanchez Genaro | IF-UNAM | Mexico |
| Tomas Ricard | U Valencia | Spain |
| Torres Manuel | IF-UNAM | Mexico |
| Torri Guiseppe | U Milano | Italy |
| Tortola Miriam | CFTP-IST | Portugal |
| Ttito Pedro | U Puerto Rico | USA |
| Uribe Cecilia | U Oxford | England |
| Urrutia Luis F. | ICN-UNAM | Mexico |
| Valenzuela Cristian | U Federico Sta.María | Chile |
| Vargas Magaña Mariana | FC-UNAM | Mexico |
| Vazquez Eric | IF-UASLP | Mexico |
| Vera Aguirre Carlos E. | U del Tolima | Colombia |
| Vergara José David | ICN-UNAM | Mexico |
| Verzegnassi Claudio | U Trieste | Italy |
| Villanueva Victor M. | IFM-UMSNH | Mexico |
| Villaseñor Luis M. | IFM-UMSNH | Mexico |
| Wagner Carlos | ANL | USA |
| Wagner Federico | UBA | Argentina |
| Wang Lin | Columbia U | USA |
| White Christopher | IIT | USA |
| Yuan C. P. | Michigan State U | USA |
| Zepeda Arnulfo | Cinvestav | Mexico |
| Zhou Wei | Ohio State U | USA |
| Zurita José F. | UBA | Argentina |

# Author Index